The Water-Powered Mills of Floyd County, Virginia

CONTRIBUTIONS TO SOUTHERN APPALACHIAN STUDIES

1. *Memoirs of Grassy Creek: Growing Up in the Mountains on the Virginia-North Carolina Line.* Zetta Barker Hamby. 1998

2. *The Pond Mountain Chronicle: Self-Portrait of a Southern Appalachian Community.* Edited by Leland R. Cooper and Mary Lee Cooper. 1998

3. *Traditional Musicians of the Central Blue Ridge: Old Time, Early Country, Folk and Bluegrass Label Recording Artists, with Discographies.* Marty McGee. 2000

4. *W.R. Trivett, Appalachian Pictureman: Photographs of a Bygone Time.* Ralph E. Lentz II. 2001

5. *The People of the New River: Oral Histories from the Ashe, Alleghany and Watauga Counties of North Carolina.* Edited by Leland R. Cooper and Mary Lee Cooper. 2001

6. *John Fox, Jr., Appalachian Author.* Bill York. 2003

7. *The Thistle and the Brier: Historical Links and Cultural Parallels Between Scotland and Appalachia.* Richard Blaustein. 2003

8. *Tales from Sacred Wind: Coming of Age in Appalachia. The Cratis Williams Chronicles.* Cratis D. Williams. Edited by David Cratis Williams and Patricia D. Beaver. 2003

9. *Willard Gayheart, Appalachian Artist.* Willard Gayheart and Donia S. Eley. 2003

10. *The Forest City Lynching of 1900: Populism, Racism, and White Supremacy in Rutherford County, North Carolina.* J. Timothy Cole. 2003

11. *The Brevard Rosenwald School: Black Education and Community Building in a Southern Appalachian Town, 1920–1966.* Betty J. Reed. 2004

12. *The Bristol Sessions: Writings About the Big Bang of Country Music.* Edited by Charles K. Wolfe and Ted Olson. 2005

13. *Community and Change in the North Carolina Mountains: Oral Histories and Profiles of People from Western Watauga County.* Compiled by Nannie Greene and Catherine Stokes Sheppard. 2006

14. *Ashe County: A History; A New Edition.* Arthur Lloyd Fletcher. 2009 [2006]

15. *The New River Controversy; A New Edition.* Thomas J. Schoenbaum. Epilogue by R. Seth Woodard. 2007

16. *The Blue Ridge Parkway by Foot: A Park Ranger's Memoir.* Tim Pegram. 2007

17. *James Still: Critical Essays on the Dean of Appalachian Literature.* Edited by Ted Olson and Kathy H. Olson. 2008

18. *Owsley County, Kentucky, and the Perpetuation of Poverty.* John R. Burch, Jr. 2008

19. *Asheville: A History.* Nan K. Chase. 2007

20. *Southern Appalachian Poetry: An Anthology of Works by 37 Poets.* Edited by Marita Garin. 2008

21. *Ball, Bat and Bitumen: A History of Coalfield Baseball in the Appalachian South.* L.M. Sutter. 2009

22. *The Frontier Nursing Service: America's First Rural Nurse-Midwife Service and School.* Marie Bartlett. 2009

23. *James Still in Interviews, Oral Histories and Memoirs.* Edited by Ted Olson. 2009

24. *The Millstone Quarries of Powell County, Kentucky.* Charles D. Hockensmith. 2009

25. *The Bibliography of Appalachia: More Than 4,700 Books, Articles, Monographs and Dissertations, Topically Arranged and Indexed.* Compiled by John R. Burch, Jr. 2009

26. *Appalachian Children's Literature: An Annotated Bibliography* Compiled by Roberta T. Herrin and Sheila Quinn Oliver. 2009

27. *Southern Appalachian Storytellers: Interviews with Sixteen Keepers of the Oral Tradition.* Edited by Saundra Gerrell Kelley. 2010

28. *Southern West Virginia and the Struggle for Modernity.* Christopher Dorsey. 2011

29. *George Scarbrough, Appalachian Poet: A Biographical and Literary Study with Unpublished Writings.* Randy Mackin. 2011

30. *The Water-Powered Mills of Floyd County, Virginia: Illustrated Histories, 1770–2010.* Franklin F. Webb and Ricky L. Cox. 2012

31. *School Segregation in Western North Carolina: A History, 1860s–1970s.* Betty Jamerson Reed. 2011

The Water-Powered Mills of Floyd County, Virginia

Illustrated Histories, 1770–2010

FRANKLIN F. WEBB *and*
RICKY L. COX

CONTRIBUTIONS TO SOUTHERN APPALACHIAN STUDIES, 30

McFarland & Company, Inc., Publishers
Jefferson, North Carolina, and London

Library of Congress Cataloguing-in-Publication Data

Webb, Franklin F.
The water-powered mills of Floyd County, Virginia : illustrated histories, 1770–2010 / Franklin F. Webb and Ricky L. Cox.
p. cm. — (Contributions to Southern Appalachian studies ; 30)
Includes bibliographical references and index.

ISBN 978-0-7864-3723-8
softcover : acid free paper ♾

1. Historic buildings—Virginia—Floyd County.
2. Historic sites—Virginia—Floyd County.
3. Mills and mill-work—Virginia—Floyd County—History.
4. Water mills—Virginia—Floyd County—History.
5. Floyd County (Va.)—History, Local.
6. Floyd County (Va.)—Antiquities.
7. Historic buildings—Virginia—Floyd County—Pictorial works.
8. Historic sites—Virginia—Floyd County—Pictorial works.
9. Mills and mill-work—Virginia—Floyd County—History—Pictorial works.
10. Water mills—Virginia—Floyd County—History—Pictorial works.
I. Cox, Ricky L. II. Title.
F232.F4W43 2012 975.5'712—dc23 2012001460

British Library cataloguing data are available

On the cover: *foreground* The Weddle/Wickham/Webb/Agee Mill on Howell Creek, late 1940s (Frank Webb collection); *inset* The second Jacks Mill on Little River, 1992 (photograph by Frank Webb)

Manufactured in the United States of America

McFarland & Company, Inc., Publishers
Box 611, Jefferson, North Carolina 28640
www.mcfarlandpub.com

This book is dedicated to Carolyn Webb, without whose patience with Frank's two great passions, trout fishing and mills, the book would never have been started; to Molly Brennan Cox, without whose steady encouragement and extensive involvement as reader, researcher, and reference librarian, it would certainly never have been finished; and to Sammie Phillips, a miller of high standards since 1938. — R.L.C.

THE WATER MILL

by Sarah Doudly Clarke

Listen to the water mill,
All the livelong day—
How the clicking of the wheel
Wears the hours away.
Languidly the autumn wind
Stirs the greenwood leaves:
From the field the reapers sing,
Binding up the sheaves;
And a memory o'er my mind
As a spell is cast—
The mill will never grind
With the water that is past.

Take the lesson to yourself,
Loving heart and true;
Golden years are fleeting by:
Youth is passing, too.
Strive to make the most of life,
Lose no happy day;
Time will never bring you back
Chances swept away.
Leave no tender word unsaid,
Love while love shall last—
The mill will never grind
With the water that is past.

Work while yet the daylight shines,
Man of thought and will;
Never does the streamlet glide
Useless by the mill;
Wait not till to-morrow's sun
Beams upon your way,
All that you can call your own
Lies in this—to-day.
Power, intellect and health
May not always last—
The mill cannot grind
With the water that is past.

Oh, the wasted hours of life
That have drifted by;
Oh, the good we might have done,
Lost without a sigh;
Love we once might have saved
By a single word.
Tho'ghts conceived but ne'r penned,
Perishing unheard.
Take the proverb to thine heart,
Take! oh, hold it fast!
The mill will never grind
With the water that is passed.

(Floyd) *Weekly News,* 12 June 1895, Page 1

Table of Contents

SECTION III: MILLS WEST OF ROUTE 8 AND NORTH OF ROUTE 221

SECTION IV: MILLS WEST OF ROUTE 8 AND SOUTH OF ROUTE 221

Acknowledgments

At least a hundred people made specific contributions to this book, whether in the form of an objective fact or a personal memory, the name or phone number of a likely contact, a loaned photograph, or skill at proofreading or editing. Equally important are the many people who encouraged us to continue by giving up valuable time to talk with us about the book. Along with the many people whose names appear somewhere in the text, in photo captions, or in the endnotes to each part of the book, thanks go to the following people (with apologies to those whose names ought to be here as well).

Margaret Harman and Wendell Peters, successive Clerks of the Circuit Court of Floyd County, and Deputy Clerks Danah Howell, Rhonda Vaughn, and Joan Slusher provided patient and courteous help with courthouse records over a span of more than twenty years. County historians Marguerite Tise and Effie King Brown contributed information about specific mills and people, as well as general knowledge of Floyd County history that helped make 140 separate stories begin to look like related parts of a single narrative. Thelma Webb Via provided contacts and made her electrical supply store Frank's Floyd County headquarters; Leslie Shelor typed his notes. Sammie Phillips, who began his ongoing milling career in 1938, answered both technical and historical questions in numerous personal and telephone interviews. Shai Cullop helped conform much of the text to *Chicago Manual of Style* standards. Debra McAlexander supplied background information and retrieved materials from Newman Library at Virginia Tech. Dale Belcher copied photographs for Frank Webb; he and Calvin Rorrer each gave guided tours of the Laurel Fork section of Floyd County. Thomas Wright and Harry Roberson provided facts about mills and millers along with anecdotes about their home territories. Aaron and Ada Altizer were consulted several times about the White Rock and Alum Ridge communities. Gene Vaughn contributed excellent photographs and personal memories about Vaughn's mill. Kathleen Ingoldsby's transcribed interviews of life-long residents of the Beaver Creek community and her archival work for the Floyd County Historical Society were helpful over and over. Lois Basham, Mrs. Paul Duncan, Nita Gray, a Mrs. Huff, Marvin Thomas, and Lucille Thompson contributed to Frank Webb's collection of mill photographs. Along with these people, we thank those who offered photographs we already had from another source, or that the publisher chose not to accept.

Barb Reninger's online compilation of transcribed census and marriage records was invaluable in exploring family connections within the ownership histories of most Floyd County mills. Dr. James M. Cox commented on the introduction and the historical overview. John N. Lovett, Jr., Ph.D., P.E. (Retired), graciously reviewed portions of the manuscript having to do with mill construction and operations. Dr. Barbara Ewell offered insightful comments on multiple drafts of the entire book, did most of the indexing, and provided steady encouragement. All are deeply appreciated.— *R.L.C.*

Preface
by Ricky L. Cox

When Frank Webb was diagnosed with cancer in the 1980s, he resolved to devote the time left to him to his three great loves—his family, fishing, and water-powered mills. He fished all over the U.S. and around the world, but limited his mill research to Floyd County, Virginia, where he was born in 1926 and to which he returned often after moving to nearby Roanoke with his widowed mother and three brothers in the early 1930s. Our mutual interest in mills and milling led to a crossing of paths in 1992 and for the next four years we spent many spring and fall Saturdays and many summer weekdays searching for and visiting mill sites. But until his death in 1996, Frank worked mostly on his own, collecting photographs, tracing deed records, and conducting interviews. Knowing that he would not live to see the project completed, Frank extracted from me a promise to "make a book" out of the ownership records (based on title deeds), local newspaper items, and oral histories he had collected for more than 140 separate mills, all but a few of which were at least initially powered by the force of falling water.

I drafted the texts of a few mill histories in time for him to comment, but most of my work was done several years after Frank's death. What seemed like a single, fairly straightforward task turned out to be many separate tasks. In addition to assembling and shaping Frank's notes and deed summaries into individual mill histories, the work included collecting additional written and oral information on individual mills; locating additional photographs; retracing deed book records with gaps or apparent contradictions; and revisiting mill sites. The single biggest job was researching and writing a historical overview that could explain, if only minimally, why so many mills existed over the course of two centuries in such a small and thinly populated place as Floyd County, Virginia. I worked at these things in fits and starts from 1997 until 2007, and more or less constantly, as time allowed, from 2007 until the book was finished in the fall of 2010. All told, Frank and I had the benefit of each other's advice and assistance for only four of the twenty-plus years this project has taken.

The liveliest sources for the individual mill histories are oral histories, collected on creek banks and front porches, over the telephone, and in kitchens and restaurants. Unattributed information within the "History" section of each mill entry—when it was built, who worked there, when it closed, what happened to it, etc.—unless found (rarely) in the deed records, comes from Frank Webb's personal notes, which were transcribed by Leslie Shelor while Frank was still able to proofread and correct the results. While they are concise and specific, these notes only occasionally name his sources or give the dates of their interviews. But each piece of information undoubtedly comes from someone who felt that he or she did know, and who was perhaps the only person who would know, the answer to a particular question.

Uncorroborated individual memories are bound to include in their details some errors, but presenting only the facts gleaned from census reports and deed records would exclude much of what is most important and interesting about the roles played by mills and millers in all farming communities. Excluding oral history would also discount the Appalachian tradition of historical folk narrative, which is, for the time being, alive and well, in spite of modern media. My only regret concerning these anonymous sources of key information is that it is not possible to name them and acknowledge their contributions. Information collected through face-to-face and telephone interviews, letters, and e-mails during the process of composing the histories and historical overview is attributed by the "Notes," which appear at the end of the book.

Oral histories are complemented and in most cases corroborated by the most reliable of sources, real estate deed records held in the county courthouse by the Clerk of the Circuit Court of Floyd County. But while the identities of the people involved—"the party or parties of the first part" (the sellers) and "the party or parties of the second part" (the buyers)—are usually plain enough, their motivations are seldom explained unless the sale was forced by heirs or creditors. The financial maneuverings and the legal tangles that often preceded or followed a sale, especially of some of the larger, more valuable mill properties, can be hard to follow. This is especially true if any of the key parties dealt regularly in real estate and therefore appears often in the Grantee and Grantor indexes to real estate transactions. Nor is it always clear exactly what property is changing hands. Boundary lines based on dead trees and set stones, with minimal reference to manmade features like mills and dwelling houses, can add to the confusion.

Some of the inevitable gaps and inconsistencies were spotted by my wife, Molly Brennan Cox, as she formatted individual entries, or by me as I composed the texts of mill histories. The lines of those particular ownership histories were retraced, sometimes with more satisfactory results, sometimes not, as we ourselves barked up several wrong trees and occasionally got into the wrong patch of woods altogether. In spite of the hundreds of hours both of us spent on this, on top of the many hundreds of hours Frank had already spent, a few such errors may still have been overlooked, since there was no point to our retracing hundreds of correct steps for the sake of correcting one or two that went astray. Upon discovering an error among the deed book records, readers are encouraged to visit the office of the Clerk of the Circuit Court of Floyd County to pick up the track one of us lost. The solution, in written form, to any legal tangle we failed to straighten out will be welcomed.

As regards references to newspaper articles, notices, and advertisements, most of those without a page number appeared in Frank's notes with only the name of the paper and the date of publication. The fact that most of the newspapers referred to consist of only a few pages prevent this from being the inconvenience it might at first seem.

Standard reference works on mills and milling, census records, local newspaper items, and additional oral history accounts provide the factual underpinnings of the historical overview. Although it grew to more than twice its intended length, the overview is intended only to provide a backdrop for the individual mill histories, and to suggest social and economic contexts in which so many mills, some only a few hundred yards from a competitor, and others in spots so obscure they have virtually reverted to wilderness, could have been necessary parts of their respective communities.

Readers familiar with old Floyd County families will spot apparent misspellings of both first and last names throughout the book. Surnames with multiple spellings include Booth/Boothe; Earles/Earls; Ekis/Ekiss; Guerrant/Guerant/Geurant; Harman/Harmon; Kropf/Kropff; Lawrence/Laurence; Quesenberry/Quessenberry/Quisenberry; Short/Shortt; Tice/Tise; and Yeats/Yeatts. Variant spellings of given names include Alfred/Alford; Cory/Cary/Cara; Jarmen/Jerman/German; and Moir/Moyer. Among the variant spellings of surnames, some are still in use as legitimate versions of the same name (e.g., Quisenberry and Quesenberry), and some are supposed to refer to distinct families (e.g., Harmon and Harman). Other unfamiliar spellings of last names may represent obsolete earlier spellings (e.g. Alterman for Alderman). Among given names, many alternative spellings are simply phonetic transcriptions, sometimes flavored by regional dialect, e.g. "Elcany" in place of Elkanah. Rather than attempt to identify the single correct spelling for every name, we chose simply to be consistent at any point with the source nearest to hand. In most cases this was a deed book record.

Actual errors in Frank Webb's laborious transcriptions of handwritten deed records are remarkable chiefly in that there are so few. Those that did escape notice are to be excused considering that he undertook a monumental and tedious task and pursued it doggedly, in spite of steadily worsening health, under the most unforgiving of deadlines. My own errors and my failures to resolve questions and consult additional sources are due simply to the limitations imposed by a job, a family, and other interests. With unlimited time and the assistance of an experienced surveyor and a real estate lawyer, one could follow every lead, reread every deed, compare all the boundaries, identify and consult secondary sources, and eventually resolve some of the ambiguities that remain. Having none of these resources, we did our best to fill gaps in ownership records and to keep track of the locations of original mill properties within tracts of land that grew or shrank as their boundaries were defined and redefined in successive transactions. Had I waited until the book was perfect to publish, it would never have been published, leaving Frank Webb's marvelous work and tenacity unappreciated and the product of his effort largely unknown.

The saving grace in this regard is that all printed records at our disposal are available to anyone, and will be for the foreseeable future. Since virtually all the unanswered questions pertain to deeds recorded before 1910, the answers to those questions are already beyond the reach of living memory. If they are to be found at all in firsthand accounts, they will be found in written records. There is no doubt that much remains to be learned and written about many of the mills in this book, but for now these things, like all grist that comes to the mill, will have to wait their turn.

Introduction

Bread may still be the staff of life, but we are leaning far less heavily on it than we used to. Our daily bread includes less and less actual bread, and most of what of we do eat comes to the table not from our ovens, but from a plastic bag. But seventy-five years ago, and in many cases much later, homemade biscuits or cornbread, made from locally grown wheat and corn, were on the table at almost every meal in Floyd County, Virginia, and in much of the rest of the rural South. To meet the demand for flour, corn meal, and animal feeds in a society that was still heavily horse-drawn and foot-powered, a mill was needed within walking distance of every community. In response to this need, over the past two centuries more than 140 grain mills were built along the banks of Floyd County's one river and numerous creeks, runs, and branches to grind various grains into foodstuffs for the dining tables of Floyd Countians, as well as for families living near the Floyd County border in the adjoining Southwestern Virginia counties of Carroll, Pulaski, Montgomery, Roanoke, Franklin, and Patrick. Along most any stream more than four feet wide and some smaller, long-time residents can point out the ruin of a building or a dam, or the shallow remnant of a race dug to channel water to a mill that was depended upon by all the people and most of the livestock of the neighborhood.

Although the term "grist mill" is often applied to a mill capable of grinding any sort of grain, it most properly applies to a mill, whatever its source of motive power, that grinds dried corn into cornmeal for cornbread, cornmeal mush, and corn whiskey; or into grits, a more coarsely textured foodstuff less widely known and consumed in the Mountain South than in the Piedmont and Lowland South. All such mills could also shell, crush, and crack corn for chicken, horse, and cattle feeds. At larger, more complex, and more widely scattered "flour mills," wheat was ground into flour for biscuits, gravy, cakes, and pies. The byproducts of these processes and custom mixes of hay, oats, corn, and wheat, sometimes combined with molasses and other supplements, were bagged as feed for hogs, chickens, horses, sheep, and cattle. A substantial number of mills were equipped to grind buckwheat, usually after hulling it, into flour for buckwheat pancakes. Other small grains— like rye, for distilling and other uses, and oats, mostly for animal feeds— were also processed. Technically, every grain mill can produce animal feeds, and is therefore a "feed mill," but at least one Floyd County mill was built exclusively for this purpose, and several one-time grist and flour mills had become feed mills, exclusively, or nearly so, by the time their doors were closed.

Not surprisingly, men and women who built and ran grist, flour, and feed mills pursued other industries that depended upon and sometimes competed for the relatively small amounts of motive power provided by Floyd County's streams. Before portable steam boilers allowed lumbermen to begin moving sawmills into the woods around 1900, a significant portion of the power extracted was used to turn the county's timber into lumber. Many 19th century and several early 20th century grain mills shared a dam, race, or water wheel with a sawmill that often predated it on the site. A much smaller portion of the county's water power was put to work carding and spinning wool into yarn, and, later, weaving the yarn into finished

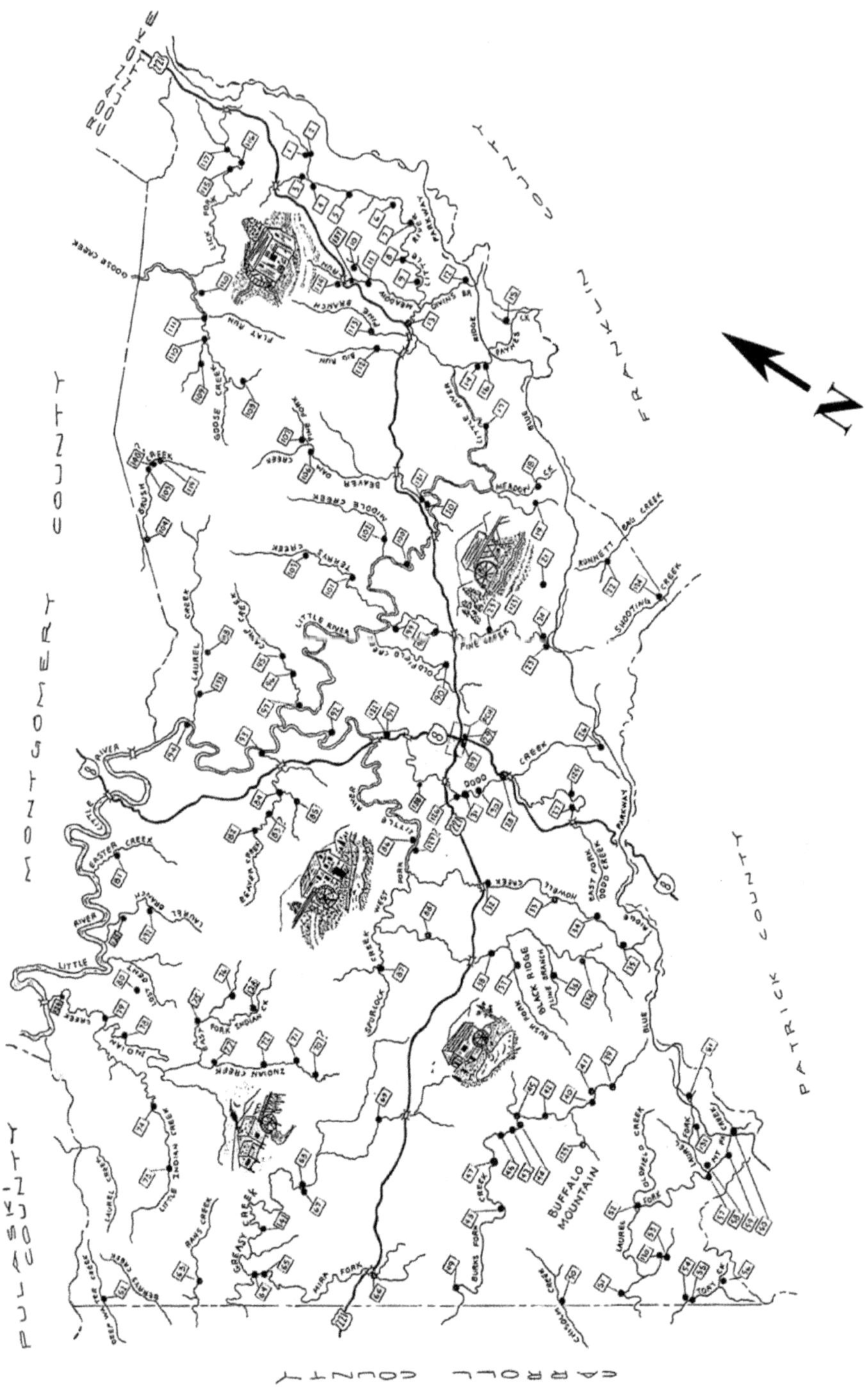

Floyd County, Virginia
Grist, Feed, Flour, and Woolen Mills
c. 1770-2010

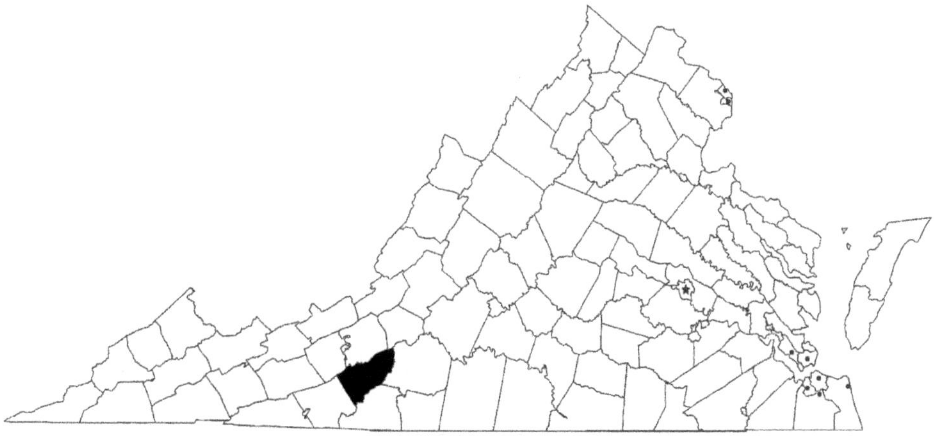

Map of Virginia showing Floyd County in black.

cloth, blankets, and coverlets. A handful of blacksmiths and furniture makers employed water power to drive trip hammers, lathes, and saws. At least two tanneries applied the force generated by a water-turned wheel to grind the hardwood bark from which they extracted tannic acid.

Before rural electrification reached up the creeks and into the hollows in the 1940s, there were even a few hydroelectric installations set up to power one or two DC light bulbs or recharge a 6-volt wet-cell radio battery in time to hear the Grand Ole Opry on Saturday night. Others illuminated the mill or a few nearby homes after the mills ceased grinding for the day. The Floyd Electric Power Corporation built by far the largest hydro-power installation at the site of Sowers Mill on Little River in the early 1920s. It supplied the Town of Floyd with electric current until its physical plant was bought out in 1937 and dismantled and its function taken over by the larger and more efficient generating facilities and transmission lines of Appalachian Electric Power Company.

The most obscure applications of water power were driving the bellows of primitive iron furnaces and grinding flaxseed for the manufacture of linseed oil. These were among the first applications to disappear, their products having been priced out of the market by the same economies of scale and geographic determinism that eventually doomed all the aforementioned industries, including sawmilling. Floyd County can grow and harvest trees as cheaply as most other places, but it has become more profitable to haul the logs twenty or thirty miles to factory-like mills in adjoining counties than it is to saw them locally and haul out the lumber.

By 1935, the county's two or three iron foundries and one or two oil mills were long forgotten, and only one woolen mill remained. One water-powered sawmill was still at work at the Flint Mill on Beaver Dam Creek and another at what was then the Murray Agee Mill on Howell Creek. The grain mills at both locations were by that time grinding only corn, having lost their flour milling trade to roller mills.

By 1940, the very last water-powered corn and flour mills had been built and many old ones were closing their doors. With the beginning of World War II, non-farm income rose rapidly, roads improved, and automobiles became more plentiful. Instead of trading a portion of their corn and wheat in exchange for having it ground into meal and flour, people could afford to buy these items at mills or stores for what amounted to a smaller and smaller portion of their cash incomes. The fields of wheat and heirloom varieties of open pollinated "bread corn" that once patchworked the hills of Floyd and surrounding counties grew fewer and farther apart, and finally disappeared. Larger and larger quantities of dairy, horse, and hog feeds, ground with powerful electric motors and diesel engines, became the primary products of a shrinking number of mills.

A cash economy, whose arrival was delayed by the Great Depression and then accelerated by the onset of World War II, along with good roads and later the interstate highway system, made the world smaller. Local mills thereby lost the only true advantage—convenience—they ever had over distant competitors, who brought in corn and wheat on barges and railroad cars instead of in wagons and pickup trucks, and sold meal and flour by the ton instead of the bag.

Today the most likely spot to find a corn shock, the familiar teepee-or tipi-shaped bound bundle of cornstalks with ears still attached, is alongside a fall season pumpkin display in a convenience store parking lot. The annual wheat harvest, for thousands of years the farmer's most important summer task, does not merit even a symbolic tribute and is instead recalled infrequently and indirectly through a reference in an old Protestant hymn ("Bringing in the Sheaves") and in dying figures of speech like a "shock" of hair or a "stubble" of beard. Even the once familiar *New Testament* analogies found in the parable of the sower and in that final separating of the wheat from the chaff, of the just from the unjust, are disappearing from daily speech. They are no longer familiar to a population that neither sows nor reaps, in any literal sense, and would recognize neither newly threshed wheat nor the worthless chaff.

Whether these changes are good or bad is a matter of opinion, one that a thoughtful person would have a hard time forming, and a harder time defending, without attaching a long list of qualifications and exceptions. Probably the only adjectives that can be applied with any objectivity are "easier" and "harder," and these only in the physical sense. It is clearly faster and easier to drive to the nearest store and buy a loaf of bread, given enough money to do so without giving up some other necessity, than it is acquire or rent a suitable field, prepare the land, sow the wheat, protect it from foraging animals until it ripens, cut and bind it, shock it, stack it, thresh it, haul it to the mill, have it ground, bring it home, mix it with other ingredients, and bake it. Merely reading this sentence aloud makes it clear why the Little Red Hen had such a hard time getting reliable help. On the other hand, she knew exactly what was in her bread, and exactly what it took to make it. Peace of mind, whatever that means to a chicken, was hers.

Either way, the grain mill as a vital part of every rural community may be gone forever. Only by selling romance and nostalgia along with meal and flour, or by serving some highly specialized niche market, could anyone extract the extra cost of doing business on such a small scale.

Whether or not doing so can illuminate an uncertain future, we ought at least to consider Floyd County's mill history, to appreciate the essential industry that once was, and to find or perhaps merely assign some purpose to what remains of its history. If these prove too ambitious, the original and still main purpose of this work remains, that being simply to record the names and locations of as many mills and mill owners as possible.

Beyond this basic goal, at least some additional information about nearly every mill was gleaned from oral histories, deed books, census records, and written histories of the county and region. Most of this material pertains to the activities carried on at a given mill, the people who owned and ran it, and its years of operation. Milling as a process is discussed in bits and pieces, but comprehensive accounts of milling machinery and mechanical processes have been left in the capable hands of other writers. In all cases, an effort has been made to include the material not likely to be found in a library or archive, whether or not it could be confirmed beyond the shadow of a doubt. As a result words like "said to have been" and "supposedly" appear more often than the authors would have liked. But much of what is most interesting about these mills, and most worthy of recording, exists or existed only in the memories of Floyd Countians who raised corn and wheat and carried them to the mill, wore clothing made from printed cotton feed sacks, and, for much of their lives, ate biscuits at every breakfast and cornbread at every dinner and supper.

As information, both oral and documented, accumulated about individual mills, the impulse arose to look more broadly at the economic and social functions of the neighborhood grist mill. These topics are occasionally treated as they arise in the discussion of why, when, where, how, and by whom a mill was built, and of the goods and services it provided. Besides storing water to power the mill, millponds, both full and drained, drew fishermen and turtle hunters. The necessity for waiting one's turn presented a natural forum for serious talk of politics and economics and for serious play like checkers, "Fox and Geese," or shooting or wrestling matches. Stores and post offices sometimes grew up in the same or adjacent buildings and were owned and operated by the same families. For people who didn't buy a great deal of manufactured goods or processed foods, the mill and the buildings and services located nearby became a rough precursor of the mall concept of today, the original home of "one stop shopping."

Also treated in bits and pieces are the applications, and implications, of an important alternative means of producing cornmeal, the small semi-portable mill that could be driven by a gasoline engine or an electric motor, or belted to a line shaft driven by water power. Examples of these self-contained mills, whose prominent manufacturers included the Meadows and Williams companies of North Wilkesboro, NC, began arriving in Floyd County around the time of the First World War, dotted the countryside until after the Second World War, and virtually disappeared by 1960. The widespread adoption of these small mills is but one of several factors that contributed to the decline of grain mills, whatever their source of power, in Floyd County.

Many of the earliest mills in the county now exist only as obscure deed book references, recorded in the spidery handwriting of a long ago clerk of the court. Others cling to life as nearly-forgotten rumors in the memories of the county's oldest citizens, as dots on obsolete postal and highway department maps, or as names or numbers in *Products of Industry* reports and gazetteers. A surprising number are still in evidence physically, albeit mostly as crumbled foundations and dam abutments or half-filled ditches along the banks of roads and streams.

Among the fifteen or so standing mills that Frank Webb and I visited before his death in 1996, a third collapsed or completely disappeared in the subsequent 15 years. Ironically, one of the very oldest and most significant Floyd County mills, Spangler Mill on Pine Creek, remained fully equipped until the early 1990s, and stood until early in 2005. Except for a scrap of the c. 1911 dam and the outline of the original dug race, no trace of it now remains. Only ten buildings that once housed water-powered mills still stood in 2010. Most were steadily decaying and some were nearing collapse. No more than three or four are likely to survive to the middle of the 21st century.

Among these survivors, only two were still at work in 2010. As the last for-profit commercial mill in the county, Vaughn's mill on Mira Fork Creek in the northwest quarter of the county is at this writing the final link in a chain leading back to the second half of the 18th century, before Floyd County was Floyd County, when the earliest permanent white settlers were planting their first crops of wheat and corn. Formerly a producer of cornmeal, wheat flour, buckwheat flour, and animal feeds, Vaughn's mill still retails meal and flour produced in North Carolina. The mill itself produces only feed for animals using electric motors and a diesel engine, having abandoned its steel overshot water wheel half a century ago.

It is not quite accurate to speak of the other still-operating mill as a survivor. It is instead what folklorists call a revenant, a returner, having been raised or at least saved from the death it would have died at least half a century ago had it not stood within the right-of-way for the Blue Ridge Parkway. The restored Mabry Mill is operated by the National Park Service as a tourist attraction and historical exhibit. Of the tens of thousands of mills that ever existed anywhere in the world, Ed and Lizzie Mabry's could be the best known; it is almost certainly the most frequently photographed, thanks to the nearly 20,000,000 people who travel the Blue Ridge Parkway, America's most visited national park, each year.

For many millions of Americans (whether or not they've actually been there and seen it) Mabry Mill is *the* mill, a splashing, creaking embodiment of pre-industrial America. But as a symbol of rural technology in the early 20th century, Mabry Mill tells only part of the story. Its basic design was already ancient when Jamestown was founded in 1607. At least a dozen mills making flour with high speed turbines and roller mill systems were already at work in Floyd County when Mabry Mill was built about 1910. Yet the fact remains that until the radio, the Model T automobile, and steam-powered sawmills became familiar, a mill like Ed Mabry's would have been the most complex example of applied technology many people in the surrounding community would ever see. Visitors to Mabry Mill expect, and in one sense receive, a distillation of the traditional culture and economy of an entire region, neatly bagged and tied with a string.

The reader is asked to bear in mind that conclusions, stated or implied, as to why an essential industry practically disappeared within fifty years are based only on conditions and examples found in one county or, at best, one small part of one region of a very large nation. A definitive treatment of this question requires a look at the history of American agriculture in general, and of small scale, subsistence farming in particular, since the collapse of the local milling industry closely parallels the decline of diversified, multi-crop farming not only in Floyd County, but in the entire nation. Kansas and California will have to be considered along with Camp Creek and Lick Fork. Henry Ford and Cyrus McCormick must be accounted for, along with Newton Hylton and Hiley Moses.

In the milling process at its most basic, a miller and his mill accept a customer's wheat or corn, separate the edible from the inedible, grind it to a digestible and hopefully palatable form, and bag what remains for the trip back home. The information gathered for this book has not made so neat a package, and will perhaps need to be run through another set of stones or a better sifter before it is fit for some tables. But it offers at least the raw materials without which the finest mill is useless.

Part One

Historical Overview

Historians usually understand and present their subjects as series of related events, each of which was both a consequence of what came before and a cause of what was yet to happen. But in writing about Floyd County's water-powered mills, a chronological approach has two major drawbacks. The first is that the history of every mill overlaps the histories of several others, with tremendous variation between the lengths of their individual working lives. Some lasted for only two or three years, while others stayed in business for a hundred or more years, with the buildings standing idle or put to some other use for many years afterward. The beginning and ending dates of many early mills can only be guessed at, making it even more difficult to keep track of which mills were running at any given time.

The second problem with a year-by-year approach is that over the two centuries covered here, there was no steady or invariable progression away from the earliest technology toward more modern machinery or methods. Without the hammer mills and bin mixers that produce bulky livestock feeds, and the diesel engines and electric motors that replaced water power, the handful of small-scale flour and feed mills operating in rural Southern Appalachia in 2010 would differ but little from mills built or rebuilt in Floyd County a century ago. In fact, although the means of storing, moving, and packaging materials have changed, and steel and plastic have replaced wood and cast iron as component materials, the flour-making machinery in the newest industrial flour mills anywhere in the country is in principle identical to the machinery found in most Floyd County flour mills in 1920.

Even fewer fundamental changes took place at the dozens of smaller mills built to make only cornmeal, rye meal, buckwheat flour, and animal feed, and at the many pre–1900 flour mills that simply stopped making stone-ground flour when customers and millers rejected it in favor of finer, whiter, roller mill flour. Two of the last water-powered mills built in Floyd County, Jack's Mill on Little River (I, 5) and the Flint Mill on Beaver Dam Creek (II, 106), were grist mills constructed in the 1930s on the sites of 19th century mills that had ground both corn and wheat. The new mills were intended to grind only cornmeal and small volumes of animal feed. Instead of being, as one would expect, the most modern mills in the county, they were in principle of operation among the most primitive. Apart from having replaced part or all of the wooden water wheels, shafts, and gearing with cast iron and steel, the builders of both mills relied on machinery and methods of grinding cornmeal that would have been familiar to a miller working in the 1830s or even the 1730s.

Mabry Mill (IV, 60), which was restored by the National Park Service in the 1940s and is still in use as a working historical exhibit on the Blue Ridge Parkway, provides a more familiar illustration of the coexistence of old and new technology. Except for the greater reliance on iron and steel in the drive train, the basic milling technology predates the 1607 settlement of Jamestown, Virginia. But Mabry Mill itself was not built until around 1910, ten years or more after the first multi-story, turbine-powered roller mills appeared in the county. Many of Ed

Mabry's regular cornmeal customers were probably having their wheat ground into flour at one of several modern, high-speed roller mills less than a day away by horse and wagon. By the time Mabry Mill ceased to run as a private business around 1935, the wooden overshot wheel had been abandoned and a one-cylinder kerosene engine brought in to drive the millstones. The few days it may have taken to install the engine and connect it to the drive train of the mill with belt and pulleys represented a technological leap of many hundreds of years.

In preparing the mill for display, restorers did away with the engine and hired Newton Hylton to build a replica of the original wooden wheel, which was already obsolete outside the upland South when it was built about 1910. The engine and other more modern aspects of Mabry's operation, including the frame house he built above the mill, were purposely excluded from the story this site tells regarding early 20th century technology in parts of Appalachia not industrialized and modernized by coal mining or large-scale timbering. Skewed as it is toward the most primitive technology in use in that era, the picture presented to visitors may be criticized as incomplete, but it cannot be dismissed as inaccurate, since at least one similarly primitive mill (the two mentioned above, and the Enoch Phillips Mill [III, 74] on Little Indian Creek) was running in each of the other three quarters of Floyd County (as defined by Routes 8 and 221) until 1920 or later. In short, the most up-to-date mill technology and the most ancient could be found at work in Floyd County, a few miles apart or a few feet apart, for most of the span of time covered in this book.

Faced with these arguments against a year-by-year arrangement, this overview focuses instead on several key aspects of mill history, with chronology receiving more or less emphasis depending upon the subject at hand. These discussions are intended as background for the histories of individual Floyd County mills, and to give readers a basic introduction to terms and concepts they will encounter there. No part of this overview is comprehensive or exhaustive.

The discussion of the machinery and methods of grain milling is deliberately brief because little if any of it is unique to Floyd County mills and because detailed treatments are already available to the serious reader. Oliver Evans' *The Young Mill-Wright and Miller's Guide* and Professor B. W. Dedrick's *Practical Milling* thoroughly describe the machinery and processes of buhr stone (or burr stone) milling, which uses the traditional massive natural millstones to make cornmeal, buckwheat flour, rye meal/flour, and wheat flour. Dedrick offers an extremely detailed explanation of the roller mill system of milling, which in Floyd County more or less eliminated stone-ground wheat flour within thirty years of its circa 1900 arrival. The construction of dams, the manufacture of millstones, and the science and history of vertical and horizontal water wheels are also the subjects of book-length studies and numerous articles. Terry Reynolds' *Stronger Than a Hundred Men: A History of the Vertical Water Wheel* stands out as both scholarly and readable. Jon Sass's *The Versatile Millstone: Workhorse of Many Industries* is a concise introduction to the history and uses of natural mill stones.

Examples of milling technology as it was applied in Floyd County are discussed as they arise within individual mill histories. References to the most detailed of these are included in the overview, and the index will point readers to shorter references within other mill histories. Treating the individual mill histories as more or less self-contained units, and referring to them at appropriate spots in the overview, allows for the fact that readers already familiar with Floyd County may look first, and perhaps only, at the entries for the mills most familiar to them. Although the mill histories were written and arranged with these readers in mind, the suggested sources, references within the overview to specific mill histories, and the table of contents should make the book equally accessible to readers who prefer to begin at the beginning or to know something about the forest before looking closely at the trees.

Old World Origins of Mills and Milling

If one accepts that permanent human settlements began to form (in grain producing climates) only after the domestication of wild grains made it possible for formerly nomadic peoples to live year-round in one place, then mills and civilization are inseparable. This is true whether the mill is one mile or five hundred miles away from the people who grow the grain or consume the meal and flour made from it; it is equally true whether growers and consumers are the same or different people.

Writing in 1924, Professor B.W. Dedrick was emphatic if not precise in declaring that the milling of grains "is as old as the cultivation of cereals by the earliest primitive man."[1] The parallel development of grain-based agriculture as a cornerstone of Western civilization, and of milling as an early industrial process, is reflected in the shared origin of the words "mill" and "molar." The back teeth of humans and other omnivores, as well as herbivores like horses and cattle, get the name "molar" from the Latin "molaris" meaning a mill, or "mola," a millstone, whose work these teeth imitate by crushing and grinding food. The French "moulin," the Spanish "molino," and the German "mühle," all of which mean "mill," reveal an obvious and ancient kinship. The American melting pot eventually converted surnames based on these words and others like them into the anglicized "Miller," which was the sixth most common surname in the United States in 2000.[2] Some of the original variety among these cognate words is preserved in the names of places as diverse as Moline, IL, the Moulin Rouge cabaret in Paris, France, and Muhlenberg County, KY.

By whatever name, the process of grinding grain to make it more palatable and more digestible has been known to agricultural societies for thousands of years. The first mills developed in the ancient world used human and later animal power to grind small volumes of grain between shaped stones, making the grains easier to cook, easier to digest, and more versatile as ingredients in baked or boiled foods. Most of the grains that originated in or were introduced to the Mediterranean, Asia Minor, and Europe were small enough and soft enough to be eaten whole, cooked or uncooked, but were far more appetizing and nutritious after milling and baking or boiling. Grain that had been boiled, whole or crushed, in water or milk, and then salted or sweetened, became the European "porridge," which, thanks to Mother Goose and the Brothers Grimm, most Americans associate with poor but honest peasants as well as the better class of talking bears. The enduring popularity of boiled grain dishes in modern day Europe and Great Britain shows that people who can certainly afford "better" continue to enjoy grain in its simplest forms. But these minimally processed foods are vastly outnumbered by breads baked from milled grains of all kinds, alone and in numerous combinations. For all of recorded history, bread baked from the products of a mill, whether a mile or a thousand miles away, has been a dietary staple among all classes of people. As a result it has provided a familiar metaphor in cultural, literary, and religious contexts, in every part of the world whose climate can produce edible grains.

Mills and Migration

The Scots-Irish, German, English, Welsh, French, and other groups of backcountry settlers came into the Blue Ridge Mountains and points west after the flat expanses of the Tidewater and the gentler slopes of the Piedmont were already settled and under cultivation. Few of them could compare pedigrees or pocketbooks with the Tidewater aristocracy. But while these pioneers may have been among the poorest in pre–1850 Virginia, they were not generally descended from the poorest Europeans, as would be the case with many middle and late 19th century immigrants.

As second or third generation Americans, if not themselves natives of Great Britain or Western Europe, the earliest white settlers in what was to become Floyd County were quite familiar with the whole-grain flours produced in thousands of mills in Europe and Great Britain, most of them water-powered, or in numerous replicas of simple European mills lining the banks of suitable colonial American streams by the mid–18th century. Blacks, whether slave or free, had little choice but to adopt diets based on European tastes and traditions, thanks to the radically different climate, the lack of familiar seeds, and the fact that slaves, at least, had no land of their own and little time to tend the plots set aside for their personal use.

The use of water power for grain milling and other basic industrial processes was quickly established in settled regions of the British colonies. Beginning with a windmill in Jamestown, Virginia, circa 1621 and a water-powered mill in Plymouth, Massachusetts, soon after,[3] mechanized mills followed the westward advance of white settlement. In an article published in the journal *Science* in 2008, Dorothy J. Merritts and Robert C. Walter surmise that by 1840, dams and water powers for various purposes were so closely spaced along some streams in the piedmont and ridge-and-valley regions of Pennsylvania that the upper ends of many millponds reached nearly to the bases of the next dams upstream. After the mills and the dams had disappeared, a tiered string of silt banks remained in the stream beds, one immediately above the location of each dam.[4] (Virginia legislators apparently sought to prevent similar overcrowding on streams by the 1705 passage of a law forbidding the construction of mills within a mile up or downstream from another "without the particular leave of the general court, or county court, in which such mill is intended to be set."[5] The law may have limited the development of water power in more densely settled parts of Virginia, but seems to have had little effect on Floyd County millers, many of whom constructed mills within a mile of an up or downstream neighbor with or without permission of the county court.)

This alteration of the colonial landscape, and the demand for motive power that caused it, was only one consequence of the overpopulation of farming people in parts of Pennsylvania prior to the American Revolution. Of far greater importance to the history of Floyd County and all of Southwestern Virginia was the pre–and post–Revolution exodus, from eastern and central Pennsylvania, of thousands of settlers of mostly Scots-Irish and German descent. As early as the middle of the 18th century they were streaming down the Shenandoah Valley, also called the Great Valley, toward Southwestern Virginia and Northwestern North Carolina, hoping to find and lay claim to better or at least cheaper land. Movement into the New River Valley was temporarily slowed by the perceived dangers of settling in a true wilderness and by the British Proclamation of 1763, which prohibited settlement beyond the crest of the Blue Ridge. Only personal reservations continued to be a factor after the American Revolution, which through its nullification of British restrictions permitted and ultimately encouraged settlement, since large tracts of western lands were granted or sold at attractive prices to Revolutionary War veterans and others willing to move west to new homes in the mountains.

As soon as the first crops could be sown on their Blue Ridge homesteads, the first arrivals must have begun wishing for mills to grind the first harvests. Temporary reversions to slow, hand-powered milling devices such as a mortar and pestle or a quern (a very small buhr stone mill whose top or runner stone is turned by hand) would have been one among many significant sacrifices of comfort and convenience made by this region's earliest white inhabitants.

As soon as land could be cleared, settlers planted the wheat and barley, as well as the hardier oats and rye, brought with them from Europe. And they had already adopted a native plant that had been utterly new to Europeans, one that could be either a vegetable or a grain, depending on the season and the form in which it was eaten. The earliest colonists had named it "Indian corn," to distinguish it from European "corn," which in British usage refers to various staple grains native to the Old World, usually wheat.

Indian corn's productivity in barely cleared "new ground" fields, and its versatility as a food for hogs, horses, cattle, sheep, chickens, turkeys, ducks, and dogs, as well as humans, made it an ideal pioneer crop. Its "patience" in waiting to be harvested was also an asset. Harried farmers had only a couple of weeks to cut and shock a fragile wheat crop, but could take months to harvest dried ear corn hanging on the stalk in its protective, umbrella-like shuck,[6] as long as foraging animals or an unscrupulous neighbor didn't find it first.

But to serve as food for humans, corn benefited even more from grinding than the small European grains. Fully matured and dried corn of the flint, dent, pop, or flour varieties (as opposed to the sweet corn that people still eat canned, pickled, or fresh in the milk stage) could be gnawed at but was essentially inedible, unless one had the jaws and teeth of a horse or a pig, or the multiple stomachs and (physical) ruminating capacity of cattle. Grains of dried field corn can be softened by parching over a fire or in an oven, but without first being crushed and sifted, as in making grits or cornmeal, or soaked in lye water to loosen and soften the protective covering, as is done in the process of making hominy, are little affected by boiling.

Native Americans had gotten along well enough for thousands of years grinding small quantities of dried corn between a rounded hand-held stone and a flat or concave bed stone, or, when suitable stones were not available, crushing it in a samp mill. In this device a heavy wooden post with a rounded end was suspended over a hollowed out stump and corn was crushed between them, a few grains at a time. The living tree limb or prepared spring-pole to which the post was attached would carry some of the weight and impart a regular rhythm to the pounding.

But native cultures in this region seldom achieved the population densities of white settlements, and most were not as heavily dependent on agriculture as white settlers would become in the 19th century. The first European settlers along the banks of the Little River and its tributaries probably did temporarily adopt native methods, or rig up querns using locally quarried stones, but the tedium and time required to grind even a few cups of coarse cornmeal would have made a mechanized grist mill a high priority for every settlement. Even if no time were saved after the walk or ride was taken into account, the opportunity to be out and about and the prospect of meeting friends or neighbors made worthwhile both the trip and the miller's toll for grinding, which was set at a maximum of one-sixth (for all grains) in 1645,[7] but reduced by the legislature to one-eighth portion of "English graine" (wheat) in 1671[8] and to one-eighth portion of wheat or corn in the legislative session ending in October 1748.[9]

Flour mills, being more elaborate, more expensive, and more difficult to operate, were even more to be encouraged, as there is no comparable makeshift method of grinding wheat into flour fine enough to make a dough or batter that will rise satisfactorily before or during baking. Moving beyond the reach of a mechanized flour mill thus meant doing without the bolted (sifted) wheat flours required to produce the soft, light, baked goods that demonstrate both the skill of the baker and the level of sophistication present in the home. Families without this luxury instead ate coarser whole-wheat flours and breads.

The First Floyd County Mills

The identity and location of the very first grain mill in what became Floyd County will never be known with absolute certainty; existing records are inconclusive, and there is little doubt that a number of early mills came and went without leaving any written record at all. That said, the reasoning behind these assumptions makes an interesting story, part of which illuminates the relative freedom from government intervention or assistance that some pioneers dreaded, some embraced, and all accepted as a consequence of leaving "civilization" behind. The effects of living for many generations in this kind of environment include the independence, self-reliance, and resentment of official meddling in one's private affairs that haven't quite died out among Floyd Countians, even two centuries later.

Existing claims to the title of "first" will be examined in the discussion that follows. But it bears repeating that the most carefully researched conclusions are, ultimately, speculative, for the simple reason that some of the best contestants were probably forgotten long ago. The mills most likely to be overlooked were naturally the oldest, the smallest, and the shortest lived, but age, size, and longevity are not the only circumstances obscuring the origins of mills, from the very earliest, c. 1770, to some that were built well over a century later. The natures of pertinent legal documents, as well as the human natures of the people responsible for creating and interpreting them, also have a part to play.

Since deed records have been the primary sources of information on where, by whom, and roughly when mills were built, those on land that did not change hands for many years are also likely to have escaped notice. In such cases, no deed would mention the mill, mill pond, or mill race until some part of the property on which these stood, or an adjacent tract that was somehow affected by one of these features (e.g., partially flooded by the mill pond or crossed by a deeded right of way for a mill race), was transferred to a new owner. This might not happen until thirty or forty years after the mill was built, by which time a dusty wooden structure, susceptible to fire and constantly threatened by flood and damp, could have been accidentally or deliberately destroyed, or simply abandoned. In the case of an adjacent property that was in some way affected by the presence of the mill, unless it belonged to the same owner, it would be nearly impossible for a researcher to identify its sale as significant to his or her research. Hence, a mill could conceivably have been built and used for half a century and still not appear in deed book records.

Even when land did change hands with a mill standing and perhaps even operating upon it, there is no guarantee that the mill or mill property would be mentioned in the deed recording the transfer. This omission occurs in a surprisingly large number of the deeds. Although puzzling and frustrating to historians, this absence is more to be blamed on long-established conventions for the conveyance of real property than on an error or an oversight by any group or individual.

For their part, surveyors had no real cause to mention a mill or its water power components unless the mill itself, the mill dam, the mill pond, or a dug (in-ground) race was useful in describing the boundary lines. The last three, in particular, were perceived by surveyors as relatively permanent features of the land, as they certainly were, compared with the dead trees, saplings, stumps, "pointers," and "set stones" frequently referred to in property boundaries. All three were regularly used as landmarks in describing the boundaries of a piece of property, sometimes years after the mill itself had disappeared, whenever they happened to follow, define, or be crossed by a boundary line.

Nor can the absence of consistent and specific references to mills in deed records be blamed on the magistrates and attorneys who actually composed and wrote out the deeds that would be transcribed and entered into the record by the current Clerk of the Circuit Court or a Deputy Clerk. Although drawn up by dozens of people all around the county, the contents of these documents are remarkably consistent in order and wording, and adhere closely to a format which tended not to enumerate or specify structures, often lumping improvements of all kinds under the term "appurtenances."

Because it cannot be taken for granted that the owner(s) named in the first deed that does include a reference to the mill is the one who built the mill or had it built, that owner's name may be of no use in searching for the legal petitions (if any were ever filed) that should have preceded the construction of a mill dam. Further, pertinent Court Order Books in Floyd County are indexed, consistently, only by the names of petitioners and other persons appearing before the court, and not according to their reasons for appearing. Hence, a search for the legal beginnings of a given mill, unless explicitly revealed in deed records or other documents, or in oral histories, means searching for the names of a string of successive mill property owners in the indexes of multiple volumes, each covering several years' worth of hand-written court proceedings.

The second main reason that early mills cannot be reliably traced through courthouse documents has to do with understandable lapses on the part of mill builders and the court officers charged with regulating them. Only those provisions of mill-related legislation that encouraged the building of mills and, by extension, the growth and development of the county, appear to have been conscientiously obeyed by persons desiring to build mills, or strictly enforced by the county court.

Beginning in 1667, prospective Virginia millers who did not own enough land to encompass both a dam and the mill pond it would create, and who could not persuade a neighbor to sell at an acceptable price (if at all), could use the power of the local court to acquire the minimal amount of necessary land as long as the transfer of property "take not away howsing, orchards, or other imediate conveniences" from a neighboring land owner.[10] This provision was later expanded to include other buildings, curtilage (fenced spaces around buildings), and gardens, as well as to address concerns of the larger community, such as impediments to navigation, blockage of the passage of fish, or endangerment of public health through the presence of stagnant water.[11] Considering the large tracts of land held by most of the county's earliest families, most probably did not need a neighbor's cooperation, voluntary or induced, to erect a dam. In cases where streams did serve as property lines, efforts were no doubt made to buy the needed land during the planning stage, in order to avoid legal proceedings both then and in the future, when the dam might need to be raised or relocated, or a race lengthened or enlarged. Then as now, for most people, legal action against a neighbor would have been a last resort.

But having exhausted personal means of reaching an acceptable agreement, and having no other means of moving forward, ambitious and determined men did engage attorneys to ask the local court to intercede on their behalf. The process seems strangely modern for a time we like to think of as having been relatively free of government interference in private affairs. By compelling one private citizen to sell property to another for use in a private business venture, it far exceeds the familiar power of eminent domain, which allows governments to seize and use private property for a supposed public good. The practice was justified with the argument that the "perverseness" of uncooperative individuals should not prevent the building of mills, since a grist mill, even though privately owned, was essential to the "public goode."

Upon receipt of the petition, the Circuit Court would issue a *writ of ad quod damnum* which led to an on-site inquest by a panel of at first two and later twelve jurors. The court could afterward compel the owner of the opposite stream bank to sell one acre for a dam abutment and pond at a price set by the panel, or to accept damages awarded by this panel for upstream flooding.[12] Among the court orders examined, only when a neighbor's land would be flooded excessively did the court limit the height of a proposed dam to less than requested, or deny permission to raise an existing dam. A complete record (application, jury's report, and court order) of one such successful application, filed and acted upon in 1874, is reproduced in the history of the Solomon Slusher/Elijah Turman Mill on Mira Fork (IV, 66).

Initially, those who already owned land on both sides of the stream, and enough upstream land to insure that no neighbors' property would be flooded or otherwise damaged by the presence of the mill pond, were not required by law to ask permission, unless (following the passage of a 1705 law) there was already another mill on the same stream within a mile in either direction.[13] But after mill legislation was expanded yet again in the legislative session ending in October 1748, every person wishing to construct a mill was required to apply for the permission of the local court, regardless of whether the property of any other landowner would be affected, and regardless of whether there were other mills nearby.[14] Had this provision of the law been consistently obeyed and enforced, court records could be used to identify and date the beginning of virtually every water-powered mill in Floyd County. But compliance seems to have been inconsistent, based on Frank Webb's records of his searches of court order books in Floyd and Montgomery Counties, and on subsequent searches related to selected mills.

Four mills built within a one-mile stretch of Burks Fork, probably between 1845 and 1888, provide convenient examples. Three of these mills—the Jacob Barringer Mill (IV, 43), the Jackson Harris/Martin Harman Mill (IV, 44), and the John Barringer Mill (IV, 45)—were on sites that had been part of a 500-acre tract bought by Jacob Barringer in 1845. While no mill is mentioned in the 1845 deed, it is conceivable that one of the mills was there at that time. The existence of two stretches credibility. The construction of the John Barringer Mill is dated at 1868 by a letter written by Henry Slusher. Yet, Floyd County court order books for the years 1843 to 1887 include no record of a petition for permission to build any of the three dams. The same is true of the William Alderman Mill (IV, 46), built on an adjoining tract a short way down Burks Fork, probably between 1878 and 1888. As will be seen, it is no coincidence that at the times these mills are believed to have been built, the land on both sides of the creek at and above the sites of the dams was owned by the presumed builders.

The implied failure on the part of the county court to compel compliance reflects the spirit of a time and place in which the public good, and therefore the duties of state and local government, seemed best served by looking the other way in some instances. Then, as now, state and local governments had three potentially contradictory primary duties: (1) to promote and encourage development of business and industry; (2) to protect existing property and the rights of its owners; and (3) to protect public health and safety. In pursuing both the emblem and the substance of progress by fostering the growth of business and industry, officials might be tempted to place less emphasis on the last two of these duties. This would be particularly likely when there was almost no existing infrastructure and when the notion that the natural environment might need protection from man (and not the other way around) was even farther in the future than the concept of a department of public health.

In the Tidewater and Piedmont regions, which were flatter than Floyd County and more densely settled, there was a real danger that a new dam and the pond it created might interfere with an existing water power site or flood valuable farm land. This was probably of much less concern in thinly populated, steeply sloped lands of western counties. And since water power was the only significant source of stationary motive power before the advent of the steam engine (which was never significant as a source of power in Floyd County, except in sawmills), the erection of a dam was a prerequisite to the establishment of mills or mechanized industries of any kind, all of which both the state and the county were anxious to promote.

In addition to the potential for flooding neighboring lands, amended state laws also addressed the concern that dams and millponds could impede navigation and block the passage of fish. Since commercial navigation was never feasible on Floyd County streams, and fishing never more than a pastime or at most a means of supplementing a diet oriented toward red meat, neither of these dangers was likely to prevent the building of a dam. The county court did typically order mill owners to incorporate "fish ladders" into their dams, as in the Hornsbarger inquest decision in 1833–36 (see Sowers Mill II, 94). The July 1874 court order granting

Solomon Slusher leave to build a dam on Mira Fork states that Slusher is to "erect over the dam a fish ladder similar to ladder required by law over all dams."[15] But neither the law nor such orders to comply with it seem to have been taken seriously, since no picture of a fish ladder on a Floyd County stream has come to light, nor has any written or oral record of one been found, aside from this and similar references in other court orders issued after the completion of *ad quod damnum* inquests.

The third duty, that of protecting public health from the threat presented by a sizeable body of potentially stagnant water, was also addressed in mill legislation, and was, by contrast, probably taken more seriously in the 18th and 19th centuries than it would be today. This fear might seem vaguely superstitious to 21st century Americans unconcerned with mosquitoes or water-borne pathogens, but as mill historian Harold Melton explained, it reflected a very real 18th century fear of typhoid fever. This was far less a danger along the cool, fast-moving streams of the Blue Ridge than in the lower, slower, and warmer streams of Tidewater Virginia, but was a reasonable concern, nonetheless.[16]

The law requiring persons who already owned the necessary land to ask the court's permission to construct a dam appears not to have been strictly followed or enforced in the several Virginia counties that at some point included the land that became Floyd County in 1831. The same is true in at least one other Virginia county, Pittsylvania, which lies to the southeast of Floyd's own southeastern neighbor, Franklin County. After an exhaustive search of Pittsylvania County court records for applications to build dams between 1767 and 1799, Harold Melton noted that "some mills were known to have been built for which no application can be found."[17] All that remains to be said of these "illegitimate" mills is that the first mill in Floyd County could as easily be among them as among those that were duly approved by a county court.

Some court orders merely grant after-the-fact legal status by giving permission to construct a dam that is already in place. The histories of the Jacob Epperly Mill (II, 98) and the James T. Lester Mill (III, 132) provide two examples. Why the owners bothered after that point is unclear. Legal status may have conferred some competitive advantage, or the owners may have been basically law-abiding people who knew that a petition filed after a significant investment had been made was almost certain to be granted. As regards early mill owners who did feel compelled to ask the Circuit Court for permission before erecting a dam, confusions and gaps exist here as well, but it is clear that mills were springing up throughout the Floyd County area before 1800. In writing that "the great influx of settlers into Southwestern Virginia and Kentucky did not begin until after 1794, when the Indians were finally subdued,"[18] Ralph M. Brown is referring to regions west of Floyd County. The frontier in Virginia had moved beyond the watershed of the New River by 1800. Settlement was well under way, and mills were being established to meet the needs of settlers, with or without legal permission.

Court order books for Montgomery County show that at least a dozen petitions to build grist mills on waters of the future Floyd County were granted between 1782 and 1828 (Appendix B). But fewer than half of those mills have been identified on a list of twelve mills that were grinding corn and wheat when Floyd County was formed from Montgomery County in 1831. Besides those twelve, as many as seven more of the mills whose histories are covered by this book might have been at work as early as 1831. All told, there may have been as many as 25 mills, for which some documentation exists, at work in that year. As far back as 1800, documents of one kind or another place a grain mill in most parts of the county, with the possible exception of the southwest quarter as defined by Routes 8 and 221. And even that potential exception is an unlikely one, since there were eventually as many mills on Burks Fork and a few tributaries as there were along the main branch of Little River, the county's longest and largest stream.

The builder of Floyd County's very first mill may well have been the sort of man who thought it better to apologize, should he ever be questioned about it, than to ask permission beforehand. But if we assume, for the sake of argument, that every mill builder *did* obtain permission, and that he did so *before* he built the dam and mill, there could still be no absolute certainty as to the identity of the first mill. The recording of the court order establishes only that the petitioner intended to erect a dam and, presumably, a mill nearby; it does not prove that he carried through with those plans. To offer a modern analogy, the application for and receipt of a building permit does not guarantee that the planned building will ever be completed.

Without supporting evidence such as deeds or wills, a list of court orders (for a partial list see Appendix B) establishes with certainty only the names of petitioners wishing to build dams across streams that would later be in Floyd County, and the order in which their petitions were granted. The very earliest of these was approved in 1770 by the court of Botetourt County.*[19] According to Lewis Summers, the court record attests that "on the motion of Thomas Goodson, leave is granted him to erect a water grist mill on his land on Pine Creek, a branch of Little River, in this County."[20] Assuming there is no other Pine Creek draining into a Little River within the very large area that was then Botetourt County, this Thomas Goodson would at first glance seem to be the same Thomas Goodson who left a grist mill and a sawmill to his son John in a will dated 1834 (see mill I, 19). But those mills were on the western fork of Meadow Creek, several miles east of Pine Creek. And according to a Floyd County genealogy Web site, this particular Thomas Goodson would have been only fifteen years old in 1770.[21] Neither Goodson's Mill nor William Bell's mill on Little River, mentioned in a court document dated 13 November 1773, has been traced forward with enough certainty to establish exactly when and where they were built, if they were ever built at all. The same is true for the William Davies mill on Beaver Dam Creek, approved by the Montgomery County court in 1782 (Appendix B).

Among the mills whose owners did petition the court for permission to put up a dam, and whose existence was afterward confirmed by a deed or will, the earliest to be approved belonged to Joshua Howell (III, 138), who received the Montgomery County Court's blessing to erect a dam across Dodd Creek (the South Fork of Little River) in 1784. In all likelihood, Howell set to work and completed his mill within a year or so, in which case his would have been the earliest among the 140-plus mill sites treated in this book whose locations and subsequent histories are known today. Though his dam across Pine Creek is not known to have been sanctioned by court order, Josiah Terry's mill (II, 99) has the distinction of being the first whose physical existence is established unequivocally by a deed book record. Ironically, Terry qualified himself for this honor when he was compelled in 1787 to deed the property that would later be known as the original Spangler Mill (II, 99) to William Logan as partial payment of debt.

A more thorough search of court orders, deeds, and wills recorded in Augusta, Botetourt, Fincastle, and Montgomery Counties could indicate more clearly which of the earliest documented mills was the first to actually grind grain. But so much early development went unrecorded that it is unrealistic to believe that this search would establish with certainty the name and location of the very first mill and miller in the county. As seen above, several mills identified by name, and almost certain to have existed, have not been located; conversely, several early Floyd County deeds mention an "old mill race" or a "mill tract" that could not be matched with any known mill, i.e., they establish the location of a mill without attaching

**The territory that is now Floyd County was part of Augusta County from 1738 until Botetourt County was formed in 1769. In 1772 that portion of Botetourt became part of the newly formed Fincastle County, and then part of Montgomery County when Fincastle was subdivided (it no longer exists) in 1776. The southern portion of Montgomery became Floyd County in 1831. A narrow strip of Franklin County was appended to Floyd in 1871.*

a name to it. Given the number of such loose ends, and the incomplete nature of records that were made well after the pioneer era, there is no doubt that a number of 18th and even 19th century water-powered mills were never heard of by either of the authors of this book.

Choosing a Spot

Knowing that they would have to build their own mills or at least encourage their construction, some of the early settlers must have chosen the sites of their homesteads with an eye toward developing a mill site or "mill seat," as a water power site was typically known once the mill had been built and more commonly after the mill itself had disappeared. The recurrence of names like Duncan, Hall, Huff, Hylton, Lester, Phillips, Slusher and Spangler in the deed records of numerous individual mills suggests an interest in mills and milling that lasted in some families for generations, regardless of whether they brought with them the necessary skills of the miller and millwright, or took up and mastered them in response to a critical need or perceived opportunity.

If these people were at all daunted by the steepness of the lands they would have to clear and cultivate, the millers among them may at least have been encouraged by the abundance of decent sized streams, by the generally high degree of "fall" (drop in elevation) over short distances, and by the narrow and readily dammed valleys through which most streams ran at regular intervals. The available "head" at a mill is the vertical distance between a point a few inches *below* the surface of the water standing behind a dam and a point a few inches *above* the surface of the water as it re-enters the stream below the dam. A high degree of downward slope or fall in the bed of the stream assures that adequate head could be obtained without having to acquire or flood excessive acreage along the stream. Perhaps more important in a time when land was abundant but labor scarce, it meant that lengthy in-ground earthen races (dug races) and wooden races—also referred to as flumes, forebays, "box" races, or sluices—would not have to be dug or constructed and maintained.

Ideal sites became harder to find as land was broken by sale and inheritance into smaller tracts, yet some mill builders, both early and late, were either diligent or lucky in acquiring a naturally favorable spot. Others who were unwilling to relocate or unable to afford more easily developed spots, or who were persuaded that an available location's economic potential outweighed its physical limitations, did whatever was required to overcome those weaknesses. Two of the best known remaining mills, the Roberson Mill (I, 27) on Dodd Creek and the Shelor/Nolen Mill (I, 123) on Pine Creek, are good examples of making the best of a less than perfect situation. The first required a very long, very tall wooden race to connect the mill and the dug race; the second stands at the lower end of the longest (by estimation) dug race in the county.

Erecting a dam at the narrowest point of a valley required the least investment of time and materials, and made the strongest dam, especially if log, stone, or concrete abutments could be anchored against natural stone outcroppings on one or both banks. Numerous Floyd County mill sites may still be identified by natural stone abutments, silt banks, the remains of stone or concrete abutments, or pieces of the dam's wooden framework, preserved by having been under water. Even without these bits of confirming evidence, these same spots continue to catch the eyes of mill buffs, and will draw the attention of future millwrights, should we ever again need these sources of energy.

Water Wheel Type and Mill Location

At some early stage a builder had to decide where to place the mill building in relation to the dam. To some degree, the practically obtainable head and the volume of the available water determined the type of water wheel to be used, vertical or horizontal. Since this choice, in turn, influenced the location of the mill relative to the dam, a brief review of the key features of each type of wheel may benefit the discussion of choosing a mill's location.

A vertical water wheel derives most of its power from the effect of gravity on the mass of the water contained in the buckets attached to the rim of the wheel. (The exception to this rule, the undershot vertical wheel, was driven solely by the impulse or momentum of flowing water, but was employed only in flat terrain with very low heads and was unknown, in any commercial application, in Floyd County.) With a vertical water wheel, whether wooden or steel, how high the water can be raised above stream level (the available head) is more important, within certain limits, than the volume of water that can be stored behind a dam or directed to the wheel by a race. Extra water allows for the use of a wider wheel and extends the time a wheel can run at full power, but the availability of more water than is required to fill the buckets of the wheel doesn't increase the power produced. Properly installed, a scientifically designed, carefully balanced steel wheel running under optimal conditions can convert potential energy to rotative motion at close to 90 percent efficiency.[22] Vertical water wheels powering Floyd County mills included wood, steel, and various wood-steel hybrids, and ranged from 8 to 25 feet in diameter, with most being between 10 and 20 feet tall. Vertical water wheels, such as the overshot wooden example at Mabry Mill (IV, 60), or the more modern steel wheels at Vaughn's Mill (III, 64) (overshot) and Pine Creek Mill (I, 123) (pitchback/high breast), are the most familiar and hence the most easily recognized, but are by no means the only types of water wheels used by Floyd County mills.

A horizontal wheel, of which the steel or cast iron turbine is the best known type, derives power primarily by reacting to the force of a solid column of water dropping through it and striking simultaneously the multiple vanes, blades, or paddles of a wheel mounted just above the level of the tail water. Having transferred its potential energy to the turbine, the water returns to the stream via the tail race. Steel turbines as small as ten inches in diameter and much cruder homemade wooden tub wheels up to four feet across were installed at Floyd County mills, in most cases immediately below the dam. While additional fall increases the impulse delivered by increasing the velocity and momentum of the falling water, the volume of available water is more important than the head, again, within certain limits. Even a few feet of fall, given enough volume, can generate tremendous power. Though less familiar to the mind's eye than a traditional vertical water wheel, the turbines coupled to hydroelectric generators inside the dams at Claytor and Smith Mountain Lakes are good examples of modern horizontal water wheels. The finned rotor of an automotive turbo charger, driven by the stream of pressurized exhaust gases, is simply a more sophisticated application of the principle behind the modern water turbine. The wooden "tub wheels" used mostly by 19th century corn mills worked on the same principle, but at much lower speeds and with far less efficiency than the 75 percent achieved by some small steel or cast iron turbines. Unlike vertical water wheels, which may rot or rust in place for decades after being taken out of service, a wooden tub wheel quickly rots, while its steel counterpart sinks into the mud of the stream bank, where it is likely to remain unless recovered and sold for scrap. At least two steel turbines are supposedly still buried along Floyd County streams and a couple more have been recovered. No wooden tub wheel from a Floyd County mill, or even a photograph of one, was found, but a tub mill documented by the Blue Ridge Parkway in 1936 was probably very similar to Floyd County mills of that type.

Whichever type of wheel was to be used, both maximum head and maximum volume could be obtained by placing the mill as close as possible to the stream. This also exposed the mill to the greatest danger of damage by flooding. Today, a building inspector or an insurance agent might calculate that risk precisely, and, based on the results, refuse to issue a building

Tub wheel installed in a corn mill, Rocky Knob Park area, Blue Ridge Parkway, 1936. The diagonal wooden spout near the center of the picture (partly hidden by the crossed poles) directs water from the bottom of the hollow wooden penstock to the angled vanes of the horizontal wooden wheel at lower left. Inefficient forerunners of steel or cast iron turbines, wooden tub wheels were built up of two or more solid pieces of wood and carved with auger and adze into a pinwheel pattern held together by wrought iron bands fitted around the outside. The true tub wheel was enclosed by a circular wooden tub which improved efficiency by containing splashed water and channeling it through the wheel. (Courtesy Blue Ridge Parkway Archives Collection, negative 6653)

permit or deny an application for insurance coverage. Under current regulations, many mills would never have been built exactly where they were. A good number would never have been built at all since locating the mill safely out of the floodplain would have been impossible or too expensive to be worth the trouble. Since no such safeguards existed in the past, a builder had to set priorities and calculate risks on his own and locate his mill accordingly.

The two distinct yet inseparable components of choosing a spot for the mill are distance from the stream, which in mountain valleys usually had a vertical as well as a horizontal aspect, and distance below (downstream from) the dam. Selecting the ideal spot for the mill would have been challenging even if it had to account for only one of the following factors: development of maximum usable horsepower; type of water wheel (vertical or horizontal); initial and long term costs; risk of flood damage; and impact on adjoining lands. The following examples are meant to illustrate the difficulty of considering *all* these factors, by showing what alternatives emerge in considering only *one* of the factors and disregarding the others, although the impact on the lower-priority factors is noted. For the sake of illustration, in the examples below it is assumed that options for placement of the mill are not limited by the topography or availability for development of the surrounding land. In reality, no single mill site would allow for all these variations.

If the first priority in locating the mill building is to *obtain maximum horsepower*, which is a function of both available head and volume of water, do one of the following: (1) Using an overshot or high-breast vertical water wheel, maximize the head by placing the mill as far below (downstream from) the dam and as close to the stream as possible. This puts the mill at greater risk of flood damage and requires long, expensive dug and wooden races. (2) Build nearer to the dam and compensate for lost fall by raising the height (and width) of the dam, increasing both fall and volume but also flooding more land behind the dam and increasing the potential for destruction downstream in case of a washout. (3) Build close to the dam at its original height and install a turbine, if an adequate volume of water is available year round. If not, raise the dam to increase reserve storage.

If the first priority is to *build and maintain the mill at the lowest cost,* eliminate dug and wooden races by building the mill immediately by the stream and as close as possible to the dam. Available head and power are reduced and the danger of flood damage to the mill increased, unless the building is raised above the level of the pond, increasing its initial cost. A turbine is initially less expensive, but requires regular maintenance. Power developed by a turbine drops precipitously if the water volume drops below optimum levels.

If the *fear of flood damage* is a decisive factor, build the mill as far from the dam and the stream as possible, both vertically and horizontally. Ordinarily this would mean as far downstream as possible, since following a contour line with a dug race is the least expensive way to acquire additional fall and simultaneously minimize flood risk. This assumes that the valley widens below the dam, which would allow for dispersal of the flood waters before they reach the mill. Build a dam of minimum height. The less water it holds, the less damage will result from a dam failure. Costs for digging, building, and maintaining races are maximized. Since storage is minimized, install a vertical wheel which turns more slowly and is initially more expensive, but maintains a constant efficiency with reduced volume of water.

If the first priority is to *minimize flooding of usable land* by the mill pond and other impacts on adjacent lands, build the dam at a point where the valley is narrow and steep-sided and the stream bed itself drops precipitously, as would be the case at a waterfall or series of short falls. Make the dam just high enough to divert water from the stream into a dug or wooden race. Build the mill far enough downstream to achieve the needed fall and close to the stream to avoid digging a long tail race to return the water to the stream. Use a vertical water wheel with a low width-to-diameter ratio since volume will be consistently low.

The more factors one attempts to account for, the more complex the process becomes. Carried out today, such a complicated risk-benefit analysis would require some portion of the skills now attributed to a civil engineer, a mechanical engineer, an accountant, a real estate appraiser, and a team of lawyers, as well as the insurance agent and building inspector. A hydrologist ought also to be consulted since the risks associated with flooding would need to be calculated based on the highest flood level likely to be experienced during the projected life of the mill.

Even if all this expert advice had been available, the result would still have been an educated guess, a highly refined and expensive one, but a guess just the same. The millwrights hired to design and build Floyd County's water-powered mills had no such human and scientific resources, but they did have tables published by manufacturers of water wheels and milling machinery and the advice found in comprehensive guides such as Oliver Evans' immensely popular *Young Mill-Wright's and Miller's Guide.* Beyond that meager help, they were left to rely on experience, intuition, and luck.

It is no surprise, then, that millwrights faced with topographically similar situations sometimes reached very different conclusions about how best to proceed, based on their assumptions about the relative benevolence of nature as well as the importance assigned to each of the priorities listed above. In fact, examples can be found to suggest that each of the four factors was actually given first priority by the builder of at least one mill somewhere in Floyd County.

At the same time, it is easy for eager historians to perceive patterns where none exist and to interpret actions of the past in ways conformed neatly to the thinking of the present. All that can be safely said is that each builder did what seemed best under the circumstances as he understood them. However much or little thought went into the process, decisions were made and plans carried out by the kinds of people who took hold, did things the best they could, and went about their business. The builders of these mills were probably not the sort of people who sat around speculating about what other people might have done a hundred years earlier, or why.

Since nature has repeatedly proven itself beyond the scope of human calculations, it may have been more a matter of luck than of good judgment that many mills, like Caleb Harman's (III, 88) on West Fork, or the Bob Lee Vest Mill (I, 13) on Little River, were washed completely out or damaged beyond repair, some after only a few years. An undocumented mill built by Jim Booth on Beaver Creek across from Dobbins' store is supposed to have washed away soon after, or even before the dam and water wheel were even completed. Boothe's grandson Russell Dulaney (b. 1923) remembered seeing iron gearing from the mill lying downstream c. 1930, but remembered no sign of the mill or a dam.[23] On the other hand, the building thought to be the second Spangler Mill (II, 99) stood by Pine Creek for over a century and a half before collapsing in a windstorm in 2005. As with assumptions made about the motives of people long dead, the perception of irony in a given situation depends much upon an observer's inclination to perceive it. That said, it at least seems ironic that the Thompson-Akers Mill (I, 14) on Payne's Creek, whose builders dug and maintained an extremely long race in order to achieve the necessary fall and place the mill safely beyond the reach of the worst imaginable flood, ran for over forty years, survived the flood of 1940 and others, and was still in daily operation when it caught fire and burned to the ground in June 1964.

Mill Dam Construction

While the locations of many mill dams can be determined by clues described later in this overview, there is often little evidence of the appearance and structure of their center portions, the parts that actually formed and retained the mill pond and directed water to the dug or wooden race or to a turbine forebay and penstock. Dams made of any material were subject to be washed out or broken by flood waters and floating debris. The upper parts of wooden dams, like wooden races, were subject to rot due to constant wetting and drying. Only the parts that were continuously submerged lasted very long. Once mill ponds were drained, the saturated planks and timbers deteriorated rapidly.

A mental picture of a traditional overshot water wheel is easily called to mind thanks to popular examples like Mabry Mill (IV, 60). Images of mill dams, on the other hand, are likely to be based on more distant, more modern, examples. Locally these would include the Claytor Dam on New River just south of Radford, Virginia, or the smaller hydroelectric installation near the mouth of Little River, which joins the New River just below Claytor Dam. Their steeply sloped downstream faces of massive poured concrete, their steel super-structures and floodgates, and their internal turbine tubes are a far cry from most Floyd County water mill dams. While some mill dams did have concrete abutments or reinforcements of some kind, only three are known to have been constructed entirely of concrete. The Sowers Mill (II, 94) dam, built farther down Little River than any other within the county, and the Irvin Huff Mill (I, 2) dam, possibly the farthest up the same stream, have both been breached, i.e. burst

Greasy Creek Mill (III, 67) dam, constructed by Mack Turman and sons c. 1930, in 1992. Most of the wooden face and upper structure of the dam have rotted and washed away, but the heavy mortised timbers forming the base of the dam and the shortest prop timbers remain. Recesses formed into the concrete western abutment indicate the size and angle of the taller prop timbers and define the angle of the upstream face. (Ricky Cox photograph)

through or broken, whether on purpose or by nature, but are otherwise still more or less intact. Both were built in the early 20th century. The third, at the Jim Perdue Mill (II, 117) on Lick Fork Creek, washed out not long after it was built in the 1940s.

Although some could have been washed completely away, earthen dams also appear to have been rare, possibly because of the difficulty of moving hundreds of cubic yards of material and of diverting the stream while the dam was built. Earthen dams would seem also to be generally inferior to wooden or stone dams on fast moving streams of much size, given the ease with which they are eroded, unless complex stone or concrete spillways are provided. What appears to be the remains of an earthen dam near the upper end of Little River may be seen near the site of the early 19th century Anthony Poff Mill (I, 4).

Dams made entirely of massive, tightly fitted stone would have been more common had there been convenient sources of quarried stone or a shortage of timber. As it is, no dams were built entirely of large, shaped stones. The only dams known to have been faced from end to end with closely fitted stacked stone are the Rakes Mill (I, 26) dam on the Blue Ridge Parkway and the Chris McAlexander Mill (IV, 53) dam, both of which span narrow valleys. Comparison of the Rakes Mill deed records with information provided at the site leaves it unclear who built the dam and when, but it is plain, considering the equally rare stacked-stone piers that once supported a wooden race between the dam and mill, that the builder or builders were skilled stone masons. The McAlexander dam, though breached near its center and not as closely fitted, has retained much of its shape, suggesting also the work of intelligent craftsmen.

A good number of Floyd County mill dams did have stacked stone abutments at one or both ends of the dam. Impressive examples remain at the sites of the German Belcher Mill (IV, 37) on Rush Fork, the Martin Weddle/Cara Simmons Mill (IV, 38) on West Fork, and the William H. Harman Mill (II, 100) on Little River. The most massive examples are at the still-standing John Epperly Mill (IV, 126) and Webb's Mill (II, 122), both on Dodd Creek.

Mill dam abutment and mill foundation of stacked stone, German Belcher Mill (IV, 37), 2008. Floyd County dam builders had few sources of large blocks of cut stone, and no easy means of moving them. Dam abutments and mill foundations were typically constructed of relatively small local stones, minimally shaped and laid without mortar. (Ricky Cox photograph)

There were probably many low diversion dams made of stone that was simply piled rather than fitted, and perhaps periodically daubed or chinked with mud, sod, or brush while in use. Rather than holding back a pond of hundreds of thousands of gallons of water, a diversion dam served mostly to turn a stream of water out of its bed and into a dug race. It was best suited for use on steeply sloping streams, where easily obtained head or fall compensated for low volume. Such dams typically fed water to tall, relatively narrow overshot wheels which are more efficient than a turbine or tub wheel in high-head, low-volume applications. Stones of a manageable size and squarish shape suitable for this type of dam might be retrieved for another use after the mill was closed. Even if left intact, dams of this type are difficult to distinguish from the natural stream bed once the profile of the dam is broken up by the displacement of a few stones. Sites likely to have used such dams include those of the Asa Pratt Mill (IV, 36) on Line Branch, the William F. Nolen/Nowlin Mill (I, 124) on Shooting Creek, and the Goodson/Howard and Lancaster/Harvey Mills (I, 19) on Meadow Creek. At all three sites, small streams run in steep, narrow, rocky beds. An important and still mostly intact exception is the piled stone dam that formed a substantial mill pond across a broad, shallow valley for the A. Newton Hylton/Walter Boyd Mill (IV, 35) on the headwaters of Howell Creek.

The majority of the larger mill dams in Floyd County, of which fragments or photographs remain, appear to have been made of wood, with stacked stone or concrete abutments. Although no documentary evidence supports this assumption, it is likely that some of the earliest dams, like the earliest homes, were constructed of logs rather than dimensioned timbers and planks, and for the very same reason: there were far more trees available than there was time to chop, hew, and saw them into planks and timbers. The most expedient method of building a log dam, in terms of time and materials, would have been to simply stack the logs, hewn on the top and bottom sides, across the stream at the narrowest convenient spot. Not surprisingly, this method would also be the least sound from an engineering perspective, since virtually all the force exerted by the constant pressure of standing water, along with the intermittent and occasionally severe pressures and blows dealt by the current and floating logs or chunks of ice, must be borne by the abutments on either end. The vertical upstream face or "breast" of the dam, being perpendicular to the motion of the stream, would be subject to the full force of the current; floating logs and rolling stones would hit it head on, instead of striking a glancing blow, as is the case with the sloped breast of a more sophisticated dam. In addition, water pouring over the top of such a dam tends to undermine the bottom log.

While poorly designed dams may have been thrown up as temporary measures, it must be supposed that men who were able to reproduce, at least in rudimentary form, the mills they had known before coming west could also re-create stable, durable dams using the most abundant and easily worked material at hand — wood. Whether they learned them from other men, by studying existing dams, or from printed sources, most would have been familiar with principles articulated in Oliver Evans' *The Young Mill-Wright and Miller's Guide.* This classic technical reference first appeared in 1795, ten years after Evans built the first automated grain mill in the world, and was already in the 8th of an eventual 15 editions by 1834. According to Evans, there are several points to be attained, and dangers to be guarded against, in building mill dams.

1. Construct them so, that the water, in tumbling over them cannot undermine their foundations on the lower side.
2. And so that heavy logs, or large pieces of ice, floating down, cannot catch against any part of them, but will slide easily over.
3. Build them so that the pressure or force of the current of the water will press their parts more firmly together.
4. Give them sufficient tumbling space to vent all the water in time of freshets.
5. Make the abutments so high that the water will not overflow them in time of freshets [flash floods].
6. Let the dam and mill be a sufficient distance apart; so that the dam will not raise the water on the mill in time of high floods.[24]

Only the last of these cautions was widely and deliberately ignored, mostly after the widespread adoption of the turbine water wheel at new and refurbished Floyd County mills around 1900, more than a century after Evans' book first appeared. As noted above, the tremendous volume of water demanded by large turbines persuaded builders, perhaps against their better judgment, to locate mills as close as possible to dams to avoid having to dig or build and maintain suitably large races.

At the Greasy Creek/Harris Mill (III, 67) on Greasy Creek the dilemma of safety versus efficiency was eventually solved, perhaps coincidentally, by conducting the water to the mill and turbine in enclosed metal pipes. The mill was originally built with a wooden race and an overshot wheel, which placed the mill out of the direct path of water released by a breached or overflowed dam. The mill is also near the point at which the valley suddenly widens, which would have allowed flood waters to spread and lose much of their force as they approached the mill. By the time the mill was converted to turbine power, a new dam had been built at the farthest possible point downstream, and moving the mill upstream was unthinkable. The traditional solution would have been a wooden forebay and penstock between the dam and the mill. The installation of enclosed steel pipe solved the problem by spanning the distance between dam and mill by the shortest and least expensive route. Since the pipe was enclosed, instead of being lost, the water's potential energy was converted to water pressure as the water dropped toward the turbine in the lowest level of the mill. The use of solid iron, steel, or concrete pipe between the dam and a mill built safely downstream would likely have become common practice, had any other mills been built, after 1925, on streams large enough to drive a turbine.* If so, millwrights could have slept more soundly, having both reduced the risk of flood damage and appeased the ghost of Oliver Evans.

Evans' most specific elaboration on his rules for mill dam construction is his contention that the breast or upstream face of the dam should form an angle of exactly 22½ degrees measured from the surface of the water.[25] At this angle, a plank "will be held firmly to its place by the force of the current, and, in this position, it requires the greatest force to remove it; and the stronger the current, the firmer it is held to its place."[26] In this expansion of his third principle Evans reveals the key to holding in place a dam made of wood, which is by nature better suited to float on a body of water than to rest, immovable, at its bottom. The ramp-like profile of the dam breast is also the key to resisting the momentum of flood waters and surviving the pounding of floating logs. The force of both is deflected upward, causing them to pass harmlessly, it is hoped, over the top of the dam. In sum, it appears that the largest part of a wooden mill dam, like an iceberg, lies below the surface of the water.

Stanley Wimmer atop the Ezra Wimmer Mill (II, 119) dam, undated. The uppermost crosswise foundation timbers, near-vertical prop timbers, and angled supports beneath the board face of a typical hollow frame dam are visible here. An "apron" of wood or piled stones at the base of the dam prevented the water that spilled over top of the dam from undermining the foundation. (Frank Webb collection)

Although inferior to a stone or concrete dam in terms of longevity and sheer mass, wooden dams did have some advantages, including lower initial cost, relative ease of construction, and ease of repair. A wooden frame dam could be constructed by anyone possessing basic timber framing skills and tools, and assembled piece by piece with less effort than would be needed to assemble and raise a timber framed building of similar proportions. Once the stone or concrete abutments were in place, no earth moving equipment was needed, nor were any means of quarrying, cutting, and placing suitably massive stones.

**A steel culvert pipe was installed between the dam and turbine at the Chris McAlexander Mill (IV, 53) about 1946.*

Schoolgirls from Greasy Creek School are assembled below Vaughn's Mill (III, 64) dam and race, 1943. Gaynor Alderman snapped this photograph of most of her female students, grades 1 through 7. In front: Bonnie Bell Marshall (lying on the ice), Lucille Cox (kneeling), and Evelyn Reece (hands clasped around ankles). Standing: Arlen Matherly, Betty Marshall, Ruby Matherly, and Elrica Quesenberry. The school, still in use as a private residence across Vaughn's Mill Road from the site of old dam, closed within a couple of years. (Courtesy Gene Vaughn)

Mill Races: Head and Tail; Dug and Wooden

A mill race is an open-topped, man-made channel of earth, wood, stone, or concrete that carries water to or away from a water-powered mill. The *head* race conveys water from a mill pond or stream to the top or to the breast of a vertical water wheel or to the turbine's penstock, a hollow column, typically wooden, or a pipe, mounted directly above the turbine. The *tail* race carries the "tail water" back to the stream after its potential energy has been converted to rotative force by the wheel. In either type of race, downward slope is kept to an absolute minimum to maximize the head available at the wheel. Unless the term "tail race" is used, all references to mill races in this book are to head races delivering water to the wheel.

Head races could be either dug into the ground or constructed of wood and elevated as required by the terrain. A few dug races have concrete and masonry reinforcements or ends, but neither of these materials accounts for the majority of the structure of any known Floyd County mill race. Alternative names for dug race are *in-ground race*, *earthen race*, *ditch*, *canal* and, *water lane* (rare). A dug race was preferred when three conditions were satisfied: the slope on at least one side of the valley was suitable; the pond-to-mill route dictated by that slope was sufficiently direct; and the ground was not inordinately rocky. Whatever its initial cost in labor compared to that of a wooden race, a dug race was ultimately less expensive. Muskrat holes and washouts had to be repaired, and silt left by flood waters or run-off from the hillside above had occasionally to be shoveled or scooped out. But there was little wood to rot, whereas every piece of an elevated wooden race would eventually have to be replaced. Although they are unimpressive to look at and could be dug by unskilled laborers, dug races were carefully planned and their construction carefully supervised. To obtain the most power, the grade had to be precisely maintained along the entire length with just enough down slope to keep the water flowing. Too much total slope reduced the power of the wheel it served by diminishing the available head. An error of only one degree in a race only 200 feet long would reduce the available head by more than three feet, a substantial amount given that steel water wheels were seldom much more than 20 feet tall, and wooden wheels rarely more than 15 feet tall. Excessive slope also wasted energy in pointless splashing if water was delivered to the wheel at a velocity exceeding the normal speed of the water wheel's rim. The absence of splash, once the wheel has reached its operating speed, is in fact one indicator of a properly set up vertical wheel.

A list of dug races, some easily identified more than a century after they were last used, may be compiled by consulting the "Visible Remains" listed for each individual mill history. Among these, the dug race at the Shelor/Nolen/Pine Creek mill (I, 123) is notable for its great length and is, like the races at Vaughn's Mill (III, 64) on Mira Fork Creek and the Charles Gillispie Mill on Meadow Creek (I, 18), easily seen from a public road. The John Epperly Mill (IV, 126) race along Dodd Creek is relatively short, but due to the demands of the mill's large turbine water wheel(s), is remarkable for a width and depth that would accommodate the full flow of a good sized creek.

Elevated wooden races are also called *box races, sluices, flumes,* or *forebays*, although the latter term is more often applied to the connection between a dam and a turbine penstock when dam and penstock are only a few feet apart. Functioning as small aqueducts, wooden races often resembled miniature railroad trestles in form and proportions. Working examples may be seen at the Pine Creek Mill (I, 123) and Mabry Mill (IV, 60). Parts of the wooden race to the Roberson Mill (I, 27) on Dodd Creek are also still standing. The engineering involved in a wooden race is apparent in the construction of the channel itself and in the precise angles of supporting posts and cross braces, as well as in the nearly level slope. Gates at the head of a race for diverting water away from it made maintenance much easier, but because of their distance (sometimes thousands of feet) from the mill, they were often not convenient for turning the water off when the miller wished to stop the wheel. Accordingly, a gate of some kind was needed in the race just above the point where water was introduced onto the wheel. In addition, including a "turnout," a low side board or a movable gate in the side of the race, at a point from which the water could find its way back to the stream, allowed the race to be drained. This was not merely a matter of convenience but one of safety as well. Even a small volume of water leaking past the gate and onto the wheel could keep the wheel turning slowly or cause it to start suddenly, presenting a danger to unwary mill workers. A slowly turning wheel or one that was stationary but "loaded" on one side was also an after-hours hazard to passersby, and a cause of off-hours damage to the mill machinery. "Chocking" the wheel with a post or wooden bar would prevent its turning, but removing the prop was dangerous, since it released a wheel loaded on one side with hundreds of pounds of water. Homer Roberson, owner of Roberson Mill (I, 27) died of injuries received when he was pulled under his water wheel after knocking out the prop in 1966.

The problem of a leaky gate was compounded in very cold weather since the water would freeze to the wheel and to the ground. Having freed the steel wheel at Vaughn's Mill (III, 64) Oden Vaughn, brother of owner Ephrem Vaughn, had to drop his sledge hammer and cling to a cross-brace while the 22-foot wheel made one complete revolution, and then rocked back and forth before settling with the heavy side down. Ephrem Vaughn, who was there but recovering from an illness, shouted, "Hold on, Oden," to which Oden replied, "I've got her!"

Vaughn's Mill, 1943. The large plume of ice marks the location of the "turn out" which could be opened to drain the race for repairs and to divert water before it reached the wheel. Ice hanging from the bottom of the race on the wheel side of the turnout marks the square hole cut by Ephrem Vaughn to drain remaining water and keep it from "loading" the motionless wheel, or freezing it in place, both potentially dangerous conditions. (Courtesy Gene Vaughn)

If the man wielding the hammer or other tools escaped injury when the wheel suddenly broke loose, there was still the task of removing enough ice to get the unbalanced wheel moving. Ephrem Vaughn solved the problem by cutting a drain hole in the bottom of the race on the downstream side of the gate, which kept water that seeped past the gate from reaching the wheel when the mill was unattended. The tapered sides of the square plug fit neatly back into the hole from which it had been sawn, and, once swollen with water, leaked minimally when the race was filled.

At the Powhatan Williams Mill (I, 25) on Pine Creek, a wooden race carried water over the heads of travelers on what is now Sandy Flats Road. Wooden races also passed over public roads at the Willis Mill on Greasy Creek (III, 69) and the Elijah Turman Mill (IV, 66) on the headwaters of Mira Fork. In an even more unusual arrangement at the Eli Sowers/Rufus Harris Mill (IV, 41), a dug race carried water along the north side of the creek to a point just above the mill. From there it was carried in a wooden race over the creek to the mill on the south side of Burks Fork, where it had been placed, apparently, for the convenience of customers traveling on what is now Conner Grove Road (Route 799). A similar arrangement was used at the A. W. Stanley Mill (IV, 130) on Taylor's Creek.

Wooden "box" races were originally employed at both ends of the long dug race at Vaughn's Mill (III, 64) to overcome imperfections in the hillside contour. At the upper end, the solid rock that underpinned the dam abutments made it impossible to dig the race all the way to the dam, so a wooden race joined to the dam carried water diagonally downstream to the upper end of the dug race. At the downstream end the race could not be extended to a point directly in line with the upper end of the mill, so here again a section of wooden race ran "catty corner" from the hillside to the race or forebay that delivered water to a turbine, originally, and later an overshot wheel. Several Floyd County Mills—including Mabry Mill (IV, 60), the Piney Fork Mill (II, 107) on Beaver Dam Creek, and Elijah Turman's mill on Mira Fork (IV, 66)—needed two wooden races drawing on separate branches of a larger stream to supply enough water.

Most mills built downstream from the dam first employed a dug race to get the water as far downstream as possible, then a wooden race to carry the water from the hillside to the water wheel. One exception to this practice was employed at the DeHart Mill (III, 136) on Indian Creek, which used only a wooden race between the dam and the mill, perhaps because of the width of the valley on the west side of the creek, where the mill, the road, and the miller's home were located. Connecting to either end of a dug race on that hillside might have required a wooden race nearly as long as the one that ran directly from the dam to the mill. Another exception was Rakes Mill (I, 26) on Dodd Creek, where a wooden race, carried on piers of stacked stone, channeled water from the dam to the mill. The hillside on the north bank was suitably sloped and only a few feet from the stream, but too rocky to be easily excavated.

As a rule, mills fitted with horizontal turbine water wheels were built very close to the dam to avoid having to dig and construct races large and strong enough to carry the great volume of water required by a modern turbine. No such installation survives intact, but good examples from the past included Webb's Mill (II, 122) on Dodd Creek and Spangler Mill (II, 99) on Pine Creek. The Spangler Mill dam was originally many yards upstream, but a new one was constructed in line with the upper side of the mill house when a turbine water wheel and a roller mill were added between 1910 and 1914. The lengthy dug race and the substantial wooden race that had fed the overshot wheel were replaced by a short wooden forebay between the dam and the turbine penstock only a few feet downstream from the dam. Traces of the original dug race remain between Spangler Mill Road and Pine Creek, nearly a century after the dug race was abandoned. Notable exceptions to the practice of locating a turbine powered mill immediately below the dam (and thereby violating a cardinal rule of Oliver Evans) are Vaughn's Mill (III, 64) and the Shelor/Nolen/Pine Creek Mill (I, 123). The fact that both are still standing and in good condition is perhaps a comment on both the foresight of their builders and the wisdom of Evans' advice.

As noted, the tail race carries the "tailwater" back to the stream after the potential energy of the falling water has been converted to rotative force by the wheel. In the case of mills located directly on the stream bank, the phrase "tail race" refers more to a function than to a structure, since the only physical distinction between the tail water and stream water was the few inches of elevation needed to prevent the tail water from pooling at the bottom of the wheel and impeding its motion. One of the key virtues of turbine water wheels, which were otherwise at a disadvantage on many high-head, low-flow streams of Floyd County, was their ability to function submerged in tail water, with little loss of efficiency. Hence, the tail water of turbine powered mills emerged at near stream level under ordinary conditions, and below stream level when the level of the stream was even a little above normal. But if floodwaters receded when the mill was not running, the turbine could be left clogged with mud and sand.

Breast or pitchback wheels, as used at the Shelor/Nolen/Pine Creek Mill (I, 123) and the Will Stanley Mill (IV, 54) could function fairly well in tail water since they turned in the same direction as the tail water was supposed to be moving on its way back to the stream. Overshot wheels, by far the most common type in Southern Appalachia, were seriously handicapped by "wallowing" when the tail water was backed up by higher-than-normal stream levels. The potential for this wallowing could easily be overlooked by a mill builder whose calculations were based on normal or below normal water levels. Even a millwright who took into account periods of high water would have to consider how much power was being sacrificed under normal conditions by installing a wheel small enough to operate efficiently when stream levels were slightly above normal. Miscalculations in both directions must have affected many mills, but only Ed Mabry (IV, 60) is known to have installed a smaller overshot wheel to eliminate chronic wallowing. Homer Roberson (I, 27) simply put up with this problem after installing a used wheel that was too tall for the head available at his location. Cab Nolen (I, 123) precluded a similar problem by reversing his oversized, used wheel and converting it from overshot to high breast or pitchback configuration.

In the case of a mill built more than a few feet from the stream, the tail race would look and function exactly as a drainage ditch. Head races, whether dug or wooden, stand out to the eye as man-made structures engineered to temporarily resist water's natural impulse to move downhill by the shortest route. Tail races, on the other hand, worked with, rather than against, nature to return the now useless water to the stream with the least possible effort or expense.

Few tail races are listed under the heading of "Visible Remains" in the individual mill histories, not because none remain, but because, while they may still be visible, they are seldom unmistakably identifiable as man-made features of the land. Once the wheel stopped turning and the water ceased to flow, a tail race grassed over and faded into the landscape more quickly than any other feature of a water-powered mill. Once the function of the ditch is no longer explained by some recognizable fragment of the mill, the tail race is easily mistaken for an old stream bed or an erosion scar. The disappearance of tail races is hastened by their susceptibility to the reclaiming power of nature. When a diversion dam is broken or a mill pond drained, a dug head race becomes instantly immune to the stream's power to erode it or fill it with silt. The tail race, by contrast, becomes more susceptible to both. Silt that would previously have been deposited in the mill pond is now left in the tail race by receding flood waters, and neither the miller's shovel nor the flow of water from the wheel pit to the stream is there to remove it. Finally, the flat creek-bottoms crossed by dug tail races often became hayfields or pastures afterward, both of which are improved by the filling of unnecessary open ditches.

Mill Construction

A substantial number of pre–1900 grain mills were built at sites where sawmills were already at work. In other cases, both were constructed at newly developed sites to make the fullest use of the rotative or reciprocating power that water wheels of various types could provide. Among sixteen Floyd County grain mill sites counted in 1881 by the compilers of a Virginia business directory (see Appendix A), ten also had sawmills. The ratio was similarly high in 1885 (22 out of 43); 1889 (15 out of 51); 1894 (23 out of 62); and 1897 (25 out of 71). The precipitous drop, reported by the same source, in the proportion of grist mills with active sawmills on-site in 1911 (2 out of 15) and 1917 (1 out of 15) reflects the en masse relocation of sawmills from creek banks to the woods following the sawmill owners' rapid adoption of the portable steam engine and boiler as a source of motive power, beginning around 1900. The equally striking drop in the number of grain mills, with or without sawmills, is due to a more complicated set of circumstances, outlined later in this overview.

Early sash type or "up and down" sawmills were usually powered by a primitive type of wheel called a "flutter wheel." Although wasteful of water, flutter wheels were apparently peculiarly suited to producing, via a crankshaft and pitman rod, the up-and-down motion of the sash saw. Though long gone from popular memory, this motion is memorialized in the seemingly nonsensical lyrics of the traditional folksong "Turkey in the Straw": "I come to the river and I couldn't get across, so I paid five dollars for an old blind hoss. He wouldn't go ahead and he wouldn't stand still, so he went up and down like an old sawmill." Flutter wheels were sometimes retained at the sawmill even after a more efficient overshot wheel or turbine was in place nearby.

Although the new mill would not have to share machinery or a roof with the sawmill, it

Asa Harman's sash type sawmill on West Fork Creek, detail from undated photograph. Flutter wheels, like the one pictured here at rest at the base of the penstock, worked well with sash type saws but were seldom used to drive a circular saw or grist mill. The vertical silhouette of the saw, with teeth filed to cut only on the downward stroke, is visible at the left end of the saw log. (Courtesy Floyd County Historical Society)

One of the last water-powered Floyd County sawmills, 1930s, sharing motive power with the grist mill at that time owned by Murray Agee (IV, 34). Standing in the foreground in a tie and flat cap is Charles Simmons Weddle, son of Andrew Weddle, who ran a water-powered sawmill on or near this spot between 1880 and 1910. (Courtesy W. Wayne Ratliff)

would have to share a limited supply of water and power. Even so, having a sawmill on site was at first a great advantage since a sawmill could by itself produce the lighter framing (floor joists, studs, and rafters) and roof sheeting for the new mill, and with the addition of a planer, could supply flooring and weatherboard siding as well as finished boards for eaves, soffits, doors, and window frames. Apart from the building itself, hundreds of board feet of finished lumber were needed for bins, hoppers, spouting, and grain elevators.

Because of the weight of milling machinery and stored grains, and the heavy stresses induced by rotating, vibrating, and reciprocating stone, wood, and steel, millwrights continued to construct large mills using traditional timber frame components and methods long after balloon framing was adopted in residential and light industrial construction. Oak sills, corner posts and cross beams 14 inches square and larger, and forty or more feet in length, were too heavy and too long for most sawmill carriages, but squaring and smoothing such timbers with hand tools would have been no great task for men capable of cutting the exact tenons and mortises by which the beams were assembled and then locked together with wooden pins. A Grayson County mill owner was told by his uncle that as long as a good roof was kept on it, a timber-framed mill would simply roll over in one piece before it would collapse.[27]

As an extra precaution against shocks and vibrations, a completely separate foundation and framework, called a *husk* or *hurst* frame, was built within the frame of the mill building to support the main drive shaft and gears and also the weight of one or two "run" or pairs of stones. Measuring up to four feet across and eighteen inches in thickness, a single stone could weigh more than 3,000 pounds. The frame of the building and the hurst frames were typically connected by wooden flooring and lightweight wood framing, but there were no rigid connections between the two. In mills with vertical wheels, which could weigh many tons when fully loaded, the hurst frame might also be made to support the inner end of the wheel's shaft. This construction further isolated the building from the vibration induced by the rotation of

Harrison Tie and Lumber Co., steam-powered sawmill, Route 760 (Quesenberry Road, SW), 1920s. Sawmillers began to abandon water power and take to the woods as soon as portable steam boilers and steam engines arrived in Floyd County around 1900. In this pre–EPA set up, runoff from the log springhouse in the background conveniently disposes of sawdust. Standing left to right are Sterling Martin, Herman Martin, Arthur Bond (sawyer), Raymond Martin, Henry Martin, and Early Quesenberry. (Ricky Cox collection)

the largest of sometimes poorly balanced gears, shafts, and pulleys. The anticipated settling these forces caused in the hurst frame could be compensated for fairly easily with wooden shims, whereas uneven settling of the building's foundation (typically of stacked stone) would cause a variety of problems that would be difficult to spot and even harder to fix.

Going to the Mill

The number of roads named for mills merely hints at the importance mills once enjoyed in rural areas. Going to the mill for the first time with a father, brother, uncle, or grandfather would have been among the initiations experienced by every farm boy and many farm girls. One or the other group would account for just about every Floyd County child at least until World War II. In August 2008, Floyd County native Jack Spangler recalled the first time he went to the Spangler Mill on Pine Creek with his father. Born in 1931, Murry Jackson Spangler was about seven years old when this took place. Most men and many women born in rural Southwestern Virginia before World War II could share similar memories of their first visits to water-powered mills.

Going to the Mill, c. 1938

Sometime around 1938 plus or minus 2 years I went with my father on one of his fall-of-the-year trips to the Spangler Mill to have some white corn ground into cornmeal and some buckwheat ground for flour. He would take two or three bushels of shelled corn and maybe a bushel of buckwheat. These quantities were sufficient to pay the toll for milling and for us to have a full winter's

> supply. We lived within a few miles of the Spangler Mill but used it primarily for cornmeal. Webb's Mill on Dodd creek, which was further away, was used more frequently for wheat flour. Each of these mills were water powered.
>
> As I recall the Spangler Mill in those days had a dam across Pine Creek a short distance upstream from the millhouse. The down-river side of the dam consisted of large timbers maybe 4 × 4s or 6 × 6s standing vertically with braces supporting heavy boards forming the back side of the dam. I don't recall how the dam was anchored to the riverbanks. The dam was not very high, maybe 6 to 10 feet. A wooden race carried the water downstream to the mill wheelhouse. I don't remember water flowing over the dam but there was sufficient leakage to keep water flowing downstream. Even though I had lots of time on my hands while the milling operations were taking place, my father was very careful not to allow me to get too close to the dam or the mill pond. But, the pond looked like a good swimming pool to me.
>
> I remember going down to the wheelhouse with my father to see what was there. The waterwheel was not an overshot wheel. I seem to recall it was either an undershot wheel but most likely it was a turbine as I think it was laying horizontally with a shaft coming upward to a 90 degree gearing arrangement to transfer power to a line shaft under the mill house. At my age at the time I did not know the difference and probably did not know enough to ask. Others have told me that they thought the mill had an undershot wheel.
>
> Cousin Walt Spangler was the miller. He had a "diverter mechanism" for the lack of a better term to channel the water either to the waterwheel or back into Pine Creek. I don't recall exactly how the water got from the race to the waterwheel. It most likely was some sort of pipe, possibly made of wood, since the water had to fall maybe as much as 10 feet to the wheel. In any case it was damp and wet in the wheelhouse. The noise level was high when the water was diverted back into Pine Creek. The noise level was even higher when the mill was in operation. There were squeaks, groans, clicking and clacking, etc. all going on simultaneously.
>
> Power from the waterwheel was transferred to the various machines in the upper levels of the mill via the line shaft with a number of pulleys underneath the mill itself. There may have been some more gears (cogwheels we called them) there also. Belts went up through the floor to these machines. I was not allowed to go into this area unless I had my father by the hand because he considered it too dangerous.
>
> The cornmeal was warm as it came out of a vertical shoot from the second floor into a bag. I remember sticking my hands into the warm cornmeal. It even smelled good too![28]

Besides a chance to spend time with a father, grandfather, or older brother, the regular chore of going to the mill presented many boys and some girls with an opportunity to assume an important responsibility and save the valuable time of an adult who had too much to do at home. Many Floyd Countians, both men and women, have told of being placed on a horse with a sack of corn and worrying, all the way to the mill and back, about how they would get the heavy sack back on the horse if it slipped off along the way. Roy Turpin's regular experience of driving his father's team and a loaded wagon to the mill, starting when he was only eleven years old, may have had many historical counterparts, but such stories are rare among men born after 1940.

Uriah Turpin was a sawyer in a sawmill five or six days a week and a farmer seven days a week. To supply "chop" for his ten or twelve milk cows, feed for his hogs, laying mash for his chickens, and horse and mule feed for the team, plus corn meal for himself, a wife, and eight children, Turpin or one of his sons made the thirteen-mile round trip to Greasy Creek or Harris Mill (III, 64) with a one-horse wagon about once a week in the winter. His team and two-horse wagon could carry enough to last two weeks. Ten years earlier, he could have gone to the Ed Strong Mill (III, 72) on Big Indian Creek, just over the hill from his farm near Duncan's Chapel Church of the Brethren, but it closed down in the late forties after the main dam washed out and then broke again soon after it was repaired. It was torn down around 1950. When the Greasy Creek Mill closed in the 1960s his horses and wagon made an eighteen-mile round trip to Vaughn's Mill on Mira Fork. Once, after Greasy Creek was shut down, and Vaughn's had sent word that they were closed for repairs, the team made the trip to Webb's

Uriah Turpin's wagon and team are parked at Greasy Creek Mill, off Indian Valley Road about three miles from Willis, in a photo made about 1956. Strawberry, the "off horse" (on the left as viewed in the picture or on the driver's right) weighed about 1,250 pounds. Lady, the lead horse, was about 100 pounds heavier. The young boy on the porch is thought to be their 11-year-old driver, Roy Turpin, standing beside an unidentified mill patron. (Courtesy Gertrude Harris family)

Mill on Route 8, just outside of town, traveling a total of twenty miles. Most of the other farmers drove pickups, one-ton trucks or even their family cars, with large trunks and spacious back seats, to the mill. A team and wagon on the highway was becoming a rarity in 1956, which is probably why these pictures were taken to begin with.

Uriah's oldest son worked with him at the sawmill and they had too much do at home on their days off to make the time-consuming trip to the mill. The second son, Roy, was eleven years old in 1956 and good with horses; both he and his father believed he could handle the chore of going to the mill. The day before the trip they loaded the wagon with baled hay, sacks of ear corn, and sheaves of bound oats to be ground and mixed into animal feeds, and a half-bushel of shelled white corn if they needed corn meal. Wheat was usually taken to be ground into plain flour in larger amounts a few times a year. Roy left as soon as the morning milking was finished. In winter, he'd have to go on a Saturday, because of school, and would leave around daylight. Days were short, and even with his early start, some of the many farmers who did "public work" during the week were lined up at the mill ahead of him.

At the end of the farm road he'd turn right and head southwest on Route 730 (Duncan's Chapel Road). If the roads were clear, he'd take more than three miles off the trip by cutting across Route 749 (Rocky Hill Road) and also save some wear on the horses' shoes and the steel tires of the wagon by minimizing his time on Route 787, the only paved road he'd travel. If he expected ice in shaded spots along Route 749, he'd take a rasp, a farrier's hammer and pliers, and a bag of ice-head horse shoe nails. When he saw ice up ahead, he'd stop at a level spot, set the wagon brake, tie the team, and go to work on the two front shoes of each horse. On each of the four front feet, he'd remove two of the eight nails and replace them with two of the ice head nails he brought along. The heads of ordinary horseshoe nails fit into rectangular recesses in the shoes to keep them from wearing away. The heads of the ice head nails stuck out about half an inch below the bottom surface of the horseshoes and gripped the ice like the cross-links on a set of automobile tire chains. On an asphalt road, they would last only two or three miles, so Roy could expect to stop and change out two other nails in each front shoe on the way back. If he anticipated too much ice on Route 749, he'd stay on 730 to its end and go left on 787. This was the main road through Indian Valley, but there wasn't much traffic besides a mailman or a milk truck and a few cars. The horses were not bothered by passing cars, but did shy away from trucks. On the back of the wagon Roy tacked a hand-lettered cardboard sign reading, "I may be slow but I'm ahead of you."

At the mill he'd pull the team and wagon out of the way and wait his turn. When he was nearly ready to grind what Roy had brought to the mill, miller Arby Harris sent Roy out to back the wagon up to the porch of the mill. When he thought Roy was out of earshot, he'd call to the other men, "Watch this boy back that wagon. He is really good." Roy turned the wagon around and eased Lady and Strawberry back against the leather harness, taking pains to center the loaded wagon in the open space and fit the tailgate of the wagon squarely against the edge of the porch. Together they unloaded the wagon through the mill's wide front door and fed the hay, corn, and oats into the hammer mill beneath the main floor of the mill. From there it was blown back up to the mixers on the main floor where molasses, soybean meal, cottonseed cake, salt, and other supplements and minerals were added to make it more nutritious and appealing to the animals. From the mixers it was dispensed into sacks weighing about 100 pounds apiece. Arby had already shown Roy the secret of the locking stitches sewn across the mouth of each sack. Start at the right spot and it all comes out in one smooth pull. Start at the wrong place, and it will have to be picked out, one stitch at a time. He had also shown Roy how to hold one finger under the string to tie a miller's knot around the mouth of a bag of corn meal or flour.

With the added weight of the molasses, supplements, and wheat middlings for the hogs, the horses would be pulling close to a ton of feed, plus the wagon, plus Roy, on the way home. The feed would still be warm from the grinder, so he'd make himself a place to burrow down among the sacks for the ride home. At a level spot near one of streams he'd cross on the way home, he'd unhitch the horses and take them to water before giving them a few ears of corn saved out when the sacks were filled the day before. While the horses rested he'd drink a Nehi bought at the store next to the mill and eat a sausage biscuit or something else his mother sent along. With luck, he'd be home by 2:00 or 3:00 p.m. If it was a busy day, or if he had to go all the way to Vaughn's Mill, he'd do well to get home by dark.

After delivering Rainbo Bakery products for 25 years, to the day, Roy Turpin is locally famous as "the bread man" and as a Church of the Brethren pastor and minister. In telling this story, he allowed that today some people would call this a hard way to treat a child, but he remembers it as a great adventure, a regular test of his resourcefulness and resolve to do the job entrusted to him. And most of the time, driving the horses and wagon was easier than the work he would have been doing back at home.[29]

The Mill as Commercial and Social Center

People like Roy Turpin could often do other errands on a trip to the mill. Depending upon the amount of power available, and the type of mill, development and construction of

Top: Spangler Mill (II, 99), 1910s. Spangler family members and neighbors take advantage of the drawn-down pond to collect stranded fish and turtles. If the smudged date stamped on this photograph is correctly read as August 5, 1911, the event was advertised and later covered in the pages of the *Floyd Press.* (Courtesy Floyd County Historical Society) *Bottom:* Waders in the Flint Mill (II, 106) pond, Beaver Dam Creek. The year is unknown, but the height of the corn in the bottom land and on the hillside beyond suggests that the picture was made in late July or August. The view is thought to be southward, which would place the hillside cornfields on the south side of what is now Route 612 (Stonewall Road, NE). (Courtesy Stewart Shank)

Mill owner Cleophas Vest, with family and neighbors, at the newly constructed Flint Mill (II, 106) c. 1933. An employee of the Norfolk and Western Railroad in Roanoke, VA, Cleo Vest had the new Flint Mill constructed on the site of the larger S. B. Lawrence/David Vest mill around 1932. A discarded runner stone, probably from the flour mill, is in use here as a temporary porch step. Most of the new structure, to the left of this view, covered one of the last water-powered sawmills in the county. Gathered here (from the left) are Jessie Sink, Effie Vest, Alwin Vest (with shotgun), Cleo Vest, Wanda Whitenack, Hazel Whitenack, Chester Whitenack, Mason Whitenack, Gracie Whitenack, Cora Martin, Rufus Sink, and Oscar Whitenack. (Courtesy Joyce Poff)

other kinds often followed the construction of a mill. The presence of a relatively simple corn-only mill of the type found in most every small community might attract a blacksmith shop. But a flour mill, to which every farmer and householder for miles around must eventually come, was the natural center of an agricultural community. Stores, post offices, distilleries, and other industries needing motive power were often built nearby. Wool carding mills were essential as long as people were spinning and weaving wool at home. A few linseed oil mills extracted valuable oil from flaxseed, a byproduct of the production of linen, which ended years before the wool carding industry. Sideline enterprises often took place in attached buildings, perhaps because of the risk of cross-contamination. Fires or stoves employed in other process would also present a greater risk of fire to the combustible interior of a flour mill.

Grain mills were also sometimes preceded by other applications of water power to an industrial process besides sawmills. A couple of very early mills were on or near the sites of iron foundries, which used water power to drive bellows. The steady blast of air was required to heat the mixture of charcoal, iron ore, and limestone above the melting point of iron. Because the iron ore was generally of such low quality, and the process so labor intensive, early forges and foundries were abandoned as soon as iron and steel from more efficient producers could be imported economically.

Ella DeHart Finch and companions at the eastern end of the Aaron DeHart Mill (III, 136) dam near the mouth of Indian Creek, undated. (Courtesy Olga DeHart Harvill)

In addition to being commercial centers, mills also served as social centers, primarily for the men and boys of a community. But the idea that millers had plenty of time to sit down for a game of checkers, "give away" or "fox and geese" is called into question by the number of times many mills changed hands. Milling seems to have been no career for a man who expected to sit around. As well as industrious, a miller needed to be tactful in his dealings

Vaughn's Mill (III, 64) water wheel at rest, late 1940s or early 1950s. The combination of the strong geometric shapes of the mill and overshot water wheel, the sparkle of rushing water, and the drama of highland landscapes has always appealed to painters and photographers. A wooden gate in the box race has been opened and the water turned out, while the wheel is stopped momentarily during the work day, for the night, or for the weekend. (Courtesy Gene Vaughn)

with customers whose personal habits might not suit him. Oscar Webb (II, 122) is supposed to have posted this notice in his mill: "If you spit on your floors at home, feel free to spit on ours." Other versions of the story add this second line: "We'll use your sacks to clean it up." The deterrent power of this threat hinges on the fact that most families kept special, tightly woven "long" sacks of heavy cotton or linen cloth for carrying wheat and corn to the mill, and the flour and cornmeal back home. These were kept spotlessly clean and were sometimes embroidered with initials to prevent mix-ups.

Individual standards of cleanliness raised more than questions of etiquette for millers as

From the left, World War II veterans Willie Showalter and brothers Arby and Marvin Harris at the Greasy Creek Mill (III, 67), probably around 1950, following a VFW parade in nearby Hillsville, Virginia. Willie Showalter's wife was the former Audrey Harris, sister of Arby and Marvin, and of the mill's owners, Noah and Harless Harris. (Courtesy Gertrude Harris family)

long as they continued to grind, in exchange for toll, the corn and wheat customers brought to the mill. As health regulations tightened it became harder to accept corn that had been exposed to rodents or was infested with corn borers, or to risk contaminating a roller mill with "weevily" wheat. Ultimately this ancient form of barter disappeared as fewer and fewer people grew their own bread corn and wheat, and millers were compelled to maintain tighter control over the raw materials passing through their equipment. Mixing even small amounts of spoiled or contaminated grain with products intended for other customers could cost far more than the trade of one customer could ever be worth, yet it was important to retain the good will of the community and a good name. At least one miller addressed this problem by maintaining a separate Meadows type corn mill exclusively for grinding patrons' corn on an exchange basis. The customer carried his portion home and the miller used the one-eighth portion taken as toll in animal feeds. The purity of flour was maintained by adding more steps and more machines, such as separators, scourers, desmutters, polishers, and cockle cleaners to the process of cleaning wheat before it was ground.

In addition to daily gossip and tale telling, mill place social occasions included community picnics and fish fries or give-aways, held when a mill pond was drained for repairs. These are mentioned in the histories of the Henry Huff (I, 1), W. H. Harman (II, 100) and Spangler Mill (II, 99). A chaperoned group of fourteen "young ladies of the town ... had a delightful time" inside the John Epperly Mill (IV, 126) after their afternoon "fishing party" was rained out in late June 1908.[30] The predecessor to Epperly's Mill, the Howard Mill, was also the site of community entertainments. A handbill from 1891 promises "Speaking, Boatriding, Fishing, Lawn Party and ... the East View Band" and invites the public to "Bring your Basket and your Family, [and] if you have no family bring your basket and your neighbor's girl. We are going for a good time."[31] Whether or not they were actually asked to come, Laurel Ridge residents anticipated being invited to a picnic at Sowers Mill (II, 94) to celebrate the completion of a new concrete dam in late 1910.[32]

Seasonal activities included swimming, fishing, ice skating, and cutting ice and storing it, packed in sawdust, in local icehouses. Shooting and wrestling matches, baptizings, several weddings and at least one funeral also took place. The county treasurer and the registrar also used mills as satellite locations for the convenience of taxpayers[33] and voters in outlying areas.[34]

And even when working mills were still common sights, the eyes of artists were drawn by the elemental appeal of wood, water, stone, and steel. Whether as backdrops to human figures or as the central subjects of paintings, poems, or photographs, mills, especially those fitted with overshot wheels, have probably gotten more attention than any other commercial or industrial structure.

The Technology of Water-Powered Grain Milling

While specific applications and combinations of milling technology may have been peculiar to a Floyd County mill or miller, the machines and the mechanical principles behind them were not. Most of the water wheels and much of the milling equipment installed in pre–1900 mills was handmade in the county by skilled millwrights, carpenters, and blacksmiths, but it was virtually all based on patterns and models widely known throughout the U.S. and in some cases already obsolete in many other parts of the country. Published sources describe the machinery and explain the processes of producing flour and various by-products from wheat, rye, and buckwheat; cornmeal from corn; and various kinds of animal feeds from grains, hay, fodder, and a wide variety of vegetable and mineral supplements. However, the following specific mill histories may be useful to the student of mill technology as illustrations of the equipment and processes that characterized Floyd County mills. Mill histories with relevant but less substantial references are noted in the index.

The pros and cons of horizontal turbine water wheels are considered in the histories of the Bob Lee Vest Mill (I, 13), Roberson Mill (I, 27), Vaughn's Mill (III, 64), and Shelor/Nolen/Pine Creek Mill (I, 123). The last of these also describes the rare (in Floyd County) "pitchback" water wheel configuration still in place at that site. The history of the Will Stanley Mill (IV, 54) also deals with the pitchback wheel, which, though never widely accepted in the Southern Appalachians, had become a favorite among American and English hydraulic engineers by the time water power was supplanted by steam in the mid to late 1800s. The history of the Henry Epperly Mill (II, 103) explores an easily overlooked drawback to the turbine water wheels typically coupled to the roller mills that sprang up all around the county in the first quarter of the 20th century. Flour making with the time honored burr or buhr stones, compared to the modern roller mill, is treated at some length in the histories of the Irvin Huff Mill (I, 2), Thomas Lester Mill (I, 121), and Roberson Mill (I, 27). The grain elevator, an essential component of multi-story mills of any type, is described along with the Spangler Mill (II, 99). Steam power, an alternative to water power that was pointedly rejected by all but a handful of Floyd County grain millers, is discussed in the history of the Peter Hatcher Mill (I, 21).

Self-contained "Meadows" type mills (Meadows was the best known of several manufacturers) are described and discussed at some length, along with the Willie Ferris Mill (I, 10), and are also treated briefly near the end of this overview, since neither of the primary sources on full sized grain mills, Oliver Evans (who was dead long before they were developed) nor B.W. Dedrick, has much to say about them.

One machine that may have been unique to the Floyd area was a mill invented by George W. O'Bryan, a native of Floyd County who was living in Cambria (near Christiansburg), Virginia, when his device was patented in 1913. An August 1912 *Floyd Press* reprint of a *Roanoke Times* item claimed that O'Bryan had invented "the first automatic crusher and grinding mill in the world" and declared it "one of the greatest inventions of its kind of this age." [35] Two years later, O'Bryan himself claimed in a *Floyd Press* advertisement that his mill "makes first-class meal, Graham flour, buckwheat flour, grinds fodder, corn and the shuck altogether, and small grain in the sheaf."[36] Interested persons were invited to inquire about purchasing mills, a half interest in the business, or the right to manufacture or sell the patented O'Bryan mill. Had it lived up to O'Bryan's claims, such a mill would soon have replaced every corn and feed mill in the country, sparing only the roller mills, since it could not, even in O'Bryan's powerful imagination, produce sifted white flour. Like most devices expected to perform many tasks, it probably did none of them very well. But one unit, at least, was installed in 1915 in Oliver Carter Bower's mill (II, 96) on Camp Creek, the history of which has more to say on this topic.

The stones in George O'Bryan's mill probably came from a mill stone quarry on Brush Mountain, also in Montgomery County, Virginia, only a few miles from Cambria and Christiansburg. The corn rocks in many Floyd County mills were quarried, cut, and dressed there and shipped by wagon. Advertisements for Brush Mountain stones appeared in a local weekly, the *Floyd Intelligencer*, as early as 1853.[37] Their manufacture is discussed in *Millstone Manufacture in Virginia: Interviews With the Last Two Brush Mountain Millstone Makers*, edited by Charles D. Hockensmith (1999)[38] and in a *Roanoke Times and World News* article by Robert Fries. Montgomery County historian Jimmie Lee Price has also researched and written on this topic on his own and in collaboration with Hockensmith.

One other connection between the manufacture of milling machinery and Floyd County existed in the Salem Foundry and Machine works, which manufactured roller mills, sifters, and various other pieces of flour milling machinery between 1890 and the early 1950s. Brothers and Floyd County natives John Emory Shank and William Samuel Shank were president and vice-president, respectively, when the company was re-chartered in 1906. Vaughn's Mill (III, 64), the Shelor/Nolen/Pine Creek Mill (I, 123) and Sowers Mill (II, 94), and probably other Floyd County roller mills as well, were fitted with Salem machinery.

The Economics of Grain Milling

Most of the machines and processes employed in the milling trade were located inside the mill, where they could be closely monitored and carefully regulated by a diligent miller. But many of the factors that determined whether or not a miller would succeed financially were, like the price of grain, beyond the miller's control. Some, like the weather, were beyond anyone's control.

If pursued beyond the most immediate causes and effects, the economics of milling becomes an even more complex topic than the technology. The smartest approach is to say even less about it. The bare outline that follows is meant only to reveal to readers the slender margin from which millers and mill owners were compelled to extract enough, at least, to live on, if not an actual profit. It might also suggest at least one reason why so many mills changed hands so often. Discussion is limited to production of cornmeal and wheat flour, one or both of which was the primary business of most mills until the middle of the 20th century, after which time the production of custom (made from customers' own grain and hay) and commercial (made partly or completely from imported feedstuffs) animal feeds became the main source of income for all but a couple of mills. Animal feeds were produced by many 19th century mills, but became a leading product only after two things had come to pass: The typical farmer had to be persuaded that his cows would give more milk, his chickens would lay more eggs, and his hogs would gain more weight if he spent the extra money for feeds mixed according to the dictates of the emerging science of animal nutrition. For his part, a miller had to invest in a hammer mill and the requisite bins, ducts, blowers, and baggers, and in a gasoline or diesel engine, or an electric motor, powerful enough to drive these machines without sapping power needed for other mill operations.

Production of rye meal or rye flour and buckwheat flour were both important processes throughout the second half of the 19th century, as evidenced by *Products of Industry* reports from 1870 and 1880. But by 1950, buckwheat milling had become a niche market specialty of a handful of mills, and the grinding of rye flour or meal a rarity. The authors understand neither of these processes well enough to venture even a rudimentary discussion. It is safe to say, however, that profit margins were no better for these grains than for corn and wheat.

The exact financial situation at a given mill at any particular time was determined by a host of factors, including debt assumed in building, equipping, or purchasing the mill and the costs of hired labor and upkeep. Customer debt carried on the books could also be a factor, but this was probably not as significant as in a retail business like a store, since most cornmeal and flour were ground on an exchange basis, and little money changed hands. Location was also important, both for the convenience of and access to customers, and for the size of the stream, which placed an absolute limit on the horsepower and production capacity. Extremes of weather, like floods and long freezes that affected the mill directly, as well as season-long weather patterns that would have an impact from one harvest to the next, were as important as they were unpredictable. More subtle but still important influences included local, regional, and even national economic conditions. In other words, millers and mill owners were affected by most every consideration faced by the owners of manufacturing or value-added businesses in all times and all places.

The balance sheet computations of Floyd County mills prior to World War II were slightly less complex than those of most modern manufacturing concerns in that they were virtually all custom mills, meaning that they ground only the grains and feed stuffs brought to them by customers. Lacking cost-effective modes of transport like rail or water, most did not import corn or wheat from outside the county (aside from the trade of customers living in the bordering districts of adjacent counties), or ship their products in quantity for sale outside the county. Transportation costs and inventory control, issues that plague most modern businesses, were therefore of less concern before the age of the automobile.

Both would, however, become significant and, ultimately, decisive factors, as feed millers began trucking in the corn, wheat, and oats that local farmers had once supplied. By the 1940s they were paying to have molasses and other bulky additives for their own mixed feeds, along

Greasy Creek Mill (III, 67) operator Arby Harris prepares to grind baled hay into cattle feed, 1962. Installation of power-hungry feed making equipment, like the hammer mill in the basement and the two mixers behind Harris, eventually cǫmpelled every Floyd County feed mill to replace water power with electric motors and diesel power units. (Courtesy Lillian Harris Van Lear)

with pre-packaged animal feeds (mostly chicken feeds) for retail sale, hauled up winding county roads. By the end of the 1940s, many millers had their own flat bed and enclosed trucks, hauling in grain, dry additives, and bagged feeds. These same trucks and their drivers delivered custom mixes in 100 pound sacks, mostly to farmers immersed in a dairy industry that had begun as early as the 1910s with an extra cow or two kept to produce butter to be packed in tubs and taken by a "huckster" by wagon to the nearest market town, like Roanoke, Radford, Christiansburg, or Pulaski. This expanded to the sale of cream in metal cans trucked to small regional creameries around the 1930s.

By the 1950s many of the larger mills were being drawn into the business of grinding and delivering multi-ton loads of dairy feed to farmers shipping whole milk to regional processing plants outside the county in eight- and ten-gallon cans. By the 1960s and 70s, feed was being delivered in specially fitted bulk delivery trucks to a growing number of Grade A dairymen who were storing milk in stainless steel tanks until it was picked up by large tanker trucks and hauled to out-of-state processors. By the year 2000, the virtual disappearance of the dairy industry in Floyd and surrounding counties ensured that a few large regional mills with large trucks could meet the milling needs of several counties. Egg, poultry, sheep, and pork operations also flourished at various times, but none of these by themselves, or together, were ever as large as the dairy feed business in Floyd County.

Although there was no home delivery involved, the rapid disappearance of small scale dairying also affected the feed milling business. The practice of keeping one or two cows to produce milk for home use and to sell for supplemental income, and to produce a calf to sell each year, was still widespread among rural families in 1960, but had disappeared only twenty years later. This was hastened by numerous factors, including a decline in family sizes, continuation of the shift toward dependence upon and identification with non-farm jobs, greater numbers of women working outside the home, expanding access to store-bought foods (places to purchase and money to purchase with), and the closure, by the dairy industry, of processing plants designed to handle tiny volumes of milk (from four to perhaps 100 gallons per day, shipped in 8 or 10 gallon cans) collected at great expense of time and labor at multiple locations.

Concern over the costs of buying, maintaining, and operating trucks was still in the future during the pre–1940 heyday of cornmeal and flour milling, but millers already had plenty to worry about. The ever present question was how to survive, let alone prosper, on the small margins of profit allowed by law and local traditions. The ancient system of grinding for toll, by which customers traded to the miller a portion of what they brought to the mill in exchange for having it ground, are outlined below, as they applied to cornmeal and wheat flour, the two most important mill products in the pre-feed mill era. Why farm feed eclipsed cornmeal and flour as the primary products of local mills is discussed later in this overview, in the history of the Farmers Exchange Mill (I, 139), and in bits and pieces throughout the book. Precisely when the balance tipped is impossible to say without examining the account books, if any exist, of the last dozen or so active mills. But this shift in emphasis from food for humans to feed for animals undoubtedly parallels the decline in small, semi-subsistence, multi-crop farms and the simultaneous increase in the proportion (but not the number) of specialized dairy, poultry, pork, or beef operations.

Although the overall financial picture is complex, the equation for computing the millers' toll on corn brought for grinding into meal could hardly be simpler. After the corn was measured, traditionally by volume rather than weight, the miller took as his share exactly one-eighth, as prescribed by Virginia state law. The process by which toll was taken in smaller, older mills is described later in the discussion of Meadows-type mills. The miller was free to sell his one-eighth portion or to grind it and sell the meal, or to trade corn or meal for some other commodity. Being naturally cleaner, after harvest, than any small grain, and also less subject to variation in quality, corn was typically not cleaned before measurement, and was cleaned only minimally, if at all, by passing it across screens and through an adjustable blast of air before grinding. The customer's portion was bagged or returned to the sacks in which it was carried to the mill if they were of suitably fine texture.

Grinding wheat is a far more complex process than grinding corn, as indicated by the much greater size of a buhr-stone flour mill, and the number and variety of machines required. The differences are even more apparent in a mill that made flour by the roller process. Wheat was typically measured by weight rather than volume, which makes necessary the first piece of equipment, a large platform scale upon which sacks of grain could be weighed or, more commonly, a scale upon which a large bin was mounted for greater convenience in weighing large amounts of loose grain. The presence of this item alone often distinguished a flour mill from a mill that ground only corn. Before it was weighed, the wheat was delivered through a trap door to the bottom end or "boot" of an elevator,* usually in the mill basement, and carried to an upper floor to be introduced to a cleaning machine called a "receiving separator." Here, reciprocating screens and a constant blast of air removed undersized and underweight grains of wheat, bits of chaff and straw not removed in the threshing process, dust and dirt, some weed seeds, and most other foreign matter. "Screenings" with enough substance to avoid being blown away, consisting mostly of swiveled (shriveled) grains of wheat, were bagged and returned to the customer, who could feed them as they were to most any farm animal, or have them ground and mixed with a batch of feed.

A speed indicator typical of water-powered roller mills. When the bell on the right side of what was essentially a tachometer warned that the RPM of a primary drive shaft had fallen below a preset limit, the miller had either to reduce the flow of material being fed into the mill or increase the flow of water to his vertical water wheel or turbine. The second option was often not available on Floyd County's small mountain streams. (Ricky Cox photograph)

The grains that survived the first cleaning were diverted into a bin, often mounted directly on a scale. The customer was credited with this amount of wheat, measured in bushels, and from that number could anticipate how much wheat and bran he would receive at the end of the process. A rule of thumb used in the water-power era was one and one-half 25 pound bags of flour per 60 pound bushel, or three bags (75 pounds) for every two bushels (120 pounds) of cleaned wheat.

After the wheat was weighed, a wooden slide was opened in the funnel-shaped bottom of the bin, feeding the wheat into the boot of a second elevator. From there it may have been run through a scourer, a desmutter, and a cockle machine, all of which were expected to remove materials that were negligible in their weight but could lower the quality of the current and subsequent batches of flour. What happened between this point and the flour bag is carefully described in other sources. The long and the short of it, in terms of the miller's share, is that for every bushel of good quality wheat, which at a percentage of moisture suitable for grinding weighs about sixty pounds, the customer would receive about 36 pounds of flour and fourteen pounds of wheat bran. The flour was bagged for the trip home and ended up in the oven. Most any farm animal—cattle, hogs, horses, chickens—would eat the bran, alone or mixed with other feeds, especially if the endosperm of the wheat had not been completely separated from the bran as it passed between the stones or through a series of rollers. While it might have been seen as a small benefit by some observant customers, consistent failure to cleanly separate the bran would be a serious problem for a miller, since success in a flour mill depended upon producing the absolute maximum amount of flour and recovering every speck of it.

For a detailed discussion of grain elevators in flour mills, see the history of the Spangler Mill, II, 99.

Jam (left) and Ham (right) with millers Ephrem Vaughn (left) and his nephew Sammie Phillips, Vaughn's Mill (III, 64), 1950. These littermates were the subjects of an experiment sponsored by the Larrowe Division of General Mills, makers of Larro Hog Concentrate, to prove the superiority of a "well-balanced hog ration" over a diet of plain corn and wheat. The results were conclusive, as Ham outgained Jam by more than 45 pounds during the nine-week experiment. Word circulated among Indian Valley farmers that a .22 short rifle bullet would not penetrate the thicker skulls of Larro-fed hogs. Hog butchers were advised to use the more powerful long rifle shells of the same caliber instead. Raising hogs was such an attractive sideline for millers that the Virginia legislature had placed restrictions on the practice before 1800. (Courtesy Gene Vaughn)

The economic fate of the miller lay within the ten pounds that remained in the mill after the customer had loaded up his fifty pounds of flour and bran and gone home. Some small part of the ten pounds would be high-grade flour. How much depended on the miller's skill and the condition of his machines. The popularity of machines such as flour scalpers and bran dusters, which actually brushed off and collected bits of flour adhering to bran particles, shows the importance of maximizing the yield of flour. The ability to do so was one of the chief attractions of the roller mills and the multi-stage swinging sifters that were typically installed along with roller mills. The fact that they made lighter, whiter flour impressed customers. The fact that they made more of it impressed millers.

Even with the best of equipment, part of the ten pounds would be "middlings," which are the product of grinding the oily and protein-rich wheat germ. Middlings are automatically separated out in the process of "bolting" the flour, passing it over or through wire screens and silk cloth in a rotating reel or swinging sifter. Left in this form, middlings usually become highly nutritious hog feed, which a miller could sell or turn into cash, eventually, by feeding it to his own hogs. Raising hogs was traditionally such an important sideline for millers that

Fordson tractor supplies supplemental power to the flour mill at Vaughn's Mill (III, 64), 1940s. Despite its location on one of the smaller streams, Ephrem Vaughn's mill on Mira Fork ran under water power, supplemented by gasoline and diesel engines and electric motors, until around 1960. Power from the tractor is transferred via flat belt to a line shaft above the second-story flour sifter. Vaughn celebrates a successful installation while Andy Branscome inspects the results. (Courtesy Gene Vaughn)

state laws prescribed the permitted proximity between the pigs and the mill yard. Several individual mill histories, including that of Logan's Mill (IV, 51), make reference to this practice. Many mills later retailed commercially produced middlings as hog feed under names like "Daisy Middlings" and "Red Dog."

Since flour costs more per pound than hog feed, the logical means of maximizing profit would be to incorporate the middlings into the flour; not surprisingly, millers and milling engineers spared no effort in their attempt to do so. Middlings grinders were used to reduce the texture, and electrical and chemical bleachers removed some of the brown color middlings added to flour. Besides making the color and texture of flour less appealing, middlings diminished the capacity of flour to rise under the influence of yeast or baking powders. Wheat germ has relatively little of the gluten that is found mostly in the starchy white endosperm, and the oil in the germ causes the flour to stick to itself. The germ was therefore routinely removed in the process of making all but the coarsest whole wheat flour. It is ironic that much of the progress in mill technology of the last century and a half has been devoted to making the whitest and lightest flour, and therefore the least nutritious and healthful, we can possibly make.

As more and more flour mills dropped out of the competition, the survivors were able to shift the rates of return to proportions more favorable to themselves. Whereas customers were receiving 36 pounds of flour and fourteen pounds of bran from a sixty-pound bushel of wheat near the end of the water-powered era, the added costs of electric and diesel power units and other costly upgrades dictated that customers be given only thirty pounds of flour and twelve pounds of bran, leaving the miller with eighteen pounds, of which half might be top grade flour. But even this improvement was not enough to save small-scale, rural flour milling. The reduced rate of return was yet another reason for local farmers to stop raising wheat and instead buy local or national brands of ready-to-use flour at a mill or grocery store. Vaughn's Mill was the last Floyd County mill to grind biscuit flour, giving it up in 1991. No other mill was even in business by then, and none of the most recently closed had made wheat flour since the 1960s. (Roberson Mill [I, 27] continued to grind buckwheat on a limited scale into the 1970s.)

Faced with such a thin margin of profit, it's no wonder that so many millers sold out within a year or two of taking over a milling business. Some made money by reselling the mill in rising markets like the middle and late 1910s, some broke even, and some limped away, wounded but still on their feet. Others held on until it was too late to break even or sell at a bearable loss. Some of these handed the property over to holders of deeds of trust. Many more faced chancery suits brought by creditors or partners and either sold their mills at bargain prices or saw them sold at public auction to the highest bidders. Sometimes the new owner was the old owner plus a partner who could infuse some cash into the business. At other times it was yet another novice anxious to prospect in the gold mine that flour mills, in particular, were apparently perceived to be as late as the 1920s. In still other instances, the buyer had no intention of running the mill at all and was instead speculating that someone who hadn't been at the auction would pay the auction price plus a handsome profit a few days, weeks, or months afterward. Samuel Brame and Winfield Scott were two such investors whose real estate ventures included other sorts of commercial properties as well. A typical advertisement for one of Brame's mill properties, which was last known as the Frazier Thomas Mill (II, 93), appeared in the *Floyd Press* on March 10, 1909: "I have on Little River, Va., a splendid three-story buhr mill with fairly good dwelling house. This property lies about six miles north of Floyd, Va., and is known as the A. A. Boothe millplace. The mill with a good miller will do a splendid business and is well suited for party or parties wishing to engage in the mercantile business. I am going to sell this property regardless of price. Call on or address S. R. Brame, Floyd, Va."[39]

Brame was also at various times the sole and outright owner of the Enoch Phillips Mill (III, 74) and the Jim Dulaney Mill (III, 78) in Indian Valley and the Shanghigh Mill property (I, 23) on the headwaters of Pine Creek. He held a deed of trust to the Ridinger Goodson Mill (II, 118) and would have become its owner had Amos Ridinger defaulted on the loan secured by the deed of trust.

Grain Milling and the Liquor Business

The production and sale of untaxed whiskey is a frequently romanticized part of the culture of the Southern Appalachians. The daring and drama that often attended the illegal manufacture and sale of whiskey, and the violence and tragedy that sometimes resulted from its consumption, are, if nothing else, sensational. As a result, this topic has been treated in hundreds of newspaper and magazine articles and depicted in numerous movies, novels, and works of non-fiction by writers whose knowledge of the subject varies at least as much as the purity of the product they were discussing. But it is undeniable that the distillers' need to buy ground corn and rye, and the millers' desire to sell these things, guaranteed a powerful and sometimes awkward economic connection between the two trades. Although he may overstate the case in *The Second Oldest Profession: An Informal History of Moonshining in America,* Jess Carr recognizes the mutual benefit in this relationship by contending that "family or commercial gristmills probably could not have survived without the subsidy of work done for stillin'."[40]

Most literature on this subject deals with the illegal trafficking in liquor motivated by the arrival of national prohibition in 1920. Less has been written about the entirely legal business of distilling that was widely practiced in the county, and, presumably, throughout the state and most of the nation, before that point. A county by county directory of Virginia businesses, published between 1877 and 1917 under various titles, named twenty-one presumably licensed Floyd County distillers in 1885, twenty-nine in 1889, and forty-two in 1894. Distillers were not counted in 1897, but the number had very likely grown at a rate that paralleled both the jump in the county's population (from fewer than 10,000 in 1870 to over 15,000 in 1900) and the steady increase in the number of corn and flour mills reported by the same source during roughly the same period — sixteen in 1881; forty-three in 1885; fifty-one in 1889; sixty-two in 1894; and seventy-one in 1897. (For complete lists of distillers and corn and flour mills in the cited years see Appendix A.) There were probably a good number of illicit distillers as well, since there was money to be made by avoiding the tax on whiskey even before distilling was outlawed.

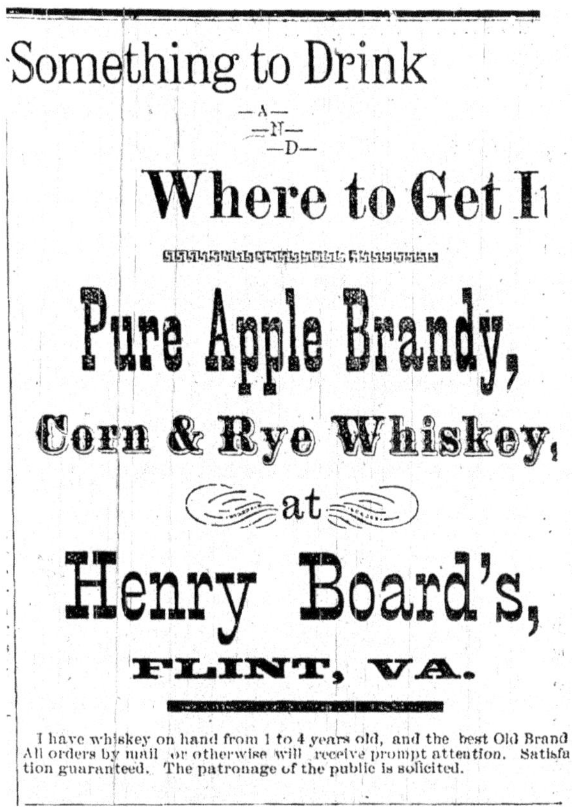

Floyd *Weekly News* advertisement. December 6, 1895. Customers of Henry Board's Flint, VA, distillery could purchase apple brandy and corn and rye whiskey by mail or in person. The Flint post office was near Double Springs School on modern day Route 612 (Stonewall Road NE). A photograph archived at the Blue Ridge Institute is identified as the "Board Mill-Post Office" in Floyd County, but no record of a Board Mill was found.

A number of millers were involved with distilling, and almost all millers knowingly or unknowingly supplied rye and cornmeal to distillers, before and after it became illegal to manufacture and distribute "spirituous liquors." Many a miller with reservations about the liquor business must have found himself on the horns of a dilemma before 1920 when a customer with only a wife and two children to feed ordered several hundred pounds of cornmeal. The miller's hesitation would have been compounded after it became a criminal act as well as a socially questionable one to make and sell whiskey.

The heat generated by this debate, locally and nationally, is frequently reflected in the pages of the *Floyd Press* between 1900 and 1920. Its editors provided a platform from which shots were fired on behalf of both sides of the national and local debate on prohibition. The anti-liquor, pro-temperance camp typically relied on direct appeals to the morals and good consciences of readers, while the anti-prohibition, pro-liquor side made its case through the more subtle financial device of advertisements concerning the buying and selling of component

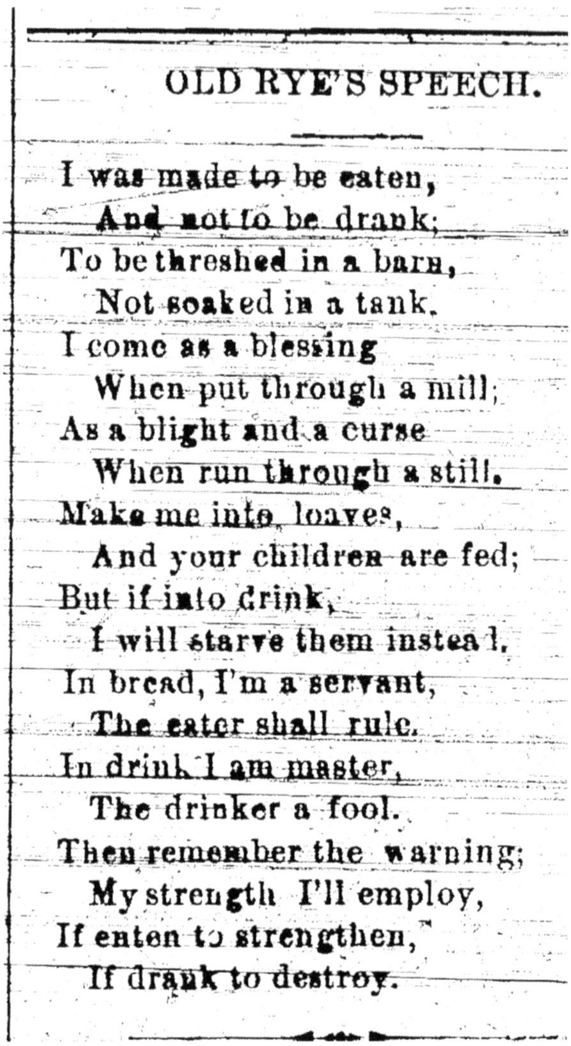

OLD RYE'S SPEECH.

I was made to be eaten,
And not to be drank;
To be threshed in a barn,
Not soaked in a tank.
I come as a blessing
When put through a mill;
As a blight and a curse
When run through a still.
Make me into loaves,
And your children are fed;
But if into drink,
I will starve them instead.
In bread, I'm a servant,
The eater shall rule.
In drink I am master,
The drinker a fool.
Then remember the warning;
My strength I'll employ,
If eaten to strengthen,
If drank to destroy.

"Old Rye's Speech," the *Floyd Press,* page 3, January 23, 1902. The temperance army enlisted poets as well as politicians in their battle against alcohol.

parts—white corn; sugar; fresh and dried berries, cherries, apples, pears, peaches, etc.—and the finished products—whiskey, brandy, and bourbon—of the distiller's art.

Although secret deliveries, nighttime raids, and high speed car chases make the most colorful stories, it was the opportunity to earn cash in a cash-poor economy that enticed most people to risk the loss, theft, or destruction of expensive distilling equipment and materials; the seizure of expensive horses, wagons, and automobiles; heavy personal fines; and even imprisonment. For men barely able to feed their families, the math could be persuasive. The price of a bushel of shelled white corn probably would not have exceeded one dollar and a half at any time in the 1930s; a pint of corn whiskey would have sold for between twenty-five and fifty cents during most of that decade. Depending upon the method and the amount of sugar used, a bushel of corn could yield from as little as one to more than ten gallons of whiskey. A modest return of four gallons to the bushel would yield thirty-two pints with a retail value of from eight to sixteen dollars. With sugar, yeast, and containers to buy, and lots of hard work to be done, the gross difference of at least $6.50 was by no means all profit, but still impressive. Using a more realistic price of 50 cents a bushel for shelled corn and more conservative estimates of both yield and price of liquor, historian Jess Carr calculated that "twenty bushels of corn might bring ten dollars: turned into corn liquor, it would bring seventy-five dollars."[41] Even a gross profit of only $3.25 per bushel had to be tempting to men who could hope to earn no more than a dollar (and not always in cash) for ten hours of labor in the fields or the woods, when there was work to be had.

The temperance army could also work the financial angle, as in this facetious account of the economic benefits of turning corn into whiskey instead of corn meal. It is interesting that the writer reveals either ignorance of the process, or a weakness for contemporary slang, by referring to the squeezing of juice from the corn. Only dried corn and cornmeal are involved in the process of making corn whiskey, and at no point does anything, except perhaps the customer, get squeezed:

Meal or Whiskey?

It is claimed that a bushel of corn will yield a larger product if the juice is squeezed from it, than if it is ground into meal. The following computation seems to prove it: Out of a bushel of corn comes four gallons of whiskey. From the whiskey the distiller gets five dollars, the saloon keeper gets seven, the government four, the railroad one, the farmer fifty cents, the consumer six months, his wife a divorce and the tax payers get the bills.

As in many other manufactures, there are also some by-products from the bushel of corn, which benefit the friends of the whiskey industry. The policeman making the arrest receives fees amounting to about two dollars, the magistrate who tries the case gets from three to six dollars, the officer who takes the drunkard to jail is allowed about 18 cents per mile for travel, the jailer gets two dollars, twenty cents per week for board and at the Supreme Judicial Court, bench and bar come in for a fifth gleaning. Finally the tax payers receive a second set of bills.

In no way can you get so much out of a bushel of corn as to make it into whiskey.[42]

Not surprisingly, the most impassioned rhetoric has to do with the physical, moral, and economic consequences of making, selling, and drinking untaxed whiskey, but the liquor trade's connection to the milling trade was generally much stronger at the other end of the process, in the acquiring and processing of raw materials. Corn liquor made in the traditional way without yeast and minimal sugar requires corn that has been dried and shelled, then wetted and allowed to sprout inside a sack or under a damp cloth cover, then dried and ground into a coarse meal. This so-called "sprouted corn" starts the process of fermentation, which turns the starch in the cooked corn mash into sugar. The addition of sugar, which was initially too expensive to be used in this way, speeds up the process and increases the yield, since it can be converted directly to alcohol, without first having to be converted from starch to sugar. The addition of yeast also makes the process more profitable by accelerating the process of fermentation during which the sugar is converted by microbial action to alcohol.

Because there is no other reason for grinding sprouted corn than to serve as "malt" in the making of liquor, a request for this service makes clear the customer's intentions, and makes the miller a knowing, if not exactly willing, participant. A group of young distillers in Indian Valley would occasionally use a mill belonging to one of their fathers for this purpose late at night. To avoid being discovered they disengaged the clutch on the steam engine that ordinarily drove the mill and took turns pulling the drive belt until the soft, sour-smelling corn was ground to the coarse texture required for their purpose. The owner would know what had happened as soon as his mill was started, and would have to grind and waste several pounds of good corn to clean his stones and push out the last of the ground up sprouted corn.

To avoid public scrutiny, a distiller could laboriously grind small quantities of sprouted corn with a hand-cranked steel-buhr mill. The next step in this direction was the purchase of a Meadows type stone-buhr mill powered, in a few cases, by water or steam, but most often by a one-cylinder gasoline engine or a four- or six-cylinder engine from scrapped Ford, Dodge, or Chevrolet car, adapted for stationary use by cutting off the frame and replacing the radiator with a barrel of water. Compared to a traditional "corn rock" (a pair of stones up to four feet in diameter with a proportionately massive drive train), a Meadows mill was portable, inexpensive, uncomplicated, and inconspicuous. With vertical stones typically 16 to 24 inches in diameter, these could grind cornmeal, as well as sprouted corn, at least as quickly as a full-sized corn rock. Their popularity in private and part-time applications throughout the rural South, for whatever purpose, no doubt hastened the decline of the smaller water-powered corn and feed mills.

As noted, a miller's degree of participation was a matter of personal choice, with the simplest and safest course being to mind his own business, and let others mind theirs. In the same way that teetotaling farm families could pick and sell gallons of blackberries and bushels of apples without concerning themselves over the potential uses for barrels and baskets of squashed, overripe fruit, millers could not be held responsible for what a customer made out of ground corn and rye. At the same time, some millers probably did express reluctance to trade with certain neighbors, perhaps with an implied threat to make public who was making large purchases. At the very least, they might insist that these transactions take place after regular hours. Customers were always free to take their business elsewhere, or to take the solution to its logical extreme and buy their own mills. The history of the Shelor/Nolen/Pine Creek Mill (I, 99) has more to say on this strategy. Other pertinent mill histories are identified under "distilling" and related index entries.

A "New South" portable mill with vertical stones, made between 1910 and 1940. New South was one of several early competitors to Meadows Mill, all of which were absorbed by the still active Meadows Company or went out of business. Mechanisms for cleaning and feeding grain into the mill, and for sifting the product, varied along with paint and trim, but in principle of operation these mills were more or less identical. (Ricky Cox photograph, 1991)

Complements and Competitors: Meadows Type Mills in Floyd County

The significance of the W. C. Meadows Mill, and a small number of virtually identical competing brands, in the history of grain milling in Floyd County is even more obscure than the story of its water-powered mills. Because of the very small physical and economic scale of businesses that ground meal or flour exclusively with semi-portable, self-contained mills of the Meadows type, these mills are virtually undocumented in legal records. And because the buildings that housed them are indistinguishable from other small, enclosed and floored farmyard outbuildings, such as granaries, workshops, or meat houses, there is little about them to attract the eye of a photographer. The few that appeared in photographs by coincidence are unrecognizable to an unsuspecting viewer unless specifically identified as mills or "mill houses."

The obscurity of the Meadows type mills is particularly ironic since they were not widely used in Floyd County until around 1920 and did not reach their heyday until the late 1930s, by which time the number of water-powered mills was in steep decline. The irony of their having been so quickly forgotten is compounded by the Meadows Mill's superiority, for grinding cornmeal, over horizontal "corn rocks" of the type that had worked so well, albeit slowly, for thousands of years. It was this traditional sort of millstone that some ancient thinker had in mind when he observed that the millstones of justice grind exceedingly slow but exceedingly fine.

Many of these old style stones had remained in service in larger mills even after the set or "run" of stones formerly dedicated to wheat-grinding was idled by the installation of a roller mill, whether in that same mill or at any other within easy reach of horse and wagon. The obsolescence of horizontal corn rocks was neither as abrupt nor as absolute, probably because the cornmeal they produced was indistinguishable from that of the Meadows type mill, in both quality (assuming proper cleaning) and quantity. In contrast, compared to burr mills, roller mills produced flour that was finer, whiter, and generally superior in appearance and baking qualities, and they made more of it from each bushel of wheat. That it was less nutritious than buhr-stone flour and much less so than whole-wheat flour was and still is irrelevant to most consumers.

While the product was more or less identical, Meadows type mills did offer distinct advantages over the traditional "corn rocks" whose jobs, and whose name, they eventually assumed in most mills. They were smaller, lighter, and essentially self-contained. A top-mounted hopper and cleaner were standard features, and a side-mounted, factory-made bolter or sifter could be added at additional cost. Both of the stones mounted vertically inside the wooden case were smaller and lighter than those of the old horizontal type. The shaft-mounted runner stone was more carefully balanced and could safely rotate at much higher speeds. Instead of a complicated drive train made up of multiple gears, shafts, pulleys, and belts, they were driven by a single pulley, mounted on the exposed end of the shaft that held the runner stone, and could be powered with equal ease by water power or by any of its more portable and more flexible successors—a steam engine, a gasoline (or kerosene) engine, or an electric motor. In construction and action, this type of mill combines the best features of pre-and post–1900 technology: the mechanism and process are easily understood and grinding is completed in a single quick pass through durable natural stones.

An excerpt from a 1983 "History of the Meadows Mill Company," written by Meadows Mill Company General Manager Edward Finley for *Old Mill News*, describes the advantages of the Meadows Mill and others like it as both complements and competitors to traditional, full sized mills:

> There were several reasons for the popularity of this type mill over the old flat or horizontal mills. The stones being smaller in diameter could turn faster thus producing a greater volume in far less time while at the same time maintaining the quality or even better since improved cleaning devices had been designed and incorporated in the machine so that much of the foreign matter which previously had been allowed to flow into the older mills could be removed before being ground. Then, it was no longer necessary to locate the mill near some stream as it could be powered by a gasoline engine or an electric motor making it feasible for small cross roads stores to put up a mill [for] custom grinding. These mills could be shipped fully assembled, making it easy to install without costly or elaborate machinery as in the past. The price of the mill was also in the range of the small operator. Records of the New Williams Mill Company [also in North Wilkesboro, NC, and absorbed by Meadows in 1924] show that in 1919 the cost per mill was $77.00 and sold for $82.79 with a profit of $5.79.[43]

It is easy to see why the owners of many large, full-time mills purchased mills of this type to supplement or replace cumbersome or worn out horizontal stones, or to circumvent aging power trains, and to see why the owners of "cross roads stores" would want to make their businesses more attractive by adding this service. Since many rural families continued to grow "bread" corn and have it ground into meal after they had given up raising wheat and began purchasing local or national brands of flour stocked in local stores, they could do all their grocery shopping at a nearby store that had a Meadows mill on hand. Since most people had wheat ground into flour in fairly large quantities only once or twice a year, those who did grow wheat could still make the longer trip to a flour mill when necessary. The miller-storekeeper could grind the toll (the one-eighth portion to which he was entitled in exchange for grinding the corn) into meal and sell it to customers with no corn of their own. Millstone historian Jon Sass noted, "Many country stores throughout the USA were able to produce

The David Vest Mill (II, 106) on Beaver Dam Creek, c. 1910. Thinly-planted, weed-free corn fields surrounding the mill illustrate the difficulty of raising enough corn to feed everything on a horse and human-powered farm. Average yields, around 25 bushels of shelled corn per acre in the 1920s, were doubled and tripled by the advent of affordable chemical fertilizers in the 1930s and after. Modern Midwest yields of hybrid varieties approach 200 bushels per acre. (Courtesy Stewart Shank)

cheap flour and meal for local consumption and many farmers invested in this type of mill so avoiding the time-consuming and laborious visits to the local custom mill. These two factors played a major part in the demise of the rural custom mill."[44]

Why an ordinary farmer would invest this kind of money, plus the cost of a one-cylinder gasoline or kerosene power unit or a salvaged automobile engine (rural Floyd County was not electrified until after World War II) is less obvious. Unlike those run by storekeepers, these farmstead outfits were typically used only one day a week to grind cornmeal and dairy and hog feeds for the owner and for neighbors living, usually, no more than a couple of miles away. Since most grinding was done on an exchange basis (one half-gallon of corn out of each half-bushel brought to the mill) there was little chance of recovering the cash outlay. Excess corn was sometimes sold, but few of these part-time millers sold corn meal.

The answer to this question lies in the constant demand for corn, both white and yellow, and the almost year-round struggle to raise, harvest, and store enough to last from one fall to the next. Before the widespread availability, and affordability, of modern fertilizers, the worn out fields of Floyd County might yield less than 30 bushels of shelled corn per acre, as compared to the harvests of at least five times that amount expected by farmers today.

With the exception of cats, every animal on a pre-tractor farm ate corn. And since cats do eat mice and rats, even their survival was only one step removed from cornfield dependency. Since a corn miller took as his toll one-eighth of the corn that was brought to him for grinding, he needed to have only eight customers to be assured of having as much bread corn, for his own use, as the average of his neighbors, without plowing, planting, cultivating, or harvesting any white corn of his own. (Meal made from white corn was and still is overwhelmingly preferred in Floyd County, but yellow is equally popular in other places.) This security, by itself, would be attractive at a time when very large families might bring half a bushel of corn to the mill every week. There would also be a toll on any feed that was ground, plus opportunities, for anyone so inclined, to grind corn for distillers who might pay in cash for the service, rather than give up the one-eighth toll. While it is unfair to generalize, it is safe to say that many owners of Meadows type mills were themselves directly involved in the production of corn whiskey.

Aside from the mill itself, and the engine, little equipment was required. Existing buildings were often modified by having holes cut in the wall through which a drive belt could pass from the mill to a pulley on a tractor parked outside, or through which an exhaust pipe could extend from an engine placed inside. No scales were needed, since the grain brought to the mill was measured by volume, most often in a wooden or metal half-bushel (four gallon) measure. At 28 pounds, plus the weight of a container, the half-bushel was a convenient amount to carry to the mill with the grain divided between both ends of a sack draped over a man's shoulder, or the back of his horse. Toll was collected by filling a "toll dish," which was in some cases simply a wooden box dimensioned so as to hold exactly one-half gallon of corn. By state law, all measures used at mills were to be "sealed," i.e., bear an official seal indicating that their volumes were as stated, but how closely that rule was followed is unknown. A well traveled joke, sometimes localized by the insertion of suitable names, told of a boy reporting to his father after "taking a turn" to the mill by himself for the first time. Having been made more observant by responsibility, he notices for the first time that the miller dips out part of his father's corn and puts it into a box of his own before starting the mill. The boy says nothing, but when the miller's back is turned retrieves what he thinks to be stolen corn, plus a little extra as punishment, and pours it into the hopper with the rest of his corn. The joke usually ends at that point, without describing the father's response.

As is often the case with faster, more productive machinery, Meadows Mills were also in some ways more dangerous than the horizontal mills they replaced. The enclosed stones and the feeding mechanism atop the mill presented few dangers when run at the recommended speed, but the flat drive belt, which could be of most any length, depending upon how far away the drive unit was located, was fully exposed and liable to catch loose clothing of an unwary observer. Aaron Altizer, who ran a Meadows Mill for himself and his White Rock area neighbors in the 1940s with a modified Model T Ford engine, had stepped away for a few seconds when he heard the note of the engine change. The broad drive belt or a pulley had caught the tail of his daughter's dress and pulled it so tightly around her that she could not breathe. She was nearly unconscious, but otherwise unhurt, when he cut the dress free of the belt with his pocket knife.

The most serious local Meadows Mill accident on record happened in November of 1926 and was reported in the *Floyd Press* on November 18, 1926, under the headline, "Young Man Killed":

> Last Wednesday afternoon, Chester W. Hall, the only son of Mr. and Mrs. Amos B. Hall, three miles east of town, was killed while working at a mill.
>
> It seems that he and his father had erected a corn mill and they were running it with a motor from a Chevrolet car and it is supposed that it was speeding too much and the rock bursted and flew in all directions. One piece struck young Hall in the head crushing his skull, killing him instantly. His father was right seriously injured but it is thought he will recover, although he is yet in a right serious condition.
>
> He was about 16 years of age and leaves a father, a mother and two sisters.

Thomas Wright recalled that part of the roof and sides of the building were also torn off by flying fragments of the mill stone.[45]

Uriah Turpin looks on as Arby Harris, long-time miller at the Greasy Creek Mill (III, 67), ties a 25-pound bag of newly ground floor. It would be difficult, today, to find a product so modestly named as "Good Family Flour." (Courtesy Lillian Harris Van Lear)

A 1991 survey showed that at least twenty of these mills ran in a 40-square-mile section of Indian Valley, or within easy reach across a county line, between 1920 and 1950.[46] If this density can be extrapolated over the 400 square miles of the county, one could guess that there were at least as many of these at work within little more than a quarter-century as there were water-powered mills in 200 years. Yet their size and portability made them almost ephemeral compared to water powered mills, and they are all but forgotten. The distribution, ownership, and output of these mills would make an interesting study on its own.

The Miller

Given that several hundred men, and in some cases women, felt persuaded or compelled to try their hands at the milling profession, it's impossible to describe a single "type" of person who would be drawn to it. It is easier to characterize the small fraction who stayed at it long enough to be counted successful.

Essential characteristics would have to include honesty, dependability, and cleanliness.

The habit of working steadily and more or less efficiently would of course be required to succeed in any business. Some measure of charity would also be important to the reputation of a man on whom every person in the neighborhood depended for the corn meal and flour that were absolute necessities of the pre–World War II rural diet, and staples for years afterward.

At least until the war economy brought non-farm jobs, mobility, and an infusion of cash, prosperity and often the very survival of farm families depended directly upon getting the most food calories—for themselves and their animals, and afterward for themselves *from* their animals—from the crops they grew. To plant, cultivate, and harvest a single acre of corn without a tractor, a farmer or some combination of family members had to walk about forty miles. Having invested so many steps already, a farmer probably wouldn't have balked at traveling a couple of extra miles to the next mill if he felt that too much toll was being taken from his grain, or that he was receiving in exchange meal, flour, or feed whose quality was lower than that of the grain he brought with him or had previously stored in the mill, as was sometimes done with wheat. The importance of giving "full measure" is lost on Americans who buy products by the box, bag, or bottle and pay little attention either to volume or weight, but the universal understanding of the phrase made it the perfect metaphor for Abraham Lincoln, who spoke at Gettysburg in 1863 of men who had given "the last full measure of devotion."

The reputation of the miller in the folk history of Floyd County is generally benevolent. Most millers seem to have been viewed as servants of their communities. Instances of farmers refusing to trade with one or another mill are known, and an anecdote that circulated around Floyd implied that at least one mill owner encouraged his hired miller to get as much toll as possible from poor farmers. But neither of the authors heard any specific claim of a particular miller deliberately cheating a customer. This stands in stark contrast to the reputations of millers in folk songs that migrated from the British Isles. In one song, known by the names "The Miller and His Three Sons" or "The Dishonest Miller" and probably others, a dying miller leaves his mill to the son who promises to cheat *all* his customers. Jessica Banks, of the Penn State University Center for Medieval Studies, states that every mill-related folksong that originated in the British Isles "paint[s] this figure as one who is known throughout society as a selfish, grasping thief who takes advantage of anyone he can."[47] Such representations are consistent with the image presented in Chaucer's *Canterbury Tales*, in which the vulgar and dishonest miller is among the least savory of the pilgrims, as evidenced by both the content of his own story, "The Miller's Tale," and the characterization of the miller in "The Reeve's Tale," the story told immediately afterward.

The reputation for dishonesty at one time attached to more or less all millers in England may stem from the once widespread practice of granting monopolies to millers as inducements to settle and remain in a given location. Requiring farmers who were in some measure attached to the land they worked, although they did not own it, to patronize a particular mill, regardless of how well or quickly their grinding was done or how fairly they were charged for the service, would certainly lead to resentment. And the knowledge that his customers could not take their business elsewhere would certainly bring out the worst in a miller already inclined in that direction. Monopolies were awarded in some industries in some British colonies in North America, including Virginia, but there is nothing to suggest that a monopoly was ever offered to a miller in what would become Floyd County. County courts certainly encouraged the establishment of mills by interpreting laws in the most favorable way, and the laws themselves were generally favorable to businesses, but it was up to the miller to sink or swim once the mill was at work.

The language of mill related laws does make clear, however, that early Virginia legislators were well aware of the potential for abuse and the specific forms it could take, perhaps because

Top: **Amos Cabel "Cab" Nolen grinding corn about 1960. Nolen also owned the Spangler Mill (II, 99), near the mouth of Pine Creek, for three decades. (Courtesy Judy Nolen Hylton)** ***Bottom:*** **Irvin Huff (I, 02) specialized in grinding buckwheat flour on the headwaters of Little River near Copper Hill after roller mills took over the wheat flour business. (Frank Webb collection)**

of what they'd seen and heard in or from England. In his study of the iron industry in Colonial America, E. N. Hartley noted that monopolies of all kinds "presumably reached their peak between 1614 and 1621" and that "in effect, the whole period from 1600 to 1685 was rich in monopolies— and in protests arising from the discrepancies between the benefits to the realm alleged by those who sought them and their concrete results."[48] As men who had found life in the colonies in some way preferable to remaining in England, early leaders may have been anxious to avoid saddling themselves with a system of such doubtful value.

The very first act of the Virginia legislature pertaining to mills or millers (1645) has as its stated intention "to rectifie the great abuse of millers."[49] As revised in the following year, the law refers to "the greate abuse of Millers in exacting excessive and illegall toll" and establishes guidelines to insure consistent measurement of grain received and toll taken.[50] A law passed in 1705 includes a similar provision "for the prevention, of abuses by evil-minded, covetous, and exacting millers, or owners of mills," but is only a small part of a law whose overarching purpose and tone are in keeping with its title, "An act for encouragement of building Water-mills."[51] The bulk of mill-related legislation passed between 1619 and 1785, when the earliest Floyd County mills were being built, is generally favorable to the milling profession, and in any case refrains, after 1705, from referring to millers as "evil-minded and covetous." Compared to their English predecessors, Floyd County millers in particular, and American millers in general, appear to have been kept honest by the competition, or the threat of it, if not by their own scruples.

Customers needed also to have meal, flour, and feeds ground in a timely fashion, although the word "timely," now a synonym for "quickly," probably had a more complex meaning in the past. Since a trip to the mill was often a social event as well as a chore, men were generally willing to wait their turns figuratively and literally in the presence of a friendly miller and congenial neighbors. The hospitality of some millers extended to their homes, when customers had to be kept overnight in order to get their grinding done. Others were known to keep mills running around the clock when a new wheat crop was ready for grinding, or when a neighboring mill was unexpectedly closed by fire, flood, or freezing weather.

On slow days, millers at all but the largest mills could attend to farm chores or other businesses like a store, post office, or blacksmith shop, but were expected to answer the bell when a customer arrived. Cast iron bells mounted under the eaves of the Spangler Mill (II, 99) and on posts in front of the Weddle/Webb/Agee Mill (IV, 34) and the John Epperly Mill (IV, 126) are thought to be typical. Unless someone, whether it be the miller himself, a retired but still-able father or grandfather, a teenage son, or, quite often, the miller's wife, appeared to grind or at least receive the customer's grain, word would soon get around that a trip to so and so's mill might be a waste of time. A story is still in circulation in Indian Valley of a miller who saw a loaded wagon approaching the mill near closing time. After the wagon rounded a curve and was temporarily out of sight of the mill, he got into his car and drove away, apparently not caring that the man was sure to find out what had happened, and not dreaming that the story of his action would outlive himself and everyone who witnessed it.

Among the essential habits of the successful miller, keeping himself, his premises, and his product clean were probably the easiest for a customer to observe, and the hardest for a miller to maintain without the cooperation of his customers. All but one or two pre–1880 Floyd County mills ground only what was carried in by customers (hence were called "custom" mills) and most continued until near the end of the milling era to grind customers' grains in exchange for a toll. The amount of weed seeds, chaff, foreign matter, mildew, inferior grain, and rodent droppings in grain must have varied tremendously from farm to farm.

Fortunately, because corn was harvested one ear at a time, it was not subject to include seeds of other plants, and foreign matter of any kind is easily spotted among the large grains. Although the feed mechanisms of the older horizontal mills were not universally equipped for this, pieces of shuck, silks, and bits of cob could be removed by passing the corn across a wire screen with holes slightly larger than a kernel of corn. Everything that passed through the first screen landed on and moved across a second, with holes slightly smaller than a normal grain. Chaff, "rat pepper," and cracked or undeveloped grains fell through and were discarded. As the grains that had passed both tests were fed into the mill they dropped through a blast of air that blew away dust and kernel-sized bits of lightweight trash, along with grains of corn hollowed out by corn borers.

Wheat grains are, by contrast, much smaller, and wheat is harvested many stalks at a time. Since wheat is never cultivated or weeded between sowing and harvest, seeds of other plants are much more likely to be mixed in with threshed wheat. To make their flour reflect the quality of the best wheat that was brought in rather than the average, flour millers were compelled to install a bewildering and redundant assortment of separators, cleaners, scourers, polishers, and de-smutters, and to add or remove moisture from wheat before it passed through the first step in the actual grinding process.

It was important also for a miller to avoid looking like the mechanic and engineer that he often had to be, since while monitoring the milling process, and especially in bagging flour and meal, he might literally have a hand in what was to be on the table that night. A Floyd area miller who powered his Meadows type mill with a one-cylinder engine, and could never seem to adjust the carburetor to his lasting satisfaction, was known for producing meal that, even after half an hour in a 425 degree oven, offered customers a faint aftertaste of kerosene.

Patience with both machinery and people would have been essential. A sense of humor would help, as would a man's ability to keep his opinions to himself regarding the particulars of religion and politics. If pressed, the safest course was to profess a general approval of the one and blissful ignorance of the other.

Prominent Floyd County Millwrights

The men who actually designed and supervised the construction of all but the most primitive water powered mills in Floyd County were referred to as millwrights. Three of the best known were John W. Epperly, Alonzo "Newton" Hylton, and Johial "Hiley" Moses. All three were also owners or part owners of at least one mill, but only Epperly was best known in that role, having built and run two of the mills still standing in 2010, the Roberson Mill (I, 27) and the Epperly Mill (IV, 126), both on Dodd Creek. Epperly was somewhat of a bridge figure in the history of mill technology. He mastered the 19th century arts of gunsmithing and blacksmithing, and his first mill, built in the 1880s, was patterned after Oliver Evans' 18th century prototype. Fifteen years later he built one of the first modern roller mills in the county and is supposed to have helped others convert to the roller mill system that is even today essentially state of the art in flour milling.

Although he outlived both the others by nearly thirty years, Newton Hylton was most often associated with the primitive type of mill represented by Mabry Mill (IV, 60). He built or supervised the building of at least two wooden water wheels for that mill, and had some part in building wheels for several others in the southwest quarter of the county, including the Will Stanley Mill (IV, 54), the Wickham/Webb/Agee Mill (IV, 34), and the Eli Sowers/Rufus Harris Mill (IV, 41). Hylton's personal charm and his mastery of many pioneer skills made him a favorite subject of photographer Earl Palmer, who saw Hylton as the embodiment of the hardiness and ingenuity that characterized the earliest white settlers in the Blue Ridge Mountains. A twelve-page section of Jean Haskell Speer's *The Appalachian Photographs of Earl Palmer* is devoted to Hylton's many talents. The dust jacket featured a picture of Hylton raking hay with a team of horses.[52]

References to Hiley Moses in newspaper accounts and deed book records suggest that he was relied upon to get complex roller mills up and running, but disinclined to stay in one place for very long at a time. He ran several Floyd County mills and was apparently working outside the county in 1900, perhaps at the Carroll County, Virginia, mill later known as the Early Marshall Mill. Moses was an early partner in the Phillippine (later Vaughn's) Mill (III, 64) and half-owner of the Greasy Creek Mill (III, 67) when he died in 1929. His wife Josie was the daughter of William Dickerson and the granddaughter of William Morricle, both millers, and was herself a competent miller. Hiley Moses' paternal grandfather, his father, and at least four of his nine brothers (Charles, Jacob, George, and John) at some point worked as millers or millwrights. His two youngest brothers became electricians after moving with their

John Epperly (front, center), master miller, millwright, blacksmith, and gunsmith. Epperly built and ran two of the ten water-powered mills still standing in Floyd County. Seated by him are sons Giles (left) and Isaac (right). Standing behind are, from the left, Fitzhugh, Levi, Calohill, and Arthur. Levi ran the John Epperly mill, briefly, after his father's death. Giles owned the Harman Mill on West Fork when it was destroyed by flood in 1943. (Courtesy Jim Altizer)

parents to Roanoke, Virginia, between 1900 and 1910. That change of surroundings would have made clear to them that water power, and the complex means of transmitting and applying its mechanical force, would be abandoned as soon as more flexible and convenient sources of motive power became available. The future prospects for the builder or operator of a water-powered grain mill would have seemed less than promising in the growing city of Roanoke.

Family Connections Among Millers and Mill Owners

The surname of most every family that was well established in Floyd County by 1900 can be found in the deed records of at least one mill, but several pop up repeatedly. In both eastern quarters of the county (Sections I and II), the Huffs were heavily involved in the milling trade, as were the Poff, Epperly, and Williams families. Though not otherwise well represented, members of the Wimmer family were long time owners of two mills in the northeast section (II). The Spanglers and the Lesters appear in the histories of mills in three of the four quarters of the county. Though they tended to hire others to do the milling, the Scott family owned mills all around the county. The Phillips and Duncan families were prolific millers in the northwest quarter around Indian Valley (Section III). The names Barringer, Hylton, Slusher, and Harris appear regularly in the deeds to mills that stood in the Burk's Fork area (Section IV).

Even among families not known as millers, once a mill came into a family it tended to stay there, as was probably the case with most rural Floyd County real estate well into the second half of the 20th century. As long as legal actions, high taxes, high land prices, or complicated estate settlements didn't force the sale of a farm to the highest bidder, one or another member of the family was usually anxious to buy out other heirs, typically at a price below market value. Even if he or she was not farming the land, crop lands or pastures could be rented and timber sold at intervals to pay taxes and maintain line fences.

Alonzo Newton Hylton, photographed by Earl Palmer on June 3, 1946. Identified variously as A. N., Newt, or Newton, Hylton appears in the oral histories of more mills, usually as the builder of the wooden overshot wheel, than any other millwright. He worked beyond the middle of the 20th century, but embodied the ways of the 19th. What others hired done by machinists, foundrymen, and carpenters, he fashioned for himself with forge and hammer, saw and axe. (Courtesy Blue Ridge Institute and Museum of Ferrum College/Earl Palmer Collection)

An important difference between a mill property and a farm is that a grain mill and the land it sits on, no matter how isolated or far from town, are commercial rather than agricultural real estate. Like a store or a small manufacturing concern, a mill must be operated to realize any economic benefit from ownership of the mill or the land immediately surrounding it. Land, left unattended, at least maintains and typically appreciates in value. A mill left unattended quickly fell into disrepair and lost the value both of its physical plant, and of its patronage, which, having moved elsewhere, would be hard to recover.

Two factors—the tendency to pass real estate to family members, whether by inheritance or sale, and the more modern capitalist impulse to amass personal wealth—played roles in the transaction when the owner of a mill was not compelled simply to sell to the highest bidder. In such cases, the seller and the relative buying the mill were heeding an agricultural society's instinct for acquiring and holding on to real property for the long-term good of an extended family to which both belonged or owed allegiance. Also at work was the imperative of a modern, commerce-based economy to convert resources into capital for the short-term benefit of the individual,

Johial "Hiley" Moses was known for his skill at setting up new roller mills. He also ran several mills in and around Floyd County and was part owner of two. His paternal grandfather and father and four of his nine brothers were millers or millwrights. (Courtesy Norma G. Sheets)

i.e., to recover, save, or make some money. These influences could work together in persuading a buyer, but might create a dilemma for a seller.

One consequence of the tendency to keep land in the family was that a considerable number of the hundreds of men and women who at some time owned part or all of a Floyd County mill probably weren't that interested in milling, and would never have considered milling as a trade or a mill as an investment, had the opportunity to run or own a mill not come along through some family connection. When the chance came, the desire for personal gain may have persuaded some to seize it, especially if they thought (in spite of ready examples to the contrary) that the millers' life, with an income seemingly guaranteed by taking a toll from each customer, would be easier than farming or sawmilling. After a couple of years of struggling to make the mill pay for itself, owners who had bought the mill for either or both these reasons realized that the "better judgment" they had ignored in buying the property had in fact been better.

Although there is no statistical or anecdotal evidence to support this idea, it seems plausible that the coincidental and sometimes almost obligatory nature of many such acquisitions, as well as misperceptions about the business side of milling, may help to explain why so many mills changed hands so many times—people were simply getting rid of business ventures that they either hadn't much wanted or hadn't fully understood in the first place. The fact that throughout the 19th and early 20th Centuries these people in turn sold the mills to people who were kin by blood or marriage, but not necessarily inclined toward or suited to that line of work, reflects a continued willingness to honor the first impulse at the expense of the second, by offering the property first to a family member, almost certainly for less money than would have been asked of someone not thought of as family.

Examples upon which the reader may base his or her own speculations can be found in the ownership history of virtually every mill that lasted for more than twenty or thirty years. Kinship connections between successive owners are explored in detail in the histories of the Fleming Duncan/Billy Dickerson Mill (III, 71), the John Duncan Mill (III, 79), and Vaughn's Mill (III, 64). These three all happen to be in the northwest quarter of the county, but this same pattern is seen in mill histories from every part of the county.

Josephine "Josie" Dickerson Moses was the daughter and granddaughter of Floyd County millers. A competent miller herself, she was married to itinerant millwright and miller Hiley Moses. (Courtesy Norma G. Sheets)

Women in Floyd County Mills

It is assumed that the wives of men selling and buying grist mills are listed among the "parties of the first part' (sellers) and "parties of the second part" (buyers) in about the same proportion as in deeds to most other kinds of real estate.* This assumption has not been tested, but the results would be of little value in any case, since deed records reveal nothing about the making of decisions, and rarely tell who actually ran or would run the mill.

One woman, Nancy Shank, was listed as the owner of a mill in the U.S. census *Products of Industry* report compiled in 1880, but no known deed names her as either grantee or grantor in the conveyance of a mill property (see Appendix C). Similar inconsistencies between courthouse documents and *Products of Industry* reports, regarding the ownership of a particular mill, tend to suggest that she was not, at least, the sole owner of the mill. In some cases, a person erroneously listed as the owner in a census record (of which the *Products of Industry* reports were a part) was a former owner, and in others, a former or current miller. There is no written or oral evidence that Nancy Shank ran the mill, but there is no doubt that she was closely associated with it in some way.

Her first husband, Joseph Gray, Sr., and her second, Lewis Shank, both gave their occu-

**The summarized deed records that follow each individual mill history do not always reproduce in full the sometimes long lists of sellers and buyers named in individual deeds. Consult the original deeds for a complete listing.*

pations as "miller" in at least one census, but no clear record has been found of either owning a mill. Whether the mill supposed by the *Products of Industry* enumerator to be owned by Nancy Shank in 1880 was left to her by Gray (who was apparently more than forty years older than she) upon his death, given to her in the settlement of her divorce from Shank between 1866 and 1879, or acquired some other way is not known. She did bring considerable but unspecified property into her second marriage, and was to retain full control of it by virtue of a prenuptial agreement recorded on 26 February 1866.[53]

Martha "Mattie" Weddle was at least part owner and the primary operator of a small Burks Fork mill (IV, 39) just up the road from where Luvenia Harris ran the Eli Sowers/Rufus Harris Mill (IV, 41) with and without her husband's help. Luvenia Harris's daughter Ella Harris Thomas did the same after she and her husband Homer bought the mill in 1936. Lacona Sutphin Cox helped her husband, "Blind Walker" Cox, run the Enoch Phillips Mill (III, 74) on Little Indian Creek and probably did the same at the Jim Dulaney Mill (III, 78).

Josephine "Josie" Moses, wife of itinerant miller and millwright Hiley Moses, may have been the most skillful of women millers, considering that her husband ran several sophisticated roller mills on Mira Fork (III, 64), Greasy Creek (III, 67), and Dodd Creek (II, 122) and at least one similar mill in Carroll County before his death in 1929. Moses' problems with alcohol, and his wife's ability to take over until he could return to work, figure in the oral histories of two of these mills.

As consumers of mill products, many women would have had distinct preferences for the flour and meal that made the best baked goods. Since farm animals were more or less indifferent to the subtle distinctions in the taste and texture of their feed, most men probably honored the preferences of their wives. Doing so was easier than it might seem in the case of wheat or buckwheat flour. A few extra miles, going and coming, was no real obstacle, since most farmers hauled these products to the mill, in fairly large quantities, only a few times a year. Cavilla McPeak recalled bringing home twenty-eight 25-pound bags of plain flour (self-rising could be had but did not keep as well) around 1950 for himself, his wife and one child, and his parents.[54] Assuming an exchange rate of nearly three 25-pound bags of flour for every two 60 pound bushels of cleaned wheat, McPeak must have taken more than 20 bushels (about 1,200 pounds) of wheat to the mill. Millers would certainly have encouraged this practice, considering the elaborate process of cleaning, weighing, and further cleaning that had to be completed before grinding the wheat of each individual customer.

Though it may have been difficult to interest the mothers, wives, and daughters of mill customers in the nutritional content of a sack of dairy feed, large regional mills and wholesalers—whose pre-mixed dairy, hog, and chicken feeds were being retailed at local mills in 100 pound bags as early as the 1930s—found another way to spark interest in feed products. Originally delivered in coarse burlap sacks, these feeds were later shipped in sacks of tightly woven cotton, printed with colorfast dyes in an infinite variety of patterns. Women who lacked the time or thought it inappropriate to visit the mill and personally select the sacks they wanted would at least ask the men to pick out sacks with the same print. These could be made into dresses, aprons, skirts, blouses, children's clothing, and curtains. Plain white cotton sacks with stenciled labels could be bleached and made into diapers, underclothing, towels, sheets, and pillowcases. Twenty-five pound flour bags might have pre-printed borders that made them suitable to serve as pillowcases without alteration, once the flour was gone.

The Decline and End of Grain Milling in Floyd County

As long as a sizable number of people in a community were growing wheat and corn to take to the mill to be ground for their own use, and as long as most depended on horses or

Sammie Phillips oversees loading of the solid runner stone from the Spangler Mill (II, 99) corn rock, August 1992. A mortised wooden shaft, probably removed from service when the roller mill was installed early in the 20th century, has been put to use as a post for the porch roof, to the right of the suspended stone. (Ricky Cox photograph)

The Weddle/Wickham/Webb/Agee Mill on Howell Creek, late 1940s. Wooden mill buildings, many with wood-shingle roofs, began to decline almost as soon as they were abandoned. Decay was accelerated by occasional flooding of the foundation, constant dampness, and insects attracted by accumulated grain dust. (Frank Webb collection)

their own feet for transportation, and as long as rough roads made it unprofitable for outside wholesalers to put their products on the shelves of country stores, grist and flour mills were needed within a few miles of every rural home. It took years for all these conditions to change, but a change in any of them was usually both a cause and consequence of changes in all the others.

Effie King Brown dated the "beginning of the end" of flour milling in the northeastern part of the county at 1927, when her family bought their first bag of Gamble's Best Flour, a regional or national brand, at a local store.[55] Within the next thirty years most people stopped growing bread corn and wheat for their own consumption, but as long as feed mills were grinding the customers' hay, corn, and oats into livestock feed, a mill was still wanted, at least close enough that a round trip wouldn't require more than half a day. The Farmers Exchange Mill (I, 139), built exclusively to grind animal feeds in the 1950s, shared this short-lived niche with several flour mills that had been refitted with hammer mills and bulk mixers. The only commercial mill left in Floyd County in 2010, Vaughn's Mill on Mira Fork Creek (III, 64), continues to retail cornmeal and flour but has ground only animal feeds since 1991.

After most rural families who were farming part-time gave up keeping milk cows, hogs, and chickens for family use between 1950 and 1970, and the Grade C dairy business was abandoned by regional milk buyers, it was only a matter of time until several related factors combined to drive small local mills out of business: The across-the-board upgrading of secondary roads—whether from dirt to gravel, or from gravel to tar and gravel, or from tar and gravel to asphalt—may be the biggest factor, though it cannot be pointed to as the original cause since, as noted, it was both a cause and an effect of the rapid increase in the number of Floyd County men and later women holding full-time jobs, many of them doing shift work at factories twenty, forty, even sixty miles away. These daily commutes took them past or near to supermarkets and other retail outlets outside the county. The same improved roads encouraged wholesale suppliers to deliver brand name goods to local stores and mills, and allowed larger mills in neighboring counties to compete for the accounts of the largest dairy farms. It was only a matter of time until economies of scale, to which Floyd County mills had been relatively immune, began to cut into already small profit margins.

Many of the men working at non-farm jobs continued to farm part-time, with substantial help from wives, children, and parents, but were forced to abandon crops not directly related to what little profit was to be made from their increasingly specialized evening and weekend operations. Heirloom varieties of open-pollinated corn had to be grown in separate patches far enough from other varieties to prevent cross-pollination. And because characteristics like grain size, stalk height, and planting density hadn't been maximized, optimized, and standardized by intensive, centralized hybridization, heirloom varieties didn't fit as neatly into the high productivity methods of planting, cultivating, and harvesting that work best when every seed, every plant, every ear is identical to every other. Hybrid varieties also matured more quickly, so could be planted later, with less risk of frostbite. And while they didn't mind feeding it to livestock, and probably knew that the name brand meal and flour they were buying were hardly "organic" as we think of that term today, some were justifiably reluctant to eat bread made from the corn they had themselves treated with increasingly potent chemical fertilizers, herbicides, and insecticides. Other factors— like the overall financial shape of a milling business, and intangibles like the personalities of the owner and employees— that delayed or hastened the decline of individual mills can be accounted for only on a case by case basis.

Although it probably had no effect, then or after, on existing businesses, it is not surprising that grain mills were neither counted accurately nor represented as particularly valuable assets in *Industrial Survey* reports of Floyd County compiled in 1930 and 1957, both of which were overtly and unapologetically intended to attract employers to the county. The compilers of both documents foresaw that the economic future of the region would depend upon industry rather than agriculture, in which case the presence of grain mills, especially small, primitive examples serving only a few square miles, might deter rather than attract employers seeking a progressive, forward-looking place to build their factories. The 1930 survey completed by the Engineering Extension Division of Virginia Polytechnic Institute includes only roller mills, and names only seven of these, although at least a dozen were then in operation.[56] At least fifteen corn meal and buhr-stone flour mills were also still operating under water power, as were dozens of very small mills powered mostly by gasoline engines and running only a day or two each week. The survey notes, "There probably are sufficient flour and grist mills in the county but bakeries are needed to supply cooked-grain foods to the larger communities and surrounding areas."[57] Mills, all of which were locally owned, were unlikely to employ any additional workers, but a bakery built with outside capital just might.

The 1957 survey compiled by the Industrial Survey Committee, chaired by Everett H. Kingrea, listed only "one meal and feed mill" under "Miscellaneous Services."[58] This is almost certainly the O. A. Webb Mill (II, 122), since this business was one of the sponsors of the survey. Several others, including the Greasy Creek Mill (III, 67), the Pine Creek Mill (I, 123), and the Roberson Mill (I, 27), were still grinding feeds and corn meal, and at least two, Vaughn's Mill (III, 64) and the Thompson-Akers Mill (I, 14), were still making flour as well. Since a primary reason for starting or relocating a manufacturing business in the rural South was to escape labor unions, the writers of the 1957 report took pains to point out, "There has been no labor trouble at any time since the inception of any factory operation."[59] Knowing that the owners of the sort of factories most likely to come to Floyd would themselves have no use for mills or mill products, and that they would prefer employees dependent on a weekly paycheck, there was no point in calling attention to the fact that even part-time farmers with easy access to a mill were only a few steps and a few years removed from an almost self-sufficient lifestyle.

The Remains of Floyd County Mills and Mill Dams

A list of mills still standing in 2010 that were powered by water initially or throughout their working lives appears in Appendix E. The general condition and completeness, as a mill, of each is discussed in the individual mill histories. Visible remnants of mill dams, ponds, and races are noted there as well. Due to their resistance to the elements, the isolated locations of some, and the sheer number of them, some of these fragments of concrete, stone, earth, and submerged wood will likely survive the very last of the mills still standing in 2010. It seems therefore wise to conclude this overview with a description of the clues that future readers may be able to see with their own eyes. Even with these bits of evidence and a photograph in hand, they will find it hard enough to imagine what once was there.

Many mill histories describe the remnants of mill dams that may be spotted by the careful observer. Even where no trace of the mill dams remain, sites may sometimes be recognized by the presence of one or more of these four features. Readers are reminded that all but one Floyd County mill site, that of Mabry Mill (IV, 60) on the Blue Ridge Parkway, are on private property. Examples that can be spotted from public roads and rights of way have therefore been singled out in the following and earlier discussions.

(1) A *silt bank* may remain indefinitely on one or both sides of the stream on the upstream side of the dam. New layers of this fine, rich topsoil, eroded from upstream fields and cut-over forests, were added every time fast-moving muddy water was slowed by its entry into the mill pond. A high, straight sided, virtually stone-free soil bank is exposed as the stream cuts its way through the silt and back to its original bed after the dam has been breached. The tops of these banks, now grassed over or even forested, were once the flat, muddy bottoms of mill ponds.
A perfectly flat field remains behind the mud sills and abutments of the Ed Strong Mill dam on Big Indian Creek sixty years after the mill was torn down. A similar plot at an even older mill site on Burk's Fork was for many years an especially productive potato patch. Easily viewed examples of silt banks are found upstream from the sites of the Duncan/Hall Roller Mill on Indian Creek (III, 79) and the Spangler Mill (II, 99) on Pine Creek. The swampy mud flats behind the Rakes Mill (I, 26) dam on the Blue Ridge Parkway show how these sites may have looked before the dams were breached and the streams began cutting their way back to their original beds. The abundance of these silt deposits behind 17th to 19th century mill dams in the mid–Atlantic Piedmont, and their lasting effects on the landscape, are discussed at length in the *Science* article cited early in this overview.[60]

(2) A *dug race* (essentially a ditch) leading downstream from the dam site along the contour line of adjacent hillsides most often supplied water to an overshot wheel, but may have served a horizontal wheel, either a tub wheel or a turbine, instead. Long dug races at Vaughn's Mill (III, 64) on Mira Fork and the Shelor/Nolen/Pine Creek Mill (I, 123) originally carried water to turbines, both of which were eventually replaced by vertical water wheels because of insufficient water and maintenance issues.

(3) Several *stacked-stone abutments* that anchored wooden dams to the banks of the stream seem likely to last for many more years. The most striking examples are at the sites of Webb and Son Mill (II, 122), the German Belcher Mill (IV, 37), and the Weddle/Simmons Mill (IV, 38). The most visible from a public road are at the sites of the W. H. Harman Mill (II, 100) and the John Epperly Mill (IV, 126). Sturdy concrete abutments remain at

A flat, virtually stone-free silt bank marked the location of the last Spangler Mill (II, 99) dam in 2008. Similar formations are the most prominent physical reminders of several Floyd County mills. The bank is formed when the stream cuts its way back to its original bed after the dam is breached. (Ricky Cox photograph)

the Ed Strong Mill dam site. The largest and best preserved concrete abutments, at the Greasy Creek Mill (III, 67), appear almost as strong as when completed in the late 1920s. Nearly intact concrete dams remain at the sites of the Caleb Sowers Mill (II, 94) and the Irvin Huff Mill (I, 2), but only the former is visible from a public road.

(4) The hardest to spot, but ironically the least susceptible to destruction of mill dam fragments, are the *wooden sills or "mud logs"* that lie across the bed of the stream. These are the remains of a wooden dam built between stone or concrete abutments, atop a grid of heavy timbers pinned, pegged, or bolted together and anchored to the bed of the stream and the abutments. These wooden timbers can last indefinitely as long as they remain below the water line. When the water is low and clear in Burk's Fork and Little River, travelers on Conner's Grove Road and Roger Road may see in the stream bed at least one mud log from the Slusher/Alderman Mill (IV, 42) dam and the William H. Harman Mill (II, 100) dam. A significant portion of the framework of the last Greasy Creek Mill (III, 67) mill dam remains, and fragments of two earlier dams are visible upstream.

Part Two

Individual Mill Histories

Arrangement

The 140-plus individual mill histories that follow are arranged in four sections, numbered I through IV. Each section represents one of the four quarters of Floyd County as defined by its two major highways, Route 8, which crosses the county near its midpoint in a roughly north-south line, and Route 221, which runs the length of the county in the other direction, from its southwestern border to the northeast corner. Because Route 221 lies for most of its length from one to two miles south of a line drawn through the center of the county, each of the two southern quarters, sections I and IV, is considerably smaller than its northern counterpart, sections II and III, respectively. But for a variety of reasons, the significant differences in the areas of the four sections are not reflected in the number of mills found in each. Although Section II (the northeast quarter of the county) is nearly twice the size of Section I (the southeast quarter), each includes about 35 mills. And although perhaps a third larger than its southern counterpart, Section III (the northwest quarter) includes only 33 mills, nearly ten fewer than Section IV.

A stream-by-stream arrangement of the mills was rejected due to the meanderings of Floyd County's many branches, runs, and creeks, and one river. The longer the stream, the less satisfactory this system becomes as a way to group mills that served the same or nearby communities. This is best illustrated by Little River, which begins in the northeast part of the county, near Copper Hill, and exits Floyd County in the northwest corner, a mile or so west of Copper Valley (the two communities are related only by name). At that point, it veers north to serve as the boundary between Pulaski and Montgomery Counties before emptying into the New River near the City of Radford. A crow flying directly from Copper Hill to Copper Valley would cover only about twenty-five miles, but a hiker following the stream bank would walk at least fifty. Even without considering its major tributaries like West Fork, Dodd Creek, and Indian Creek, the Little River passes through or touches on three of the four quarters of the county, as defined above. Patrons of the Wertz Mill, by no means the farthest up Little River, may never have heard of the regular customers of the Sowers Mill, which was also powered by the flow of Little River, but lay many miles downstream. But Wertz Mill customers could regularly have eaten with friends and neighbors who had their meal and flour ground at mills on Big Run, Meadow Run, Pine Branch, or Divins Branch, all of which lay within easy reach by horse and wagon.

Although mills with successive individual numbers, (e.g., 3, 4, 5, and 6) are in many cases close to each other, this is not necessarily the case, since Frank Webb learned of the existence of many additional mills after assigning the first numbers. Hence, a mill numbered above 100 may be only a mile or two from a mill numbered below twenty. There seemed no point to renumbering, since mill locations are identified by stream (which, as noted, is not always helpful), and more importantly, by road number and road name, and can easily be found on a county map.

Primary Sources

Mill histories were constructed, primarily, from four types of sources: undocumented oral histories, local newspapers, various historical and scholarly sources, and deed book records. Memories are fallible and oral histories often cannot be confirmed by other sources, but as long as they did not directly contradict a documented source, the memories of millers, their descendants, and their customers have been included. The abundance of truth and human interest they do include cannot be sacrificed for the sake of excluding a few errors. The newspaper items were mostly brief, objective notices of events or legal actions; a few report injuries to millers or damage by fire or flood; and a couple are feature stories about particular mills. They are in all cases accepted as accurate, to the extent that their own sources were trustworthy. Published books and magazine articles were presumably subject to some form of review, and unless otherwise noted in the text of this book, are assumed to be reliable. By far the most important documentary sources for this book are the real estate deed records maintained by the Clerks of the Circuit Courts of Floyd and Montgomery counties. While the motives and actions that result in the transfer of property from one owner to another may be subject to question, title deeds themselves represent what is certainly the most jealously guarded privilege of American society, that of private property. Hence, deeds are among the most familiar and most trusted of public documents.

Additional Sources for Late 19th and Early 20th Century Mills

Two sources consulted but not heavily relied upon in the writing of the mill histories are transcribed in part or completely in the tables and appendices that follow the individual entries. Tables A, B, and C present information collected for the United States Census Products of Industry Reports (sometimes referred to as the industrial census) conducted by the federal government in parallel with the population censuses of 1850, 1870, and 1880. Some background and an explanation of why these are, for some purposes, not particularly reliable precede the tables. The reports are nonetheless interesting, and for some purposes quite valuable. Since only a fraction of the statistical information they contain is incorporated into the texts of corresponding mill histories, readers are referred by note to the pertinent tables.

Parts of a Virginia business directory published under various titles between 1877 and 1917 are reproduced in Appendix A. The excerpted lists include the names and locations (by nearest post office) of corn and flour mills reported in the county* in 1881, 1885, 1889, 1894, 1897, 1911, and 1917. Asterisks indicate grain mill owners also listed as the owners of sawmills. References to Appendix A entries appear only rarely in the corresponding mill histories, and because it includes so little additional information, readers are not directed to it by notes in the corresponding mill histories. Listed mills that could be matched to documented mills are identified by section and mill number (e.g., III, 64; I, 19). The fact that several of

**The means of collecting this information is unknown. One plausible method would have been to mail a survey to all Floyd County postmasters, whose names would have been on file in a Washington, D.C., office. This would explain cases in which the same mill, located between two post offices, was sometimes listed at both.*

the listed mills could not be matched to a known mill is consistent with the fact that the lists in Appendix A are known to include errors and redundancies. More importantly, it shows that the work of documenting Floyd County's water powered mills is by no means complete.

Reading the Mill Histories

Each history is formatted using identical headings, though not all are used in every case. A word of explanation on each will help the reader get the most from each part of an individual mill history.

Mill Names. While Floyd County mill owners were not above boasting about superior meal and flour in newspaper advertisements, most were refreshingly modest in naming their businesses. Out of about 140 mills, fewer than ten were known by the names of streams and communities, and the remainder simply by the names of the owners. Even the exotic sounding "Phillipine Mills" (built about 1900, and still running as Vaughn's Mill, III, 64) was probably a play on the surname of two of the original partners and possibly a reference to the recently ended Spanish American War. The only overtly immodest example is William H. Harman's "Palace Roller Mill," a name that probably only Harman himself ever used. It's hard to imagine a farmer, unless he was given to sarcasm, telling his wife that he was loading the wagon for a trip to the Palace Roller Mill or to the Eureka Mill, a name supposedly employed by Matthew Scott for his Dodd Creek mill.

While the practice of calling a mill by the owner's name is admirable in its simplicity, it has flaws, one being that changes in the name by which a mill was commonly known did not always keep pace with changes in ownership. Several mills listed in the *Products of Industry* reports for 1860, 1870, and 1880 (Tables A, B, and C) were reported to be property of men who had sold them years before. The names by which mills are known might also vary with the age of the person asked, since it would be quite natural for a person to remember a mill by the name of the man who owned or ran it when he or she first traded at or knew about the mill. Family connections can also play a part, since family-oriented people might assign unmerited importance to the name of a relative or family friend who played even a minor role in the history of the mill.

In the absence of a widely agreed upon name for a given mill, it seemed reasonable to use the name(s) of the owner(s) who lasted the longest, even though this method risks glorifying a silent partner who seldom if ever set foot in the mill or of ignoring the man who sacrificed everything to build a mill but soon sold out and moved on. In the later case, Johile (or Johial) "Hiley" Moses comes to mind. A highly skilled miller and millwright, Moses was a one-time partner in what is now Vaughn's Mill (III, 64) and in the Greasy Creek Mill (III, 67), but in the first case soon sold his part, and in the second, died a few years after acquiring it. He is among the best remembered millers and millwrights and worked in several other mills but his name is attached to none of them. The folly of attempting a logical explanation of a naming process that was itself far from logical is best illustrated in the discussion of Jack's Mill (I, 5) on Little River.

Upon failing to find a mill listed under its (to them) familiar name, readers are asked to consult the index, since the names of all known millers, and the names of all owners for whom a deed was found, will be included there. If a mill was known to have been operating in a particular year, Tables A through C and Appendix A may be consulted, bearing in mind that none of these is absolutely complete and that both A and C are decidedly incomplete.

Note that distinctly different situations are indicated by subtle differences in the formatting of mill names. A mill listed as the "Jones-Smith Mill" or the "Jones and Smith Mill" was at one time owned by partners Jones and Smith. Real examples include the Thompson-Akers

Mill and the Kropf and Custer Mill, both in Section I. Use of the name "Jones Mill/Smith Mill" (with a forward slash) would indicate that the mill was owned by both Jones and Smith, but not at the same time. The Anthony Poff/Hosea Wimmer Mill is one example. The name "Jones Mill and Smith Mill" (the word "Mill" appears after each of two names joined by "and") means that Jones and Smith owned two entirely different mill buildings, one of which succeeded the other on or very near the same spot. The "Peter Geurrant Mill and John O. Jack Mill," also from Section I, is one example. Ideally, each separate mill building would have been given a separate number, but several sites were discovered to have had more than one building only after numbers were assigned.

Location. Every mill in Floyd County could at one time be reached by a road passable to horse and wagon. As mills closed and isolated farms nearby were abandoned, a number of these one-time public roads became private drives or were completely abandoned. The Irvin Huff Mill (I, 2), the Quinter Lucas Mill (II, 108), the Crockett Duncan Mill (III, 77), and the Cara Simmons Mill (IV, 38) all stood on roads that are now maintained privately or not at all. Yet most mill sites still lie within half a mile of a public road. Hence, even those mills that are now accessible only on foot are located by references to the nearest roads, rather than by the distance to the nearest mill up or down the same stream. Both road names and road numbers, as used on the *2007 County Road Map* distributed by Floyd County, are supplied, since neither system is by itself foolproof. Along drivable roads, distances were measured with an automobile odometer. Cross-country distances were estimated by scaling various maps, including the one just mentioned.

Power refers to the type of water wheel(s) supposed to have been used at a particular mill. Vertical wheels used in Floyd County were almost all overshot, with a handful of high breast and pitchback wheels. Most of the overshot wheels were of wood, but since these were generally older and vulnerable to rot, no good examples aside from the reconstructed wheel at Mabry Mill are known. Among the steel overshot wheels, all those for which a maker has been identified were built by the Fitz Water Wheel Company of Hanover, PA. The Savage Co. of Knoxville, TN, also sold steel overshot wheels in the southeastern United States, but no example has been reported at a Floyd County mill. Among horizontal wheels, steel turbines had probably replaced wooden tub wheels by 1920, but it is not clear that record keepers always distinguished between the two, which are similar in principle but quite different in construction and in operation. When the stated horsepower ratings or wheel types are taken from the reports on the *Products of Industry* which, beginning in 1810, were completed on the same ten-year cycle as the U.S. population census, the date of the report (1850, 1870, or 1880) is indicated in parentheses. (Mill by mill enumerations for these three years are included in Tables A, B, and C.) Horsepower ratings are in every case estimates, but whether they were made by the miller or the enumerator is not known. Those that seem unrealistic are noted.

Stones. Mill stones necessarily function in pairs, traditionally mounted horizontally. The stationary bottom stone is called the bed stone. The rotating top stone is the runner. Each pair is traditionally called a "run" of stones. In this usage, "run" is both singular and plural, e.g., a matched top and bottom stone make up one run of stones; two such pairs of stones are counted as two run of stones. The more familiar and grammatically regular "set" and "sets" are used to avoid confusing readers who overlook these explanatory notes. As a rule, a mill with one set of stones ground mostly corn. A mill with two sets typically ground cornmeal on one set and flour on the other. A third set might be for buckwheat flour. While it was, evidently, common for large commercial mills outside Floyd County to have two sets of stones for grinding wheat, so that one could be used while the other was sharpened or otherwise repaired, there is no certain example of this in a Floyd County mill.

The terminology of roller type flour mills is also potentially confusing, since one "stand" of roller mills houses two sets or pairs of rollers. Two stands of roller mills, the practical minimum, contain four pairs of rollers, allowing for four distinct breaks or reductions of the wheat, as it passes between rollers of successively finer corrugations with successively smaller spaces between them.

Meadows type mills enumerated under this heading used smaller stones (typically 16 to 24 inches in diameter) mounted vertically on a horizontal shaft. These essentially self-contained mills were used mostly to grind cornmeal or crack corn. Since they were relatively portable and could be driven by a gasoline engine or electric motor, they were frequently set up in farmyard outbuildings and at country stores. By 1920 they were cutting into the corn grinding business that smaller water-powered mills had retained after flour milling was completely taken over by larger but often more distant roller mills.

Also noted under this heading are other water-powered machines such as sawmills, planing mills (planers), wool carding mills, and generators.

Visible Remains that could not be visually confirmed, or that were not visited in 2008 or 2009, are followed by the date of the last confirmed observation. A blank in this space indicates uncertainty rather than confirmed absence of remains, which is indicated by "None."

History. The narrative history of each mill comprises both written documentation and oral history. The plainest are merely summaries of public documents; the most elaborate incorporate discussions of evolving mill technology and of changes in the local, regional, and national economies. The authors have not completely resisted the temptation to speculate about motives, causes, and consequences, but have tried at least to acknowledge the speculative nature of such conclusions.

Deeds and Other Public Records include summaries of pertinent title deeds, court orders, and wills. Occasionally, a chancery suit or other legal action such as the filing of a mechanics lien is referred to. Although they provide only a skeletal outline of a mill's ownership history, title deeds are the single most valuable source of reliable information, and the most frustrating due to the frequent omission of specific references to mill property. Along with the elaborate formal phrasing that characterizes legal documents of any kind from any time, the ornate handwriting of clerks and deputy clerks (until the Circuit Court clerk's office purchased a typewriter c. 1917) can be difficult to decipher. Readers interested in a particular mill are encouraged, without embarrassment or apology on the part of the authors, to retrace the trail of its ownership, and to confirm independently that documents were not overlooked, and that the documents summarized were not misinterpreted or simply misread.

Section I. Mills East of Route 8 and South of Route 221

1. Henry Huff Mill

Little River

LOCATION: East side of Route 795 (Huff Mill Road SE) 1 mile south of Route 221 (Floyd Highway North); POWER: Tub wheel (1870), Breast wheel (1880); Stones: 1 set (1870), 2 sets (1880); VISIBLE REMAINS: None. This mill may have stood on or near the site of the Irvin Huff mill (I, 2)

History. Henry Huff (1786–1877) built his mill on land bought in 1818 from Jacob and Annie Sowder. At one time there was a legal distillery and small storehouse where the brandy was kept and sold. In 1848 Huff sold the mill and land to his son Byrd (1820–1890), who sold it to his younger brother Robert (1828–1895) in 1867. Descriptions of a Henry Huff mill in the 1870 *Products of Industry* report and a Byrd Huff mill in the 1880 report could refer to other, now-forgotten mills, since both men are supposed to have sold the mill at least a decade before these census dates. This discrepancy might also reflect a community's tendency to refer to a business by the name of a long-time owner years after it changed hands, in the same way that a farm continues to be "the Smith place" long after the Smiths are gone. If these records do refer to the same business, the mill's source of power and its capacity were substantially upgraded between 1870 and 1880. Henry Huff's great-grandson Irvin operated a mill on or near this same spot. Henry Huff's father, Phillip, had a mill on nearby Lick Fork Creek in the late 1700s (II, 116). The following memoir was written by Valentine M. Sowder (1866–1956), a prominent Floyd County lawyer, around 1940. He lived near the Huffs as a young man. The brick house at the "Jack's Mill" place still stands. Frank Webb's source is unknown, but the meeting must have taken place around 1875. Quoted verbatim.

Deed for the Huff Place. The Huff place, 234 acres, was conveyed to Henry Huff by my great grandfather, Jacob Sowder and wife, by deed dated August 29, 1818, and recorded in the Clerk's Office of the Circuit Court of Montgomery County, Virginia, in Deed book K, page 85.

Peter Geurrant. About one-half mile on the road going south from the Henry Huff house is another large brick house, built up on a hillside above the road. This place in my recollection, was known as Jack's Mill, as there was also a large mill there on the river near his house. My brother, A. T. Sowder, lived there from about 1893 to 1916.

Henry Huff. I never had occasion to contact him but one time. On this Huff farm is a mill and the water to operate the mill is collected in a dam. This is the nearest mill to the headwaters of Little River. Sometimes they would draw off the mill dam to catch the fish. On one occasion, Byrd Huff, who had charge of the mill and the dam, drew off the dam. He invited in the neighbors to share in the fish. My father was one of those invited. When he left home he told me to come at a certain hour of the morning and bring a basket and get his part of

Map of Mills in Section I

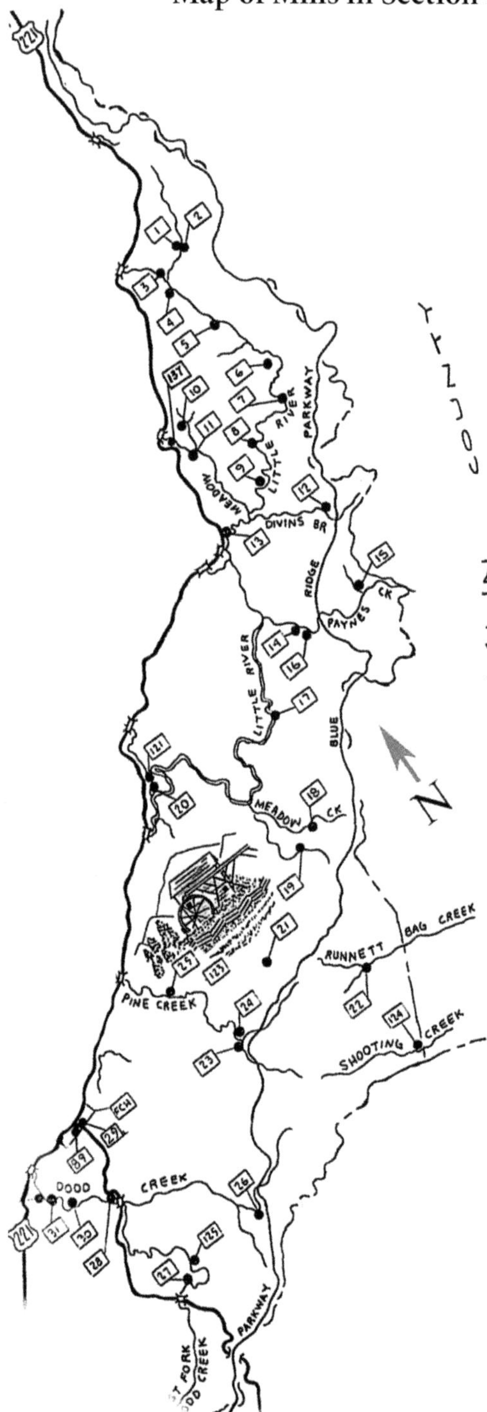

the fish. I took the basket and went to the mill and they put my father's part of the fish into the basket. I started home with it. When I got about 100 yards from the mill, but out of sight of the mill, I met Henry Huff going towards the mill, walking with his cane. He stopped me and asked me what I had in the basket. I showed him that I had the fish, and he wanted to know where I got the fish and I told him they were drawing off the mill dam and these were the fish that my father got. He said that was his mill, that he built that mill and the dam, and now they were drawing off the dam and catching the fish and didn't even let him know that they were planning to do so, and tears came into his eyes. I looked down in the basket and the largest fish in it was right on top. I reached down and took out this large fish and gave it to him. He took the fish and went on towards the mill and I went home. There that day, his life and my life touched and the two lives reached back at least to a time before the Constitution of the United States was adopted, and the two put together cover administrations of all the presidents of the United States.

Before the war, the Huffs had a still house and made brandy near the mill, and they had a small house there where they kept their brandy for sale, with accommodations for the person in charge to sleep. My mother said that the rough element of the community gathered there almost nightly and drank and gambled. A Frenchman by the name of Marello, it appears, was put in charge of the business. How long he was there I never heard, but one night the building in which he slept was burned and he was burned up in it. The theory was that he had been murdered and robbed and the building burned to cover up the crime. It was never known who robbed him, killed him and burned the building.

For further history of this mill site, see Irvin Huff Mill (I, 2).

Deeds and Other Public Records.

Montgomery County Deed Book "K," page 85, 29 August 1818. Jacob and Annie Sowder to Henry Huff, 234 acres.

Floyd County Deed Book "E," page 68, 6 March 1848. Henry and Rachel Huff to Byrd Huff, 2 tracts of land on Little River, containing 234 acres.

D.B. "M," page 137, 7 March 1867. Byrd Huff to Robert Huff, 234 acres on Little River.

D.B. "N," page 345, 15 August 1874. Byrd Huff to Robert Huff, 2 acres known as the Byrd Huff Store House.

D.B. "N," page 600, 28 August 1875. Robert Huff and Mary Huff to Jackson Huff, 35 acres known as part of Henry Huff old tract.

D.B. "Y," page 258, 31 December 1896. Robert Huff, deceased, to Penelope Ann "Nellie" Huff, 199 acres of land and home place in the division of Robert Huff estates.

See Table B, *1870 Products of Industry.* Name of corporation or individual: Huff, Henry

See Table C, *1880 Products of Industry.* Name of corporation or individual: Huff, Byrd

2. Irvin L. Huff Mill

Little River

LOCATION: East side of Route 795 (Huff Mill Road SE) 1 mile south of Route 221 (Floyd Highway North); POWER: Overshot wheel; STONES: 2 sets of French burrs; 1 set of Brush Mountain stones. Sawmill and planing mill; VISIBLE REMAINS: Part of concrete dam, about 100 yards upstream from Route 795

History. A Huff family friend says that this mill was built by N. T. (Napoleon Tazewell) Fowler (1842–1921) and his second wife, Penelope "Nellie" Huff Fowler (1848–1937) in the early 1900s, and was first known as Fowler's Mill. A granddaughter of Henry Huff (I, 1), Nellie Huff Fowler was first married to Giles Huff (1843–1893), with whom she had one son, Irvin L. Huff (1882–1969). N.T. Fowler, her second husband, a miller, millwright, and carpenter, was almost certainly connected to a mill known in 1880 as the "Fowler and Huff" mill on the South Fork of the Roanoke River. The only written documentation of the Fowler and Huff Mill is its enumeration in the Products of Industry report that accompanied the 1880 United States census (see Table C). N. T. Fowler and his first wife, Maggie, did own the H. W. Carr Mill (II, 115) on Lick Fork Creek, one of the headwaters of the South Fork of the Roanoke River, between 1886 and 1891. The possibility of a connection between those mills is discussed in the H. W. Carr Mill history.

N. T. Fowler is supposed to have built what became known as the Irvin Huff mill expressly for his stepson and to have taught him the milling trade. Irvin Huff's buckwheat flour was a local favorite, and his mill pond was known to locals and people from Roanoke and surrounding areas as a place to ice skate. Pictures and articles about skating on the Huff Mill pond appeared in the *Roanoke Times* in the late '40s and early '50s. The mill stopped running in the early to mid '50s. Part of the concrete dam, one of only three in the county (the others were at

Skaters on the Irvin Huff Mill pond. Accounts and photographs of skaters on Floyd County mill ponds are rare. The group in this undated snapshot includes, second from left, Ron Francis; second from right, Kitty Coxe; and far right, Neale Huff. Irvin Huff is said to have tested the ice by driving his car onto the frozen mill pond. (Frank Webb collection)

Sowers Mill [II, 94] also on Little River, and the Siner/Perdue Mill [II, 117] on Lick Fork Creek), is all that remains of the mill. There were also a blacksmith shop and a planing mill at this site.

This mill was unusual in having had two sets or "runs" of French burr/buhr stones, which were used exclusively for grinding flour. Highly prized for centuries, these and all other flour milling stones were rendered suddenly and utterly obsolete (in Floyd County) by the introduction of the roller mill system of making flour between 1900 and 1930. One set at the Irvin Huff mill may have been used exclusively to grind the buckwheat flour for which Irvin Huff was well known in the east end of Floyd County. The Homer Roberson mill on Dodd Creek (I, 27) specialized in this same niche market after being cut out of the wheat flour trade by mills outfitted with the high-speed, high-yield roller system of grinding. It was either not technically feasible, or simply not worth the trouble to grind relatively small quantities of buckwheat flour using the elaborate roller mill system.

Irvin Huff's great-grandfather Henry Huff built and ran a mill at or near this site in the first half of the 19th century (I, 1). Henry Huff's son Byrd Huff also ran the first mill on this site, before selling the property to his brother Robert, father of Nellie Huff Fowler.

Deeds

Floyd County Deed book "Y," page 258, December 31, 1896. Robert Huff, deceased, to Penelope Ann "Nellie" Huff, 199 acres of land and homeplace in the division of Robert Huff estate.

D.B. 54, page 287, 13 January, 1933. Penelope A. Huff to Irvin Linden Huff, 199 acres of land, homeplace and all possessions, "Nellie" Huff to retain control of said land during her lifetime. This land willed to Penelope Ann Huff per Will Book "F," page 366, 13 February 1892.

D.B. 80, page 205, 5 July 1958. Irvin L. Huff to Paul G. Willett and wife, Nellie Virginia Huff Willett, 193¾ acres land and all his possessions.

3. Morgan O. Poff Mill

Little River (Wilson's Branch)

LOCATION: On Wilson's branch east of Route 221 (Floyd Highway North) and northwest of where Route 649 (Bear Ridge Road SE) crosses Little River; POWER: Overshot wheel, 8-foot; STONES: 1 set

History. Morgan Poff (1883–1967) erected his water wheel to drive a generator which recharged wet-cell, direct current batteries, probably to power light bulbs or a radio. A single run of stones was added later to grind for his own family as well as neighbors. This mill was built in the early 1930s and torn down in the early 1940s, by which time 115V alternating current from a commercial utility was available, or soon would be.

Deeds

Floyd County Deed Book 71, page 471, 9 February 1934. Nellie Miles to Morgan Poff, 50 acres.

4. Anthony Poff Mill/ Hosea Wimmer Mill

Little River

LOCATION: Route 649 (Bear Ridge Road SE) about .5 mile east of Route 221 (Floyd Highway North); POWER: Unknown; RUN STONES: Unknown; VISIBLE REMAINS: Part of what appears to have been an earthen dam

History. The documented history of this mill is limited to one reference in a Montgomery County will and four Floyd County deeds. The remains of the earthen dam are distinctly visible from both up and downstream. The site is unusual in that there is little evidence of the silt that typically built up behind mill dams and remained as a level, fertile patch of land after the dam was broken.

Deeds and Other Public Records

Montgomery County Will Book 5, page 21, 14 September 1830. Henry Poff to Anthony Poff, Montgomery County, Virginia.

Floyd County Deed Book "A," page 110, 13 August 1832. Anthony Poff to Hosea and Christenor Wimmer, Anthony Poff mill tract. Eighty-one acres adjoining Henry Poff and Henry Huff.

D.B. "E," page 387, 20 November 1849. Hosea and C. Wimmer to Charles B. Reynolds, part of Benjamin Huff tract, being land the mill is on.

D.B. "G," page 390, 25 January 1854. Charles B. Reynolds to Henry Poff. "Supposed to be 8 acres." Mentions Old Poff Mill dam, but may not be the actual mill property.

5. Peter Geurrant Mill/ John O. Jack Mill and Jack's Mill

Little River

LOCATION: Route 648 (Jack's Mill Road SE) 1.4 miles southeast of Route 221 (Floyd Highway North); POWER: Breast (first mill, 1880); steel overshot (second mill); **STONES:** Originally 2 sets of stones. Second mill, 1 set; VISIBLE REMAINS: Ruins of building, steel wheel and gearing. Dug race clearly visible parallel to Route 648

History. Peter Geurrant (born c. 1800 in Buckingham County) petitioned the court of Montgomery County in 1823 to build a water grist mill on Little River.[1] Geurrant ran a legal distillery at this site as well, and in 1850 reported having made 1,900 barrels of flour and 4,000 gallons of whiskey in the past year (see Peter Geurant, Table A). Partners Staunton R. Aldridge and Charles B. Reynolds bought Geurrant's estate of some 1,375 acres at public auction on October 24, 1859, after his death earlier that year, for $11.00 an acre. The mill was known for a while afterward as the Aldridge and Reynolds Mill, and then as the Aldridge Mill, after court-appointed commissioners divided the land between Reynolds (815 acres) and Aldridge's heirs (the mill and, presumably, the remaining acreage) in November 1866. Part of the race wound up on Reynolds' side of the line, but the rights of the Aldridge heirs and subsequent owners to the water required for the mill were guaranteed in the Circuit Court's March 1869 confirmation of the partition. They were also guaranteed the right to access and maintain the race and a road across the Reynolds property. The November 1870 deed recording the partners' 1859 purchase repeats the terms of this agreement. Meanwhile, John O. Jack had bought the mill and 531¾ acres from Aldridge's heirs at public auction on December 2, 1868; the deed confirming his purchase was recorded on February 2, 1872.

After John O. Jack sold the mill to James Smith in 1878, it changed hands fairly regularly until being torn down or destroyed sometime after 1935, in which year C.W. Janney bought the mill Geurrant had erected more than a century earlier along with a tiny fraction of the original acreage. Janney hired Troy Conner to build a smaller, less elaborate mill on the same spot. Janney's mill, with its single set of stones, probably ground only corn meal and cattle feeds, and was one of the last two (see IV, 53) water-powered mills built in Floyd County. The Fitz overshot water wheel was bought used in North Carolina and installed with help from Kenneth Walton, who was badly injured while putting the wheel together. The building was still standing in the mid-1990s but much of the metal roof was gone and the back gable end and floor had collapsed. One run of stones and a single line shaft were on the ground beneath the mill. Stairs to the second floor ran along the east wall. The building was fully collapsed when photographed by Dot Howery of Floyd, VA, in August 2001.The Henry Huff Mill (I, 1) history mentions Peter Geurrant and Jack's Mill.

Although he may have operated it for a much longer time, John O. Jack owned "Jack's Mill" only from December 1868 until March 1878 — not quite 10 years out of this mill site's 180-year history. Yet the original mill kept his name for 60 or more years, through at least 14 subsequent owners. Even the mill that replaced it was known as Jack's Mill, and Route 648 is

The second Jack's Mill on Little River, 1992. Built about 1936, this was the last water-powered mill built in Floyd County. The new mill took the name of the one it replaced, even though John O. Jack had sold the property more than half a century earlier. (Frank Webb photograph)

now officially Jack's Mill Road. The absence of any logic behind the persistent attachment of Jack's name illustrates the difficulty of assigning a single owner's name to any of the many mills about which the voice of the people—in this case so loud, so clear, and so illogical—is conflicted or silent.

Deeds

Floyd County Deed Book "M," page 230, 10 November, 1870. Peter Geurrant heirs to Charles B. Reynolds and Staunton R. Aldridge, 1375 acres including mill (land was actually bought at auction in 1859). Deed refers to 1866 partition necessitated by Aldridge's death. Guarantees right of "S. R. Aldridge Heirs and those claiming by through or under them the full complete and entire privilidgs of all notes[?] privirlegs necessary for the Mill aforesaid and the privirlige of repairing and keeping in repair the Mill race."

D. B. "L," page 368, 1 November 1866. Partition of land between Charles B. Reynolds and heirs of Staunton R. Aldridge. Confirms mill owners' right to water supply separated from mill by partition line.

D. B. "M," page 312, 2 February 1872. Staunton R. Aldridge's heirs to John O. Jack. 531¾ acres, including mill. Deed records Jack's purchase of this property at auction on 2 December 1868.

D.B. "O," page 645, 22 March 1878. John O. Jack to James W. Smith, land and mill property.

D. B. "Q," page 326, 12 December 1881. James W. Smith to John H. Lemons, land, brick office, and mill known as the mill and office tract.

D. B. "Z," page 537, 24 April 1894. W. D. Vaughn, commissioners' sale to John T. Price, 1½ acre land on which is situated a flouring and manufacturing mill, a brick dwelling and outhouse known as Jack's Mill.

D. B. 36, page 386, 21 February 1910. J. T. Price to John S. Keaton, ¾ acre and old Jack's Mill property.

D. B. 37, page 122, 23 February, 1910. J. S. Keaton to G. W. Hundley, ¾ acre and Jack's mill property.

D. B. 38, page 107, 20 October, 1911. George Washington "Tood" Hundley to Posie C. Poff, mill, mill house, blacksmith's shop, house and all water rights.

D. B. 38, page 110, 27 January, 1912. Posie C. Poff to Leonard B. Stump, mill and etc. as above.

D. B. 39, page 352, 13 November 1913. L. B. Stump to William D. Craighead, ¾ acre land and mill.

D. B. 39, page 507, 13 November 1913. William D. Craighead to G. W. Whorley, one acre of land and mill known as Jack's Mill, including mill house, water rights, dwelling house and blacksmith shop.

D. B. 41, page 433, 15 March 1916. George W. Whorley to William D. Craighead, same.

D. B. 43, page 541, 7 April 1919. W. D. Craighead to Samuel Zachariah Dewitt, same.

D. B. 44, page 460, 24 February 1921. S. Z. Dewitt to Curtis Conner, "Jack's" old mill and 1 acre land.

D. B. 44, page 564, 10 May 1921. Curtis Conner to O. Blaine Ferris and wife, "Jack's Mill" store and dwelling house.

D. B. 47, page 499, 16 April 1926. O. Blaine Ferris to

Fletcher Claytor, water mill, blacksmith shop, store, dwelling house and water rights.

D. B. 54, page 408, 13 September 1935. Fletcher Claytor to Charles W. Janney, same.

D. B. 80, page 365, 10 September 1958. C. W. Janney to Norman E. Knowles, Jr., same.

D. B. 86, page 431, 26 January 1963. Norman E. Knowles to Allen W. Janney, same.

D. B. 90, page 497, 21 July 1965. Allen W. Janney heirs to James Donald Pugh (still owned mill in 1996).

See Table C, *1880 Products of Industry.* Name of corporation or individual: Smith, Jas. W.

6. Dewey Swanson "Tood" Hundley Mill

Little River

LOCATION: Just north (upstream) of Route 610 (Hummingbird Lane SE) 2.4 miles east of Route 221 (Floyd Highway North) and .4 mile east from Route 648 (Graysville Road); POWER: Turbine; STONES: 2 sets and Midget Marvel flour mill; VISIBLE REMAINS: Dug race

History. The August 1875 deed conveying a right of way for William O. Carr to dig a race across William H. Poff's land is remarkable for its lack of restrictions on either party. The owners of property through which rights of way for mill races had been granted frequently specified the size and exact location of the race. In some cases, the deeds spelled out privileges reserved by the miller, such as accessing the race as needed to maintain it, and by the land owner, such as crossing the race with roads, bridges, and fences and cultivating the ground around it. The lack of detail in this agreement suggests that neither the race nor the mill had yet been built, and supports the supposition that Carr erected a sash type sawmill here between 1875 and 1878 and had added a grist mill by the time he sold the property and both mills to J. H. Lemons in 1882.

Unless Carr had acquired land on Little River through some unrecorded transaction, his mills were on land purchased from Charles B. Reynolds a week before the right of way for the race was deeded to Carr by Wm. H. Poff. One of the boundaries described for the 135 acre tract bought from Reynolds "cross[ed] canal and rive[r]," raising the possibility of an even earlier water power site nearby. It seems likely, however, that William O. Carr did build the grist mill eventually known as the Hundley Mill.

In a May 1993 interview with Frank Webb, Howard N. Hundley stated that this mill was last run by Dewey Swanson "Tood" Hundley (1898–1989), his wife, Rachel, and son, Howard. There were also a sawmill, a blacksmith shop, and a store near the mill. The mill ceased operating in the mid–1950s as the building and equipment fell into disrepair. Dewey Hundley did not use the Midget Marvel flour mill. Dewey Swanson "Tood" Hundley was the son of George Washington Hundley, also a Floyd County miller and mill owner, also nicknamed "Tood." Weaver Argabright ran this mill for George A. Cannaday (1869–1949), who preceded the Hundleys as owner.

Deeds

Floyd County Deed Book "O," page 173, 31 July 1875. Charles B. Reynolds to Wm. O. Carr, 31 July 1875, 135 acres on headwaters of Little River. One boundary line "crossing canal and rive[r]." This transaction was made with the consent of W. H. Poff. *Why Poff's consent was required is not stated.*

D. B. "O," page 173, 7 August 1875. William H. Poff to William O. Carr, right of way for "a race or canal through my land on the waters of Little River ... for purpose of running any machines the said Wm. O. Carr may see proper to run on his own land."

D. B. "P," page 78, 25 November 1878. William O. Carr to Emaline Carr, sash type sawmill on west side of Little River, 4 acres more or less.

D. B. "Q," page 438, 21 February 1882. W. O. and Emaline Carr to J. H. Lemons, 2 acres, plus or minus, on which is located a saw and grist mill. Headwaters of Little River where Pine Spur Road crosses river. "A water way which was conveyed to us by William H. Poff in D. B. 'O' page 173, 7 August 1875 is conveyed to J. H. Lemons. Sold by us to Fannie Poff and Adam H. Poff, but not deeded or transferred to them."

D. B. "S," page 313, 2 October 1885. John H. Lemons to William H. Poff, 2 lots, one has 2 acres and mill, one has ½ acre and race.

D. B. 35, page 100, 21 September 1908. W. L. and B. G. Howard, Special Commissioners, to Minnie Cannaday, in settlement of chancery suit of *W. J. Poff vs. Ira W. Poff*, tracts totaling 161¼ acres, including land deeded to Wm. H. Poff by John H. Lemons in D. B. "S," page 313.

D. B. 54, page 558, 31 July 1937. George A. and Minnie Cannaday to Dewey S. Hundley, 161¼ acres, "including the mill on said premises and the fixtures used in connection therewith."

D. B. 175, page 391, 12 October 1990. Mary W. Hundley (Mrs. D. S. Hundley) to Howard N. Hundley (son), Salem, Virginia, Tract "A" of 161¼ acres. Same tract as in Deed Book 54, page 558.

7. J. S. Dewitt and S. Z. Dewitt Mill

Little River

LOCATION: Route 648 (Graysville Road SE) .5 mile south of Route 610 (Hummingbird Lane SE); POWER: Turbine; STONES: 2 sets and chop/feed mill; VISIBLE REMAINS: None

History. George Washington "Tood" Hundley is believed to have built this mill 10 or 11 years before selling it to Blaine Ferris in 1919. James Strickler Dewitt (1886–1966) and Samuel Zachariah Dewitt bought the property two years later; J. S. ran the mill and his brother S. Z. ran the store. The mill was damaged by high winds in the 1940s, by which time J. S. was the sole owner, and torn down in the late 1940s by S. Z. Dewitt.

Deeds

Floyd County Deed Book 44, page 383, 20 March 1919. G. W. Hundley to O. Blaine Ferris, mill, 30 acres land and water rights.

D. B. 44, page 382, 15 April 1921. O. Blaine Ferris to James Strickler Dewitt and Samuel Zachariah Dewitt, 30 acres land, mill and other buildings (including, apparently, a store), known as the G. W. Hundley Mill tract.

D. B. 49, page 434, 4 April 1929. S. Z. Dewitt to J. S. Dewitt, his ½ interest in land and mill.

The Dewitt Mill on Little River stood for only about forty years, and may already have been in disrepair when this undated photograph was taken. The wooden structures along the near gable end of the building appear to be a wooden race and a turbine penstock. Part of the wall of the mill above them has been hastily covered with random length weatherboarding. (Frank Webb collection)

8. Kropff and Custer Mill

Little River

LOCATION: Route 642 (Meadow Run Road SE), about 2 miles east of Route 221 (Floyd Highway North); POWER: Breast wheel; STONES: 2 sets

History. As late as the 1990s older residents of this community remembered the ruins of this mill and the millstones lying on the bank of the river. The race providing water for this mill ran under the road. The mention of a penstock, as the word is now understood, in the first deed (below) suggests the mill was driven by a turbine or tub wheel in 1857. The breast wheel listed in the 1880 census *Products of Industry* report may therefore have been installed by Kropff and Custer. The purchase of two tracts of land totaling 52.5 acres adjoining the mill property in 1876 could be coincidence, or could have to do with a Virginia law placing restrictions on millers who wished to fatten hogs on mill byproducts and spoiled or surplus grain. Passed in 1748, the law stated, "No owner or occupier of a mill shall keep any hogs at his or her mill, except in inclosures, unless such owner or occupier shall have fifty acres of land at the least, adjoining to such mill; and if any hogs, belonging to the owner or occupier of such mill, shall be found running at large, it shall and may be lawful for the proprietors of the land adjoining to such mill, to kill, or cause to be killed or destroyed, all such hogs."[2] Deeds connected to several other mill properties record mill owners' purchases of tracts of 50 or slightly more acres "adjoining the mill tract" or "mill lot."

Deeds

Floyd County Deed Book "J," page 331, 19 October 1857. Gary Gray to William Riley Gray, a tract of land on Little River containing 49 acres. Metes and bounds of this deed mention the mill race and penstock.

D.B. "N," page 422 and 423, 27 January 1875. W. R. Gray to Mark Custer and R. W. Kropff, 2 tracts of land and mill, one tract of 24½ acres and one of 31¼ acres and mill.

D. B. "O," page 98, 31 March 1876. James M. Gray and Nancy Gray, his wife, to Mark H. Custer, their entire interest in 52½ acres land on Little River adjoining mill property of Kropff and Custer.

D. B. "T," page 262, 2 December 1889. R. W. Kropff to M. H. Custer, 30¼ acres land and mill on Little River.

D. B. 50, page 9, 16 June 1934. J. E. Proffitt, commissioner for M. H. Custer, deceased, to Claude C. Custer, two tracts of land, one of 74¾ acres and another of 30¼ acres.

See Table C, *1880 Products of Industry.* Name of corporation or individual: Kropf and Custer

9. John W. Wertz Mill

Little River

LOCATION: Route 641 (Timberline Road SE) 1.5 miles east of Route 221 (Floyd Highway North); POWER: Turbine; STONES: 3 sets and wool carding machine; VISIBLE REMAINS: Part of dam and foundation (1992)

History. Dr. Amos Wood's *Floyd County: A History of Its People and Places* notes that "James Howard Gray built the first old-fashioned water wheel mill in that part [the east end] of the county and he was interested in the smelting of iron."[3] Grey (or Gray) bought and sold a number of tracts along Little River and Roanoke River, one of which was known as the Old Forge Tract. John Wertz had similar interests, buying the Grey tract and several other tracts for mineral rights. It is not certain that Wood's reference is to this mill, but Grey did own the land on which it stood, and he does not appear in any known record as the owner of another mill tract in Floyd County.

The Wertz mill is said to have been in ruins in the early 1920s. An auction sale bill found at the old Wertz homeplace in 1994 shows that one fan mill, one upright corn mill, three sets of mill burrs, and "about 30 barrels of flour" were to be sold on March 21, 1917, by George C. and C. H. Wertz. The appraisers' inventory that preceded this sale lists other mill related equipment among the personal property of Eliza Lewis Wertz.[4]

Chataigne's Gazetteer for 1880 lists J. W. Wertz and R. O. Harvey mill at Graysville, Virginia. The same source lists J. W. Wertz as merchant and miller at Graysville in 1885, but in 1889 says only that R.O. Harvey was a merchant, making no mention of a mill (see Appendix A). R. O. "Red" Harvey was better known as the owner of a mill (I, 19) and store in the Pizarro community.

Deeds

James H. Grey, James M. Grey, and John W. Wertz all owned large tracts of land, resulting in complicated settlements of their respective estates. Gaps and inconsistencies in the following record reflect the difficulty of tracing the ownership history of a single small tract.

Floyd County Deed Book "L," page 19, 18 June 1862. James H. Grey to John T. Jett, 300 acres on Little River, known as the Old Forge Tract.

D. B. "L," page 338, 7 July 1866. James H. Grey and John T. Jett and wife, Elisa, to James M. Grey, 300 acres plus or minus, known as the old Forge Tract on Little River.

D. B. "N," page 128, 26 October 1870. J. T. Jett, administrator (for J. H. Grey?), to Rhoda Grey, 52¼ acres as her dower land. (Mill is on this tract of land.)

D. B. "N," page 135, 30 January 1873. J. T. Jett, administrator for J. H. Grey, deceased, and M. H. Custer and wife, Charles H. Reid and wife, transfers to Charles H. Reid, 12¼ acres on Little River, which is all their interest in the estate of J. H. Grey.

D. B. "N," page 156, 12 November 1873. J. T. Jett, administrator for James H. Grey, 160 acres of land to James Grey.

D. B. "N," page 176, 6 January 1874. J. T. Jett and wife to James M. Grey, their ½ interest in lands of James H. Grey, deceased, and Rhoda Grey.

D. B. "O," page 91, 31 March 1876. James M. Grey to Mark Custer and wife, his interest in dower of Rhoda Grey, being 52½ acres adjoining mill tract of Mark Custer.

D. B. "Q," page 349, 8 February 1882. J. L. Thompkins, commissioner in causes of *Poole and Hurt vs. William Riley Gray and Others, Poff and DeWitt vs. W. R. Gray and Others*, and *J. T. Jett vs. W. R. Gray and Others*, to John W. Wertz, 334½ acres of land. R. O. Harvey and J. W. Wertz had bought the land together at public auction, then Harvey sold and transferred his interest to Wertz.

D. B. 27, page 511, 4 May 1907. R. L. Wertz to Elizabeth L. Wertz, his interest in personal and real estate of his father, J. L. Wertz, deceased, consisting of some 300 acres in Franklin County, and 500 acres in Floyd County, also 19 head of cattle, 76 head old sheep and 35 lambs.

D. B. 42, page 509, 18 April 1919. Estelle Wertz Himes to George C. Wertz, her entire interest in her father's estate.

D. B. 42, page 510, 12 May 1919. Garnett Wertz to George C. Wertz, his interest in his father's estate.

D. B. 43, page 427, 16 April 1919. Willie Elizabeth Wertz Sterling and husband to George C. Wertz, her interest in J. W. Wertz lands.

D. B. 75, page 589, 14 September 1954. George C. Wertz (deceased 11 July 1954) and First National Bank of Roanoke, executors, to Dixon Lumber Company, 392 acres plus or minus, three tracts.

See Table C, *1880 Products of Industry.* Name of corporation or individual: John W. Wertz

10. Willie Ferris Mill

Laurel Branch of Meadow Run Creek

LOCATION: About halfway between Route 610 (Hummingbird Lane SE) and Route 642 (Meadow Run Road SE); POWER: Overshot wheel; STONES: One set (Meadows Mill)

History. Willie Ferris ran this small mill on Laurel Branch with a small wheel (said to be 8 feet in diameter) and a 16-inch Meadows Mill, a semi-portable, self-contained unit with vertical stones manufactured in North Wilkesboro, NC. Meadows Mills, and the nearly identical units marketed by their competitors (Williams, New South, etc.) were widely used in Floyd County and were frequently powered by gasoline engines. They could be set up in any convenient building, safely away from flood-prone streams. Although capable of grinding whole wheat or buckwheat flour if well maintained, most Meadows type mills ground only corn meal. Owners of Meadows or competitor mills seldom made a living at milling, and often ground on only one day each week, typically a Friday or Saturday. When grinding corn, full and part-time millers typically accepted a toll of ⅛ by volume as payment. The miller received no money for grinding, but neighbors with no corn of their own could buy or trade for meal or surplus corn taken in as toll from other customers. Part time millers like Ferris were usually farmers who needed the extra corn for themselves, and did not sell either corn or meal.

To get wheat and buckwheat flour, customers had to visit a full-time mill with stones adapted for grinding flour. Like corn, wheat could be ground on an exchange basis, but the system was more complex and followed a variable, weight-based formula requiring scales and printed tables. But millers with debts and wages to pay did depend on cash sales of raw and processed products.

The Meadows Company still manufactures stone-buhr grist mills. Aside from having frames and cases of metal in place of wood, and improved bearings, lubrication, and pulleys, modern versions are much the same as models made eighty years ago. Mill stones are now fashioned from North Carolina granite, but at one point were imported from quarries near Brush Mountain in Montgomery County, VA. Many traditional, horizontal "corn rocks" from larger county mills came from Brush Mountain, as well.

Restored Meadows portable mill with 24-inch stones, 2010. The use of castings identified with the New Williams mill, a cross-town competitor to Meadows in North Wilkesboro, NC, suggests that this example was manufactured not long after Meadows absorbed the New Williams Company in 1924. Mills with 16- and 20-inch stones were common in Floyd County. (Ricky Cox photograph)

Deeds

Floyd County Deed Book 29, page 371, 10 May 1907. Martha King to Mary J. Ferris, wife of William H. Ferris, 80 acres of land on the waters of Little River.

D. B. 48, page 72, 1 May 1928. J. H. Custer to W. H. Ferris and wife, a road 13 feet wide from Ferris' land to Pine Spur Road.

11. Zadock "Zed" Conner Mill and Gus Beamer Mill

Meadow Run Creek

LOCATION: Route 642 (Meadow Run Road SE) around .8 mile from Route 221 (Floyd Highway North) near the confluence of Laurel Branch and Meadow Run Creek; POWER: Tub wheel (1880); STONES: 1 set and sawmill

History. Deeds made in 1822 when Zadock Conner or his heirs sold this land do not mention a mill, but an 1880 deed refers to the property as the "Zed" Zadock Conner mill site, indicating the existence of an earlier mill near where the Beamer mill stood. The assignment of a single number to two different mill buildings usually means that one succeeded the other at the same location. But it is unlikely that these two stood on exactly the same site or "mill seat" since Gustavus "Gus" Beamer purchased a right of way to bring water to a mill pond in 1878, two years before purchasing the 5/8 acre Zadock Conner mill site, along with an adjoining tract, in 1880. Whether the same dam or race served both mills is not known.

The terms "old Beamer mill property" and "old mill tract" in deeds dated 1898 and 1918 suggest that the mill was no longer running when J.P. Conner (1864–1949), who was also a tombstone maker, bought the Gus Beamer mill tract in 1898. The Beamer mill was one of a handful, including Mabry Mill (IV, 60) and the Slusher/Turman Mill (IV, 66), powered at some point in their histories by the water of two separate branches of the same stream. A source of extra water may have been routed to the mill only after the original source proved inadequate. Or, though expensive to build and maintain, double in-ground or wooden races could have been part of the original plan if a suitable parcel of land could not be had below the confluence of the two streams.

Deeds and Other Public Records

Montgomery County D. B. "H," page 356, 27 December 1822. Zadock Conner to Andrew Conner, 100 acres on Muster Ground Branch (Laurel Branch) and Meadow Run.

D. B. "H," page 357, 27 December 1822. Zadock Conner to William Conner, 50 acres on Muster Ground Branch. This land willed to Zadock by David Conner, deceased. No consistent record found of transfer to Gus Beamer before 1878.

Floyd County Deed Book "P," page 10, 26 July 1878. Peter Conner to Gustavus Beamer, right of way to turn Laurel Creek into mill pond on Meadow Run Creek.

D. B. "P," page 516, 14 April 1880. Mildred Price, Peter Conner, and Mary Conner to G. A. Beamer, Zed Conner mill site containing ⅝ acre and adjoining tract containing 11⅛ acres.

D. B. "P," page 493, 13 May 1880. Edward H. Price to G. Beamer, 4 acres land on Daniel's Run Road and Meadow Run Creek.

D. B. "S," page 534, 15 November 1886. G. A. Beamer to Jacob H. Poff, 14 acres of land, grist mill, sawmill and all water rights. Peter DeWeese was living on the property at this writing.

D. B. "Z," page 376, 2 December 1898. Jacob H. Poff to James Preston Conner, 7 acres and old Beamer Mill property.

See Table C, *1880 Products of Industry.* Name of corporation or individual: G. H. Beamer

12. N. C. Thomas Mill

Divins Branch

LOCATION: Route 651 (Stuart Road SE), 2 miles east of Route 221 (Floyd Highway North) and north of intersection with Route 641 (Ponderosa Road SE); POWER: Overshot wheel; STONES: 1 set, 1 Meadows Mill, 1 feed mill, and sawmill; VISIBLE REMAINS: Dug race

Marvin Thomas at his father's mill, undated. The massive wooden cribbing supporting N. C. Thomas's water wheel is perfectly adequate, but has a temporary look compared to the stone and concrete piers at other mills. The clipped gable roof (not shown) was and is a rare feature on any Floyd County building, and was unknown on any other mill. (Frank Webb collection)

History. Nathaniel Clayborn Thomas continued to be involved with this mill after he sold it in 1926 and is said to have torn it down in the early to mid–1950s. The mill was powered by an automobile engine in periods of low water.

Like a number of other millers, Thomas was also a licensed distiller before Prohibition outlawed the production and sale of whiskey. From a business perspective, milling and distilling are complementary trades, given that whiskey makers needed large amounts of corn meal or rye. Grinding one's own corn, whether raised or purchased from local farmers, would save the standard toll of one-eighth (one gallon out of one bushel), yielding an instant 12.5 percent increase in revenue. After the 1919 passage of the Eighteenth Amendment outlawed the manufacture of whiskey, many miller-distillers gave up what may have been the most lucrative aspect of their businesses. Others continued to be involved to varying extents. At one extreme were those who merely asked no questions when asked to grind sprouted corn, which has no other use than aiding the fermentation of the mash that will be distilled to make straight corn whiskey. At the other extreme, at least one miller (I, 123) is said to have sold untaxed liquor on the premises to carefully screened customers.

A "Legal Notice" from J. D. Honaker, Jr., deputy collector, states that he "will sell at public auction 16 gallons of apple brandy and about 12 gallons of corn whiskey, seized for violation of the Internal Revenue Laws." The auction was to be held at the residence of N. C. Thomas, but whether Thomas himself or some unnamed local colleague in the distilling trade had failed to pay the required taxes on these spirits is left unsaid.[5]

Deeds

Floyd County Deed Book 31, page 314, 13 March 1905. J. W. Gray to Sarah Board, land and store, but no mill mentioned.

D. B. 40, page 443, 28 March 1914. G. W. "Belly" Board and Sarah Board to N. C. Thomas and E. P. Radford, 62 acres land, water grist mill, store house and granary.

D. B. 45, page 237, 5 January 1922. E. Peach Radford to N. C. Thomas, 24 acres and grist mill.

D. B. 47, page 510, 28 April 1926. N. C. Thomas to Judia A. Thomas, 3 tracts land. Number 1 was conveyed by G. P. Thomas to N. C. Thomas, D. B. 26, page 590, 25 Acres, 2 P. Number 2 Tract, 24 acres and mill from E. P. Radford, D. B. 45, page 237. Number 3 Tract, G. M. C. Ingram to N. C. Thomas, D. B. "V," page 203, 61½ acres, and D. B. "Z," page 311, 31 acres.

D. B. 69, page 599, October 10, 1949. Judia A. Thomas to Minnie E. Bower.

13. Robert "Bob Lee" Vest Mill

Little River

LOCATION: Route 651 (Stuart Road SE) .1 mile east of Route 221 (Floyd Highway North) just upstream from the bridge, behind Shady Grove Primitive Baptist Church; POWER: 2 Turbines; STONES: 1 set, 2 Meadows Mills, hammer mill, sawmill, and planer; VISIBLE REMAINS: None

History. In 1921 Charles Daniel Vest and Robert Lee Vest bought three acres of land from A.K. Smith with the express purpose of building a mill. It was run by Charles Vest, Robert Lee Vest, Harvey Vest, and Sam Vest. One of the two turbine water wheels, which are smaller, lighter, and less expensive than vertical wheels, may have driven the sawmill and planer, and the other the grist and hammer mills. Aside from the one set of what are assumed to be conventional, low RPM millstones, the machinery on hand was compatible with the higher shaft speed of turbines. The unforeseen weakness of turbines in many Floyd County applications was their huge consumption of water. The Little River on its lower reaches, well below this mill, was one of few Floyd County streams large enough drive turbines reliably, but just how far upstream this was true depended upon the needs of a particular application and upon the storage capacity of the mill pond.

The Bob Lee Vest mill dam washed away around 1940 and was never rebuilt. *The Floyd Press* did not name this or any other mill as having been damaged in its reports on flooding within the county, but record breaking floods and extensive property damage are known to have occurred all along the New River and tributaries in August 1940.

Deeds

Floyd County Deed Book 45, page 89, 23 April 1921. A. K. Smith to Charles Daniel and Robert Lee Vest, 3 acres of land. Parties of the first part contemplate erecting a mill on the land and building a mill dam. If the water from said dam should back more than 10 feet on the land, the parties of the first part shall pay $50.00 per acre for land covered outside the 10 feet. The land for the mill was purchased for $150.00, and 10,000 feet of lumber to be sawn at the mill for A. K. Smith.

D. B. 51, page 397, 21 January 1932. C. D. Vest and Lydia Vest to Harvey Vest, ½ interest in 3 acres and mill, to include ½ interest in all machinery and appurtenances.

D. B. 62, page 271, 6 July 1944, J. H. Poff, Commissioner in Chancery cause of *Lucy B. Vest vs. Bessie Craighead and J. L. Craighead*, her husband, to Wallace C. King, 3 acres land on Little River known as the Bob Lee Vest Mill property with all water rights.

14. Thompson-Akers Mill

Paynes Creek

LOCATION: Route 639 (River Ridge Road SE) 2.1 miles south of Route 221 (Floyd Highway North); POWER: Overshot wheel; STONES: 1 set and Midget Marvel flour mill; VISIBLE REMAINS: Dug race with concrete reinforcement, concrete base of molasses tank, and concrete mill foundation (filled and converted to truck shed)

History. J.D. Akers and B.O. Thompson began acquiring land and water rights in 1917, and by 1922 had added a third partner, E.K. Thompson, and completed a new roller mill. In the meantime, the George Ingram Mill, which stood on or near the same site, had been destroyed, likely taken down to make room for the new mill. Besides its owners, the Thompson-Akers mill was operated by hired millers Eli Board, Luther Cannaday, and Luther Holdern. Ercil Austin, who bought a one-third interest from E. K. Thompson's widow in 1937, stayed on as miller for five more years after selling his interest to B. O. Thompson, one of the two remaining original partners, in 1941.

One disadvantage of shared ownership became apparent when J. D. Akers, the other original partner, died still in possession of the remaining ⅓ interest. A chancery suit filed by his heirs resulted in a Circuit Court decree that the mill be sold at public auction on December 20, 1944. An auction notice in the November 30, 1944, edition of the *Floyd Press* concludes with the assurance that "this mill is in good state of repair and in full operation with ample water power, and is considered one of the best mills in Floyd County." B. O. Thompson himself advertised the mill for sale in the same edition of the paper. He evidently preferred to negotiate a private sale rather than take his chances at a public auction, but the content of the ad suggests that he was ready to get out of the mill business anyway: "Water power, flour, meal and feed mill, known as the Thompson-Akers Mill, eight-room bungalow, water and lights. All in good condition. Price is very reasonable. Reason for selling ill health." The December 20 sale was cancelled when Moyer

The Thompson-Akers mill as seen from the air, c. early 1960s. The house directly in front of the mill still stands but the road has been relocated to the viewer's right. (Frank Webb collection)

Thompson's private offer of $4,500 was accepted by the Circuit Court of Floyd County. Though unconfirmed, it seems likely that the new owner of the Thompson-Akers mill was the "Mr. Thompson in the Paynes Creek area" who bought the steel water wheel from the A. A. Booth/Frazier Thomas Mill (II, 93) on Beaver Creek after that mill was dismantled in the mid–1940s. Moyer Thompson sold his mill and an adjoining tract to O. T. Conner in 1956.

A concrete reinforcement of the in-ground race bears the date 8–18–59 and the initials of Shirley A. Peters, who bought the mill from O.T. Conner in December 1958. The Thompson Akers mill was still producing feed, flour, and cornmeal when it burned on the night of Saturday, June 13, 1964.

The molasses tank, which was mounted outside the building on a concrete foundation and survived the fire, was half full when Glen Peters bought the property later that year. His cows enjoyed the molasses spread on hay. When Route 639 was widened and paved, the mill foundation was filled with excess dirt, then covered and enclosed as a truck shed. The Fitz steel overshot water wheel was sold for $200 and moved to Cockerham's Mill at Meadows of Dan, VA.[6] It was mounted below the dam in mid-stream, apparently for a purpose more decorative than functional, and was crushed when part of the concrete dam across Dan River collapsed in 1990.

For earlier history on this mill site, see the George M. C. Ingram Mill (I, 16).

Roof painters on the Thompson-Akers Mill on Paynes Creek, date unknown. The chimney-like projection near the center of the ridgeline provides access to the roof for repairs and painting. Some mills simply closed in the coldest weather, when grinding with water was difficult and business slow anyway, rather than risk a fire inside a dusty wooden mill. (Frank Webb collection)

Deeds

Floyd County Deed Book 43, page 265, 7 August 1917. George M. C. Ingram to J. D. Akers, mill, blacksmith shop and dwelling house.

D. B. 43, page 442, 7 August 1917. E. T. Ingram to B. O. Thompson, 8¹⁄₁₀ acre on Payne's Creek.

D. B. 45, page 245, 27 October 1922. J. D. Akers and Della M. Akers, B. O. Thompson and Addie Gaye Thompson, to E. K. Thompson, ⅓ interest in land and mill property. "J. D. Akers, B. O. Thompson, and E. K. Thompson have built and are operating a roller mill on said land, and each owning ⅓ interest in said mill and land."

D. B. 54, page 354, 30 January 1937. Mary Thompson, wife of E. K. Thompson to Ersel Austin, ⅓ interest in mill.

D. B. 59, page 327, 17 March 1941. Ersel C. Austin to B.O. Thompson, his ⅓ interest in mill.

D. B. 62, page 509, 23 December 1944. J. H. Poff, Special Commissioner in Chancery Cause of *E. J. Akers vs.*

R. W. Akers, to Moyer Thompson mill and ½ acre and all water rights.

D. B. 63, page 102, 6 February 1945. Moyer F. Thompson and wife to J. H. Poff, deed of trust for mill and ½ acre. Paid off 23 February 1946.

D. B. 78, page 220, 6 October 1956. Moir F. Thompson to O. T. Conner, Tract 1: 24½ acre and mill machinery and fixtures, Tract 2½ acre and mill and machinery buildings and water rights.

D. B. 80, page 541, 15 December 1958, O. T. Conner to Shirley Peters, land, mill and fixtures.

15. Joseph Turner Mill

Paynes Creek

LOCATION: Route 640 (Franklin Pike SE), 1 mile east of Blue Ridge Parkway; POWER: Overshot wheel; STONES: 1 set and sash type sawmill

History. This mill site is supposed to have been developed by members of the Lewis Payne family. The circular-saw sawmill built by N. T. Fowler for J.B. Payne is only peripherally related. It was supposed to have been a replacement for the more primitive sash type sawmill sold to Joseph Turner (1841–1931) along with the Payne grist mill. The mill was gone when the land changed hands in 1934.

The tiny stream at this spot today seems hardly capable of powering any sort of machinery. One of the two deeds summarized mentions Paynes Mill Creek, but neither deed includes a reference to a mill of any kind. Frank Webb's placement of a mill at this spot is based on oral sources not cited in his notes.

Stock car racer Curtis Turner was a grandson of Joseph Turner.

Deeds

Floyd County Deed Book "M," page 481, 31 December 1872. E. W. Graham to Joseph Turner and Nancy M. (Kelley) Turner, his wife, 130 acres land on Paynes Creek, being part of the Lewis Paynes, dec., and J.B. Payne, deceased, lands.

D. B. "O," page 33, 20 August 1875. Bond. John B. Payne to N. T. Fowler. "I, J. B. Payne, promise to pay said Fowler the sum of $98.00 for building a circular saw sawmill."

D. B. 51, page 195, 2 February 1934. J. E. Proffitt, Commissioner in Chancery Cause of *James Turner, deceased, vs. Sallie Young*, to A. K. and Eliza J. Smith his wife, 126 acres land on headwaters of Paynes Creek. Reserved is about ½ acre containing graveyard and right of ingress and egress to and from same for Turner family members and others.

16. Ingram and Radford Mill/ George M.C. Ingram Mill

Paynes Creek

LOCATION: Route 639 (River Ridge Road SE) 2.1 miles south of Route 221 (Floyd Highway North); POWER: Overshot wheel; STONES: Number unknown, blacksmith shop; VISIBLE REMAINS: The Thompson-Akers Mill (I, 14) was built on or near the same spot. The foundation and a dug race for the later mill remain.

History. It is unclear exactly who built this mill or when. Frank Webb's last notes say it was built in 1891 or 1892, but deeds recorded in 1889, 1890, and early 1891 transfer full, one-half, and one-half interest, respectively, in an existing mill (presumably a grist mill) and water rights. A sawmill, probably of the sash type, stood on or near this site before there was a grist mill.

The mill changed hands many times in its relatively short history. George M. C. Ingram owned part or full interest in it three separate times, selling it for the last time in 1917, apparently without water rights, to J.D. Akers. Akers and two partners did away with the old mill and put up a new one in its place, having acquired additional land on Paynes Creek, possibly to secure the needed rights to water. For later history of this mill site, see the Thompson-Akers Mill (I, 14), which is the newly built (in 1922) roller mill referred to in the last deed below.

Deeds

Floyd County Deed Book "Q," page 72, 3 May 1881. Commissioner for Deceased, Lewis Payne, to Thomas Sarles, Lot # 3, 36 acres and sawmill on Paynes Creek.

D. B. "T," page 502, 4 November 1887. T. L. Sarles and wife to S. R. Miles, 101½ acres on Paynes Creek.

D. B. "T," page 569, 23 October 1888. Mary Akers to Otho Cannaday, 3½ acres plus or minus on Paynes Creek.

D. B. "U," page 661, 7 January 1889. Sam R. Miles to Otho D. Cannaday, 21¼ acres on Paynes Creek. Mill and water privileges up to dam. Dam not to be raised.

D. B. "V," page 205, 15 November 1890. Otho D. Cannaday to G. M. C. Ingram, ½ interest in mill property on Paynes Creek, ½ interest in water rights and ½ interest in yard rights to mill.

D. B. "V," page 16, 7 February 1891. Otho D. Cannaday to Edward A. Radford, ½ interest in land and mill on Payne's Creek. On 21¼ acre tract bought from S.R. Miles.

D. B. "W," page 58, 10 March 1893. George M. C. Ingram to J. M. Cannaday, ½ interest in mill known as the Ingram and Radford Mill. One-half interest sold by Ingram to T. Marion Radford (no deed), and T. M. Radford to J. M. Cannaday, ½ interest (no deed) in ½ acre land, mill, blacksmith shop and dwelling.

D. B. "Y," page 17, 25 March 1893. Edward A. Radford to J. W. Thurman, 24 acres (21¼ acres and 3½ acres) on Paynes Creek and his ½ interest in mill tract with all rights. This is the same land sold by E. A. Radford to John B. Cannaday (no deed).

D. B. "W," page 580, 10 May 1893. George M. Ingram and wife and T. M. Radford and wife to J. M. Cannaday, ½ interest in mill property. Mill formerly known as the Ingram-Radford Mill.

D. B. 26, page 526, 26 March 1900. James Marshall Cannaday, J. W. Thurman and John T. Smith to George M. C. Ingram, mill property on ½ acre lot.

D. B. 29, page 334, 17 March 1903. George M. C. Ingram to C. W. Janey, same. Same as sold on 31 January 1903 by George M. C. Ingram to Peter Cannaday and Peter Cannaday to C. W. Janey (no deed).

D. B. 30, page 134, 5 September 1903. C. W. Janey to John Moran, same and right of water way up to Jefferson Davis Akers' spring.

D. B. 30, page 373, 2 March 1904. John Moran to George M. C. Ingram, same, including right of water way as far as J.D. Akers' spring.

D. B. 43, page 265, 7 August 1917. George M. C. Ingram to J. D. Akers, "a Water Grist Mill, Dwelling house, and Blacksmith shop" on ½ acre. No mention of water rights.

D. B. 43, page 442, 7 August 1917. E. T. Ingram to B. O. Thompson, 8⅒ acre on Payne's Creek.

D. B. 45, page 245, 27 October 1922. J. D. Akers and Della M. Akers, B. O. Thompson and Addie Gaye Thompson to E. K. Thompson, ⅓ interest in land and mill property. "J. D. Akers, B. O. Thompson, and E. K. Thompson have built and are operating a roller mill on said land, and each owning ⅓ interest in said mill and land."

17. James Ferguson Mill

Little River

LOCATION: About .75 mile northeast of Cannaday School; POWER: Unknown; STONES: Unknown

History. James Ferguson petitioned the courts of Floyd County in October of 1834 for permission to build a mill on his own property on Little River. The only proof that Ferguson carried through with his plans is a mention of an "old mill house" in the last deed summarized below. No public road passes near Little River to the northeast of Cannaday School, but an unnamed private road appearing on U.S. Geological Survey and other detailed maps leaves Cannaday School Road just northeast of the school and reaches Little River about a half mile to the northeast. Tracing deed records forward from A. B. Ross's 1873 purchase of part of the Ferguson lands could reveal more about this mill.

Deeds

Floyd County Deed Book "J," page 295, 25 September 1857. James Ferguson to Thomas Mosley, Deed of Trust, the following property, to wit: Slaves Jimmy, Mary and her three children, Erasmus, Tazewell, and Jimmy, Amy, Pleasant, Bill, Octa, Mit, Tom, Wise, Henry, Adaline, Olive and William, and a tract of land containing about 300 acres where James Ferguson now resides.

D. B. "N," page 190, 11 November 1867. J. L. Tompkins, Commissioner in cause of *Huffs Administration vs. James Ferguson*, to Charles Thomas, 139 acres on Little River. 11 November 1867, J. L. Tompkins to A. B. Ross, 113 acres land at public auction.

D. B. "N," page 165, 30 December 1873. J.L. Tompkins, Commissioner, to A.B. Ross. Mentions "old mill house" in property bounds.

18. Charles Gillispie Mill

Meadow Creek

LOCATION: Route 661 (Smarts View Road SE) .5 mile south of Route 681 (Franklin Pike SE); POWER: Overshot wheel, probably wooden; STONES: 2 sets; VISIBLE REMAINS: Lengthy dug race visible before and after it is crossed by Route 661

History. Charles Gillispie (1872–1951) and his wife, Betty Moore Gillispie, both ran this mill, which was built sometime around 1900 and stopped operating in the early 1930s. This is according to Charles Gillispie's grandson Ernest, who also recalled in a 1992 interview with Frank Webb that the mill and water wheel, which must have been wooden or a wood-steel hybrid, were built by Fanton Smith.

Deeds

Floyd County Deed Book 30, page 112, 30 November 1903. W. J. Robertson to Charles Gillispie, mill and water rights, tract known as old sawmill site, containing 1 acre plus or minus.

D. B. 30, page 500, 10 June 1904. Thomas L. Robertson to Charles Gillispie, land adjoining mill site containing 2½ acres.

D. B. 33, page 429, 14 March 1907. T. L. Robertson to C. H. Lancaster, 105 acres land. T. L. Robertson retains water rights to Gillispie mill. Water was to be taken from creek at the same point it was taken for the old sawmill and run in the old race, and any dam erected across the creek was not to damage any land.

19. Goodson Mill/Howard Mill and Lancaster Mill/R.O. Harvey Mill

West Fork of Meadow Creek

LOCATION: Route 793 (Cannaday's Gap Road SE), .1 mile East of Route 681 (Franklin Pike SE); POWER: Overshot wheel, 3 feet wide; STONES: 1 set and a sawmill (Howard Mill), 1 set (Lancaster/Harvey Mill); VISIBLE REMAINS: Dug race along west side of Route 793 (Cannaday's Gap Rd NE)

History

In 1834 John Goodson inherited land that included a sawmill and a grist mill owned and probably built by his father, Thomas Goodson, Sr.* The elder Goodson included with this gift "all of my saw mill irons" and stipulated that John Goodson was to grind the grain of the family free of charge as long as his mother lived. Another large tract, adjoining on the north and also lying along Meadow Creek, was bequeathed to Thomas Goodson's unmarried daughters. In subsequent transfers, the bounds of both tracts include references to the race or races of the early Goodson mills, but the deeds have little to say about a later and probably larger mill built on the northern tract, apparently between 1859 and 1880. The last mill was known as the Lancaster Mill, after Washington C. "Jeff" Lancaster (1819–1893), who deeded the land and mill to his daughter and son-in-law Delilah (1860–1938) and Robert O. Harvey (1853–1916) in 1880. Harvey appears in the U. S. Census *Products of Industry* report for 1880 as the owner of a mill on Meadow Creek. *Chataigne's Virginia Gazeteer and Classified Business Directory, 1880–81* lists a W. C. Lancaster mill in Pizarro. The 1885 edition lists the R. O. Harvey mill in Pizarro.

Amos Wood's *Floyd County: A History of its People and Places* states on page 21 that the mill that came to be known as Lancaster's Mill was built for Major Howard in 1870 by millwright

**In the earliest written record of a Floyd County mill, a Thomas Goodson was granted permission to build a mill on Pine Creek near Little River by the Botetourt county court in 1770 (Summers, 98). No clear link between that mill and this Thomas Goodson has been found.*

Christopher "Stuffle" Bower and that it was "for many years a landmark for that section of the country being remarkable for its great height." The mill could have been built in 1870, and the builder may well have been Christopher Bower since Bower owned at that time the same adjoining tract that John Goodson had inherited from Thomas Goodson, Sr., in 1834, and upon which the original Goodson mills had stood. But if the mill was built in 1870, it wasn't built for Major Howard, since he died in 1866, the same year that Bower bought 96 acres of what had been the Goodson land from Howard's heirs.

Deeds are notoriously inconsistent in their mention of mills, hence unreliable for establishing when mills were built. Bearing this in mind, deeds for the land on which the Lancaster Mill stood do suggest that Joshua Underwood had the mill built on land he bought from Thomas Goodson's daughters in February 1859. He conveyed land and mill to his daughter Laura and her husband, Jeff Lancaster, in January 1860. Peter Terry ran the mill for R.O. Harvey, who also owned a store on Franklin Pike.[7] The mill is said to have been abandoned by 1910 but still standing when the mill tract was sold by Delilah Harvey in 1923.[8] Natives of the Pizarro community born c. 1923 say the mill was gone before 1930.[9]

Deeds and Other Public Records for the Goodson/Howard Mill

Floyd County Will Book "A," page 94, 8 May 1834. Thomas Goodson, Sr., to John Goodson, lands including grist mill and sawmill.

Floyd County Deed Book "C," page 257, 9 May 1843. John Goodson to Major Howard, 162 acres land on Meadow Creek, including the old sawmill, water grist mill and all water rights.

D. B. "L," page 374, 8 December 1866, Peter L. Howard, Tazewell M. Howard, Thomas H. Howard, Malinda Howard, Monroe Howard, Madison S. Helms, and Sarah A., heirs of Major Howard, deceased, to Christopher Bower. Acreage unspecified.

D. B. "P," page 394, 4 December 1879. Christopher Bower and Susannah, his wife, to Delilah E.A. Harvey, 96 acres adjoining mill lands. Mill (on the conveyed property) not mentioned.

Deeds and Other Public Records for the Lancaster/Harvey Mill

Floyd County Deed Book "E," page 431, 29 January 1850. Elizabeth Goodson to Mary Goodson Sandifur, adjoining mill lands willed by Thomas Goodson, deceased, to them jointly.

D. B. "K," page 128, 24 February 1859. Mary Sandifur and Jacob and Elizabeth Goodson Helms to Joshua Underwood, 90 acres.

D. B. "K," page 277, 10 January 1860. Joshua Underwood to Laura Lancaster (his daughter), wife of Washington C. Lancaster, tract of land on both sides of Mill Creek (Meadow Creek), containing mill and 90 acres.

D. B. "P," page 408, 28 January 1880. Washington C. "Jeff" Lancaster to Delilah E. A. Harvey, his daughter, wife of R. O. Harvey, land and mill tract of 90 acres.

D. B. 45, page 523, 20 October 1923. Lila E. Harvey, et al., to Arthur Peters, 111⅓ acres (mill tract, all of Lots 1 and 4).

See Table B, *1870 Products of Industry.* Name of corporation or individual: Bower, Christopher.

Because no stream name is supplied, the relevance of this record to this mill is uncertain. In 1870 Bower did own the land on which the old Goodson/Howard mill had stood, and may have been running it, although no mill is mentioned in the 1866 deed conveying the land to Bower from Major Howard's heirs. Bower is said to have built the Lancaster Mill, but the fact that the mill Bower was running in 1870 had two sets of stones, and the Lancaster mill had only one in 1880, make it unlikely that they are the same.

See Table C, *1880 Products of Industry.* Name of corporation or individual: Harvey, Rob A.

20. William J. Williams Mill

Beaver Dam Creek

LOCATION: Beaver Dam Creek, .2 mile downstream from Route 221 (Floyd Highway North); VISIBLE REMAINS: See Thomas S. Lester Mill (I, 121)

History. William J. Williams was the first miller of record to advertise in a Floyd County newspaper. That his ad is for "Dry Goods" rather than mill products is not surprising, since many millers aspired to offer an early version of "one-stop shopping" with a variety of goods and services. In addition to general stores, many millers ran sawmills; a good number operated legal distilleries, and several had wool carding machines in the same or adjacent buildings. In his advertisement Williams

> begs leave to say to his friends, and the public generally, that he is now receiving a full and splendid

assortment of Fancy and Staple DRY GOODS [including] Hardware, Queens Ware, Drugs, Medicines, Dye Stuffs, Groceries, Boots, Shoes, Hats, Bonnets of the latest style; and in fact everything usually kept in a country store; which he will sell low for Cash or exchange for Country Produce or sell on time to punctual customers.

He wishes to buy three thousand pair of large heavy Socks, for which he will pay the highest market price in Goods. WM. J. WILLIAMS

Beaver Dam Mills, Floyd Co., April 22, 1854[10]

Who among the other subscribers to the *Floyd Intelligencer* might have had "three thousand pair large heavy Socks" to sell is an interesting but unanswered question.

The Thomas Lester Mill (I, 121) may have been built later on the same site as this mill. Also, a dam and mill approved by the Montgomery County court in 1782 in response to a request by William Davies must have been near this spot, if there are no other Beaver Dam Creeks on Little River, and if Davies actually built his dam and mill after receiving permission to do so.[11]

Deeds

Floyd County Deed Book "F," page 239, 16 January 1851. John Helms to William J. Williams, 146 acres land on Beaver Dam Creek and Little River.

No record found of sale of mill property by William J. Williams to George and J.M. Akers

D. B. "K," page 409, 22 August 1860. Trust Deed, J. M. Akers and George Akers to A. J. Kirby, 146 acres land and mill property bought of William J. Williams.

D. B. "O," page 415, 5 May 1877. John Merritt, Commissioner for William J. Williams, to Lorenia S. Lester, wife of John W. Lester, 146 acres land and mill site.

See Table C, *1880 Products of Industry.* Name of corporation or individual: Geo. Akers. This mill is the most likely of those on Beaver Dam Creek to have been referred to as the George Akers mill in 1880, but this is an uncertain match since the land and mill site (this phrasing suggests that the mill itself may have been gone) were sold to Lorenia Lester in 1877.

21. Peter Hatcher Mill

LOCATION: Route 680 (Starbuck Road SE), about 1.9 miles east of Route 860 (Shooting Creek Road SE); POWER: Portable steam engine, 15–20 HP; John Deere Model B farm tractor; STONES: 1 set and Hammer mill for grinding feed; sawmill; VISIBLE REMAINS: None

History. An undated *Floyd Press* item, supposed to have appeared c. December 1912, reported that "Mr. Peter Hatcher has moved in his new mill and is ready for business." Nearly 40 years later, on July 13, 1950, the *Floyd Press* printed notice of the "Public Sale of land and personal property of Peter F. Hatcher, deceased, 2 tracts of land consisting of 130 and 19½ acres respectively on tributary water of Little River, (Pine Creek), and 10 room house, a good corn and feed mill, a saw mill and land [?] mill and about 50,000 ft. timber. Sale 21 July 1950 at 10 o'clock."

Peter Hatcher's brother Henry worked as a miller here. Ivan Hatcher, a grandson, also worked at the mills and recalled that firing the steam engine was "hot in the summer, just right in the winter." After Peter Hatcher's death his son Tiamon bought his father's land from the other heirs along with personal property from the estate, including a gristmill for $30.00, 1 sawmill for $40.00, and 1 set of scales for $10.00.[12] Another son, Wayne Hatcher, last ran the mill.[13]

Thomas Wright, a boyhood neighbor, remembered the Hatchers as being interested in mechanical innovations. Besides building one of the county's only steam-powered grist and saw mills, Peter Hatcher was among the first to own a tractor, an Avery.

National Park Service/Blue Ridge Parkway photos of the mill show a tall, weatherboarded structure of two and one-half stories, with a door and three windows in the gable end, including one in the attic. In a 1998 interview Thomas Wright stated that the mill, which was built for Hatcher by Charles Underwood, stood 300 to 400 feet from the house, downhill and to the west, and was visible from the road. (This estimate could be refined using archived NPS/BRP photographs showing both the house, which still stands, and the mill.) It had one set of stones of 30 to 36 inches in diameter and a large corn sheller powered by a Maytag washing machine motor. Feed was ground with an International Harvester hammer mill. Hatcher's original

Peter and Antonia "Toni" Hall Hatcher, undated. Hatcher employed the latest technology in his steam-powered mill and on his farm on the headwaters of Pine Creek with mixed results. (Courtesy Thomas Wright)

intention to install a flour mill explains the unused second and partial third stories which were unneeded in a mill producing only cornmeal and coarse feeds for livestock. In a flour mill, these upper stories would have stored cleaned wheat and housed the machinery, spouts, and elevators it took to make the highly refined flour that Floyd Countians began at first to prefer and eventually to demand, after the first roller mills arrived around 1900.

Wright was told as a boy that the Hatchers' "Geiser" circular-saw, rack-and-pinion carriage sawmill had first operated across from a quarry site on Shooting Creek Road, north of the current intersection with the Blue Ridge Parkway. Hatcher bought the mill after three employees were killed by a boiler explosion, a danger which, along with the increased threat of fire, probably kept many millers from abandoning water power until the arrival of gasoline or diesel engines and finally electric motors. These more modern forms of motive power were even more mysterious than steam power, and far from trouble free, but the consequences of inexpert or careless operation were rarely fatal. The one known fatality caused by over-speeding a mill driven by a gasoline engine occurred within a short distance of the Hatcher Mill, when 16-year-old Chester W. Hall was killed by an exploding mill stone in November 1926.[14]

According to Wright's sources, the boiler and engine purchased by Hatcher to drive his grist and saw mills came by rail from Pennsylvania to Franklin County and were brought from there up Cannaday's Gap or the Runnet Bag Road using a low-slung rented wagon with steel wheels pulled by two yoke of oxen. Boiler and engine were set on flat rock foundations near the mill. The boiler flues were replaced at the beginning of World War II, but by the end of the war the boiler was once again in need of repair. It and the engine were sold and replaced by a John Deere model B farm tractor.

Because the Hatcher Mill was never water powered, it is technically beyond the scope of this book. But it merits inclusion as a sort of window on the world beyond the Southern Appalachians, a world in which the vertical water wheel was already an emblem of a primitive past, eighty years before the last water-ground Floyd County cornmeal was bagged and tied by Artice Nolen (I, 123) in the 1970s. Peter Hatcher's 1912 application of steam power to grain milling was an anomaly in Floyd County, but the use of steam power had by then become the rule rather than the exception among contemporary industrial applications in England, whose topography was not well suited to hydro-power development. The same was true in those regions of the United States where the demand for power was rapidly outgrowing substantial but finite and immovable water-power resources.

While the topography of Floyd County, with its numerous streams and narrow valleys, worked against the adoption of steam power by grain millers, by 1912 these same characteristics were persuading sawmillers to abandon water power sites and take to the hills. The compelling difference was in the relative difficulty of transporting the raw materials to the mill. Newly cut, sap-heavy logs could only be limbed and cut into lengths of not less than eight feet, whereas the weight and bulk of wheat and corn had already been substantially reduced by drying, threshing, shucking, and shelling before they left the farm. A sawmill and a steam engine could be moved to the middle of a boundary of timber, and the sawn lumber hauled out, far more easily than logs could be loaded and transported from the woods to the nearest water-powered sawmill. On top of that, even a small portable steam engine was more powerful than all but the largest water wheels in the county, and would be unaffected by low stream levels or by low temperatures as long as the boiler was drained or kept warm during a hard freeze.

Since Peter Hatcher didn't own a suitable mill seat, he took the next logical step by locating his grain mill away from the stream bank and its attendant hazards. Whether he intended from the beginning to add a sawmill is unknown, but its inclusion would have helped the engine earn its keep when there was no cornmeal or feed to be ground, and to get the most return from the time, fuel, and effort required to bring the boiler up to working pressure. At the very least, the sawmill operation provided fuel for the engine in the form of slabs, which would have been considered a waste product at a water-powered mill. Hatcher's purchase of the sawmill therefore counterbalanced two of the operating advantages—"free" energy and almost instant readiness—that water power had over steam as an industrial power source. Sparks from the flue of the steam engine would have increased the risk of fire, but the more constant danger of flood was virtually eliminated.

The greatest drawback of steam power might have been the constant need for a second worker with two important duties. He had first to be the "fireman," whose job it was to maintain a good head of steam by feeding slabs, edger strips, and other scraps of wood to the boiler. At the same time he had to be the "engineer" and keep the level of the water inside the boiler safely above the minimum and the steam pressure safely below the maximum. Although it would not always be so, labor was cheap and abundant in Floyd County in 1912 and for most of the next thirty years, especially for a man with family close by.

The truth of Peter Hatcher's thinking could be much less sophisticated—he may simply have been unable to find or afford a suitable stream bank site on which to build his mill. But nearly a century later, the plans he made look like those of a forward thinking man.

Deeds and Other Public Records

Floyd County Deed Book "W," page 12, 20 July 1892. A. L. Cannaday to P. F. Hatcher, 130 acres on waters of Pine Creek.

D. B. 27, page 520, 8 May 1901. D. F. Hall and B. H. Hall to Peter F. Hatcher, their interest in 130 acres land where P. F. Hatcher resides.

D. B. 54, page 97, 6 February 1936. P. F. Hatcher to Commonwealth of Virginia, 2.29 acres.

D. B. 56, page 209, 15 July 1937. P. F. Hatcher to Commonwealth of Virginia, two tracts land, one of 5.77 acres, another of .84 acres, total of 6.61 acres.

D. B. 71, page 207, 9 August 1950. W. F. Hatcher and wife, Myrtle Hatcher Hall and husband, and Tiamon M. Hatcher, heirs of Peter F. Hatcher, deceased, to Tiamon M. Hatcher, the lands of Peter Hatcher containing two tracts, one of 130 acres and another of 19½ acres.

D., B. 98, page 207, 13 October 1969. T. M. Hatcher to Peter Hall, two parcels of land containing 149½ acres.

22. Asa Shortt Mill/Jake Brogan Mill

Runnet Bag Creek

LOCATION: Approximately 3 miles upstream from the point in Franklin County where Runnet Bag Creek meets Route 793 (Cannaday's Gap Road NE). This road is state maintained at both ends, but a lengthy stretch between is impassable to all but foot traffic; POWER: Overshot wheel; STONES: 1 set; VISIBLE REMAINS: Wheel pit and stacked stone foundation

History. Asa D. Shortt was a Baptist minister and member of the Floyd County School Board as well as a miller. He rebuilt his mill with help from various skilled and unskilled laborers after it burned in 1893. His detailed account of the cost of materials and labor mentions several millers and millwrights working in Floyd and Franklin counties (see Appendix D). Couples married by Shortt at his mill included Arkansas "Kansas" Underwood and Andy Jarrels on May 28, 1903. Mrs. Jarrels was later married to Jacob "Jake" Brogan, by whose name the mill was last known. Brogan was never the owner of this mill, but is supposed to have run it from 1923 to 1937, for both John Peters and Sammy J. Turner. Each of these last two owners also ran the mill, but not long enough for either of their names to replace Brogan's in the oral history of this community. This same photograph of the Jake Brogan mill appears on page 24 of *Franklin County Life and Culture: A Pictorial Record*, published by the Franklin County Bicentennial Commission, Rocky Mount, VA, 1986, but any doubt

The Asa Shortt/Jake Brogan mill on Runnet Bag Creek, date unknown. The wooden race is sturdy and surprisingly free of leaks, but the ladder up to it inspires less confidence. The mill was completely rebuilt by Asa Shortt after it burned in 1893. (Courtesy Blue Ridge Heritage Archive/Franklin County Bicentennial Collection)

about the location of this mill would seem to be settled by the recording of the deeds to the mill property in Floyd County.

Deeds

Floyd County Deed Book "S," page 86, 10 June 1885. James P. Martin and Martha Martin, his wife, to Alexander Graham, 3 or 4 acres and grist mill on Runnet Bag Creek. *Whether James and Martha Martin purchased the mill or built it is not known, since no deed to them was found in Floyd County records. That deed could conceivably be recorded in Franklin County if the mill tract was at that time part of larger tract lying mostly in Franklin County.*

D. B. "S," page 614, 12 February 1887. Andrew J. Graham, executor for Alexander Graham, deceased, to Sarah C. Short, wife of Asa D. Short, 3 or 4 acres and grist mill.

D. B. 39, page 576, 26 November 1913. Sarah C. and Asa D. Short to Walter G. Houchins, 5 acres plus or minus, saw and grist mill.

D. B. 43, page 33, 26 February 1917. Walter Houchins to General W. Boyd, 4 or 5 acres, saw mill and grist mill.

D. B. 43, page 74, 15 September 1917. General Washington Boyd [1894–1983] to W. C. Webb, 5 acres plus or minus and mill property. (Webb died before completing the terms of payment, resulting in a sale by a court commission.)

D. B. 43, page 565, 4 November 1919. B. G. Howard, Commissioner, to Henry Atkins, about 5 acres land and mill.

D. B. 47, page 360, 18 April 1925. Henry Atkins to P. W. Nolen, 5 acres more or less and grist mill.

D. B. 46, page 381, 18 April 1925. P. W. Nolen and wife to J. W. Peters, 5 acres more or less, the Shortt Mill property.

D. B. 48, page 590, 11 January 1930, heirs of J. W. Peters, deceased, to S. J. Turner, 5 acres more or less and mill property.

D. B. 101, page 26, 15 February 1971, S. J. Turner to T. E. Hall and Carnell Hall, his wife, the Shortt Mill property, containing 5 acres more or less.

23. Shanghigh Mill

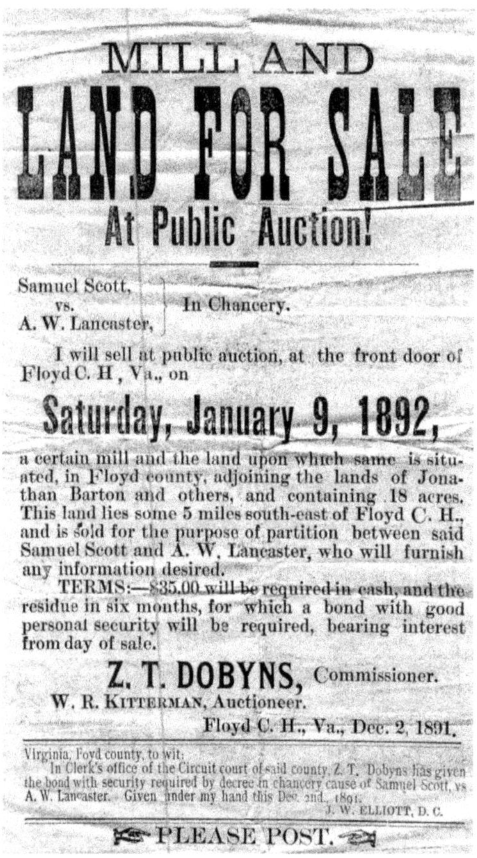

MILL AND

LAND FOR SALE

At Public Auction!

Samuel Scott,
vs. In Chancery.
A. W. Lancaster,

I will sell at public auction, at the front door of Floyd C. H, Va., on

Saturday, January 9, 1892,

a certain mill and the land upon which same is situated, in Floyd county, adjoining the lands of Jonathan Barton and others, and containing 18 acres. This land lies some 5 miles south-east of Floyd C. H., and is sold for the purpose of partition between said Samuel Scott and A. W. Lancaster, who will furnish any information desired.

TERMS:—$35.00 will be required in cash, and the residue in six months, for which a bond with good personal security will be required, bearing interest from day of sale.

Z. T. DOBYNS, Commissioner.
W. R. Kitterman, Auctioneer.
Floyd C. H., Va., Dec. 2, 1891.

Virginia, Foyd county, to wit:
In Clerk's office of the Circuit court of said county, Z. T. Dobyns has given the bond with security required by decree in chancery cause of Samuel Scott, vs A. W. Lancaster. Given under my hand this Dec. 2nd., 1891.
J. W. Elliott, D. C.

PLEASE POST.

Pine Creek

Location: Route 860 (Shooting Creek Road SE) near intersection with Route 637 (Thomas Farm Road SE); Power: Overshot wheel; Stones: 1 set; Visible Remains: None

History. In 1874 partners A.W. Lancaster, J.R. Lee, and William A. Lee paid John Young thirty dollars for three acres on Pine Creek with the express purpose of "erect[ing] water grist mill and other machinery. Water rights included, dam to be not more than two feet high." In 1883 Lancaster bought out his two partners for $53, and the next year sold one-half interest in an 18-acre corn mill tract to Samuel Scott for $53.72. Disagreement between the partners resulted in a chancery suit and the advertisement of a public sale of the property on Saturday, January 9, 1892. Ultimately, Scott acquired the remaining half interest from Lancaster in 1893 and willed it to Winfield Scott, who took possession in 1899. Although the mill was apparently still standing,

Chancery suits were frequently filed against mill owners to recover bad debts. But in this case, co-owners Scott and Lancaster apparently could not agree on a way to dissolve their ten-year partnership. Scott became sole owner of the mill in June of 1893. (Courtesy Gerald Via)

it was not mentioned in the deed when the property changed hands again in 1909, possibly at a public sale advertised on June 17, 1909, in the *Floyd Press:* "Tract of land including the old Shanghigh Mill, containing about 42 acres. This lot includes part of the Tom Lancaster land, balance of Shanghigh mill tract."[15]

Deeds and Other Public Records

Floyd County Deed Book "O," page 202, 18 June 1874. John Young to A. W. Lancaster, J. R. and William A. Lee, 3 acres to erect water grist mill and other machinery. Water rights included, dam to be not more than two feet high.

D. B. "Q," page 429, 25 March 1882. J. L. Thompkins, Commissioner, to A. W. Lancaster, J. R. and William A. Lee's ⅔ interest in mill and property.

D. B. "R," page 344, 28 April 1883. A. W. Lancaster to Samuel Scott, ½ interest corn mill tract, 18 acres.

D. B. "W," page 320, 6 July 1893. A. W. Lancaster to Samuel Scott, 18 acres and mill on Pine Creek.

Floyd County Will Book "F," page 577, 17 October 1899. Mill and rest of land willed to W. Scott by Samuel Scott, deceased 1899.

D. B. 36, page 10, 21 July 1909. V. M. Sowder, Commissioner for W. Scott, to S. R. Brame.

24. Phillip Williams Wool Carding Mill

Pine Creek

LOCATION: Between Route 680 (Starbuck Road SE) and Blue Ridge Parkway; POWER: Unknown; STONES: None. Wool carding and wool picking machines

History. Phillip Williams (1773–1834) willed to his daughter Sarah Williams Lancaster his wool carding machine and picker, per records for May court 1834, Will Book "A," page 22.

Although it was not known to have shared a water wheel or turbine with a grist mill, the Phillip Williams Wool Carding Mill is included as an example of the application of water power to a process that was essential before 1900. By that time cheaply produced cotton fabrics were replacing home-spun and home-woven linen and wool for everyday clothing. Because cloth lasts longer than flour and cornmeal, there were relatively few wool processing mills countywide, even during their heyday. Carding mills like this one, which could only clean and straighten the wool fibers for customers to spin, dye, and weave at home, were of little use after spinning wheels and looms were relegated to attics and barn lofts.

Home-spun fabrics, whether knitted or woven, required considerable skill and enormous amounts of time to produce. By the beginning of the 20th century, bolts of inexpensive fabrics stocked by every general store and mail order house were replacing homemade cloth. Once farmers were persuaded that it paid them to purchase enriched animal feeds in 100 pound sacks, sacks of bleached white cotton became a major source of fabric for bed "linens," diapers, towels, and underclothing. Sacks printed with an infinite variety of bright, colorfast patterns supplied fabric for girls and women's clothing, curtains, aprons, and quilt squares through the Great Depression and into the 1960s.

Aside from simple carpets and more complex coverlets woven by people of unusual skill or passion for this craft, home production of textiles had disappeared by the early 20th century. Only those mills that could complete the process of spinning, dying, and weaving wool into finished products like blankets and patterned coverlets survived beyond.

The most successful example was the Vaughn-Dunn Woolen Mill on Burks Fork Creek (IV, 48) where J.T. Dunn was still inviting Floyd County sheep farmers to send him "WOOL, to make into BLANKETS, COVERLETS and knitting yarn" as late as 1942.[16]

Deeds

None found.

25. Powhatan "Port" Williams Mill

Pine Creek

LOCATION: Across Route 690 (Sandy Flats Road SE) from Pine Creek/Nolen Mill (I, 123); POWER: Overshot wheel; STONES: At least one set and sawmill of unknown type; VISIBLE REMAINS: Almost certainly used long dug race now serving Shelor/Nolen/Pine Creek Mill

History. An 1874 deed to this property conveys the land on which a dam, and probably the race as well, were already in place, along with the right to cross the land of Isaac and Peter Huff to maintain the race and dam. The deed is quite detailed in its specification of the rights, obligations, and prohibitions agreed to by both parties. The sawmill that was supposed to saw lumber for Isaac Huff "at half the public rate" was probably an "up and down" sash saw which cuts only on the downward stroke. The limiting of the dam to its existing height of 2½ feet suggests that it served as much to divert flowing water into the race as to hold water in reserve for use during peak demand.

This mill site is one of many whose histories illustrate the very logical practice of reusing a proven water power site. Good locations were abundant but not unlimited even in Floyd County. The difficulty and expense of gaining permission to dam streams and flood upstream acreage, and of digging and maintaining in-ground races, were additional reasons to build a new mill on or near the site of an obsolete or damaged mill building. The deed confirming conveyance of water rights from J.J. Sweeney and wife to George W. Shelor (transcribed below) highlights the fact that a mill is of little value without the necessary water rights. The extreme length of this race would have made this a more serious cause of concern at this mill than at most others.

Powhatan "Port" Williams (1813–1893) and Elizabeth "Betsy" Dickerson Williams (1818–1899), about 1875. (Courtesy Floyd County Historical Society)

The Powhatan Williams (1813–1893) Mill was one of several whose wooden races crossed over public roads, a circumstance that posed less of a problem before the appearance of tall truck bodies and other large or wide machinery on rural roads. In the same way that some horses are reluctant to cross running water, some could not be compelled to walk beneath dripping wooden raceways without first having their heads covered. People were either reluctant or delighted to pass beneath, depending upon the season, their ages, and the way they were dressed.

George W. Shelor built the current Shelor/Nolen/Pine Creek Mill (I, 123) in 1913 across the road from the Powhatan Williams Mill and used at least one set of its stones. Whether the old mill was then still standing, and if so, how long it remained thereafter, are unknown.

Deeds

Floyd County Deed Book "N," page 573, 3 November 1874. Isaac and Peter Huff to Powhatan Williams, David E. Williams, G.M.D.

Williams, W.T. Williams and John H. Williams, land for mill race and dam and all water rights. Dam not to be raised higher than it is at the time of the deed, being 2 ½ feet. *No record found of transfer from Powhatan Williams et al., to William T. Williams, son of Powhatan Williams.*

D. B. 32, page 353, 1 January 1906. William T. Williams to J. J. Sweeney, 2 acres and mill property, being grist mill and sawmill and shop.

D. B. 34, page 507, 7 April 1908. J. J. Sweeney to Josie and J. W. Peters, ½ interest in grist and sawmill and shop on Pine Creek.

D. B. 35, page 12, 14 October 1907. J. J. and Elizabeth Swinney and Morton and Ida Radford to J. W. Peters, "½ interest in the mill and mill property known as the Williams mill on Pine Creek." *According to D. B. 36, page 595 (see below), J. J. Swinney/Sweeny sold a ½ interest in the mill to Morton Radford. No deed was found recording that conveyance. The listing of both Sweeney and Radford and their wives as the parties of the first part in this deed may reflect the unofficial status of that transaction.*

D. B. 35, page 577, 29 June 1909. Josie and J. W. Peters to L. P. Spence and Amos D. Spence, their interest in 2 acres and mill on Pine Creek.

D. B. 36, page 579, 24 or 29 July 1910. G. W. Shelor and wife to A. D. and L. P. Spence, ⅔ and ⅓ interest, respectively, in 50 acres at Wills Ridge. *This deed conveys the 50 acres traded to the Spences along with $100 in exchange for the mill in the next deed. It does* not *convey the Powhatan Williams Mill property, but is included because it clarifies that the water rights, which were apparently overlooked in the writing of the deed (immediately below) conveying the mill to G. W. Shelor, were conveyed by the Spences as follows: "Grist Mill and appurtenances with full right of way for water power."*

D. B. 36, page 594, 27 July 1910. L. P. and A. D. Spence to G. W. Shelor, for one hundred dollars and fifty acres of land "in the Low Gap of Wills Ridge," the property known as the W. T. Williams mill place and 2 acres on Pine Creek.

D. B. 36, page 595, 29 July 1910. J. J. Sweeney to G. W. Shelor. "This deed, made and entered into this 29th day of July 1910 between J. J. Sweeney and his wife, parties of the first part, and Geo. W. Shelor, [party] of the second part. Witnesseth, that whereas J. J. Sweeney sold to Morton Radford one half interest in what is known as the Powhatan Williams Mill and J. J. Sweeney and Morton Radford sold entire interest in said Mill to Josie Peters and Josie Peters sold said mill to Amos A. Spence and L. P. Spence and said Amos L. Spence and L. P. Spence sold said mill to George W. Shelor and there being a question as to whether said J. J. Sweeney conveyed the Mill Race and all rights to Josie Peters and Martin Lanford. This deed be made by the parties of the first part to quit the title to said water rights and mill race.

"Now therefore in consideration of the premises the parties of the first part doth grant sell and convey to George W. Shelor all of his right[,] title[,] and interest in the water rights and Mill race belonging to what is known as the Powhatan Williams Mill and the parties of the first part convey the same right and title to said water rights and mill race as is given them in deed from W. T. Williams and wife dated Jan. 7th, 1906 and recorded in Clerk's Office of Floyd County to which reference is here made. Given under my signatures and seal. J. J. Sweeney (seal)."

See Table C, *1880 Products of Industry.* Name of corporation or individual: P. Williams and Sons

26. German (or Jarman or Jerman) Rakes Mill

Dodd Creek, South Fork of Little River

LOCATION: North side of Blue Ridge Parkway at Milepost 162.4; POWER: Overshot wooden wheel; STONES: 2 sets; VISIBLE REMAINS: Rebuilt dam is intact and maintained by Blue Ridge Parkway; parts of mill foundation and stone piers for wooden race remain on north side of stream. Extensive mud flats above the dam illustrate the original appearance of silt banks remaining behind some long-gone mill dams.

History. A photograph (unsuitable for reproduction) of this mill appeared on page 3 of the August 12, 1937, issue of the *Floyd Press.* The accompanying caption places the mill "on the south fork of Dodd's Creek, five miles south of town on Route 716" and states that John Howell built the mill in 1837. Ira Compton, the last miller as well as the last private owner of the mill, stated that the mill was in active service and capable of grinding forty bushels a day when it was purchased as parkway property. John Howell, German Rakes, Monroe Yates, Reed Agee, W. A. Compton (Ira's grandfather), and J.W. Ferguson were named by Compton as previous millers. Others said to have worked here included Carl Roberson (in 1925–26); Ira Compton's father, Bill; and his sons Eldridge and Ashby.

Deed records indicate that the mill was not built until 1859 or soon after, but the 1837 date could still be significant in the history of this mill site, if not of this particular mill. Early deeds refer to this fork of Dodd Creek as "Old Mill

The Rakes Mill dam on the Blue Ridge Parkway, 1992. Parkway planners chose not to restore what was by the 1930s the Compton Mill, but have maintained the wood and stone dam. The extensive mud flat above the dam was formed by years of silt deposits, left when mud-laden storm runoff slowed as it entered the pond. (Ricky Cox photograph)

Creek," and it was common practice to reuse existing water power sites.

Until it was torn down by the Blue Ridge Parkway in the early 1940s, the German Rakes Mill stood an estimated 25 yards below the dam on the north side of the stream, across from the parkway. The creek bank between was too steep and rocky for a dug race, so a wooden race, supported in part by stone columns, connected dam and mill. These stacked stone supports, a rare feature for a Floyd County mill, are still evident, as is part of the mill's stacked stone foundation. The dry-laid stone face of the downstream side of the dam is also unusual in that most Floyd County dams of that height appear to have been of the wooden frame type flanked by stacked stone or concrete abutments.

According to a Blue Ridge Parkway sign at the site, the dam "was built early in the 19th century by one Jarman Rakes." The facts that the mill still standing here when the land was deeded to the Commonwealth of Virginia in 1937 probably was not built before 1859, and that Rakes did not own it until 1868, don't necessarily rule out that possibility, since the dam could predate the mill. But the German Rakes by whose name the mill and mill pond were known was born either in 1828 (calculated based on 1860 census) or 1825 (based on 1870 census) and is unlikely to have built anything of much consequence before 1845 at the very earliest. Hence, either the sign is incorrect, or a different German or Jarman Rakes built the dam. The interpretive sign goes on to say that Rakes' "operation was notable for a scheme of advertising that would credit a much later day. Rakes we are told allowed his customers the sole privilege of fishing for brook trout in his pond while they waited for their grist."

Deeds

Floyd County Deed Book "K," page 144, 23 March 1859. John W. Howell from John Howell, land where mill is to be built.

D. B. "M," page 444, 20 January 1868. John W. Howell to Jerman Rakes, land and mill tract.

D. B. "P," page 617, May 8, 1880. German Rakes, deceased, to Mary Rakes, dower land, 19¼ acres and mill and dwelling house.

D. B. "V," page 33, 26 November 1890. Mary Rakes, widow of J. Rakes, to W. Reid Agee, mill and land known as the Jarman Rakes mill property, 37 acres.

D. B. 29, page 361, February 5, 1903. W. Reid and Louisa Agee to Thomas Monroe Yeatts, 2¾ acres land and mill property on South Fork Little River (Dodd Creek).

D. B. 33, page 297, 23 February 1907. Thomas M. Yeatts to John W. Ferguson mill, land and water rights on headwaters of Dodd Creek. Dam not to be high enough to cover spring.

D. B. 43, page 20, 12 May 1917. James Wesley Ferguson to Ira M. Compton, mill property and all water rights. Not to build dam so high as to interfere with spring.

D. B. 56, page 80, 15 July 1937. I. M. Compton and Cassie Ann, his wife, to Commonwealth of Virginia, 2.37 acres land on Dodd Creek.

27. Old J. W. Epperly Mill/Homer Roberson Mill

Dodd Creek

LOCATION: South side of Route 714 (Roberson Mill Road SE) about .25 mile east of Route 8 (Parkway Lane S); POWER: Wooden overshot wheel, replaced by Fitz steel overshot wheel in 1956; STONES: Two sets and buckwheat huller, sawmill; VISIBLE REMAINS: Mill still standing with overshot wheel in place, both in poor condition. Dug race and remnants of the long (⅓ mile) wooden race

History. John W. Epperly, one of Floyd County's most accomplished millwrights, built what is now known as the Roberson mill in the 1880s on land that may already have been developed as a water power site. Epperly was also a skilled blacksmith and gunsmith, and according to his great- grandson Jim Altizer, had a blacksmith shop across the road from the mill.[17] Stuart Cannaday is said to have worked as a miller for Epperly and for the mill's third owner, Rufus Abraham "Abe" DeHart, who bought the mill from Harrison Scott in 1910. DeHart ran the store while Cannaday looked after the mill, but was himself a miller, having learned the trade from his father, Stephen DeHart, on Rock Castle Creek in nearby Patrick County, Virginia. Abe DeHart later owned the second John W. Epperly Mill (IV, 126) and, at some point, McDonald's Mill on the North Fork of the Roanoke River in Montgomery County, Virginia.

Cleve Smith bought what was by then known as John Epperly's Old Mill from J. W. Brammer in 1922 and stayed on as miller after selling the property to John Smith in 1926. The mill has been in the Roberson family since John Smith sold it to Homer Roberson in 1931. Current owner Harry Roberson operated the mill under water power until the wooden race or flume collapsed about 1970. Electric power was used to grind buckwheat flour on a part-time basis until 1984. Phillip Gettel noted in a 1985 survey of Floyd area mills that a single poplar tree provided the weatherboard siding for one side and both ends of the 2½ story Roberson mill.[18]

The Roberson Mill was one of the last two commercial mills to operate under water power in Floyd County and one of only two existing Floyd County flour mills that was neither designed to incorporate a roller mill nor modified afterward to accommodate one. The other is the William Allen Simpson Mill (II, 113) which is older but far less complete, having been used

An early view of the west end of the "Old John Epperly Mill." The wooden race was among the longest and tallest in the county. (Courtesy Annie Mae [Roop] Gardner)

The Fitz steel wheel was installed in 1956, replacing the last wooden water wheel in use at a commercial mill in Floyd County. The race collapsed soon after this picture was taken in 1972 but the mill ran until around 1984, grinding buckwheat part-time using electric power. (Courtesy Floyd County Historical Society)

for other purposes since closing its doors around 1941. The Roberson mill represents, therefore, the county's most authentic picture of flour milling as it was done in the United States through most of the 19th century. At the same time, it is the most distinctive among mills (that were still standing in 2010) in design and machine layout, and also demonstrated some of the finest craftsmanship in wood and metal components. The pride Epperly took in what he built, evidenced by the stamping of his name on the forged steel arms of the stone hoist, was well justified.

Unusual features include a split-level main floor; no basement; main gearing exposed to and accessible from the main floor; elevated stone platform; elevators with individual drive shafts, oriented at right angles to the peak of the roof; fully floored third level; stairs placed in the center of the mill (most are along a wall or in a corner); and minimal shafts and belting. Power is carried to upper floors by a vertical shaft extending upward from the main drive gear. This last may be characteristic of mills that pre-dated the arrival of roller mills and their attendant line shafts, drive belts, and pulleys; a similar system was used in the William Allen Simpson Mill, the only other remaining flour mill that was never converted to the roller system.

Why a mill whose owners resisted a powerful trend toward modernization outlasted many newer mills both as a structure and a business may have as much to do with personalities as it does technology, but nonetheless invites speculation about the impact of two interesting circumstances on the history of this and other Floyd County mills.

The first of these circumstances is that the mill was built in the 1880s, during which decade the flour milling industry in the United States was undergoing the most significant technological change in its entire history—the transition from the time honored method of crushing and cutting grain between two millstones, one revolving and the other stationary, to the roller mill process, which abrades grain between as many as eight separate pairs of corrugated steel rollers rotating in the same direction, like the rollers in a clothes wringer, but at much higher and slightly different speeds. Had John Epperly built a roller mill on this site in the 1880s, it would have been the very first in the county. As it turned out, his was among the very last Floyd County flour mills* to incorporate the old Oliver

**The Irvin Huff Mill (I, 2) and the William Jackson Stanley Mill (IV, 54) were built after 1900 to make flour using the Evans system, and both were on streams so small that a roller mill system may have been unworkable. The Irvin Huff Mill was, like Roberson's, known for its excellent buckwheat flour.*

Evans system of flour milling, the chief components of which (in Floyd County flour mills) are one set of conventional millstones specially dressed for grinding wheat and a simple reel-type sifter. Additional needs include shafting with pulleys and belts or gears to transmit power, elevators and spouts for moving wheat and flour between levels, and one or more machines for removing foreign matter from the wheat before it is ground. But even in its most complex form, the Evans system is simpler and less expensive than the roller mill system that replaced it. It is also slower and less efficient at producing the highest return per bushel of the ultra-white, ultra-fine flour Americans had been persuaded to prefer over the slightly coarser, slightly darker flour produced by buhr-stone flour mills. Had he waited even ten years, John Epperly probably would have planned from the outset to install a roller mill system, and having made that choice, may not have built a mill here at all since the relatively low volume of water probably couldn't have reliably generated enough power to run a high speed roller mill. He was, in effect, correcting both shortcomings—an obsolete flour milling system and an inadequate source of power—when he built a large, modern roller mill (IV, 126) several miles downstream, and below the confluence of Dodd Creek's two main branches, fifteen years later.

The second circumstance that helps explain why the Roberson mill better represents 19th century flour milling than 20th is the fact that the men who owned it between 1900 and 1930 chose not to refit the interior and modify the water power to accommodate a roller mill system, although they no doubt knew this was happening at other older flour mills, like the Spangler Mill on Pine Creek and the Harman Mill on West Fork Creek. The owners of still other flour mills decided it was impractical to rework an existing mill and water power, and chose instead to tear down an obsolete building and build a new one. While they were at it, several replaced (with a turbine or a factory made steel water wheel) or upgraded (with a new or raised dam) the power system. This was the case with the flour mills then or later known as Greasy Creek, Vaughn's, Ed Strong, and Thompson-Akers, among others. Only about thirty years old at the 1915 peak of the roller mill mania, the "J. W. Epperly Old Mill" may have seemed too new for its owners to consider razing and replacing it, considering, especially, how well constructed it has in the meantime proven itself to be. When it came to refitting the mill for a roller system, they may have perceived, as some of their counterparts on the headwaters of other streams did not, that there simply wasn't enough power to be had from the available water.

Whether the men who owned the mill during these decades consciously chose not to remodel or rebuild for practical (lack of water), personal (lack of energy), or financial (lack of money) reasons, by the time Homer Roberson bought the mill in 1931, the opportunity had passed, along with the thirty-five-year flurry of mill building and refitting that would be the high-water mark of Floyd County grain milling. By the late 1920s, forward-looking Floyd County investors may have been deterred from putting money into a small-scale, agriculture-based business by the economic trends of the teens and twenties, which predicted an economic future more and more dependent on a centralized, industrialized national economy. They would certainly have been demoralized by the cash-poor reality of the 1930s. As a result of either or both circumstances, no new roller mills were built, and no old mills refitted to accept the roller system after 1930 in Floyd County. Roberson either decided, or was left with no choice, to specialize in the grinding of buckwheat, a process that was either too difficult to adapt to the roller process, or that was thought too exotic to be worth the trouble. Roberson also occasionally ground rye flour for customers who had acquired a taste for rye biscuits, which Harry Roberson described as the color of brown sugar and good right out of the oven, but not so good after they got cold.[19]

According to Harry Roberson, Homer Roberson's son, the only major change made by Roberson was the replacement of the 16-foot wooden wheel with Fitz steel overshot wheel. The salvaged 18-foot wheel was bought for $90 from P.L. Shelor, Jr., about 1948 and came from the Huffville Roller Miller/Ezra Wimmer Mill on Brush Creek (II, 119). It was not installed until 1956 by which time the ⅓ mile long wooden race or flume was also in need of repair. The steel wheel was an improvement but the extra two feet of diameter sometimes interfered with the tailwater, thereby wasting some of the wheel's power.[20]

Considering the number and variety of exposed gears, shafts, pulleys, and belts in a water-powered mill—or any pre-OSHA industrial plant in which every machine and mechanism was driven by a single power source—accidents were inevitable. In the last fatal accident at a Floyd County mill, Homer Roberson was caught and dragged beneath the wheel on October 30, 1966, while removing a prop from the water wheel which had partially filled due to a leaking flume. Roberson was able to pull himself out of the water but died before he could be gotten to a hospital. During the 1940s Roberson's mill had been the scene of a funeral after a procession of mourners passing by the mill on its way to the cemetery was caught in a downpour. The funeral was conducted inside the mill and the casket kept there until it could be hauled over the muddy roads and to the cemetery on the back of a flat bed truck.[21]

The setting, longevity, and rustic appearance of the Homer Roberson mill, combined with its nearness to a major county highway (Route 8) and to the Blue Ridge Parkway, place it among the county's most frequently photographed mills. A series of photographs taken by Roanoke Photographer W.H. Bratton around 1945 is but one proof of its appeal.

Deeds

Floyd County Deed Book "P," page 524, 5 May 1880. John Merritt, Special Commissioner to John W. Epperly, 55¾ acres of "the land on which Eden Epperly resided." South fork of Little River. "Balance of tract to John T. Agee at request of John W. Epperly." Bounds include "white oak, on north side of the Mill race."

D. B. "R," page 318, 9 February 1884. John T. Agee to J. W. Epperly, rights for race and to spill water from same.

D. B. "U," page 258, 13 December 1889. Sarah E. Epperly to J. W. Epperly, land on which mill race now exists.

D. B. 30, page 82, 30 September 1903. John Epperly to Harrison F. Scott, mill and all rights.

D. B. 37, page 536, 29 August 1910. H. F. Scott to Rufus Abraham "Abe" DeHart. 2 November 1921 (no deed), J. W. Brammer purchased mill, land, and water rights.

D. B. 45, page 37, 3 April 1922. J. W. Brammer to G. C. "Cleve" Smith and wife, mill, land, and water rights.

D. B. 46, page 30, 3 July 1926. G. C. Smith to John S. Smith, mill.

D. B. 49, page 463, 31 August 1931. J. S. Smith to Homer A. Roberson, mill, land and rights.

D. B. 164, page 166, 10 June 1988. Dale Profitt, Gino W. Williams, and Kendall O. Clay, Special Commissioners, to Harry A. and Mary F. Roberson, three tracts including the same conveyed by J.S. Smith to H. A. and Annie Roberson, 31 August 1931.

29. Floyd Milling Company/Peter Dickerson Mill

LOCATION: East side of Route 8 (S Locust Street/Parkway Lane S), Town of Floyd, across from former Skyline Sportswear Building, currently (2010) Winter Sun/Sun Music Hall; POWER: Gas or diesel engine, then electric motors; STONES: Midget Marvel flour mill; VISIBLE REMAINS: None

History. Numerous small mills were built in Floyd County after 1920 to produce cornmeal and cattle feed using gasoline or electric power, but by the time these alternatives to water power were widely available and perceived to be reliable, flour milling was in decline. The Peter Dickerson Mill was one of only three Floyd County mills built expressly to make flour using a source of power other than falling water.*

Millwright Joe Terry helped Peter Dickerson build this mill and may have worked there as well. Dickerson's choice of the Midget Marvel Mill, a self-contained unit that combined cleaning, grinding, and sifting functions in a single machine, allowed for the use of a building not radically different from other industrial or commercial structures. As a result, the building was more easily converted to retail space after the mill closed than would have been the case with a traditional multi-story flour mill with all the attendant spouts, elevators, and shafting.

A restaurant is supposed to have operated in an attached building while the Dickerson mill was running. Posey Dickerson was living in the mill when it was sold to Albert P. Smith and Fred Sweeny in 1940. Afterward, Fred Sweeny ran a furniture store in the mill building for several years.

The two-story wood-frame building was covered with squares of patterned tin made to look like blocks of hewn stone. It stood very near to the street on the left as one drives south on Route 8/Locust Street, across from the brick

**The Burdine Reed flour mill (III, 80) used a gasoline engine. Peter Hatcher intended to make flour at his steam-powered mill (I, 21) but the flour mill was never installed.*

The Peter Dickerson mill was powered by gas or diesel engine, or by electric motors throughout its short working life. The building served as a furniture store, an antique store, and a warehouse for the Southern States Cooperative across the street before it was torn down in 1987. (Courtesy Kathleen Ingoldsby)

building that for many years housed the Skyline Sportswear garment factory. The Dickerson Mill building was home to Perkins' Antiques c. 1970–75, and was last used as a warehouse space for Southern States Feed and Seed. Owners Freeman and Randolph Cockram had the building taken down by James Link, who used some of the material to remodel the dairy barn, called at that time "Old Jacksonville," and known in 2010 as the Jacksonville Center.[22] The back of the photograph taken as the building was being dismantled is dated "1987" by the processor and by hand-written notation.

The Farmers Exchange Feed Mill (I, 139) operated in a building directly behind the Dickerson mill in the 1950s before relocating to Route 221 on the east side of town.

Deeds

Floyd County Deed Book 49, page 24, 1 March 1929. J. M. Thomas to Peter Dickerson, lot on Locust Street.

D. B. 57, page 527, 20 February 1939. Peter Dickerson to Posey Guy M. Dickerson, one-half interest in land and mill on Locust Street.

D. B. 57, page 528, 20 February 1939. Peter and P. G. M. Dickerson to J. E. Proffitt, deed of trust, land and mill.

D. B. 59, page 200, 29 November 1940. P. G. M. and Peter Dickerson to A. P. Smith and Fred Sweeney, land and mill. Mill machinery room and room occupied by P. G. M. Dickerson, deferred until May 1, 1941.

121. Thomas S. Lester Mill

Beaver Dam Creek

LOCATION: .2 mile downstream from Route 221 (Floyd Highway North); POWER: Overshot wheel; STONES: 2 sets and roller mill; VISIBLE REMAINS: Abutment; possibly part of dam (1992)

History. Exposed interior woodwork, flammable lubricants caked or puddled around unsealed bearings, and bone-dry dust in the air and on every flat surface made fire a constant danger in any flour mill. And since no amount of human care could either predict or prevent them, periodic floods posed an even greater threat to every water-powered mill and to the dams that supplied them with water. Thomas Lester (1869–1931) had his share of trouble with both. His mill was heavily damaged by fire in April 1904.[23] His 1908 ledger refers to a washout around that time, and the dam was completely washed out by heavy rains in 1921.[24] He chose not to rebuild when the mill he'd constructed after the 1904 fire burned to the ground on the night of Saturday, October 30, 1926.[25]

With losses estimated between $7,000 and $8,000 and only $3,000 insurance, the second fire was disastrous. Lester was then 57 years old, his first wife had died the year before, and he himself would live only five more years. But rebuilding the mill at least once between 1904 and 1908 made sense for a man still in his thirties, living in a county of over 15,000 citizens, most of whom lived on farms. Rebuilding was especially practical if the mill that burned in 1904 was the one built by William J. Williams on that site in the 1850s.

The four year span between the date of the fire and the 1908 ledger entries raises the question whether the mill actually took four years to rebuild, or was rebuilt twice, soon after the fire and again, perhaps from the ground up, in 1908. In any case, spending $7,124.83 to build a roller mill made sense at that point since a buhr-stone flour mill built in the mid–19th century could not have competed with the roller mills that were then beginning to dominate flour milling in Floyd County.[26] Two brand new roller mills on Dodd Creek, the John Epperly Mill (IV, 126) and Scott's Roller Mill (II, 122), were already operating to the southwest and to the north of the Town of Floyd. On the east side of town, and much closer to Thomas Lester's mill site on Beaver Dam Creek, George W. Shelor would complete his new roller mill (I, 123) in 1913. At about the same time, Spangler Mill (II, 99), down Pine Creek from Shelor's and even closer to Lester, would be reworked inside and out and refitted to make flour with a roller mill system.

The fact that Lester did not replace his overshot water wheel with a turbine in 1908 put him at some disadvantage compared to the mills just mentioned, all of which used (at least at first) turbines. A turbine water wheel is well suited to a roller mill system because of its high shaft speed, but also demands more water than a properly set up overshot wheel. Lester was probably able to produce flour of as good quality, but in less quantity, than his larger competitors. Effie King Brown recalled that her father, who had his corn ground at the William Allen Simpson Mill (II, 113) on Pine Branch, made the trip to Lester's mill for the superior flour, meaning, probably, that it was whiter, lighter, and finer than Simpson's buhr-stone flour.[27]

Men who worked as millers for Thomas Lester included J. M. Richards (1921–24) and Elliot Wood (September 1909 to March 1915). P. L. Shelor rented Lester's Mill for a year before buying his father's mill (I, 123) on Pine Creek in 1919. G. R. Smith and Walter Conner were blacksmiths for T. S. Lester.

See the William J. Williams Mill (I, 20) for earlier history of this mill site.

Deeds

Floyd County Deed Book "O," page 415, 5 May 1877. John Merritt, Commissioner for Wm. J. Williams, to Lorenia S. Lester, wife of John W. Lester, 146 acres land and mill. *Earlier purchasers of this land had apparently defaulted, resulting in a chancery suit and sale of this land to John Lester, who directed the court to record the deed in his wife's name.*

D. B. 29, page 158, 25 October 1902. Lorenia S. Lester to T. S. Lester, 146 acres land and mill. Thomas Lester, party of the second part, promises to "take care of and support the party of the first part [his mother] during her life."

123. Shelor Mill/Nolen Mill/Pine Creek Mill

Pine Creek

LOCATION: East side of Pine Creek, east side of Route 690 (Sandy Flats Road SE) near its intersection with Route 860 (Shooting Creek Road SE); POWER: Originally turbine; now high-breast/pitchback, 20' by 4' Fitz riveted steel wheel; STONES: 1 set, Salem Roller Mill, hammer mill, Meadows mill; VISIBLE REMAINS: Mill is structurally sound and restored to near-operable condition. Dug and rebuilt wooden races usable. The house at the intersection of Routes 690 and 860 was the miller's house. The larger house immediately behind the mill on Route 690 was the miller's house for the Powhatan Williams Mill.

History. The Shelor/Nolen/Pine Creek mill was the last Floyd County mill to grind grain for a paying customer using water power. When the mill ground its last turn of corn within a couple of years after Artice Nolen sold the mill in 1974, it marked the end of a tradition that began shortly after the first white European settlers arrived in the 18th century, and which extends in Western civilization at least back to the first century B.C.[28] Aside from Mabry Mill, which was restored and is now maintained and operated by the Blue Ridge Parkway as a historical exhibit, the old Nolen Mill is the only Floyd County mill still capable, with only minor additional restoration, of operating under water power. Its pitchback wheel and race configuration is the only remaining example of the handful known to have been used in the county. This mill is also unusual but not unique among Floyd County mills in having experienced an apparent regression in hydro-power technology when its original high-speed horizontal turbine was taken out in 1938 and replaced with the existing vertical Fitz wheel.

Like nearly every other water-powered mill built in Floyd County in the 20th century, this mill was erected on a site or "mill seat" already proven effective by the existence of earlier saw or grist mills. The Powhatan Williams mill may still have been standing on the other side of Route 690 when George W. Shelor put up his new roller mill in 1913 using the stones from the Williams Mill and the same c. 3,000-foot in-ground or "dug" race, which is thought to be the longest in the county.* The wooden part of the race, alternatively called a flume or sluice, which had previously carried water from the in-ground race and over the road to the Williams mill was shortened and partly or completely rebuilt to deliver water to the new wooden penstock. This square box was constructed both to house the steel and cast iron turbine and to supply it with a vertical column of water. The momentum of the water was imparted to the turbine's curved steel blades as it dropped from the height of the "head race" down to the tailrace a few inches above normal stream level, thence back to Pine Creek and on downstream.

The original race and vertical wooden penstock are still in place in this pre-1938 view of the Shelor/Nolen/Pine Creek Mill. The one-story addition on the gable end was at one time a woodworking shop. The size and arrangement of its doors and windows changed several times before it was removed in the 1960s. (Frank Webb collection)

Salem Foundry in Salem, VA, supplied the roller mill, probably as a complete installed system including cleaners, line shafts, spouts, elevators, sifter, and cast iron roller mill "stands." Each stand held two independently functioning pairs of corrugated steel rollers, with each pair providing a single "break" or reduction in the wheat milling process. A minimum of two stands, providing four separate breaks, were required. Three or even four stands, providing six or eight reductions, could recover more high-grade flour from each bushel of wheat, but cost more money and required more space; more belts, spouts, and elevators; more auxiliary cleaning and separating machinery; more upkeep; and more horsepower.

Pelham "P. L." Shelor, George W. Shelor's son, bought the mill in 1919 and owned it until

**Frank Webb estimated its length at 3,000 feet; Phillip Gettel believed it to be almost 4,000 feet (three-fourths of a mile) long.*

The unusual "pitchback" configuration places the water on the wheel between the 12 and 3 o'clock positions, causing the wheel to turn in a direction opposite the flow of the water in the race. Objects in the right foreground appear to be segments of another steel water wheel. Though somewhat more weathered, in 2010 the mill looks much as it did in this 1960s photograph, the race having been reconstructed and wheel restored to operating condition by Phillip Gettel. (Courtesy Vernon Harris family)

financial troubles resulted in the sale of the mill at public auction in 1936. According to a history of the mill compiled by Judy Nolen Hylton, her grandfather Amos Cabel "Cab" Nolen entered the winning bid through a double proxy while he himself was at Camp Lee, Virginia, serving a nine-month sentence for operating, or at least having been caught at, a moonshine still. For some reason the deed between the court's commissioner, J. E. Proffitt, and Nolen was not recorded until 1939. Upon his return Nolen put the idled race, turbine, and mill back in order and set them to work under miller Joe Terry, working only part time himself until moving in 1939 into the modern home Pelham Shelor had built in 1928 and furnished with electricity and running water, all powered by the mill's turbine. In the meantime Nolen sometimes walked the eight miles to the mill and stayed for several days before walking back home.[29] Cab Nolen was already the owner of the Spangler Mill (II, 99), a larger and much older flour mill near the mouth of Pine Creek.

In 1938, with help from sons Louis and Artice, Cab Nolen made two trips to a North Carolina mill to dismantle and retrieve the 20' by 4' Fitz riveted steel wheel; it was still in place at the mill in 2010.[30] The turbine required too much water and was prone to silting up. Because used steel water wheels in good condition weren't that easy to come by, and this one was taller than the available head at his mill site, Cab Nolen mounted the salvaged wheel in an unusual "pitchback" configuration. In the familiar overshot setup, the race extends slightly beyond the very top of the wheel. The weight of water delivered to the rotating wheel at this point is acted upon by gravity for the maximum time before spilling from the downward moving buckets. In the pitchback setup the race does not extend across the top of the wheel, but instead approaches the wheel and introduces water to it at a point between the level of the axle shaft and the top of the wheel, causing the wheel to turn in the direction opposite that of the water flowing in the race. The Shelor/Nolen/Pine Creek wheel might also be classified as a "breast" wheel, since the race approaches the wheel at breast level, instead of "overshooting" it, as is the case with most vertical water wheels in mountainous areas where high heads are usually easy to obtain. When water is introduced to a true breast wheel near or even below the middle, a shroud may be fitted closely against the face of the wheel to keep the water from spilling until the last possible moment. And since they were typically employed on streams with low head but high volume, wheels manufactured for use in a true breast configuration were often wider, proportionally, than overshot or pitchback wheels.

The main disadvantage of the pitchback wheel, compared to the much more common overshot wheel, is that the momentum of the flowing water is lost by having the direction of its flow reversed before it is introduced to the rotating wheel. Hence, the weight of the water, alone, acts to turn the wheel. On the other hand, an advantage of a pitchback wheel is that the "head" or vertical fall of the water need not equal the height of the wheel, which can be a real benefit in flatter terrain. A second advantage is that the motion of the wheel is not impeded by the "tail water" when the stream level is above normal, since the wheel is rotating in the same direction as water flowing from underneath it and back to the stream.* Judging from Floyd County mills, these advantages plus the extra leverage and torque offered by the taller wheel were not enough to offset the mechanical drawbacks or the extra cost of building, installing, and maintaining a wheel of diameter greater than absolutely necessary. Only a handful of pitchback or breast wheel installations were used in Floyd County, the most impressive being at the Will Stanley Mill (IV, 54) on Tory Creek, with a 22- to 24-foot diameter wooden wheel. While pitchback wheels were never popular in this region, they did gain favor with leading hydraulic engineers in England before steam power completely displaced water wheels. Several of the largest, most powerful, and most efficient water wheels ever built, most of them in Great Britain, were of the high breast or pitchback configuration.[31]

It's possible that Cab Nolen was actually looking for an oversized wheel to replace his turbine, but the most likely explanation is that it was simply less expensive to purchase and adapt (by mounting it "backwards") this particular used overshot steel wheel than it was to buy a new one of the correct height, or to raise the wooden race to the height necessary to overshoot the taller wheel, then dig a new race to match it. This last was probably the deciding factor, considering the tremendous length of the existing dug race and the complexities of renegotiating water rights that had been in place for more than 60 years.

A less significant change to the mill was the removal by Artice Nolen in the late 1960s of an A-roofed addition between the wheel and the road. It partly covered the water wheel added in 1938 and at one time housed a complete woodworking shop. The room itself or the outline left when it was removed appears in many photographs of the mill. A still-existing room built on to the other end of the mill housed a small stove and was referred to as "the warming room," a name that reflects the fact that mills were not well heated due to the risk of fire. The danger posed by walls, floors, and framing; and spouts, elevators, and machine cabinets all of seasoned wood, was compounded by layers of highly combustible dust on every horizontal surface. The small chimneys emerging from the corners of mill roofs, if they are seen at all, are tiny compared to the large brick and stone chimneys seen in the roof centers and gable ends of contemporary homes, schools, and factories. Some millers minimized the discomfort by closing during the coldest months when customers would be apt to stay at home and water power became less dependable due to freezing.

At some point during his years as miller Cab Nolen was also involved with the less reputable but almost certainly more lucrative trades of making and selling whiskey. As evidenced by the numbers of pre-prohibition millers who also ran legal distilleries, the combination of the two enterprises is a natural one, since a distiller needs a steady supply of ground corn or rye to be made into "mash" and converted to alcohol,

**The original wooden overshot wheel at Mabry Mill (IV, 60) apparently performed poorly because it was too tall to clear the tailwater. But in that case it was deemed easier to build a new, smaller wheel (wooden wheels required periodic replacement anyway) than to convert the existing wheel and drive train to a pitchback setup. An oversized wheel installed in the standard overshot configuration at Roberson's Mill (I, 27) in 1956 barely cleared the tailwater under normal conditions. The wheel's power was reduced by "wallowing" when high waters in Dodd Creek raised the tailwater high enough to interfere with the motion of the wheel.*

By the time this picture was made c. 1995 owner Phillip Gettel had stabilized and weatherproofed the structure, including the "warming room" on the back side of the mill. (Ricky Cox photograph)

and a miller needs a market for his excess grain. As a place where trucks might come and go without notice at most any hour, and as a regular destination for most men in a farming community, a mill could also serve as a drop off point or even a convenient retail outlet for a discreet distributor. According to a family history, a soldered copper tank that happens to fit neatly inside a wooden grain elevator inside the mill was actually one of several identical containers made to be hidden in or somewhere around the bed of a 1925 Dodge truck.[32] A trap door in the floor of the warming room, ordinarily covered by the box of sand in which the stove sat, gave access to another potential hiding place.[33]

Cab Nolen ran the mill until 1962 and died the following year. He had also owned the larger, older Spangler Mill near the mouth of Pine Creek from 1930 to 1961. Artice W. Nolen acquired the Nolen Mill from the rest of his father's heirs soon after and ran it until 1974, but was less active toward the end, as both his health and the milling business declined. An oral history compiled around 1976 by Stefan Klosowski states that the mill was reworked to grind only corn in 1965, and that it was open only on Thursdays in its last years.[34] The Salem roller mills were removed at that time and probably sold for scrap. An unnamed source believed they ended up in a restored Kansas mill, but probably confused the fate of this machinery with that of several items from the Spangler Mill which did end up in a re-created flour mill in Meriden, Kansas. After Ed Molesky bought the mill in 1974 it was run very little, if at all, and had definitely ceased operation by 1976.[35]

Phillip Gettel bought the mill and an adjoining tract in 1980 and still owned it when he died in September 2005. In summer 1995 the mill was equipped with a "corn rock" (one set of stones); a Fairbanks scale; nine Sprout-Waldron elevators, oriented at a right angle to the peak of the roof; various cleaners and separators; and a self-contained Meadows mill, marketed by (and bearing the name of) International Harvester. A devoted fan of old mills, Gettel had by the late 1990s stabilized and restored the building and water wheel, and reconstructed the wooden race that carried water from the in-ground portion of the race to the wheel. His daughter owned the mill in 2008. For the earlier history of this site, see the Powhatan Williams Mill (I, 25).

Deeds and Other Public Records

D. B. 37, page 594, 27 July 1910. L. P. and A. D. Spence to G. W. Shelor, property known as the W. T. Williams mill place and 2 acres on Pine Creek.

D. B. 36, page 595, 29 July 1910. J. J. Sweeney to G. W. Shelor. *This unusual deed conveys and confirms all water rights to Shelor. It was written and recorded to remove all doubt whether full and complete water rights had been conveyed by Sweeney to Josie and J. W. Peters (See D. B. 34, page 507, 7 April 1908) and thence to Shelor through subsequent owners. The full text is reproduced with the history of the Powhatan Williams Mill.*

D. B. 43, page 584, 26 December 1919. George W. Shelor to P. L. Shelor and Ada Shelor, his wife, parcel of land and mill and all water rights known as the Williams grist mill property.

D. B. 48, page 3, 25 February 1928. P. L. Shelor, J. E. Proffitt and K. M. Weeks, trustees, three tracts of land, one of two acres and grist mill with all water rights, second ⅛ acre, third 1.7 acres to secure note due A. C. Nolen.

D. B. 74, page 90, 15 February 1939. J. E. Proffitt, Commissioner in Chancery Cause of *Floyd County Bank vs. P. L. Shelor*, to A. C. Nolen, 3 tracts land, one of 2 acres and mill property, one of ⅛ acre and one of 1.7 acres.

D. B. 90, page 50, 24 October 1963. A. C. Nolen heirs to Artice W. Nolen, 3 tracts on Pine Creek, one of 2 acres and mill property and water rights, one of 1.7 acres and one of .8 acres.

D. B. 111, page 570, 5 July 1974. Artice W. Nolen and Fern T. to Edward J. Molesky, 2 acres plus or minus and mill with all water rights, tools, etc.

D. B. 135, page 176, 9 December 1980. Edward W. Molesky to Philip Gettel and Sylvia C. Beyen, 2 tracts land, one being the mill tract and 2 acres with all water rights, the other tract being .554 acres land.

124. William F. Nowlin (or Nolen) Mill

Shooting Creek

LOCATION: On Route 860 (Shooting Creek Road SE), about .2 mile west of Franklin County line. North side of road, 1000–1,500 feet uphill from the house (barely standing in 2010); POWER: Overshot wheel; VISIBLE REMAINS: Dug race barely visible (1992)

History. There is no mention of this mill in any known deed to the land on which it stood, and little if any physical sign of it remains. But Lester and Ed Nolen recalled that a mill was running at the W. F. Nolen homeplace in the 1930s. The mill was operated by Larken Thomas and later by Isaac Trail and was gone by the early 1940s. Marvin Nolen recalled that Ike Trail lived in the house still standing above the road on the north side just above the Franklin County line, and that Trail was the last to operate the mill, which had been abandoned by the time Nolen was old enough to explore the neighborhood on his own.[36]

Neither the mill nor any other business was mentioned when the property changed hands in 1911 and again in 1925, in spite of the grand intentions with which it all began. The "County Line Items" column on page 2 of *The Floyd Press*, 11 June 1903, informed readers that "Mr. W. F. Nowlin is aiming to do a bustling business. He is aiming to run a grist mill, sawmill, store, furniture shop, blacksmith shop, and canning factory." Which of these ambitions, besides the grist mill, were ever realized is unknown.

Marvin Nolen remembered the mill as having stood on the same side of the road as the Ike Trail house, as much as 1000 or 1500 feet up the hill. He did not recall how the water was conveyed from Shooting Creek to the mill but the state road must have crossed over a dug race, via bridge or ford, or beneath a wooden race.[37] The former seems most likely since the steady and rapid drop of Shooting Creek makes it easy to achieve the required fall, at about any point on the stream, with a fairly short in-ground race following a sidehill contour line. Also, wooden races are more complex to construct and more expensive to maintain than in-ground races. On the other hand, along with being some of the steepest terrain in the county, the Shooting Creek valley at this point is also some of the rockiest, making excavation an equally daunting proposition.

Frank Webb wrote in 1992 that the remains of a dug race were visible, but this could not be absolutely confirmed in 2009. What could be taken for the lower end of a steeply pitched (as opposed to nearly level, as was typical) dug race runs for a short distance on the north side of road just above where the mill is supposed to have stood. This ditch, with a lower bank of piled stones, might also have been intended to channel runoff from the steep hillside, although there is no obvious purpose for doing so. It might even have been "dug" by the runoff turned downhill by what was originally a low stone fence or merely a rock pile spread along the side of the hill. Neither are there unmistakable remains of a dam, but this absence is less puzzling under the circumstances. With so much available fall, placing a large volume of water on the wheel would not have been a concern to the planners of this mill. Only a low dam would have been needed to divert a portion of the stream into the race, as opposed to a higher, stronger dam required to impound a large reserve of water. Hence, strongly anchored abutments or a thick base would have been unnecessary.

The foregoing speculation is merely and only that — speculation. Changes made over the years to both road banks and possibly even to the course of the stream in the process of widening, straightening, and paving Shooting Creek Road have obliterated the evidence required for more than a guess.

Deeds

Floyd County Deed Book 37, pages 273, 274. 7 February 1911, W. F. Nowlin, deceased, and Mollie M. Nolen, wife, appoint W. C. Lester attorney in fact to sell 199 acres on Shooting Creek and Pine Creek.

D. B. 37, page 412, 27 May 1911. W. C. Lester, attorney in fact for W. F. Nowlin heirs, to L.A. Thomas and Effie L. Thomas, his wife, 193 acres known as the W. F. Nolen land (Effie and Larken Thomas appear to have been the daughter and son-in-law of W. F. Nowlin).

D. B. 47, page 135, 18 February 1925. Larken A. Thomas and E. L. Thomas, his wife, to Eula M. Boyd and G. W. Boyd, 158 acres on Shooting Creek.

D. B. 47, page 215, 19 March 1925. Eula Boyd to Isaac Trail, 56 acres on Shooting Creek.

125. John Kitterman Mill

South Fork of Little River (Dodd Creek)

LOCATION: Route 714 (Roberson Mill Road SE), approximately 1 mile northeast of Route 8 (Parkway Lane S); POWER: Overshot wheel STONES: 1 set, also sawmill.

History. The Agnew brothers, Samuel, Jr., Joel, and Booker, and a fourth partner, David Owens, built a sawmill on Dodd Creek on a 9½ acre tract they bought from Samuel Agnew, Sr., in 1852. John Kitterman bought the land and their sawmill in 1859 and bought an adjoining tract of 17½ acres in 1873. His 1875 request for permission to build a mill dam 15 feet high was granted, but the height of the dam was limited to 12 feet, 3 inches, probably to diminish the size of the mill pond and prevent flooding on the land of his upstream neighbors. When both tracts were conveyed to new owners in 1896 and again in 1898, the larger tract is referred to as the mill and foundry tract, but it is unclear whether a mill was actually there. No mill was mentioned when Joseph Proffitt sold most of the combined tracts in 1909. The deed to a smaller Dodd Creek tract sold by Proffitt in 1903 likewise makes no mention of a mill.

Local people interviewed in the late 1980s remembered having seen an old water wheel and millstones at the site of Kitterman's mill.

Deeds and Other Public Records

Floyd County Deed Book "F," page 441, 27 November 1852. Samuel Agnew, Sr., to Samuel Agnew, Jr., Joel W. Agnew, Booker P. Agnew and David Owens. 9½ acres of land on South Fork, Little River.

D. B. "K," page 262, 20 December 1859. Samuel Agnew, Jr., Joel Agnew, David Owens and Booker Agnew to John A. Kitterman, 9½ acres land lying on the south fork of the Little River. Includes a "sawmill built by the parties of the 1st part."

D. B. "M," page 547, 7 April 1873. George Young to John A. Kitterman, 17½ acres and 35 poles South Fork, Little River. Bounds refer to a "bluff near said Kitterman's sawmill pond."

Floyd County Court Order Book, 1870–1877, December term 1875. John A. Kitterman, request to build dam 15 feet high at his mill on South Fork, Little River.

D. B. "N," page 525, 12 January 1875. John A. Kitterman, request to build a dam (apparently for an existing mill) on the South Fork of Little River, 15 feet high. The dam allowed by the court is 12 feet, 3 inches high.

D. B. "Y," page 545, 22 June 1896, Ellen Kitterman and others to Headen Morricle, 2 tracts of land on South Fork of Little River, one of 17½ acres and 35 poles and known as the foundry and mill tract, and a second tract of 9½ acres.

D. B. "Y," page 546, 3 January 1898. Headen Morricle and wife to Joseph Proffitt, 2 tracts of land on South Fork of Little River, one known as foundry and mill tract of 17½ acres and 35 poles, and second tract of 9 ½ acres.

D. B. 30, page 226, 9 December 1903. J. P. Proffitt to J. T. Agnew, 8½ acres on South Fork of Little River.

D. B. 39, page 269, 2 December 1909. J. P. Proffitt to S. G. Proffitt, 21 or 24 or 27 acres of land on South fork of Little River, being part of land sold to J. P. Proffitt by Headen Morricle, D. B. "Y," page 546, 25 June 1896.

137. Joseph M. "Josie" Custer Mill

Meadow Run Creek

LOCATION: Route 642 (Meadow Run Road SE), about .25 mile east of Route 221 (Floyd Highway North); POWER: Overshot wheel; STONES: 1 set and sawmill, planing mill, steam heated lumber kiln; VISIBLE REMAINS: None

History. Joseph M. "Josie" Custer built this small mill between 1907 and 1919. He sold the property to Johnson Poff in 1919 and, according to Basil "Poff" Simpson, Sr., worked for Poff as a miller and blacksmith before moving to Oklahoma, where he lived the remainder of his life. Johnson Poff was operating this mill and another larger mill (II, 114) in the village of Check, one-quarter mile up Meadow Run Creek, when he died in 1933.[38] The Josie Custer mill is sometimes referred to as Johnson Poff's "lower mill." R. L. Poff never operated this mill, but the machinery and buildings were still in place when Mrs. J. J. Poff deeded the property to him in 1946.

Deeds

Floyd County Deed Book 34, page 554. 9 December 1907, J. L. Poff to J. M Custer, 2 acres land on Meadow Run Creek.

D. B. 37, page 355, 4 August 1910. J. J. Poff to J. M. Custer, 3¹⁄₂₀ (three and one-twentieth) acres land on Meadow Run Creek.

D. B. 44, page 117, 17 November 1919. J. M. Custer and wife to J. J. Poff, a parcel of land (and mill) containing about 6⅓ acres and known as the J. M. Custer mill.

D. B. 64, page 337, 22 June 1946. Mrs. J. J. Poff to R. L. Poff, mill, mill house, planing mill and all water rights (old J. M. Custer mill).

139. Farmers Exchange Mill/ Floyd Exchange Mill

LOCATION: Originally on South Locust Street, behind the Peter Dickerson mill building (I, 29) then on Route 221 (Floyd Highway North), opposite intersection with Route 615 (Christiansburg Pike NE), on a now (2010) empty lot between Harris and Baker Furniture and the Dollar General store, 50–75 feet from Harris and Baker; POWER: Electric motors; STONES: No stones or other machines for grinding meal or flour. Hammer mills and other feed grinding and mixing equipment; VISIBLE REMAINS: None

History. This mill was never powered by water and was used exclusively for processing and manufacturing feed for livestock; no products were made for human consumption. It is included in this history of water-powered grist mills because its appearance in the 1950s reflects a fundamental change in the age-old experience of "going to mill." Following World War II there was a steady decline in the number of people bringing a few sacks of threshed wheat or half a bushel of shelled "bread corn"* to be ground, on an exchange basis, into flour and cornmeal. Instead, the men who continued to farm,

**The traditional "bread corn" of most Floyd Countians was one of several varieties of open-pollinated, white "dent" corn and not the hybrid yellow dent corn that was introduced to Floyd County before World War II and which dominates U.S. and world corn production today. Yellow cornmeal is favored in other parts of the country. The mapping and explanation of regional preferences would make an interesting study in itself.*

Powered entirely by electric motors, the Farmers Exchange Mill was the last commercial mill of any kind built in Floyd County. The decline of the feed business and rising commercial real estate values led to the removal of the mill to make room for more lucrative development. (Courtesy Dot Howery)

whether full or part time, were bringing yellow corn, hay, and oats to be pulverized with a hammer mill and mixed with molasses and other supplements into various livestock feeds which they hauled home in hundred pound sacks.

A few had even begun grinding their own feed at home following the arrival of electric power throughout the county in the late 1940s. On September 11, 1947, the *Floyd Press* reported that "a large number of farmers are installing the electric driven mills in the county, because they have proven to be very economical to operate and a time and labor saver."[39] The fact that Mr. L. C. Angle, the only source named in the article, was the agricultural engineer for the Christiansburg office of the Appalachian Electric Power Company, may explain what seems like a bit of an overstatement of the trend. Angle noted, "So many mills have been installed in Floyd County during the past few years that complete records are not available."[40] By way of example, the article offers the names of only ten farmers, but it is nonetheless correct in observing that the needs of Floyd County farmers, along with the resources available to them, were changing rapidly.

The Farmers Exchange Mill, whose focus on livestock feeds anticipated these changes, was first located in a building directly behind the old Peter Dickerson mill on South Locust Street/ Route 8 (Parkway Lane S). In 1956 or 1957 the Farmers Exchange Mill moved to its last and best known location on land also owned by Albert Smith east of town. Smith bought half interest in the property in 1944 and became sole owner by buying the other half from C. L. Gallimore in 1947. The business changed hands several times during the lifetime of this mill, but Smith and his heirs retained ownership of the land and building until 1986, by which time the mill was no longer running.

An April 1950 newspaper advertisement stated the intentions of Wiley Webb, proprietor of the Farmers Exchange Mill when it was still located on South Locust Street. Neither Kenneth Sewell nor Kyle Lee, two subsequent owners, recalled Webb's involvement in this business, so it may have been short-lived.[41] The ad read, "Attention Farmers, we are now opening in Floyd for custom grinding and mixing. Bring hay or anything from the farm. We also have the latest equipment in molasses mixing. We can put molasses on all feed. Open 8 a.m. to 5 p.m. Monday thru Saturday."[42]

In November 1951 Robert A. Moore, who ran a feed mill supply business in Salem, Virginia, formed a partnership to operate the Farmers Exchange Mill with his son F. C. Moore, who left the partnership in 1953. Culver Gardner ran the mill for Moore. The mill was still running on Locust Street as late as 1956, but had moved out of town by the time Kenneth Sewell bought it in October 1957 and changed its name to the Floyd Exchange Mill. Sewell remembered that there was plenty of hard, heavy work, but not much profit, and that he was glad to give it up and become a rural mail carrier in 1965.[43] Sewell was succeeded as owner by Kyle Lee, who bought the business at auction in October 1965 in a short-lived partnership with Richard Moran. Eventually Lee, who went back to the original name of Farmers Exchange Mill, sold half the business to Linwood Belcher, who became sole owner in 1976 or 1977. The last owner, William B. Wickham, bought the business from Belcher, around the time Belcher purchased the O. A. Webb Mill (II, 122) in 1979. The Farmers Exchange Mill building was torn down by Blair Clarke, one of a group of investors who bought the land and building in 1986.

Along with owners Albert Smith, Kenneth Sewell, Kyle Lee, Lynwood Belcher, and Bill Wickham, others said to have worked here include Culver Gardner, William Clower, and Kenneth Belcher.

Deeds and Other Public Records

The deed of trust included below is one of three, two by Kenneth Sewell (1960 and 1964) and one by Kyle Lee and Richard Moran (1965), made to secure loans that were duly repaid. These documents have no bearing on the record of ownership of this business, but the 1960 inventory of the mill's most valuable equipment shows clearly the emphasis on handling large volumes of coarsely ground animal feeds. The other two inventories are much the same. One difference is that the seed cleaning machine does not appear in the 1965 inventory, but "1 candy machine" and "1 cola machine" were on the premises by that time. Though insignificant in themselves, and completely unrelated to each other, the removal of the one machine and the addition of the other two might be seen as emblematic of the rapid changes occurring in the economic life of Floyd County in the 1960s.

Floyd County Deed Book 62, page 298, August 11, 1944. Elvira Dickerson, widow, to A.P. Smith and C.L. Gallimore, 4 lots, lying on the waters of Old Field Creek, just east of the town of Floyd, Lots 10–11–22–23 conveyed to parties of the 1st part from Charles M. Jett and wife, by deed dated February 16, 1932, D.B. "50," p. 53.

Floyd County Deed Book 65, page 549, 6 June 1947. C. L. Gallimore and L. S. Gallimore, his wife, to A. P. Smith, their one-half interest in lots conveyed in D.B. 62, page 298.

Floyd County Partnership and Ownership Book 1, page 13, 27 November 1951. The Farmers Exchange Mill was incorporated in Floyd, Virginia, by Robert A. Moore and F. C. Moore of Salem, Virginia.

Partnership and Ownership Book 1, pages 17 and 18, 30 December 1953. F. C. Moore has no interest in Farmers Exchange Mill. R. A. Moore is sole owner.

Partnership and Ownership Book 1, page 63, 22 October 1957 "Farmer's Exchange Mill name will be Floyd Exchange Mill," Kenneth Sewell, sole owner. (Sewell had bought the business on 11 October 1957, according to the deed of trust referred to below and one other recorded in D. B. 88, page 634, 2 April 1964.)

D. B. 83, page 443, 21 December 1960, Kenneth Sewell to Curtis A. Sumpter and R. B. Jarrett, Deed of Trust. "Items of personal property ... upon the premises generally known as the Farmer's Exchange Mill," one 20 foot Jay Bee Hammermill with 75 HP motor and starter, one 24 foot Fords Hammermill with motor and starter and dust collector, two 3,000 pound duplex batch-mixers with motors and starters, one 3,500 pound floor scale, one 15 ton Howes truck scale, one #2 Puritan molasses blender with pump and motor, one H-100 Hance Seed Cleaning Machine with elevator and motor, one Elton conveyer loader complete with motor and starter, and one 3000 gallon molasses tank. All other machinery in building excepting desk and adding machine.

Floyd County Partnership and Ownership Book 1, page 339, 30 August 1983. William B. Wickham, sole owner, is trading under the name of Farmers Exchange Mill.

D. B. 156, page 381, 24 October 1986, Ethel F. Smith, widow of A. P. Smith, to Emerald Investors, two parcels of land in Floyd County, Virginia, containing 1.951 acres and 2.309 acres.

Section II. Mills East of Route 8 and North of Route 221

90. George Phlegar Mill

Oldfield Creek

LOCATION: Route 817 (Phlegar Hill Road NE) just above where road crosses creek; POWER: Overshot wheel (1870); STONES: 1 set and sash type sawmill (1870)

History. George Phlegar built a mill on Oldfield Creek sometime between his 1802 arrival in what would become Floyd County. and his death in 1839. By 1845 his son Benjamin had bought his siblings' interest in their father's property, including the saw and grist mills. The 1870 Industrial Census lists Benjamin Phlegar (1812–1892) as the owner of a sash type sawmill (driven by a 10 horsepower water wheel) and a gristmill (with an 8 horsepower wheel) at the same Oldfield Creek location. The different power ratings could be a simple error in describing a single wheel or there may have been separate water wheels for grinding and sawing. Some millwrights found it easier to transmit power from one operation to the other with belts or shafts, but dual wheeled sites are not unheard of, especially when the sawmill preceded the grist mill, and when it was of the up-and-down sash saw type, as this one was. The flutter wheel, an inefficient type of wheel with little to recommend it otherwise, was apparently peculiarly suited to providing the up and down motion via a crank and pitman rod.

Alexander Gardner ran the mill for both George Phlegar and Benjamin Phlegar. The mill was in operation at least long enough for a granddaughter born in 1872 to remember the mill and the miller. A house built by George Phlegar as early as 1816 still stands at the edge of the Floyd County Industrial Park on Route 615 (Christiansburg Pike NE). In a description of the house and farm, Mattie Susan Phlegar Brown (b.1872) recalled that "a private road runs directly in front of the house from the Christiansburg highway more than a mile through the farm to the old mill. This mill was a popular center of the community and corn was ground on the big mill stones—also hominy. Mr. Alex Gardner, who lived just across the highway, was the miller. He drove a rattling one-horse wagon and was so regular in his morning hour we called it the morning freight. The mill pond was full of fish and turtles and we often feasted on both."[1]

Deeds

D. B. "B," page 467, 26 April 1841. David Phlegar and Nancy, his wife, to Benjamin Phlegar, "one undivided sixth part of a tract of about 374 acres," home of George Phlegar, deceased

D. B. "D," page 58, 18 November 1844. Joseph and Nancy Phlegar to Benjamin Phlegar, their interest in George Phlegar property.

D. B. "D," page 137, 5 May 1845. Lidia Phlegar to Benjamin Phlegar, her interest in George Phlegar property.

D. B. "D," page 161, 4 October 1845. Elizabeth and Rhoda Phlegar to Benjamin Phlegar, their interest in George Phlegar property.

D. B. "V," page 164, 25 May 1891. Benjamin and Sarah Phlegar to son Willie S. Phlegar (1869–1953), all property.

See Table B, *1870 Products of Industry.* Name of corporation or individual: Phlegar, Benj.

Map of Mills in Section II

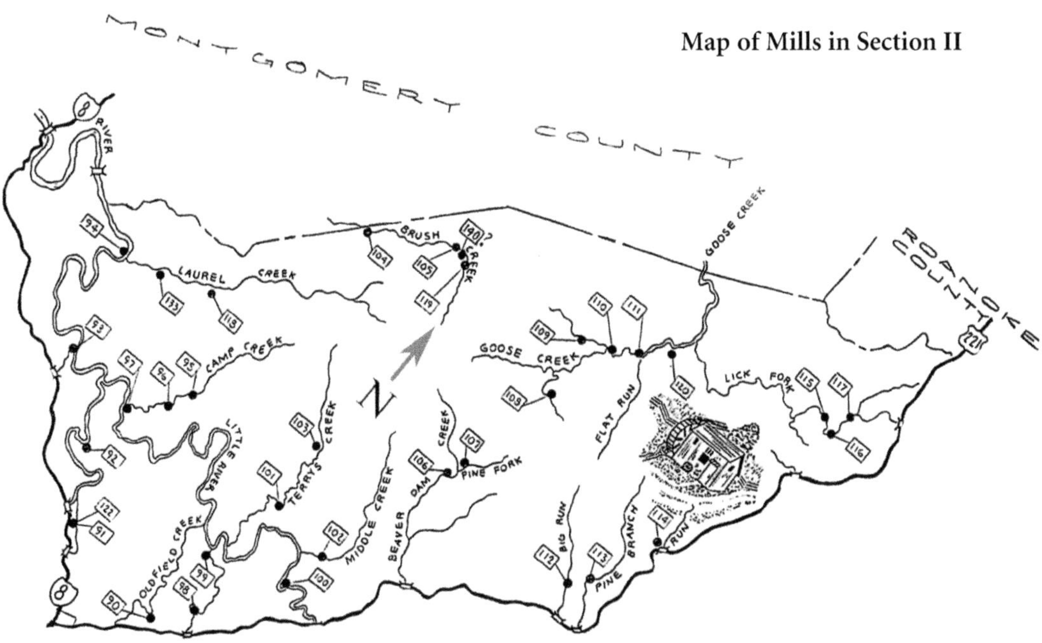

91. Matthew Scott Mill

Dodd Creek

LOCATION: Route 8 (Webb's Mill Road N), 2 miles north of Town of Floyd; POWER: Turbine (1870, 1880); STONES: 2 sets (1870, 1880); VISIBLE REMAINS: The O.A. Webb or Dodd Creek Mill (II, 122) now stands on or near the site of this mill.

History. The long history of one of Floyd County's best known mill sites began in December 1835, when William Morricle was granted permission by the Circuit Court of Floyd County to build a dam and a sawmill. A gristmill was added before Morricle sold 20 acres of land and his two mills on South Fork Little River (Dodd Creek) to Matthew Scott in 1863.

The *Products of Industry* report compiled as part of the United States Census of 1870 reported that Matthew Scott's mill included a sash type ("up and down") sawmill driven by a 16 horsepower water wheel of unspecified type and two run of stones powered by a 10 horsepower turbine. On 8 August 1874, Scott was ordered to pay Harry Dickerson $19.58 for damage done to Dickerson's land by flooding when Scott's new 12-foot, 11-inch high mill dam was closed and the water backed up behind it. The greater fall and increased storage provided by the taller dam may account for the increase in horsepower noted in the 1880 *Products of Industry* report as well as the impressive reported production of wheat flour, buckwheat flour, and cornmeal.

In the June 6, 1896, settlement of Scott's estate, ownership of the mill was transferred to his son, Winfield "Field" Scott, who built a new mill on the same site in 1903, after the original mill burned to the ground around 1902.

Winfield Scott was a partner in several mills, including the Buffalo Mountain Roller Mill (IV, 42), and other business ventures, but is best remembered for having disappeared, never to return, along with the funds of the Farmers' Bank of Floyd in 1908. The story of Winfield Scott's financial dealings and sudden departure is told, in somewhat sensational fashion, in "In Search of Winfield Scott," an article written by Joan Schroeder for the February-March 1993 issue of *Roanoker Magazine*.

For further history of this mill site, see the O. A. Webb Mill (II, 122).

Deeds

Floyd County Deed Book "L," page 98, 17 April 1863. William Morricle to M. Scott, 20 acres land and mills on South Fork Little River.

D. B. "N," page 310, 8 August 1874. Judgment, *Harry Dickerson vs. Matthew Scott.* "The jury finds for H. Dickerson $19.58 [?] damages for overflow of water from new dam built by M. Scott which is 12 feet 11 inches high."

D. B. "Z," page 236, 6 June 1896. Samuel Scott, Executor for M. Scott, Deceased, to Winfield Scott, a part of the Morricle tract and mill containing 20 acres.

See Table B, *1870 Products of Industry.* Name of corporation or individual: Scott, Matthew. (Listed under grist mills and under sawmills).

See Table C, *1880 Products of Industry.* Name of corporation or individual: Scott, Matthew.

92. Bird's Mill/Amos Graham Mill/Whitlock Mill

West Fork of Little River

LOCATION: Route 810 (Gallimore Lane NE) about 1.1 miles east of Route 8 (Webb's Mill Road N). State maintenance ends at bridge. Mill was upstream from the bridge on east bank; POWER: Tub wheel (1870), overshot wheel, then turbine; STONES: 1 set (1870), later 2 sets. Oil mill, wool carding mill, distillery

History. Although it stood beside what is now a dead-end road, this mill was part of an ambitious business in its heyday. In 1860 Daniel and B.V. Bird were operating a grist mill, a wool carding mill, an oil mill (producing linseed oil by crushing, heating, and pressing flax seed) and a distillery. An 1883 deed mentions a sawmill as well, making this one of only a few locations where almost all the manufacturing processes associated with 19th century water power were carried on. Early references to the mill property as "Hill's Old Forge" raise the possibility that this water power might once have driven an iron furnace bellows. In modern usage "forge" usually refers to the device used by blacksmiths to heat and prepare work pieces for welding and shaping. At one time it referred to the manufacture of wrought iron by "forging" reheated iron "pigs" with heavy trip-hammers, often actuated by water power.

The Whitlock Mill pond was the scene of suicide by drowning in 1888. The body of W. J. "Worth" Graham, who had left home without explanation on November 27, was discovered on December 16 "about 200 yards above the dam." Searchers were slowed in their efforts to drag the pond by having first to remove 2 or 3 inches of ice from the surface.[2] J. A. Whitlock is believed to have been the last to run this mill. Jack Lawson had two mill stones from this mill in 1992.

Deeds

Montgomery County Deed Book "G," page 203, 30 September 1818. Robert Correll to Ebinezer Watkins, mill and ten acres land on West Fork Little River at mouth of a branch above H. (Hezikiah) Sumner house. Mill on east side of river.

Floyd County Deed Book "B," page 74, 15 July 1837. Owen Sumner to John Bird, 3 acres on West Fork Little River.

D.B. "C," page 226, 4 March 1843. Eli Phlegar, Commissioner in Cause of *John Bird vs. Ebinezer Watkins*, to John Bird, land and mills known as Hills Old Forge. *No record was found of Robert Correll's purchase of the land, or of an owner named Hill.*

D. B. "D," page 106, 15 May 1845. Owen Sumner to Benjamin V. Bird, 5¾ acres, West Fork, Little River.

D. B. "E," page 405, 5 January 1850. A. J. Kirby, in Chancery Cause of Abijah Booth, to Daniel H. Bird, 40 acres land and mill known as Bird's Mill, delivered to B. V. Bird, July 1854.

D. B. "G," page 362, 25 February 1854. Benjamin V. Bird, son of John, to Robert W. Whitlow, 10 acres and Daniel H. Bird's mills, known as Bird Mills, West Fork Creek.

D. B. "J," page 174, 22 November 1856. Robert W. Whitlow to William C. Black, David Bair, and Abram Bair, Roanoke County, ten acres and mill near Owen Sumner land, known as Bird's Mill.

D. B. "M," page 325, 19 June 1869. William C. Black to David Bair, his ⅓ interest in Bird's Mills.

D. B. "N," page 393, 5 March 1873. David Bair to Abraham Bair, his entire interest in Bird's Mills and 10 acres land.

D. B. "N," page 416, 10 February 1875. Abraham Bair to Amos Graham, Bird's Mill and 10 acres, West Fork Little River.

D. B. "R," page 141, 31 July 1883. Amos Graham and wife to Asa T. Whitlock, ½ interest in grist and sawmill, dwelling house and other buildings known as the Bird or Bear (spelled Bair earlier) mill place conveyed to Amos Graham by Abraham Bear.

D. B. "R," page 172, 22 August 1883. Amos Graham and wife to Dennis and Joseph A. Whitlock, his ½ interest in Birds Mill and sawmill, South Fork Little River, formerly purchased of Abraham Bear.

D. B. 30, page 417, 10 April 1904. C. P. Graham, son of Amos to Joseph A. Whitlock, ½ interest in mills and land known as the Amos Graham Mills.

D. B. 44, page 54, 31 January 1920. J. A. Whitlock to W. G. Cockram, adjoining lands of Roscoe Graham, South Fork, Little River. Mentions Graham Mill Place. Sold in 4 tracts, the fourth including house, outbuildings, and mill.

See Table A, *1850 Products of Industry.* Name of corporation or individual: D. H. Bird.

See Table B, *1870 Products of Industry.* Name of corporation or individual: Bair, David.

93. A. A. Booth Mill/Frazier Thomas Mill

Beaver Creek

LOCATION: Route 705 (Slusher Store Road NE) .7 mile east of Route 8 (Webbs Mill Road N); POWER: Overshot wheel; STONES: 2 sets and roller mill

History. The Tugg Thomas mill was owned by only two families, the Booth(e)s and the Thomases, for all but a few of its 110 years. It was built sometime after 1833 by Daniel Booth(e) in partnership with his son William C. (sometimes referred to in deeds as William Lee) Boothe (1814–1856), to whom Daniel Boothe deeded his one-half interest in the one-acre mill tract in 1836. A tract of 44 acres, including the mill, was allotted to William's son Armstead A. when William Booth's land was divided between his widow and children in 1877. When A. A. Booth sold his grandfather's mill to S. R. Brame in 1905 it had been in the family for 70 years. It changed hands two more times before settling with Jude (or Juda) Tuggle "Tugg" Thomas in 1914. His son Frazier Thomas went to work in the mill then, bought it in 1918, and ran it until 1944. Items listed for sale along with the mill at public auction show that even a full-time miller of corn and wheat was still very much connected to the rural way of life in 1944. Along with his "water-powered roller mill and good 7-room house" and other items, Thomas offered for sale "two 8-gallon milk cans, 2 extra good Guernsey cows, 3 and 8 years old, and a lot of White Leghorn hens."[3] The mill dam is said to have been washed away in 1928 by high water in Beaver Creek. While most mills were threatened by the waters of only one watershed and stream, the Booth/Thomas mill was doubly susceptible to flooding due to its standing only a short distance from the meeting point of Beaver Creek and Little River.

Nothing in the deeds indicates conclusively that the mill run by Frazer Thomas was not the original Booth mill, but a drawing made by the surveyor for the 1877 partition is inconsistent with the location of the Thomas mill as seen on modern topographical maps. If a new mill was built nearer the mouth of Beaver Creek, it probably happened after Tugg Thomas bought the property in 1914, around the time a number of old flour mills were modernized by the installation of roller mills, and even more were built from the ground up, usually on existing mill sites. The mill is described as a "three-story buhr mill" in a 1910 real estate advertisement, so was significantly reworked at some later time if not completely replaced by a new roller mill. The wording in Deed Book 39, page 567, which grants Thomas the right "to go to work *on* [instead of *in*] mill at any time," could be a subtle clue to his intentions.[4]

Frazier Thomas moved to Alexandria, Vir-

The Thomas Mill's location at the mouth of Beaver Creek made it susceptible to flooding from both Little River and Beaver Creek, with no obvious advantages to compensate for the added risk. This photograph is probably from the late 1950s or 1960s. (Frank Webb collection)

ginia, after selling the mill to Ottis H. Pugh, who ran it for a short time before tearing it down. Alvie P. Allen bought part of the mill machinery. The water wheel was sold to a man named

The original of this photograph is dated November 1959 by the photofinisher, but when the photograph was actually taken was not recorded. (Frank Webb collection)

Thompson in the Paynes Creek area, possibly Moyer Thompson, who had recently bought the Thompson-Akers mill (I, 14). In 1992 Harriet Allen had one of the stones from this mill.

Deeds

Floyd County Deed Book "A," page 200, 26 April 1833. Daniel Booth, and Rachel, his wife, to William See (C.?) Booth, 100 acres plus or minus on Beaver Creek and Little River. "Said Daniel Booth reserves one acre for mill seat with privileges of a pond and timber to build with." Gives Wm. See Booth (son) "½ interest in one acre mill seat by his bearing ½ of expenses to build with."

D. B. "A," page 406, 1 May 1836. Daniel Booth to William See Booth, "one acre including the mill seat that I the said Daniel Booth deeded to my son William See Booth."

D. B. "O," page 499–500, May 1877. W. C. Booth, deceased, to Armstead A. Booth, lot number 2, 44 acres on Beaver Creek. Division of W. C. Booth's Land.

D. B. 31, page 561, 13 September 1905. A. A. Booth and Mattie Booth, his wife, to S. R. Brame, 44 acre tract "known also as the mill tract." Brame to receive "all buildings on said land and the Mill that is on said land."

D. B. 36, page 478, 30 March 1910. S. R. Brame and his wife, Daisy, to C. W. Harman, 32 acres. This land has upon it what is known as the A.A. Boothe Mill and all water rights."

D. B. 36, page 491, 10 May 1910. C. W. Harman and Alice Harman to G. W. Smith, 31 acres "and the Mill property situated there on known as the A.A. Booth Mill."

D. B. 39, page 567, 10 April 1914. Geo. W. Smith and Vila, his wife, to Jude Tuggle Thomas, "31 acres more or less and the mill property situated thereon known as the A. A. Boothe Mill." Smith to "have possession of dwelling house and Garden the remainder of this present year" and the right to use the mill until Thomas was ready to "go to work on the mill," which he had the right to do at any time. A telephone and interest in the party line were also conveyed to Thomas.

D. B. 43, page 221, 5 January 1918. J. T. Thomas (1860–1935) and Jennie A. Thomas, wife to F. G. Thomas, 19 acres, tract of land and mill on the waters of Beaver Creek.

D. B. 62, page 183, 31 May 1944. F. G. Thomas and Ada E. Thomas, wife, to O. H. Pugh, 19 acres, with all water rights.

94. George Hornbarger Mill/Caleb Sowers Mill and Floyd Electric Power Corporation

Little River

LOCATION: Route 663 (Sowers Mill Dam Road NE), 1.7 miles East of Route 617 (White Oak Grove Road NE); POWER: Turbine, at least after a roller mill was installed. Floyd Electric Power Company used two turbines with an oil engine (probably semi-diesel) for backup or supplementary power; STONES: 2 sets (at least), sawmill; after 1900, 3 stand roller mill. VISIBLE REMAINS: Concrete dam, largely intact. Concrete foundations of Floyd Electric Power Corporation hydroelectric plant.

History. While the mill itself is long forgotten, the Sowers Mill site is a Floyd County landmark. According to Marguerite Tise, the mill was built by George Hornbarger, who petitioned the Floyd County Court in March of 1833 for permission to build a dam and mill on Little River.[5] For some reason, the writ of *ad quod damnum* (to determine damages) subsequently issued by the court was not ruled upon until three years later, when the jurors declared, not surprisingly, that no damage would be done to the land or property of Hornbarger's neighbors and that "ordinary navigation" would not be obstructed. The decision required that a "slope" or fish ladder, 30 feet long and 10 feet wide, be built "across the slope in the main part of the channel,"[6] which seems entirely reasonable considering the number of upstream tributaries of this, the county's longest and most important stream.

Hornbarger ran the mill until 1863, after which it and the surrounding property went through a series of complex transactions involving, for the most part, Hornbarger's daughter and son-in-law Sarah and Charles Lester and their heirs. An entry in the 1870 *Products of Industry* report describing a wool carding mill operated by Lester McAlexander may have to do with this mill, since McAlexander, one of the Lester family heirs, became the mill's owner in 1872. On the other hand, there is no other reference to a wool carding mill connected with this mill.

Caleb Sowers and granddaughter, Nancy Sowers, around 1923. (Courtesy Nancy Sowers Mutter)

The various liens and loose ends were sorted out in an 1879 deed that summarized nine earlier deeds in the process of conveying undisputed ownership of the land from Harvey Trail and numerous relatives to Caleb Sowers, James A. Simmons and Daniel R. Simmons.[7] Each was the sole owner of a separate part of the property.

The new owner of the mill was Confederate veteran Caleb Sowers, who had served with the Floyd Riflemen, Company A, 24th Virginia Infantry, and was at Gettysburg as part of Kemper's Brigade during Pickett's charge. In 1888 Sowers acquired as an equal partner D. W. Williams by giving him an interest in the mill in exchange for $1,000 worth of new machinery. Sowers resumed sole ownership in 1894 after resolving a chancery suit filed by Williams, and at some point added a general store. Sowers also ran a post office for at least part of his time at the mill.

In 1900 Caleb Sowers and Sons hired Salem Machine Works of Salem, Virginia, to install a roller mill system for $1,445. In a mechanics lien filed on October 27, 1900, the suppliers spelled out the terms for payment of the remaining $925 and listed the machines installed under the terms of the contract (punctuation added): "1 Rotary Sifter; 3 Dawson Roller Mills; 1 Reel Flour Dresser; 1 Bran Duster; 1 Wheel; [2 words unreadable]; all necessary belting, boots, and heads; [word unreadable]; pulley cups; bolts; shafting; chains; and all other necessary furnishings for putting in a first class mill for making flour by what is known as the Roller Process."[8] The phrase "1 Wheel" may refer to a new turbine water wheel installed at this time, a practice followed at virtually every other new or retrofitted roller mill in the county. Or Sowers or a predecessor may already have installed one, since no other site in the county was better suited for a turbine, due to the low head and large volume of water.

Caleb Sowers sold the mill to a son, Thurman D. Sowers, in 1907. Another son, Cabell/Cabble V. Sowers, is said to have worked at the mill as well. Thurman Sowers hired A. Niel (or Neal) Keith to run the mill in March 1911, shortly after a "wonderful concrete dam across Little River" was completed in late 1910 or 1911.[9] In December 1911 the mill, land, and business were sold to H. L. Lawson. Less than two months later, Lawson sold it to the Floyd Electric Power Corporation, which was formed in 1912 with the intention of producing power for the Town of

The Floyd Electric Power Company power house under construction, about 1922. Water was channeled from the refurbished Sowers Mill dam to two turbines. (Courtesy Floyd County Historical Society)

Floyd, Virginia, and the City of Radford, Virginia.[10] H. L. Lawson was the first treasurer of the Floyd Electric Power Corporation and was the vice president when the company was dissolved and its assets bought out by Appalachian Electric Power Company in June 1937 and the hydroelectric plant taken off line.[11]

Nearly ten years passed after the formation of the corporation in 1912 before much visible work was done. In the meantime, the Sowers roller mill was leased to A. Niel Keith, who ran it until it burned about 1913. Whether this was the same building erected by George Hornsbarger in the 1830s, or a new one put up by Sowers in 1880 or 1900, when major upgrades were made to the machinery, is unknown. Construction of the power plant and the line to Floyd began in earnest as soon as a franchise was granted by the Town of Floyd on September 6, 1921.[12] On September 21 the *Floyd Press* published an advertisement for "poles, peeled and placed along the line ... delivered by October 25." The dam was also refurbished.[13] Limited service began in September 1922, and on November 22 "the Floyd Electric Power Corp. began to give a continuous service day and night with the exception of one hour per day from 12 to 1 o'clock and on Sundays from 8 a.m., to 4 p.m."[14]

The service in Floyd was welcome but not utterly reliable, even though the two turbines were backed up by what was referred to in one instance as an "oil engine" and as a diesel in others. When the lights flickered, Floyd residents were apt to comment that a fish had gone over the dam. The *Floyd Press* apologized to its readers in July 1926 for delays caused by the fact that "on account of work on the dam of the electric light plant, we have not had current this week."[15]

The concrete dam still known as "Sowers' Mill Dam" is notable for its size and construction. One of only three mill dams made entirely of concrete, it spans the county's largest stream at a point farther down than any other Floyd County mill, and was therefore a logical site for the introduction of commercial hydroelectric power to Floyd County. The dam was breached with dynamite in the summer of 1947 by the newly formed Christiansburg rescue squad in order to recover the body of a drowning victim and to prevent future swimmers from being pulled under by the suction of the old turbine intake. Squad member Frank Shelton photographed the explosion from a hilltop on the Montgomery County (north) side of the river. His 1946 Plymouth, one of the first new cars released to the public after World War II, was damaged by debris from the blast.

Deeds

Floyd County Deed Book "L," page 165, 166, 8 January 1864. George Hornbarger to Sarah Lester and heirs, 75 acres land and mill.

D. B. "M," page 454, 11 November 1872. Charles S. and Sarah Lester and heirs to Lester McAlexander, mill and land in two tracts and all rights.

D. B. "N," page 112, 27 August 1873. Lester McAlexander to Harvey W. Trail, Hornbarger Mill and 20 acres of land.

D. B. "N," page 389, 17 December 1874. Harvey W. Trail and wife and Charles S. Lester and wife to William Trail, Hornbarger Mill and 20 acres of land.

D. B. "P," page 566, 17 June 1879. Harvey W. Trail, et al., to Caleb Sowers, James A. Simmons and Daniel R. Simmons property "known as Hornbarger Mill property." *This complex deed summarizes 9 earlier deeds in the process of conveying uncontested and exclusive ownership of three separate tracts to Sowers, Simmons, and Simmons. The mill property was transferred to Caleb Sowers.*

D. B. "T," page 412, 28 June 1888. Caleb Sowers to D. W. Williams for $1000.00 in machinery furnished for said Caleb Sowers, ½ interest in mill and 3 acres.

D. B. "X," page 143, 19 May 1894. W. L. Howard and W. H. Morgan in the cause of *D. W. Williams vs. Caleb Sowers,* to Caleb Sowers said D. W. Williams' ½ interest in mill and lot. *Williams had petitioned the Circuit Court of Floyd County to force a sale of the mill property, the proceeds to be divided equally between himself and Sowers. Sowers apparently avoided a sale by purchasing Williams' ½ interest. The complete text of Williams' petition is below.*

D. B. 35, page 235, 1 November 1907. Caleb Sowers to Thurman D. Sowers [son], mill and land. T. D. Sowers to have possession of home 1 February 1908.

D. B. 38, page 23, 11 December 1911. Thurman Sowers and wife and Caleb Sowers to H. L. Lawson, 35 acres and milling business situated on the premises, along with good will of same beginning on January 1, 1912.

D. B. 38, page 108, 29 January 1912, H. L. Lawson to Floyd Electric Company, 35 acres, mill and water rights.

See Table A, *1850 Products of Industry.* Name of corporation or individual: George Hornsbarger.

See Table B, *1870 Products of Industry.* Name of corporation or individual: Lester McAlexander (Wool Carding Mill). Though uncertain, this seems the most likely site of McAlexander's wool carding operation.

The following is the complete text of D.W. Williams' petition to force sale of the Caleb Sowers mill property. The exact date of the petition was not recorded, but must be between the formation of the partnership between Sowers and Williams in 1888 and the settlement of the suit in May 1894. The document is interesting for its language, and because it represents the kind of problems that could emerge from partnerships. Some were no doubt necessitated by the huge expense of building, upgrading or purchasing a large, automated flour mill.

TO THE HONORABLE HENRY E. BLAIR, JUDGE OF THE CIRCUIT COURT OF FLOYD COUNTY:

Humbly complaining showeth unto your honor your orator, D. W. Williams, that on and prior to the 28th day of June, 1888, Caleb Sowers was seized and possessed of a certain tract of land situate and lying in the County of Floyd on Little River, containing ____ acres, upon which, was and is still, situate valuable mill property, consisting of wheat, corn and saw mills, and also a storehouse, said mills being known as the old Hornbarger Mill property, and on the 28th day of June, 1888, said Caleb Sowers and wife conveyed by deed dated the day and date aforesaid to your orator, in consideration of One Thousand ($1,000.00) dollars in machinery furnished for said mills by your orator, one undivided half interest in said mill property and three (3) acres of said tract of land on which said mills and storehouse are situated and adjacent thereto, which said deed is recorded in the Clerk's Office of the County Court of Floyd County, in Deed Book T. page 412, a copy of which is herewith filed marked (A) as part of this bill; and that since that time the said Caleb Sowers and your orator have owned said land and mill property as joint tenants, each owning one undivided moiety thereof.

Your orator further represents unto your honor that he now desires to have the said mill property and lot of land partitioned between the said Caleb Sowers and your orator, and that the same not being capable of partition in kind, your orator desires that it may be sold as a whole and the proceeds of sale divided between the said Caleb Sowers and your orator, you orator believing that such sale and division of proceeds of sale will be to the advantage and interest of both, the said Caleb Sowers and your orator.

Your orator further represents unto your Honor that some time in the latter part of the year 1884, and prior to your orator becoming the owner of one moiety of said lot of land and mill property by the purchase and conveyance aforesaid, your orator, by and with the consent of the said Caleb Sowers, built upon said land a storehouse at the cost of Two Hundred and Nineteen ($219.00) Dollars, under a contract between the said Caleb Sowers and your orator, in which it was agreed and stipulated that your orator was to build said storehouse on said lot at the cost and expense of your orator, and that your orator was to have the use, occupation and benefit of the same at a rental of Twenty-Five ($25.00) Dollars per annum, the rent to go to reimburse your orator the expenditure he should make in building said storehouse, and that your orator should hold, use and occupy said storehouse for sufficient length of time for the rental of Twenty-five ($25.00) Dollars per annum to exhaust or consume the price and cost of building the same.

Your orator further represents unto your Honor that he has had the use, control and benefit of said storehouse since it was built under said contract up to the present time, except that for about two years of the time your orator did not use or occupy it, the said Caleb Sowers occupied and used it as a post office, and that for these two years your orator received no rent or compensation therefor, and now although your orator according to the stipulations of said contract would be entitled to hold, use and occupy said storehouse for a longer time than the present, nevertheless, in order and to the end that there may be a speedy sale of said mill property, and that the same may be unencumbered with any claim or title of your orator to the use and occupation of said storehouse, your orator is willing to relinquish all further claim or right to use and occupy said storehouse under said contract; for any longer time than a sale may be made, and will, when such sale is consummated, surrender and release any and all claims he may have under said contract.

And now in as much as your orator is without remedy save and alone by the aid of a court of equity, when matters of this kind are alone and properly cognizable, your orator prays that the said Caleb Sowers be made a party defendant to this bill and required to answer the same, but an answer under oath is waived; that said mill property, lot of land and storehouse may be sold, and the proceeds of sale, after the payment of all proper costs, charges and expenses, be equally distributed between the said Caleb Sowers and your orator, giving to each one-half, that your Honor will fix and allow reasonable attorney fees, that proper process may issue; and that such other, further and general relief may be afforded your orator as the nature of his case may require, or equity and good conscience may demand, and your orator will ever pray, etc.

95. James G. Lester Mill/Wolford Scott Mill

Camp Creek

LOCATION: Uncertain. Possibly on the south side of Route 689 (Huckleberry Ridge Road NE) near a fork in Camp Creek, about .25 mile northeast of Route 615 (Christiansburg Pike NE). POWER: Overshot wheel, 10 H.P. (1880); STONES: 1 set and sawmill

History. James G. Lester and his wife had already been part owners of a mill on Easter Creek (III, 81) when they bought 5 acres, more or less, from Isam Shelor in 1871. The inclusion of water rights in the deed makes clear James G. Lester's intention of building a mill. This site included both saw and grist mills when in 1875 it was sold along with another tract of 10-plus acres for $2,200, a substantial amount given that the United States was in an economic depression brought on by the Panic of 1873. Marshall A. Davidson made up the purchase price by trading $200 worth of furniture, a tract of land in Franklin County, land and a mill in Patrick County, and a $1,000 interest in the Central Hotel at Floyd Court House. (This might explain why Lester, who is listed as a millwright in the 1870 census, gave his occupation as "hotel keeper" ten years later.) The balance was made up with a note for $350.

Four years later, Davidson sold the mill property and thirteen acres for only $1,500. The buyer was Wolford Scott (1827–1908), a grandson of Matthew Scott, who settled in southern Montgomery County (now Floyd County), Virginia, about 1770. Wolford Scott lived near Laurel Fork in Carroll County before moving to the Camp Creek section of Floyd County in the 1870s. He moved to Patrick County in 1886.

At what date the mill was abandoned is unknown, but a reference to "the old mill race" in a 1909 deed suggests that by this time the mill was gone or of so little value that it need not be mentioned as other than a convenient landmark.

Deeds

Floyd County Deed Book "M," page 347, 28 September 1870. George Winter, Commissioner, *C. Bower vs. J. V. Graham*, to Isham B. Shelor.

D. B. "M," page 320, 3 July 1871. Isham B. Shelor to James G. Lester, 5-plus acres land and water rights on forks of Camp Creek.

D. B. "N," page 616, 4 December 1875. James G. Lester, wife Mary A. Lester, and son D. D. Lester to Marshall A. Davidson, 5 acres and 58 poles, saw and grist mill and another tract of 10¼ acres. For the sum of $2,200.00, including $350.00 in a tract of land in Franklin County, $300.00 in land and mill property in Patrick County, $300.00 in cash and $1,000 interest in the Central Hotel property at Floyd Court House.

D. B. "P," page 214, 1 April 1879, M. A. Davidson to Wolford Scott, 13 acres and mill property. $1,500.00.

D. B. "T," page 37, 12 July 1887. Wolford Scott to Lucy A. Howell, 13 acres, mill and water rights.

D. B. "W," page 118, 9 December 1892. Lucy A. Howell to James H. Booth, 13 acres land, mill and water rights.

D. B. 27, page 393, 12 December 1900. James H. Booth to Samuel W. Moore, 5 acres and mill, including water rights from both creeks.

D. B. 35, page 280, 7 January 1909. S. W. Moore and Eliza E. Moore, his wife, to G. R. Hungate, 20.5 acres on Camp Creek. Boundaries mention "old mill race."

See Table C, *1880 Products of Industry.* Name of corporation or individual: Wohlford Scott

96. Oliver Carter Bower Mill

Camp Creek

LOCATION: Route 615 (Christiansburg Pike NE) about .5 mile downstream from Camp Creek Church; POWER: Overshot wheel; STONES: 2 sets

History. Neither of the deeds (below) from Deed Books 38 and 41 conveys ownership of the Oliver Carter Bower mill or dam, but their contents show how a mill owner protected essential "water rights" when he did not own all the land affected by his dam or mill race. When a race passed through adjoining property, or a mill pond backed water onto a neighbor's land, it was wise to include the terms of the agreement in the deed to the land in case of a conflict when the land changed hands or a friendship went sour. Both deeds stipulate that "the water right along the mill pond is hereby retained [by the mill owner] to the present height of the dam." The rights of the two buyers, Hungate and Sumpter, are likewise protected, with each deed specifying further that "should the present or future owner of the mill desire to raise the dam, consent must be obtained from the owner or owners of the property herein conveyed." The potential for disagreement over the interpretation of arrangements of this kind, and the potential liability posed by the presence of running or standing water crossing or adjoining another's land, help explain why mill tracts were usually larger than the fraction of an acre required for the mill building and mill yard. To completely preclude water-related disputes with neighbors, a mill tract needed to include both banks of the stream, from a point below the dam, upstream to the farthest extent of the pond when completely filled, and the entire lengths of any dug or wooden races.

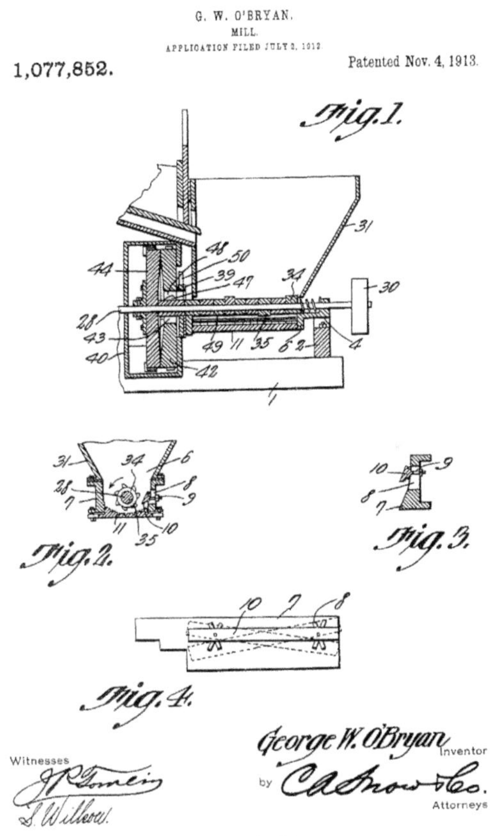

Patent drawing 1,077, 852, George O'Bryan's mill. Patent examiners evidently agreed that an adjustable blade (part no. 10 in figures 2, 3, and 4) made O'Bryan's machine sufficiently different from similar devices to merit a patent. Whether or not the mill was, as claimed, distinctly better than competitors has been forgotten.

For a mill owner who didn't own all the affected acreage and thus had to depend on agreeable neighbors, placing the dam and pond well upstream and connecting them to the mill with a lengthy contour-line race was the riskiest arrangement, from the legal perspective. Ironically, this configuration was the safest in terms of protecting the mill from floods, since it allowed for placement of the mill farther from the stream bank.

A page 3 item in the *Floyd Press* of 28 January 1915 stating that O. C. Bower has installed a George W. O'Bryan mill raises a curtain on the story of a Floyd County native who involved himself in the milling business not as an installer, maintainer, or operator of milling technology, but as an inventor. O'Bryan enjoyed as much success as most inventors, in that at least one of his machines was actually manufactured and sold. O'Bryan applied for a patent on his design for a "Mill" on 2 July 1912, and on 4 November 1913 was granted United States Patent No. 1,077,852. The improvement or unique feature that merited the awarding of a patent seems to have been an adjustable steel blade, the position of which relative to a set of rotating teeth affected the texture of the product and the speed with which it moved through the mill. But the claim that would have made O'Bryan's machine most attractive to millers and prosperous farmers was that this one machine could take the place of two by grinding shelled corn into cornmeal and also crushing and grinding ear corn and other farm products for animal feed. Cornmeal and animal feeds were among the most common products of every mill, and they were virtually the only products of smaller and older mills that did not produce flour.

On August 8, 1912, a month after O'Bryan's patent application was submitted, the *Floyd Press* reprinted an undated but presumably recent *Roanoke Times* notice that

> Mr. George W. O'Bryan, of Cambria, Va., the inventor of the first automatic crusher and grinding mill in the world, is now in the city. Mr. O'Bryan will be here two or three days conferring with Mr. H. B. Rockhill of the Rockhill Machine Works, relative to the material for the building of fifty of these mills.
>
> The automatic crusher and grinding mill is one of the greatest inventions of its kind of this age. The mill will crush corn on the cob into meal for chop with one operation. The machine will also make first class table meal, graham flour and buckwheat flour."[16]

The article closed with the notation that "Mr. O'Bryan is a Floyd man, being the son of Mr. and Mrs. A. T. O'Bryan, near town."[17]

How many Floyd Countians besides O.C. Bower put their faith, and money, into one of O'Bryan's multi-function mills is unknown. But the number of single-function mills and grinders still occasionally seen at country auctions and antique engine rallies implies that most farmers and millers chose to spend their money on the two separate machines, having concluded that one machine supposed to produce food for farm families as well as farm animals would not completely satisfy either group. To the knowledge of the authors, no single machine that could do both jobs satisfactorily has yet been invented.

In a 1994 interview with Frank Webb, Hugh Bowman recalled that the O. C. Bower mill house washed away in a flood and was never rebuilt.

Deeds

Floyd County Deed Book 30, page 9, 7 September 1903. M. F. Scott to O. C. Bower, 68 acres on Camp Creek.

D. B. 38, page 544, 14 November 1912. O. C. Bower to C. D. Hungate, 3 acres, including dwelling house, store house, barn and other outbuildings. "The water rights along the mill pond are hereby retained to the present height of dam, and should the present or future owners of the mill desire to raise the dam, consent must be obtained from the owner or the owners of the property conveyed." Did not convey mill property.

D. B. 41, page 143, 20 January 1915. C. D. Hungate to T. E. Sumpter, 3 acres more or less, dwelling house, store house, barn and other buildings. "Also water rights by mill pond are hereby retained to present height of dam." Did not convey mill property.

D. B. 45, page 517, 7 May 1924. O. C. Bower, deceased, to J. J. Sumner, 165 acres land on Camp Creek.

97. Abajiah Booth Mill/Joseph Howard Mill

Camp (Chestnut Camp) Creek

LOCATION: Mouth of Camp Creek and Little River; VISIBLE REMAINS: None

History. As with many of the first mill sites in Floyd County, the developers of this one applied water power to other important tasks in addition to grain milling. An 1826 deed refers to "mills" instead of a single mill; Joel Sumner is said to have been carding wool here in 1846–47. The Booth family, who also ran mills on Beaver Creek, and the Howards owned this mill for most of its working life and afterward. Whether Ira or Darius Howard actually operated it after 1866 is unknown. The fact that no Howard mill on Camp Creek was listed in the industrial census for 1870 or 1880 suggests not, but is not conclusive since only mills whose products were valued at over $500 were supposed to have been included. While there must once have been a public road to this mill, none now comes within half a mile.

Deeds and Other Public Records

Montgomery Court Order (?) Book. 3 July 1798, Abajiah Booth granted leave to build a mill at the mouth of Chestnut Camp Creek.

Montgomery County Deed Book "I-J," page 493, 23 December 1826. Abajiah Booth to Riley Booth, 110 acres land, mills and orchards on Chestnut Camp Creek.

Floyd County Deed Book "C," page 174, 6 October 1847. Riley Booth to Owen Sumner, 110 acres on Camp Creek and Little River.

D. B. "F," page 185, 7 September 1850. Owen Sumner to Joseph Howard, 50 acres plus or minus on Little River and the mouth of Chestnut Camp Creek.

D. B. "L," page 305, March 1866. Colonel Joseph Howard to Ira Howard, 90 acres (lot no. 4), division of Joseph Howard property.

D. B. "L," page 281, 22 March 1866. Ira Howard to Darius Howard, 90 acres (lot no. 4), at mouth of Camp Creek known as the Joseph Howard mill place.

D. B. "L," page 717, 26 July 1869. Darius Howard in bankruptcy sale to Ira Howard, 90 acres and mill place.

D. B. "Q," page 250, 21 November 1881. Ira Howard to William and James A. Martin, 90 acre tract on Camp Creek.

98. Jacob C. Epperly Mill

Pine Creek

LOCATION: About .75 mile downstream (north) from Route 221 (Floyd Highway North); POWER: Tub wheel 10 HP (1870); VISIBLE REMAINS: Part of dug race

History. Although this mill is not among the very earliest in Floyd County, its history illustrates especially well two of the difficulties of determining where and when the first mills were established. The first stems from the fact that, in spite of a law requiring them to do so, early millers didn't necessarily ask the county court's permission before erecting a mill dam. Considering that these pioneers were less accustomed than we are to following government rules and regulations, they may have felt about the need to obtain the court's permission much the same way the average person, today, feels about applying for a building permit. Despite its purpose of protecting the public's health and the owner's investment, a building permit can seem like a needless intrusion into an individual's private business and a source of frustrating delay, not to mention yet another expense in an already costly business. Whether or not fees had to be paid to the court or to the men whose time was taken up with the inquest, the applicant for the court's permission had at least to hire an attorney to represent him in court. He had also to adhere to the court-appointed panel's recommendation on the height of the dam, which might be lower than requested in the original motion. At least some landowners apparently believed that it was better to ask forgiveness, should they ever be questioned, than to ask permission.

This must have been the case when Jacob Epperly asked the county court in May 1850 for permission to dam Pine Creek, since he had already built the dam, as indicated by the stipulation that he was to "keep up the dam to the present height." As to how long the dam had been in place, we may safely assume that it was erected after Epperly was in possession of the land on both banks of Pine Creek from some point at or below the proposed dam to the farthest upper reach of the filled mill pond. Otherwise, before building he would have had no choice but to file a motion asking the county court for a writ of *ad quod damnum*. The county sheriff would have then impaneled a twelve-man jury to determine, among other things, a dollar amount to be paid by Epperly to his neighbors as compensation for one or both of these financial losses: (1) an acre on which to build the abutment for one end of his dam; (2) damage to land or buildings that would be overflowed by the millpond. The law that created this process, called "An Act for Encouragement of Building Water Mills," was passed in October 1705.[18] An amended version, passed in 1748, created a panel of twelve jurors in place of the two commissioners specified in the original legislation.[19]

A second difficulty in finding a beginning date arises when, as in this case, a mill owner bought, at different times, multiple tracts of land on the stream in question. Epperly bought three tracts on Pine Creek, the first of unspecified acreage, the others of 10 and 29 acres. As long as it included land on both sides of the stream, any of the three could have supplied sufficient land for his dam and pond. The first deed refers only to existing property lines, and supplies neither a physical description nor a length for each of the four sides of the tract. Both the other two tracts lie along Pine Creek, but neither crosses the stream, nor do they have in common any boundary lines, at least none that are easily recognized as such. But either or both could adjoin the first tract to provide the required land on both sides of Pine Creek. Hence, one is compelled to say that the mill could have operated as early as 1832, yet it may not have existed until nearly twenty years later.

Unless the destruction of a mill is reported in a newspaper or deed, the date at which it ceased to operate is even harder to determine, since a mill tract may be referred to in deeds many years after the mill stopped running. At the other extreme, specific references to a mill building aren't always included in pertinent deeds, even when the mill is still in operation. We do know that Jacob Epperly's mill ran at least long enough to be included in the *Products of Industry* report of the 1870 United States census.

Deeds and Other Public Records

Floyd County Deed Book "A," page 54, 17 April 1832. Jacob Graham to Jacob C. Epperly, a boundary of land on Pine Creek (acreage unspecified).

D. B. "A," page 244, 16 June 1833. Mathias Bower to Jacob Epperly, 29 acres land on Pine Creek.

D. B. "B," page 339, 18 March 1840. Issac Phlegar to Jacob Epperly, 10 acres and 41 poles on the waters of Pine Creek.

Floyd County, Virginia, Court Order Book, 1850–1855, May Court, 1850, page 28. "On a motion by Jacob C. Epperly the court doth establish said mill with authority to the said J. C. Epperly to keep up the dam to the present height and take the toll allowed by law." (Dam and grist mill already erected.)

D. B. "M," page 136, 1 January 1871. Jacob Epperly to John A. Epperly, 82 acres on Pine Creek. Meets and bounds of this deed refer to the Jacob C. Epperly Mill. *A deed with John A. Epperly as grantor of this property was not found.*

See Table B, *1870 Products of Industry.* Name of corporation or individual: Epperly, J. C.

---

99. Spangler Mill

Pine Creek

LOCATION: Route 682 (Spangler Mill Road NE), about 2 miles west of Route 221 (Floyd Highway North) and .6 mile southeast from Route 686 (Moore Road NE); POWER: Overshot wheel (1870), 1 turbine (1880), and later 2 turbines. Four cylinder Wisconsin gasoline engine used for power after water power fell into disrepair; STONES: Two sets (1870, 1880), 1 Brush Mountain and 1 French buhr. Two-stand Salem roller mill added about 1914. International Milling Company feed mill. VISIBLE REMAINS: Several pieces of the lengthwise wooden timbers that formed the base of the frame dam remain in the stream bed; the bottom ends of planking protrude from the stream bed to normal water level on the east side of the stream. Many large stones are scattered in the water around the dam site, but no clear outline of either abutment or of the mill foundation is visible. Remnants of the in-ground race from the upstream dam that served the mill before 1914 are visible

The finery of the women among this unidentified group suggests a date before 1910. The picture was definitely made before the new dam was built in line with the upper side of the mill and the old dug and wooden race seen here was abandoned in the early 1910s. The identity of the mounted soldier and the occasion for his saber-drawn, front-and-center pose are lost. (Courtesy Dot Howery)

between Spangler Mill Road and Pine Creek immediately above the mill site and perhaps farther upstream as well. The most visible marker of the site is a concrete pillar or pier, with two steel bolts protruding, on the stream bank about 20 yards downstream from the mill site. It was probably unrelated to the operation of the grain mill and may have been the foundation for one end of a foot bridge across Pine Creek.

History. The original Spangler mill on Pine Creek was built no later than November 1787 on a site that was continuously occupied by a mill longer than any other in the county. While it was probably not the first mill in what is now Floyd County, it was one of the two earliest mills whose exact locations are known today. Joshua Howell was granted permission by the Montgomery County court to build a mill on Dodd Creek in April 1784, but how soon afterward his mill was completed, or whether it was actually begun or in use before permission was granted, is unknown. Neither is it known for how long the original Spangler mill on Pine Creek had been running when it was sold in 1787. While these are certainly among the earliest mills, it's unlikely that either is actually the first mill, the identity of which will never be certain, regardless of what records come to light. Men of an independent turn of mind were erecting mills on their own land without bothering to ask for permission to build dams throughout the 1800s, and would certainly have done so in the pioneer days of the late 1700s.

The documented history of this mill site begins with a deed between Josiah Terry and William Logan, dated 23 November 1787, recording the conveyance of "forty eight acres together with the Mill and the Appurtenances there unto" along with an additional 200 acre tract. Terry owed Logan 400 pounds and was conveying the mill and land to satisfy that obligation. Logan signed the property over to Daniel Spangler on 11 August 1792.[20]

While this mill site was continuously occupied from its very beginning and in use for most of two centuries, the building that collapsed in early 2005 was probably not the original mill. In "Memories of My First 25 Years," Jack Spangler writes: "It is believed that Samuel [Spangler], Sr.

rebuilt the mill and probably 'modernized' it sometime after he inherited the property in 1840. The single bit of evidence found to support this is a letter written by one of his granddaughters to her children which states in part: 'Your great grandfather, Samuel Spangler was a highly respected citizen in the county in which he lived. He was a miller by occupation. He built the first large mill in the county. It was three storied [*sic*] high.'"[21]

The belief expressed in 1944 by the writer of the letter, Lillia Adaline Spangler Bowling (1864–after 1944), that Samuel Spangler (1795–1875) built an entirely new mill is probably more accurate than the idea that he merely rebuilt and modernized the existing mill. This assumption is based on the fact that tall, multi-story flour mills like Spangler's became the standard only after American inventor Oliver Evans developed the automated flour mill in Wilmington, Delaware, around 1785, employing two key innovations: the vertical elevator, made up of an endless belt with attached cups, for moving loose grain and mill products to upper floors; and several variations of what we now call conveyors and augers, for moving materials horizontally or over slight inclines.[22] A two and one-half story mill would have been practical only after the widespread adoption of the Evans system and its key feature, the grain elevator.

At the heart of each of Evans' grain elevators was an endless loop or belt of sturdy 4- to 6-inch-wide fabric with wooden or sheet metal cups fastened to it a regular intervals. The belt ran on a pair of flat pulleys of about 20 inches diameter, one at the very lowest point of the mill, typically in the basement, and the other directly above the bottom pulley at the highest possible point, usually near the peak of the roof on the very top floor. Both the ascending and descending halves of the elevator were enclosed in individual square wooden cases which were fitted at their top and bottom ends into wooden shrouds, called "boots" at the bottom and "heads" at the top, which enclosed the pulleys and the belts as they passed over the pulleys and reversed direction at the top and the bottom of the loop.

In most installations, each pulley, top and bottom, was mounted on a line shaft running parallel to the ridge line of the roof. With it on the same shaft were as many more identical pulleys as there were additional elevators in the mill. The pulleys on the top floor and basement line shafts were spaced along the shaft identically and as compactly as could be done without causing the elevator belt housings to interfere with machinery on lower floors and without requiring the removal or weakening of key pieces of the mill building's wooden timber framework. The latter was a concern only when adding elevators to an existing mill, as in the process of retrofitting a roller mill system.

As the filled cups reached the top of the elevator and rolled over the upper pulley they were inverted and the contents dumped into a wooden chute or "spout" attached to the very top of each elevator. Placing the elevators near the center of the mill made it easier to direct materials to various bins and machines in any corner of any lower floor through the main spouts or through a network of additional connecting spouts fitted with diverter paddles at strategic points. The inverted and now empty cups descended in the other half of the elevator case and were refilled as they rounded the bottom pulley. The wooden housing or "boot" that enclosed the bottom pulley acted as a gravity-fed collection point for the material that had just completed the preceding step in whatever milling process was underway. Newly received grain could also be fed through a trap door in the floor of the main level directly into elevator boots to be raised to a cleaning machine or storage bin on an upper floor.

Evans' simple (from our perspective) belt and cup elevators and his wooden pegged and rope-wrapped augers eliminated the backbreaking task of carrying sacks of grain and flour, and materials at all stages between, from place to place on the same level or up a set of stairs or a ladder so that the materials could be fed by gravity into the next step in the milling process. By eliminating the need to collect and carry tons of loose material, Evans' innovations enabled millers to carry on the processes of cleaning, grinding, and sifting continuously and simultaneously, without having to handle the product between the time it was received in the mill and the time it was sacked and handed over to a customer. In addition to requiring additional bins in which products could accumulate before moving to the next step, Evans' process included or led to the addition of devices for cleaning, cooling and sifting, all of which took up some of the space on the upper floors, which could now be used to advantage with little extra labor.

It is highly unlikely that Evans' design could have reached southwestern Virginia in time to

Spangler Mill viewed from across Pine Creek, around 1914. The original dug race, no longer needed after a new dam was built at the mill, is clearly visible below the public road. Road and race appear to merge just above the mill. The sawmill pictured below the mill in earlier views has disappeared. (Courtesy David Spangler)

influence the design of a mill built no later than 1787, but it is very likely that an 1840s millwright would have adopted Evans' ideas. By that time, his book *The Young Mill-Wright and Miller's Guide*, which was first published in 1795, had been through several editions, and Evans himself had spent many fruitless years trying to collect royalties from owners who had happily adopted his patented system but were not willing to pay for that privilege.

Strangely, the Spangler Mill, which was probably the largest in the county at that time, is not among the seven grist mills described in the 1850 *Products of Industry* report. Yet it is the only Floyd County mill included in the report submitted for 1860.[23] According to the first column heading, only mills that produced more than $500 worth of products in the year ending on June 1, 1860, were to be included, but conclusions to be drawn here probably have as much to say about flaws in the reporting system as about the state of mills and milling in mid–19th century Floyd County.

Like a handful of other county flour mills built before 1900, the Spangler mill was remodeled (most were torn down and replaced) between 1900 and 1920 and fitted with a high speed roller mill system, in this case supplied by the Salem Foundry in Salem, VA, which manufactured the flour making machinery still in place in Vaughn's Mill in Indian Valley (III, 64). The completion of what is believed to be an upgraded rather than a completely new mill was announced in the *Floyd Press* on July 6, 1911: "W. H. Spangler has erected a roller mill at his old stand, and is doing good work." The word "erected" suggests a new building, but this is unlikely for two reasons: The first is that all known photographs of the mill show the same building that stood until 2005. While none of the photos seen by the authors is positively dated pre–1911, some certainly appear to be at least that early. The second and most persuasive reason is that the mill was equipped with a pair of expensive French buhr millstones for grinding flour using the Evans process. This would have been a waste of space and money in a mill built to employ the roller process.

Inside the mill, conversion to the roller mill system required additional spouts radiating from several basement-to-attic elevators clustered near the center of the building and line shafts throughout the mill. A minimum of two "stands" (each containing two pairs of corrugated steel rolls) of cast iron roller mills were installed on the main floor. Various cleaners, bran dusters, and scalpers might also have been installed to maximize the amount and quality of flour. But the key device of the roller mill system, after the rollers themselves, was the swinging sifter which was mounted at the highest possible point in the mill so that the bran and various grades of flour it separated could fall through a maze of square wooden spouts angling down to the next step in the process, whether that would be another pass through the roller mills, a storage bin, a reel type separator, or flour packer. A turbine water wheel, which accompanied many roller mill conversions, had already been installed in place of the overshot wheel between 1870 and 1880. The second turbine may have been added when the roller mill was installed.

Outside, the water power was upgraded by building a new dam directly behind the mill. The new dam took the place of the smaller upstream dam and rendered obsolete the dug and wooden races that linked the mill and the original pond. This change increased the risk of flood damage, but provided greater water storage capacity and made it much easier to deliver the maximum volume of water to the two high speed turbines mounted at the bottom of an enclosed wooden penstock.

The date of the new dam's construction is uncertain. When Walter Spangler bought the mill from his brother Glen in 1914, the deed confirmed his right to build a dam at the west corner of the mill. But the writer may have meant only to reserve the right to rebuild or perhaps raise the dam already in place, since a photograph (see Historical Overview) of a gathering at the mill, which appears to be dated August 5, 1911, clearly shows the dam and pond already located by the mill.

It may have been in the spirit of celebrating the renovation of the mill and the water power that Walter H. Spangler joined Florence Graham in posting an invitation on August 3, 1911:

> Notice of Picnic
>
> Saturday, Aug. 5th 1911, there will be a fish fry and picnic in the Graham Grove, near Spangler's Mill. Everybody is invited to come and bring your baskets well filled with good things to eat, and have an enjoyable day with us. Spangler's pond will be drawn off early in the day, and the pond and river also will be seined for fish. Everybody welcome.
>
> Walter H. Spangler,
> Miss Florence Graham,
> Committee

A lengthy write up in the same newspaper one week later reported that between 300 and 400 people were filled, and plenty was left. Photographer D. L. Cole and Rev. J. D. Utt made pictures of the crowd surrounding the table, just before eating, and then one after they had eaten. "On the whole," the writer concluded, "we think this was the nicest outing we have had for a long time and was enjoyed very much. It was one of the most orderly and good natured crowds that has met for some time. Everybody was in a good humor and went in for a good time and they had it. We extend thanks to the Misses Graham and the Spangler boys for the good time they gave us and hope it may be their pleasure to give us another good time next year."[24]

The photograph referred to above, if in fact dated August 5, 1911, evidently was not taken by either of the named photographers, since it shows men and boys wading the half-drained pond, apparently seining fish in preparation for the picnic.

After 130 years in the hands of Daniel Spangler (1747–1823) and his direct descendants, the mill property was sold at auction in 1924 to Hestel Sowers, who had bought a half interest in the mill from Walter Spangler (1874–1955) in 1917. Though ownership had passed out of the family, "Walt" Spangler ran the mill at least into the 1930s. A newspaper notice of the 1924 sale referred to "that splendid merchant and custom roller mill, locally known as the Spangler mill property" and declared that it "is one of the best mills in this section of the country and has long been in the enjoyment of a liberal and profitable patronage."[25] Hestel Sowers immediately sold all his interest in Spangler Mill to Thurman Hall with all rights.

Miller and millwright John W. Epperly is said to have helped rebuild this mill in 1931, but the nature or extent of repairs or improvements is unknown. The mill had been purchased the year before by A. C. "Cab" Nolen, who was also a long-time owner of the Shelor/Nolen/Pine Creek mill, which still stands farther up Pine Creek. Betty and Joseph Hoffman, who purchased the mill in 1972, were the last to operate any of the machinery, using an improvised system of belts and a 4-cylinder gasoline engine to grind corn meal or perhaps whole wheat or buckwheat flour on one set of stones.

The mill was almost fully equipped when Sammie Phillips bought it in 1990, but the building was deemed beyond repair due to water damage to the floor and parts of the timber framing. Also, the dam was gone and neither acreage for a mill pond nor water rights were sold with the mill. In August 1992 Bob Shelor, Kathleen Ingoldsby, and the authors of this book videotaped a tour of the mill's three and one-half story interior, with identification of equipment and functions by Sammie Phillips. By July 2000, most of the milling equipment, excluding line shafting, pulleys, and some spouts and grain elevators, had been removed and stored or sold. A Salem Foundry swinging sifter, a flour dresser, a bran duster, and a grain scouring machine were shipped to Meriden, Kansas, for installation in a replicated flour mill. A report on the Kansas mill in *Old Mill News* names the Spangler mill as the source of several pieces of equipment. Both authors of this book took part in disassembling and removing two sets of mill stones, two stands of Salem roller mills, and various other pieces of equipment. Sammie Phillips still has the stones and roller mill stands in storage in 2010.

The Spangler mill collapsed in a wind storm in February 2005. By 2008 the remains had been

Spangler Mill in 1992, as it looked for much of the second half of the 20th century. By the time this picture was taken, the frame and foundation on the corner nearest the creek on the upstream side of the building had been severely damaged by a leaking roof and dampness beneath the mill. (Ricky Cox photograph)

piled and burned and the mill basement filled in, so that nothing remains of what was one of the two or three very oldest mill buildings, and little is left to identify one of the very oldest mill seats in the county.

Deeds and Other Public Records

23 November 1787. Josiah Terry to William Logan (see text, above), 48 acres and mill, and second tract of 200 acres.

11 August 1792. William Logan to Daniel Spangler, rights and title to the mill (complete text of deed unavailable).

Floyd County Deed Book "B," page 363, 5 April 1840. Daniel Spangler to Samuel Spangler and William Spangler (his sons) in a division of land bought from Daniel Spangler; Samuel Spangler to get mill and William Spangler to get rest of land.

D. B. "N," page 289, 28 July 1874. Samuel Spangler and Pheba, his wife, to Samuel Spangler, Jr., a parcel of land known as Samuel Spangler Mill tract. Samuel Spangler, Sr., is to have ⅓ of proceeds from mill. (Samuel Spangler Sr. died the following year.)

D. B. 30, page 260, 27 March 1902. Adaline Spangler, widow of Samuel Spangler, Jr., deceased, to W. G. Spangler (his son), mill property of Samuel Spangler, Jr.

D. B. 40, page 317, 11 August 1914. W. Glen Spangler to Walter H. Spangler (his brother), mill and 16½ acres land and rights to build mill dam not more than ten feet high. Dam to be built at west corner of mill.

D. B. 43, page 94, 11 September 1917. W. H. Spangler to Hestel A. Sowers, ½ interest in Spangler Mill.

D. B. 47, page 16, 28 October 1924. J. E. Proffitt and B. S. Pedigo, Commissioners in Chancery for W. H. Spangler, to H. A. Sowers, his ½ interest in Spangler Mill.

D. B. 47, page 18, 30 October 1924. Deed of Trust, Thurman L. Hall to J. E. Proffitt, Trustee, to secure Spangler Mill property.

D. B. 49, page 236, 26 July 1930. J. E. Proffitt and Kyle M. Weeks, Trustees for Thurman L. Hall, to A. C. Nolen, lot or parcel know as Spangler Mill. The phrase "property with the water rights connected therewith and appurtenances there unto" refers back to DB 47, p. 18. (Thurman Hall defaulted.)

D. B. 85, page 78, 20 December 1961. Heirs of A. C. Nolen, and Sallie E. Nolan, his wife, to Howard C. Haines, Spangler Roller Mill property, by estimate, 1 acres or less (refers back to DB 49, p. 236 and "more definite description from Walter H. Spangler, to H.A. Sower, DB 43, p. 94).

D. B. 105, page 343, 1 August 1972. Howard C. Haines and Minnie E. Haines to Betty H. Hoffman, mill and property.

D. B. 173, page 59, 15 April 1990. Betty Hoffman and Joseph George Hoffman to D. W. Spangler.

D. B. 174, page 882, 4 September 1990. David Wilburn Spangler to Sammie Phillips, mill property.
See Table B, *1870 Products of Industry.* Name of corporation or individual: Spangler, Saml.
See Table C, *1880 Products of Industry.* Name of corporation or individual: Spangler, Sam.

100. W. H. Harman Mill/Tice's Mill

Little River

LOCATION: Route 683 (Roger Road NE), 1.1 miles northwest of Route 221 (Floyd Highway North), east side of river; POWER: Turbine; STONES: Roller mill, corn rock, and French buhr; VISIBLE REMAINS: Remains of wooden timbers that made up the base of the wooden dam lie lengthwise in the stream and are presumably still spiked to heavier crosswise timbers, or "mudsills," buried beneath. The dry-laid stone abutment or mill house foundation is visible on the east bank of the river.

History. William H. "Bauley" or "Bawley" Harman (born c. 1832) bought the Blackburn Akers mill on Middle Creek from Lewis Bower in 1865, and is named as the owner of a mill in each of the next two industrial censuses, conducted along with the population tallies of 1870 and 1880. Despite discrepancies in the two records, both are probably describing the old Akers mill. In any case, W. H. Harman had been a mill owner for over thirty years when in 1903 he completed a new roller mill and a new road to give his customers easy access to it from the south side of Little River.

Harman's new mill was as thoroughly and glowingly publicized as any ever built in Floyd County, with notices of his progress and eventual completion of the mill appearing regularly in the *Floyd Press*. The author of the "News From Fred" column noted in the March 26, 1903, edition that "Mr. W. H. Harman is progressing nicely with his Roller mill, which will probably be completed some time next summer."

In a page 3 advertisement that appeared on October 3, Harman himself employed more ink and less restraint in describing his soon-to-open mill:

> NEW ROLLER MILL. Notice to the People of Floyd and Adjoining Counties
>
> I take this method of notifying you that my new Roller Mill will be ready next Thursday, Oct. 1st, to grind for the public. It is given up by experts that I have the finest mill in the state of Virginia, and all I ask of my brother farmers is to give me a trial to convince you that it will be to your interest to patronize my mill. My buckwheat plant cannot be surpassed anywhere and if you want the finest rye flour in the world I am prepared to make it. I also have a separator to clean your corn so I can make you superior meal. My famous old bench burr [probably French burr/buhr, a type of millstone], known all over for making fine flour, will have a place in my new mill.
>
> I have employed J. P. Ekiss and Jacob H. Moses to attend my mill. Mr. Ekiss is one of the finest millwrights in the whole country and also a very fine grinder. It is not necessary to comment on Mr. Moses, as everybody knows that he is a fine miller. Thanking the people for past patronage, I hope to have increased support in the future.
>
> Persons coming to my mill from South side of Little River will please come by my bridge at John Brammers. I am making a fine road on South side of river and will complete it soon and I will then notify the public. WM. H. HARMAN, Fred, Va., September 28, 1903.

Harman's installation, in a brand new roller mill, of a set of antiquated stones for grinding wheat flour was unusual if not unique, since stones were slower and less efficient than roller mills, and could not produce the fine white flour that patrons would soon be demanding. Only the fact that in 1903 many patrons had not yet seen roller mill flour can explain his having gone to the trouble and expense to install these stones in his new mill.

Readers of the October 29 edition discovered that Harman had once again taken pen in hand to apprise them of the lengths to which he had gone on their behalf:

> Notice to the Public. This is to notify the people that I have completed the new road on the South side of Little River, and have opened up a splendid ford across the river below my Roller mill. You leave the tern [*sic*] pike at Mr. John Brammers it is only

about ¾ of a mile by the new road to my mill. It has an excellent grade and there are no hills to pull. I had to build the road with my own funds but it will be a great benefit to the public generally. As the road cost me considerably, I hope the people will reward me for the enterprise by bringing me their grain to grind. Respectfully W. H Harman, Fred, Va. Oct. 27, 1903.

An unsigned notice on page 3 of the February 25, 1904, *Floyd Press* invited customers to "GO TO HARMANS PALACE Roller Mill if you want fine work done. His flour, buckwheat flour, and bolted cornmeal cannot be surpassed anywhere. Give him a trial and be convinced." Harman's "Notice to the People, Customers and Others of my Palace Roller Mill" published in the same paper on May 20, 1909, advised that rye flour, which could be had at $7.00 a barrel, "makes the best bread that you can use for indigestion. I know this for a fact. Dr. Harman advised me to use rye mush and milk for dyspepsia which I have been doing for same and have been greatly benefited by using it.... I want all the dyspeptics in this section to have some of this flour."

Harman also ran a legal distillery and a general store, at which he was "licensed to sell by Retail and not to be drank at the place of sale ... Spiritous Liquors."[26] In the population census of 1910, the W. H. Harman household included a 24-year-old "salesman [in a] dry goods store, for wages."

Whether or not it was, as expected, overrun with customers when it opened, Harman's mill did become a part of the social as well as the economic life of his community. On July 10, 1913, the *Floyd Press* reported, "Quite a number from town went to Harman's mill on Little River yesterday to enjoy a fish and picnic. The pond was drawn off and a good lot of fish caught."[27]

Only sixteen months later Harman's heirs sold his land and mill to Fraser Epperly, who had worked as a miller for Harman, and William Blackwell. In 1922 Blackwell sold his half interest to Will Tice, who was a blacksmith and wagon maker as well as a miller. Within a month Epperly had sold his half of the business and property to his son-in-law John Henry Vest, thus forming a "Tice and Vest" partnership that lasted until Tice became sole owner in 1930. In 1948 the mill and 66.5 acres were sold to William Burke in the process of settling Tice's estate. The machinery was removed and sold for scrap by the next owners, O. R. Sweeney and Roger Short,

A barrel stencil from one of William H. Harman's two flour mills; 196 pounds was the nation-wide standard weight for a barrel of flour. The labeling of almost all Floyd County mills as "custom" rather than "merchant" mills suggests that most mill products stayed in the county. Little is known about how, or how much, surplus wheat or flour was shipped out of the county. (Ricky Cox photograph)

before they sold the mill building and part of the land to Perry Cox, who later tore down the mill house. According to Phillip Gettel's 1985 summary of Floyd County mill histories, the mill he knew as the Tise or Tise and Vest mill last ran in 1945 and was torn down in 1958. At the time of Gettel's writing, the approximate location of the stone foundation on the river bank across from Route 683 was marked by a swinging bridge, the remnants of which were still visible in 2009, at a distance Gettel measured to be 1.2 miles from Route 221.[28]

Deeds

Floyd County Deed Book "L," page 455, 22 August 1865. Lewis Bower, son of Mathias Bower, to William H. Harman, 100 acres on Middle Creek, and 126 acres and mill of Mathias Bower.

D. B. 40, page 323, 22 November 1914. V. M. Sowder and B. G. Howard, Commissioner in Chancery Cause of *Frank P. Harman and others versus William H. Pendleton*, to Peter Fraser Epperly and William Isiah Blackwell, real estate of William H. Harman, deceased, and the Harman roller mill. $4,660.00.

D. B. 45, page 91, 24 August 1922. William Isiah Blackwell to William S. Tice, his ½ interest in land and custom roller mill.

D. B. 45, page 112, 16 September 1922. Peter Fraser Epperly and wife conveyed his ½ interest in mill to John Henry Vest, his son-in-law.

D. B. 49, page 329, 15 November 1930. John Henry Vest to William S. Tice, his ½ interest in land and mill.

D. B. 69, page 172, 2 December 1948. J. H. Poff and Curtis Sumpter, Commissioner for *Pallie Vest vs. Edna Vest, Administratrix of William S. Tice, deceased*, to William H. Burke, 66½ acres and roller mill on Little River.

D. B. 70, page 422, 27 December 1949. William H. Burke to O. R. Sweeney and Roger Short, 66½ acres and roller mill on Little River.

D. B. 74, page 310, 25 April 1953. O. R. Sweeney and wife, Daisy M. Sweeney, and Roger Short and Margie S. Short to Perry F. Cox, 27 acres land and mill on Little River.

101. Eli Mercer Williams Mill

Terry's Creek

LOCATION: Route 684 (Dobbins Farm Rd NE), .5 mile south of Route 685 (Red Oak Grove Rd NE); POWER: Tub wheel, overshot wheel, turbine; STONES: 2 sets of stones, a wool carding machine and a sawmill; VISIBLE REMAINS: Some sign of mill race (1992)

History. The presence of George Winter's (1827–1888) mill on Terry's Creek in 1861 is confirmed by a deed made to Winter in that year by neighbors Jacob and Elizabeth Helms. An 1854 deed made to his brother David refers to a mill race but is inconclusive since it names neither the mill nor its owner. George Winter was trying to sell the mill as early as 1862 but wound up with it back in his possession after a series of buyers, including George Spangler, Elisha Anderson, and Jacob Yearout, defaulted. No deed was recorded for any of these transactions.

Grain milling was a sideline for wool carder and wood worker Eli Mercer Williams. The structure at the back right corner resembles a wooden penstock, the presence of which would be consistent with Mary Williams Willis's memory of a flutter wheel providing power for her father's mill. But its replacement before the mill was sold in 1888, long before this photograph is likely to have been taken, raises questions about the accuracy of that source or of the identification of this photograph. (Frank Webb collection, undated)

Deed records seem to indicate that Eli Mercer Williams (1829–1899) bought the mill in 1886, but boundary references imply that he owned land on Terry's Creek even before his earliest appearance as a grantee for Terry's Creek land in 1883. In 1888 Williams sold out and followed a son to Texas, where, according to Williams' great-granddaughter Anna Hale, all but one son eventually settled. Hale writes that before leaving the county Williams and his family "operated a grain mill and a sawmill; made furniture and coffins, and processed raw wool."[29] An unpublished typescript written by Mary Williams Willis echoes the impression that the family was well established on Terry's Creek and states that Eli Mercer Williams owned or ran another mill and woodworking shop before buying the farm and mill on Terry's Creek, but no deed records to that effect have been found.[30] Willis also states that the wool carding machinery was driven by a flutter wheel until a more powerful and efficient overshot wheel took its place.[31] Williams's father, also named Eli (1792–1857), owned and ran a wool carding mill on Dodd Creek (IV, 30).

Charles Emmitt Williams (1870–1965), the only one of Eli Mercer Williams' sons who stayed in Floyd County, bought the property and ran the sawmill and gristmill for a short time. He was a millwright and miller and also made furniture and coffins. A family historian gave 1913 as the date of his purchase but he does not appear as a grantee in Floyd County deed records after 1909. It appears more likely that he bought the mill tract from

Ballard Dobbins in 1896. All of the equipment is said to have been in the mill house when it collapsed between 1913 and 1918.

Deeds and Other Public Records

D.B. "K," page 584, 3 August 1861. Jacob P. Helms and Elizabeth J. Helms to George Winters, 2½ acres on the waters of Terry's Creek. Boundaries of said land "beginning on a water oak near the aforesaid Winter's Mill." *This deed establishes the existence of George Winters' mill on an adjoining tract of land, but does not convey the tract on which it stood.*

D. B. "M," page 531, 20 February 1873. Zac T. Dobyns, Commissioner, in cause of *George Winters vs. George Spangler*, to George Winter, 12 acres on Terry's Creek, "with a carding machine and grist mill thereon."

D. B. "S," page 354, 15 April 1886. George Winters, and Martha J. Winters, his wife, to Eli M. Williams, 75⅜ acres "lying on the waters of Terry's Creek, a branch of the Little River."

D. B. "T," page 178, 14 January 1888. Eli M. Williams and Adeline, his wife, to Z. T. Dobyns, trustee, 2 tracts of land. 77½ acres and 75¾ acres. The grantors "convey unto said Z. T. Dobyns, Trustee 2 certain tracts or parcels of land," 77½ acres and 75¾ acres. Land was previously conveyed to Eli M. Williams by George Winters and wife in DB "S" p. 354 and 355, 15 April 1886. Also includes 2⅕ acres conveyed to Eli M. Williams by Z. T. Bowers and others in DB "R," p. 182, 14 March 1883. To pay off debts and any left over money goes to Eli M. Williams. Debts are to H.M. Williams, J. L. Williams and S. T. Williams.

D. B. "T," page 391, 18 May 1888. Eli M. Williams and Adeline, his wife, to J. L. Williams, H. M. Williams and S.T. Williams, 2⅕ acres, 75⅜ acres. Amount of 2 bonds described in a deed of trust from Eli M. Williams to Z. T. Dobyns, trustee, recorded in DB "T," p. 178 and cash. Land is "lying on the waters of Terry's Creek."

D.B. "U," page 125, 21 September 1889. J.L. Williams, H.M. Williams, and S.T. Williams to B.C. Dobbins, 2⅕ acres and 75⅜ acres on the waters of the Little River consisting of what was formerly two tracts. According to deed this land was conveyed earlier in DB "T," p. 391. "There is upon this land a carding machine which the parties of the 1st part do not sell with the land but reserve the right to remove said machine from the building in which it is now operated." *The mill seems to have been included with the larger tract, the smaller having been acquired in 1883, years after the mill was built.*

D. B. "Y," page 167, 27 August 1896. Ballard C. Dobbins, and wife, Sarah, to Charles E. Williams, 9 acres.

102. Blackburn Akers Mill

Middle Creek

LOCATION: About midway between Routes 679 (Bethlehem Church Rd NE) and 683 (Roger Road NE). Flaming Cedar Road, now a private driveway, must once have given access to this mill. POWER: Overshot wheel (1870, 1880); STONES: 2 sets (1870, 1880); VISIBLE REMAINS: Evidence of dam and mill foundation (1992)

History. The written record of the Blackburn Akers mill, like many others, hints at a time past when documentation was scant. In granting permission to build a grist mill in 1837, the Floyd County court added the condition that Akers not raise the dam more than 18 inches higher than the original dam, perhaps to prevent the enlarged mill pond from flooding land belonging to Akers' upstream neighbors. The questions of who built the "original dam" and for what purpose are left unanswered. Frank Webb thought that this might have been the site of the Daniel Shelor grist mill, the existence and location of which he was never able to firmly establish.

In 1865 William H. Harman bought the Akers land and mill on Middle Creek, presumably in running condition. No stream names are supplied in the *Products of Industry* report for 1870, but the W. H. Harman mill described therein is probably the Akers mill, since Harman is named as a owner of a mill on Middle Creek in the 1880 report. Significant differences between the two reports in the stated horsepower and capacity of the mill are puzzling, but both reports do describe an overshot water wheel and two runs of stones.

The mill was presumably still running when Harman petitioned the Floyd County court to alter a road running by his mill c. 1887, but it must have been abandoned or torn down between that time and the 1903 opening of his new roller mill (II, 100) on Little River, upstream from the mouth of Middle Creek. In an October 3, 1903, *Floyd Press* advertisement for the new mill, Harman wrote, "My famous old bench burr [probably French burr/buhr, a type of millstone], known all over for making fine flour, will have a place in my new mill," indicating that the old mill had by then been at least partially dismantled.

Deeds and Other Public Records

Montgomery County Deed Book "I-J," page 81, 28 March 1825. George and William Shelor to Blackburn Akers, 67 acres on Little River and Middle Creek.

Floyd County Deed Book "A," page 175, 25 August 1832. John Terry and wife to Blackburn Akers, 120 acres on the waters of Little River.

D. B. "E," page 260, 9 February 1849. Blackburn Akers and wife, Elizabeth, to Samuel Harris, 71¾ acres on Little River. Bounds mention bank of mill pond.

D. B. "H," page 191, 3 April 1853. Samuel Harris to Mathias Bower, mill and 71¾ acres.

D. B. "L," page 455, 22 August 1865. Lewis Bower, son of Mathias Bower, to William H. Harman, 100 acres on Middle Creek, and 126 acres and mill of Mathias Bower.

See Table B, *1870 Products of Industry.* Name of corporation or individual: William H. Harman

See Table C, *1880 Products of Industry.* Name of corporation or individual: W.H. Harman

103. Henry Epperly Mill

Terry's Creek

Location: Just below intersection of Routes 679 (Bethlehem Church Road NE) and 673 (Cole's Knob Road NE); Power: Overshot wheel (1870); Stones: 2 sets (1870); Visible Remains: None

History. The *Products of Industry* reports that accompanied the decennial United States censuses are the only sources of information on this mill. The specifics of these reports must be taken with several grains of salt, since enumerators had no choice but to rely upon their own and the millers' off-the-cuff estimates of horsepower, as well as on the millers' self-reported statistics on a wide range of questions. Millers probably could give an accurate estimate of average daily capacity, but questions concerning annual totals and averages varied from census to census and were asked, at most, only once every ten years. Since the miller himself would have little need for calculating, for example, his exact annual output of rye flour, his responses were apt to be very general estimates. This supposition is supported by the abundance of zeros in the first two, three, or even four right hand columns in the amounts of one or another mill product reported in the 1880 industrial census. Millers who maintained detailed daily ledgers had the raw data on hand, but with no compelling reasons to do so, would not likely have done the sorting, adding, and dividing needed to arrive at the requested statistics.

As a mechanical, rather than economic, measurement of a mill's capacity, no figure was more significant than the number of horsepower supplied by the water wheel, of whatever type. Nor did the accuracy of any figure depend more upon the enumerator's or the miller's skill at estimating. An accurate calculation would have required complex and time-consuming measurements of the usable head and average volume of the water as well as knowledge of the *efficiency* of a water wheel (its ability to convert potential energy to usable motive power). Water wheel efficiency varies absolutely between various types of wheels (overshot, breast, tub, or turbine), and can vary considerably within those absolute limits depending on a wheel's design, construction, and installation. In the case of turbines, efficiency and output vary widely *within* the type, depending upon the volume and flow of water.

Ignorance of this fact was a source of regret for several Floyd County millers who installed turbines in place of vertical wheels around 1900. The power produced by a vertical wheel diminishes with the flow of water, but its *efficiency* remains unchanged. The efficiency of the turbine, in stark contrast, is very high at full flow, but drops precipitously as the volume of water pouring through the turbine is reduced.[32]

Despite the margins of error in some areas, several key features of a mill could be ascertained even by an untrained or hurried observer, including the type and dimensions of a vertical water wheel, and at least the presence of a horizontal turbine wheel inside a wooden penstock or submerged beneath it. We can be confident, therefore, that between 1870 and 1880 Henry Epperly did in fact replace his overshot wheel with a turbine and that he increased his mill's daily capacity by 25 percent. We know also that he reduced the number of months of full-time operation from 10 to 8. Though it is too great a simplification to be taken seriously, and may be a disservice to Henry Epperly, we might infer that he had the very healthy perception of progress as finding a way to get the same work done in less time, as opposed to the more widely held view of progress as getting more work done in the same amount of time.

The only puzzling fact is the reported drop in the number of stones from two sets to one, which typically means that a mill grinds only corn meal, since a different type of stone is required to grind first class wheat flour. Yet Epperly continued to grind wheat, buckwheat, and rye flour, though in relatively small quantities. When the takers of the population census reported that x number of people of whatever ages lived in a certain home, there is little reason to doubt any part of that. The *Products of Industry* reports, while interesting and extremely valuable, cannot be assumed to be as objective or as trustworthy.

Deeds and Other Public Records

Floyd County Deed Book "E," page 352, 19 January 1849. David Spangler to Henry Epperly, 37 acres.

Floyd County Will Book "F," page 536, 17 April 1899. Henry Epperly willed to his son George W. Epperly this mill.

D. B. 38, page 559, 8 January 1913. George W. Epperly to Peter J. Martin, 92 acres.

See Table B, *1870 Products of Industry.* Name of corporation or individual: Epperly, Henry

See Table C, *1880 Products of Industry.* Name of corporation or individual: Henry Epperly

104. Brush Creek Mill/Lawrence Roller Mill

Tributary of Brush Creek

LOCATION: On the south side of Route 612 (Daniels Run Road NE) near or on the Montgomery County Line; POWER: Overshot wheel; STONES: Roller mill; VISIBLE REMAINS: None

History. With the exception of the undocumented Jim Booth mill on Beaver Creek, which is supposed to have washed away around 1920, perhaps before it was even completed, the Lawrence mill may be the shortest lived water-powered mill in the county. Harvey Weaver, one of the three original partners, quickly sold his one-third interest to S. F. Lawrence and Jacob E. Lawrence. The "Walcot News" column on page one of the October 9, 1902, *Floyd Press* noted that "Messers. S. F. and John Lawrence have erected a new store at the roller mill, which will be a great help to our community." The Walcott Post Office was also run by the Lawrences at or near the same location. Silas F. Lawrence sold his interest in the store to his partner early the next year, and on August 11, 1904, the *Floyd Press* reprinted a *Montgomery Messenger* notice that "W. B. Palmer, superintendent of machinery at the Arsenic Mines ... has purchased the Lawrence roller mill and will move the same from Walcot to Childress in a short time."[33] The "Arsenic Mines" referred to were near Huffville, only a few miles from Walcott, and were worked from about 1903 until the end of World War I.[34] Gold mining also took place on Brush Creek in the immediate vicinity of the Lawrence Mill in Floyd and Montgomery Counties, from the 1870s until sometime early in the 20th century. Roderick Lucas included a brief history of the "Gold Rush at Brush Creek" in *A Valley and Its People In Montgomery County, Virginia* (Blacksburg, VA, Southern Print Co., 1973).

Exactly how much of the mill Wyatt B. Palmer intended to move to Childress in neighboring Montgomery County is open to speculation. Locals say the building itself was relocated, but whether at the same time or to the same place as the machinery is unknown. One precedent for moving a mill building along with its contents exists in the Early Marshall roller mill in northeastern Carroll County, VA, which was completely disassembled and moved to a safer, more convenient location when the mill was converted from water power to gasoline engine power.

Whatever the fate of the mill building, the flour milling equipment was moved and set to work on or near the site of the S. G. Palmer and Sons mill in Childress, which by 1904 was owned solely by Wyatt B. Palmer. The Palmer Mill stood on the west bank of Mill Creek just above where Route 693 (Childress Road) crosses the stream in the community of Childress.

Another unusual feature of the Lawrence Mill, as operated in Floyd County, was that it was powered by water from a single spring, only about seventy-five yards from the mill. Nearby Brush Creek is much larger, but is at this point meandering through a relatively flat stretch of land which would have necessitated a very long race or a very high (and very wide) dam to achieve adequate fall for an overshot vertical wheel. The second deed (below) is unusually detailed in its description of the mill's location and is reproduced verbatim.

The rest of the story belongs to Montgomery County history, but it is worth noting that customers came from as far away as Horse Ridge in the Indian Valley section of Floyd County to have their wheat ground into flour at W.B. Palmer's mill in Childress. Truman Quesenberry (b. 1901), who accompanied his father to the mill, recalled that customers sometimes had to spend the night in busy times, and that Palmer or his miller was an accommodating host. *Virginia State Gazetteer and Business Directory* for 1897 listed S. G. Palmer and Sons Mill, Childress, Virginia.[35] The 1917 edition (renamed the *Virginia Business Directory and Gazetteer*) listed the W. B. Palmer Mill, Childress, Virginia.[36]

The Palmer mill is said to appear in a view of the village of Childress included in a 1995 calendar produced by the Riner (Montgomery County) Elementary PTA. Industries at the site included a large sawmill, and perhaps a planing mill, along with the grist mill. A flue pipe emerging from a building near the mill indicates that the saw and planing mills were steam powered. The Lawrence mill machinery is supposed to have been disassembled and moved again to rebuild the Shanks-Jewell mill at the mouth of Goose Creek after that mill burned.

Deeds

Floyd County Deed Book 71, page 452, 17 December 1878. Noah Walters to S. B. Lawrence and S. F. Lawrence, 23 acres land on Brush Creek.

D. B. 29, page 21, 9 September 1902. S. F. and Eugenia Lawrence to S. F. Lawrence, Harvey Weaver, and Jacob E. Lawrence, one half acre. Text of deed reproduced verbatim:

> This deed made and entered into this the 9th of September 1902 between S. F. Lawrence and Eugenia his wife of the one part and said S. F. Lawrence[,] Harvey Weaver and Jacob E. Lawrence of the other part and Harvey Weaver and Biah A. his wife of the one part and James E. Lawrence of the other part witnesseth the following[:]
>
> First that for and in consideration of the sum of ten dollars cash in hand paid by said S. F. Lawrence[,] Harvey Weaver and Jacob E. Lawrence to said S. F. Lawrence before the building of the roller mill herein after mentioned and the further consideration that the said S. F. Lawrence, Harvey Weaver, and Jacob E. Lawrence should erect a roller mill on said land and which has been done and of which roller mill said S. F. Lawrence is the owner of one third, Harvey Weaver the owner of one third, and Jacob E. Lawrence the remaining third including the land upon which same stands and the water right now used in the operation of said roller mill. The said S. F. Lawrence and Eugenia his wife grant sell and convey with generally warranty to said S. F. Lawrence, Harvey Weaver, and Jacob E. Lawrence a parcel of land in County of Floyd but on or near the Montgomery County lines on waters of Brush Creek on the south side of the Franklin and Montgomery road and bounding thereon containing about one half acre upon which said Brush Creek roller mill now stands bounded by said road and other lands of said S. F. Lawrence together with water right now used in operating said roller mill from a spring of said S. F. Lawrence on his land on opposite side of said Franklin and Montgomery road from said mill and some seventy five yards distant with the right of ingress and egress to said spring to keep water power in operation and repair the same when necessary.
>
> And for and in consideration of five hundred dollars in cash paid by said James E. Lawrence to said Harvey Weaver the receipt of which is hereby acknowledged by said Harvey Weaver they the said Harvey Weaver and Biah A. his wife grant sell and convey with general warranty to said James E. Lawrence their entire interest in said land[,] said roller mill and water right and privileges being one undivided third
>
> Witness the following signatures and seals
>
> S. F. Lawrenceseal
> Eugenia A. Lawrenceseal
> Harvey Weaverseal
> Biah Weaverseal

D. B. 71, page 453, 12 October 1904. Jacob E. Lawrence to S. F. Lawrence, his ⅓ interest in mill and water rights and ½ acre land on Brush Creek.

Deeds pertaining to the Wyatt B. Palmer Mill, Childress, Montgomery County, VA

Montgomery County D. B. 43, page 96, 26 July 1892. A. A. Phlegar, Commissioner in Chancery Suit of *Garnand vs. Childress*, to Samuel G. Palmer, H. S. Palmer and W. B. Palmer, 1¾ acres land, mill and store property at Childress, Virginia, on old Christiansburg Snowville Road.

D. B. 45, page 343, 26 November 1896. Wm. M. Howard to Wyatt B. Palmer, 12 acres plus or minus on Mill Creek at Childress.

D. B. 49, page 312, 16 June 1902. S. G. Palmer and H. S. Palmer to Wyatt B. Palmer, their ⅔ interest in roller mill lot on Mill Creek containing 11/16 acre.

D. B. 56, page 461, 16 June 1902. H. S. Palmer and W. B. Palmer to S. G. Palmer, their ⅔ interest in mill property known as the Childress Mill property.

D. B. 87, page 506, 1 October 1926. W. B. Palmer and Lucy Palmer, his wife, to R. L. Jordan, O. P. Jordan, Jr., and C. D. Glenny, roller mill lot of 11/16 acres.

105. Flemon B. Wimmer Mill

Brush Creek

LOCATION: About .25 mile downstream from the Hosea/Ezra Wimmer Mill (II, 119), just below which Route 672 (Laurel Creek Road NE) crosses Brush Creek. The road that passed by this mill is no longer open to the public. POWER: Wooden overshot wheel approximately 12 by 3.5 feet wide; STONES: 1 set; VISIBLE REMAINS: Dug race approximately 700 yards long; 1 set stones, wheel pit and support column (1992)

History. The site of Flemon Wimmer's mill on Brush Creek was still clearly marked in the early 1990s by a pair of mill stones and fragments of concrete foundations in what was then an open pasture. The history of this mill is complicated by the presence of one and very likely two other mills, all within a half- mile stretch of Brush Creek, and all running at roughly the same time. While the deed records for this mill (below) are thought to be complete and consistent, overlap among the names of adjoining landowners and ambiguous references to the "Kinser mill" and "Kinsey's mill" have left some loose ends regarding O. T. and William R. Kinsey, father and son. William owned this, the Flemon Wimmer mill, and appears to have built it or had it built. A separate history of O. T. Kinsey's mill (II, 140) is based on a short string of deeds that fit more logically together than they do in the records of either this mill or the nearby Hosea Wimmer Mill and Huffville Roller Mill/ Ezra Wimmer Mill (II, 119).

William R. Kinsey was 22 years old when he bought a 13-acre tract in 1874. Despite his age, the fact that no mill or anything to do with one (race, dam, water rights) is mentioned in the

The wooden overshot wheel and the wooden flume or box race have been abandoned and are nearing collapse in this undated photograph. The concrete pier supporting the right (upstream) side of the wheel is still in place in 2010, as is Flemon Wimmer's house, which stands upstream and across Bush Creek from the site of the mill. His general store also remains, a few yards downstream with its back to the creek. The road by the mill and store joined what is now Huffville Road near Huffville Church cemetery. The wooden bridge in the foreground spans the overgrown tailrace but not Brush Creek, which is seen in the lower right hand corner. (Frank Webb collection)

deed and that the price had risen to $660, nearly double, when he sold it 28 months later support the assumption that he built the mill. Kinsey's 1876 purchase of another tract of land and the right of way to a spring "from which [Huff and Weaver's Old Still House] was supplied with water" illustrates the common pre-Prohibition association between grist mills and distilleries. The implication is that Kinsey intended to re-establish that business using his own corn meal and rye flour and the Dangerfields' good spring water. It is also conceivable that he was striking a blow for temperance by ensuring that no one else could resume the production of liquor at that location.

Before settling more or less permanently in the hands of the Wimmer family in 1882, the mill was owned for at least three years by a Johnny Epperly. His sale of the mill to Burwell Wimmer, for which no deed or date was recorded, was confirmed in an 1882 deed from Wm. Kinsey and wife to Burwell Wimmer, but Epperly kept the mill long enough to appear as John Epperly, owner of a mill on Brush Creek, in the *Products of Industry* report for 1880. Neither that report nor the deeds provide enough information to rule this out, but he is probably not John William Epperly (born 1856) who later built two mills still standing on Dodd Creek. John W. Epperly is listed in the 1880 population census as a 23-year-old farmer living in the Jacksonville District.

In 1893 a commissioner for Burwell Wimmer, deceased, conveyed the land and mill to Flemon B. Wimmer (1862–1953) who owned the mill for the remainder of its working life. The store that Wimmer ran was still standing just downstream from the mill site in the 1990s.

Deeds

Floyd County Deed Book "N," page 401, 31 December 1874. Burwell Wimmer, and wife and Mary Weaver, wife of John Weaver, to William R. Kinsey, 18 acres on Brush Creek. Property boundary included "Dangerfield corner hence with Kinser [probably Kinsey] Mill Tract."

D. B. "Q," page 420, 7 April 1877, Wm. R. Kinsey, and Emma, his wife, to Johnny Epperly, 18 acres "land and mill property known as Kinsey's Mill, lying in the waters of Brush Creek ... being a part of the land belonging to said Mill on a part of land bought of John Weaver."

D. B. "O," page 457, 7 April 1877. William R. Kinsey and wife to Hosea Wimmer, 11¼ acres. "The party of the first part Reserves the Race with 20 feet of land clear through. The water is not to be moved from the channel where it now runs. The right to run in with wagon when necessary to repair race is also reserved. The two roads leading from Wimmers Mill and Brush Creek church to Kinsey mill is reserved."

D. B. "Q," page 490, 15 March 1882. W. R. Kinsey, and Emma, his wife, to Burwell Wimmer, 18 acres and mill. "Lands sold to Johnny Epperly and by him to Burwell Wimmer ... known as Kinsey's Mill lying on the waters of Brush Creek ... also including the water race and roads as reserved in Hosea Wimmer Jr. deed."

D. B. "Z," page 549, 9 April 1893. John Weaver, Commissioner for Burwell Wimmer, deceased, to Flemon B. Wimmer, 18 acres and mill on Brush Creek.

D. B. 48, page 299, 15 May 1922. F.B. Wimmer and Mollie Wimmer to Stanley G. Wimmer, 6 acres. "A right of way for the mill race through this tract of land where it is Now located is hereby reserved for benefit of parties of the 1st part of sufficient width to carry water and keep race in repair."

D.B. 72, page 282. 10 July 1951. F.B. Wimmer, widower, to W.S. Walters and Mary Ann Walters, 3 tracts of land. One tract (18 acres) mill site.

D. B. 74, page 89. 29 October 1952. W. S. Walters to F. B. Wimmer, widower. Walters defaulted on said note and land (3 tracts) was returned to Wimmer.

D. B. 74, page 181, 26 March 1953. S. G. Wimmer, et al., heirs of F. B. Wimmer, to Raz L. Poff, 1 tract (18 acres) mill site.

D. B. 78, page 183, 10 October 1956. Raz L. Poff to Winston Wimmer.

See Table C, *1880 Products of Industry.* Name of corporation or individual: Epperly, John

106. Silas Lawrence Mill/ David Vest Mill and Flint Mill

Beaver Dam Creek

LOCATION: Route 612 (Stonewall Road NE) 2.8 miles west of Route 221 (Floyd Highway North); POWER: Wooden overshot wheel, wood/steel hybrid wheel, Fitz all-steel wheel. STONES: 1 set and sawmill; VISIBLE REMAINS: Mill building in fair condition, lengthy dug race clearly visible on north side of Route 612, which now passes between the race and mill, but originally ran below the mill

History. A map drawn by C. M. Shockley around 1890 shows this mill, the Piney Fork Mill

Owner David Vest at the Lawrence/Vest mill, around the 1920s. The wood and iron hybrid water wheel, with wooden sides, buckets, and arms, bolted to cast iron hubs and axle illustrates the reluctance to abandon traditional forms all at once. The iron or steel axle provided better bearing surfaces, was more durable, and transmitted power more conveniently, through keyed gears or pulleys, than its bulky octagonal predecessors. The wooden components, though they could be made on site by local craftsmen using locally available materials, were subject to rot, freezing, and imbalance, due to the differences in density resulting from the constant cycle of wetting and drying. (Courtesy Joyce Poff)

(II, 107) and the Double Springs School, which still stands a short distance up Stonewall Road. The assumption that Silas B. Lawrence built the original mill on this site after 1863 is separated from certainty by two circumstances. The 169 acres Lawrence sold along with his mill in 1876 had been purchased in two tracts. The deed for the largest, bought from John Tinsley in 1863, does not mention a mill, but there are many similar instances, especially when large tracts are conveyed, in which operable mills are not mentioned. The second tract, supposed to be only 4 acres (actually 9 acres seem to be unaccounted for), was bought from Bird Cole, but no record of this transfer has been found, leaving open the possibility that the mill was already standing on those four acres when Lawrence bought the land. An 1860 deed between Bird Cole and Henry Light (II, 107), in which Cole reserves water rights for a dam and mill pond on the Piney Fork branch of Beaver Dam Creek, shows that he had at least entertained the idea of building a mill in this vicinity.

On January 22, 1876, Silas Lawrence sold "the Beaver Dam Tract and Mill property" with 169 acres to David Hall. Hall paid Lawrence about $200 of the agreed upon price of $3,650 before assigning the title bond to Jesse Hall, who settled the debt to Lawrence and was granted title to the mill and land in 1881. During the 1890s the tract of land conveyed with the mill shrank from 169 to 52 acres, and eventually dropped to 3 acres, but the 1892 conveyance by G. L. Lawrence to Samuel Scott of water rights, which could be conveyed to each subsequent owner, guaranteed that at least this portion of the dug race would be usable no matter who owned the land through which it ran. The complete text of that deed is included below.

The mill changed hands fairly often until brothers John Henry and David W. Vest bought it together in 1899. David Vest (1860–1932) became sole owner in 1909. In 1932 his widow deeded the mill tract to their son Cleophas "Cleo" Vest, who lived in Roanoke, VA, and worked for the Norfolk and Western railway. By then the mill was called the Flint Mill in honor of the nearby Flint Post Office. Only three acres

Nearing completion in the early 1930s, Cleophas Vest's "Flint Mills" was a mixture of old and new. The modern exterior housed a corn mill of a type more primitive than the 19th century corn and flour mill it replaced, alongside one of the last water-powered sawmills in the county. The overshot wheel is mounted directly behind the corn mill in the back right corner of the building. Note workmen in white shirts and neckties and the discarded mill stone on the right. (Courtesy Joyce Poff)

of land came with the mill, but the water rights conveyed by G. L. Lawrence to Samuel Scott 49 years before were included in the sale. A series of c. 1910 photographs of the old mill show a 1½-2 story building about 20 feet long sitting more or less where the mill that replaced it is now. The well-made wooden overshot wheel is 14 or 15 feet tall and sits on a framework of heavy wooden timbers. Two views show four young women, fully clothed, wading in the mill pond. The thinly planted, weed-free cornfield in the foreground of one panoramic photograph illustrates both the labor intensive nature of farming in that era and the meager crop yields before the arrival, twenty or more years later, of synthetic fertilizers. The field is also much more steeply sloped than any that could be cultivated with a tractor. A later picture (p. 153) shows an elderly David Vest in front of a hybrid water wheel, with steel or cast iron hubs and axle, and wooden arms, rim, and buckets.

Cleo Vest had the existing mill built, retaining the wood and steel wheel perhaps because it was relatively new. The dam was rebuilt and the race cleaned out and repaired at the same time. The water wheel is supposed to have been replaced at some point with a more efficient all-steel wheel made by the Fitz Company in Hanover, PA. The rebuilt mill was and still is a handsome structure with tapered concrete columns, the number of which is minimized by elaborate wooden trusses supporting the roof over the part of the building that housed the sawmill. The concrete foundation of the grist mill portion is reinforced with scrap iron and worn out farm tools, including the blade of a mowing scythe and a mowing machine sickle bar. Cleo Vest continued to work in Roanoke while the mill was run by Oscar Whitenack.

A 2002 *Roanoke Times* article states that the Flint Mill was still grinding corn and sawing lumber in the late 1930s under the care of owner and miller Oscar Whitenack.[37] Whitenack did not in fact own the mill, but did own land adjoining the mill tract, including the land on which the dam and millpond stood, roughly a quarter-mile upstream. Whitenack's granddaughter Wanda Whitenack (Vest) earned nearly five dollars carrying water for ten cents a bucket to the men repairing the race.[38] Benny Conner, whose father, Claude, bought the mill from Cleophas Vest in 1937, recalled that the wooden dam was 8 to 10 feet tall and the pond behind it perhaps 100 yards long. The grist and saw mills last ran in the 1940s and were taken out in the 1950s, at which time Claude Conner sold the steel water wheel, shafts, and pulleys for scrap and converted the mill to a dairy barn.[39]

Deeds

Floyd County Deed Book "L," page 130, 2 February 1863. John W. Tinsley to Silas B. Lawrence, 160 acres on Beaver Dam Creek.

Deed not found. Perhaps not recorded. 22 January 1876. Silas B. Lawrence to David Hall, 169 acres on Beaver Dam Creek known as Beaver Dam Tract and Mill property, "about four acres of which was conveyed to said Lawrence by deed from Bird Cole and residue by deed from Tinsley and wife."

D. B. "R," page 273, 13 June 1881. Silas B. Lawrence to Jesse Hall, 169 acres on Beaver Dam Creek known as Beaver Dam Tract and Mill property. (Jesse Hall paid off the debt David Hall had assumed when he bought the same property from Lawrence on 22 January 1876.)

D. B. "R," page 341, 19 January 1884. Jessie W. Hall, and Emeline, his wife, to Samuel Scott, 80 acres, on the waters of Beaver Dam, branch of the Little River.

D. B. "W," page 70, 11 October 1892. G. L. Lawrence to Samuel Scott, water rights to said mill. The deed entered into on October 11, 1892, between G. L. Lawrence, party of the first part, and Samuel Scott, party of the second part, follows:

> Witnesseth that for and in consideration of the fact that the party of the second part is the Owner of a Mill situated on Beaver Dam Creek in Floyd County and that said mill is a convenience to the party of the first part and whereas the race or ditch through which the water is conveyed to said mill is on the lands of the party of the first part for a distance of about fifty yds. that the party of the first part grants and conveys to the party of the second part and his assigns the right of way for said race or ditch where it now is without further charge or consideration. The dam from which the water passes through the race or ditch is not to be raised higher than is necessary to convey water through said race or ditch where it now situated, said land on which race is situated is on Beaver Dam Creek. Witness the following signature and seal,
>
> G. L. Lawrence

D. B. "W," page 69, 2 November 1892. Samuel Scott to Asa Poff, mill on Beaver Creek "known as the Silas Lawrence Mill near Flint Post Office."

D. B. "Z," page 42, 1 March 1898. B. G. Howard, Commissioner, in suit between Asa Poff and O. W. Vest, to Samuel Scott, 3 acres of Mill property, and 52 acres.

D. B. "Z," page 484, 26 May 1898. Samuel Scott to James Daniel Ekiss, mill and water rights.

D. B. 27, page 23, 17 January 1899. J. D. Ekiss to David W. Vest and John Henry Vest, mill and property on Beaver Dam Creek.

D. B. 27, page 495, 8 January 1909. John Henry Vest to David Vest, his entire interest in mill property on Beaver Dam Creek.

D. B. 50, page 229, 22 January 1932. Viola E. Vest, widow of David Vest, to Cleophas E. Vest, 3 acres land and mill and water rights to same, Tract No. 1.

D. B. 54, page 540, 1 April 1937. Cleophas E. Vest to Claude Conner, 4 tracts, including Tract No. 1.

D. B. 173, page 440, 5 May 1990. William B. Conner, Executor for Claude and Claudine Conner, to Murrie V. Poff and wife, Lot No. 4, including (mill) tract containing 4.428 acres.

See Table B, *1870 Products of Industry.* Name of corporation or individual: Lawrence, S. B.

See Table C, *1880 Products of Industry.* Name of corporation or individual: David Hall

---

107. Piney Fork Mills

Beaver Dam Creek and Piney Fork of Beaver Dam Creek

LOCATION: Just west of Route 669 (Wilson Cemetery Road NE) near Double Spring School and Route 612 (Stonewall Road NE); POWER: Overshot wheel; STONES: At least one set and sawmill. VISIBLE REMAINS: Part of old dug race from a pond on Beaver Dam Creek and mill site

History. In 1878 Bird Cole sold four partners two acres "for sawmill seat or any other mashinery it is the wish to put up." The fact that no grist mill was described at this location in the *Products of Industry* report that accompanied the 1880 Census may mean that it was not yet running or that the estimated value of production between the first day of June 1879 and the last day of May 1880 was less than $500, below which point mills were not included in the report.

With some mills that drew water from two streams, the second race was added when the water delivered by the original race proved inadequate to power the mill. That the builders of this mill planned from the beginning to use two races is confirmed by the inclusion in the deed

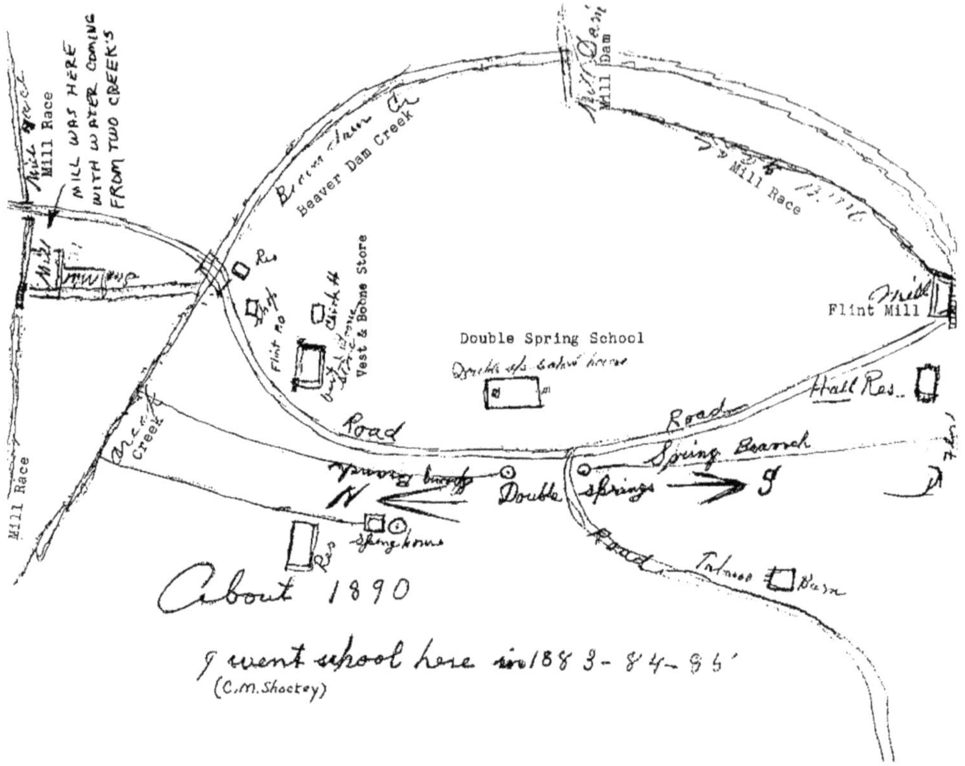

Along with the Fork Mill (center left), this c. 1890 map by C. M. Shockey locates the Flint Mill (center right) and the Double Springs School, both of which still stand along Route 612 (Stonewall Road). Modern day Route 669 (Wilson Cemetery Road) crosses Beaver Dam Creek below the Fork Mill. (Courtesy Mason Whitenack)

of their right to "cut races and dam to each race, eight feet high if necessary." A map drawn by C. M. Shockey depicting the neighborhood circa 1890 indicates that the grist mill and a sawmill immediately below it were in fact served by two races from two different streams. One is Beaver Dam Creek, and the other a tributary that today joins Beaver Dam just below where Beaver Dam is crossed by Route 669 (Wilson Cemetery Road). The second stream is referred to as Piney Fork in deeds to the mill property but is unnamed on recent maps, among them the U.S. Geological Survey topographical map dated July 1, 1981. This and other maps assign the name "Pine Fork" to a tributary that joins Beaver Dam Creek about one quarter mile downstream from the Fork mill site, very near where the pond for the Flint Mill (II, 106) was located throughout the lifetime of the Fork mill. It could be that the two streams were given similar names—Piney Fork and Pine Fork—and that the first was eventually forgotten.[40] Or a simple error concerning which of the streams was known as Pine and which was Piney in the 19th century could have been accepted as truth after going unnoticed and uncorrected for many years.

Besides employing two separate races, this mill property is remarkable for having had nearly a dozen full or part owners in only 28 years (1878–1906). After buying out the other three original partners, John Link lost the mill as a result of four chancery suits, none of which involved a previous mill partner. As might be expected for a mill that changed hands often, the Fork Mill was known by several other names, including the Bird Cole Mill; Link, Zink, and Lawrence Saw and Grist Mill; the Mills of Link, Laurence, and Zink; J. W. Roberts Mill; John H. Link Mill; the Fork Mills; and the Forks Mill(s). Among the men named, two never actually owned the mill. J. W. Roberts is supposed to have run the mill for partners Zink, Link, and Lawrence. Why Bird Cole's name would be attached is less clear, since he sold the land before the mills were built on it. But Cole had at one time at least contemplated building a mill of his own on Piney Fork, as seen in the first deed summarized below, although there is no compelling evidence that he ever acted on that intention. The advent of the Civil War, 14 months later, could have affected any plans he had made along that line.

The final chapter of the Fork Mills' history, spelled out in the text of the 1906 deed transcribed below, is by contrast neither complicated nor dramatic. After buying out two partners, I. J. Vest removed and sold for unstated amounts the grist and saw mills and their buildings. The $200.00 C. D. Smith was to pay Vest for the land was to go toward the $182.44 still owed to the previous owner, Moses Martin. The balance of $17.66 was Vest's, to do with as he pleased.

Deeds and Other Public Records

Floyd County Deed Book "K," page 295, 24 January 1860. Bird Cole to Henry Light, 100 acres on the Piney Fork branch of Beaver Dam Creek. Property conveyed is not the mill tract, but in this deed Cole "doth reserve the right to himself to as much of the above named land as the water would cover in building a mill dam and a road through the same where it now is." *Whether Cole actually built the contemplated mill or was merely retaining that option for the benefit of potential future buyers, such as partners Link, Zink, Zink, and Lawrence, is unknown.*

D. B. "P," page 341, 13 November 1878. Bird Cole to John H. Link, A. J. Zink, S. W. Zink, and Gideon P. Lawrence, 2 acres. "It be for sawmill seat or other machinery it is the wish to put." Grantees have the "right to cut races and dam to each race, eight feet high if necessary."

D. B. "V," page 466, 12 April 1882. G. P. Lawrence to Harriet Overstreet, his interest in land and mill (no deed recorded).

D. B. "Q," page 595, 26 May 1882. Bird Cole to Zink, Link, and Lawrence, "½ acre more or less ... on the waters of Beaver Dam, Branch of the Little River ... at the fork of Beaver Dam and Piney Fork."

D. B. "T," page 379, 23 May 1888. Harriet Overstreet to John H. Link, her interest in mill and land.

D. B. "V," page 405, 17 February 1892. Elijah Light to W. M. Vest, 53⅜ acres land, Pine Fork of Beaver Dam. This deed does not convey the mill property, but does reserve the right to the bed (as opposed to a dam across or a race alongside) of Piney Fork Creek "which leads to the mill known as J. Links better known as J. W. Roberts Mill."

D. B. "V," page 467, 13 April 1892. According to this Amended Memorandum of *Lis pendens*, A. J. Zink and S. W. Zink had sold their interests to John H. Link, but "if the deed or deeds has been made they have not been recorded [as of] April 14th, 1892."

D. B. "V," page 463, 11 April 1892. Memorandum of *Lis pendens* (lawsuit pending), *Isaac F. Martin vs. J. H. Link*, for an interest in 2½ acres and mills.

D. B. "V," page 464, 11 April 1892. *Lis pendens, J. R. Coffer vs. J. H. Link*, same.

D. B. "V," page 466, 12 April 1892. *Lis pendens, Sidney Sheetman vs. J. H. Link*, same.

D. B. "V," page 467, 13 April 1892. *Lis pendens, J. H. Blackwell vs. J. H. Link*, same.

D. B. "X," page 427, 16 December 1893. J. L. Tompkins, Commissioner (suit involving Isaac Martin, J.R. Coffer, Sidney Sheetman and J.H. Blackwell against J.H. Link), to Moses Martin, 2½ acres, land and mills known as Fork Mills, or mill of Link, Zink, and Lawrence.

D. B. 26, page 219, 8 October 1899. Moses Martin to William M. Vest, 33½ acres land and mills (adjoins mill lot; pertains to next deed).

D. B. 31, page 394, 26 April 1903. W. M. Vest and wife to Moses Martin of Montgomery County, race right through land deeded to Vests in D. B. 26, page 219. Specifies rights of each party.

D. B. 32, page 503, 11 June 1906. Full text of deed:

> This deed made and entered into this the 11th day of June 1906 by and between Moses Martin, a widower, and I. J. Vest, Ina Vest, his wife, parties of the first part, and C. D. Smith, party of the second part.
>
> Witnessed that whereas at a sale made by Moses Martin on the 14th day of August 1902 said Moses Martin sold to C. F. Eans, B. H. Eans, and I. J. Vest, a tract of land containing about 2 acres more or less on which there was a grist mill, saw mill, dwelling home and other improvements. Said property is situated in the County of Floyd, Virginia, on Beaver Dam and Piney Fork branches of Little River and known as the Link Mill property.
>
> Said Martin presented no deed of conveyance and was not to do so until all of the purchase money had been had but he gave said parties a title bond. Whereas C. F. Eans and B. H. Eans assigned all of their right and title in said property to I. J. Vest and whereas the said I. J. Vest has paid all of the purchase money to the said Moses Martin except $182.44 (one hundred and eighty two and 44/100 dollars) as of June 6, 1906 and whereas the said I. J. Vest has sold and removed from said land the two mills and the mill buildings and he is now selling the said 2 acres of land with other improvements thereon to the said C. D. Smith for and in consideration of $200.00 (two hundred dollars) which is to be applied as far as necessary to pay the balance due Moses Martin.

108. Quinter Lucas Mill

Goose Creek (Dashing Run)

LOCATION: Between Route 660 (Goose Creek Run NE) and Route 634 (Level Bottom Road NE) on a tributary of Goose Creek. Road which formerly joined these two roads and passed by the mill is now a private road, Quiet Drive, NE. POWER: Overshot wheel; STONES: 1 set and shingle mill

History. Quinter Lucas outfitted his mill with equipment from the Marion Hall mill (II, 109) which is supposed to have been torn down by its last owner, Moses Poff, between 1912 and 1914. Whether Lucas erected his mill on land he had owned since 1903 or waited until he purchased 31 additional acres in 1915 might be determined by a close study of the boundaries described in these and subsequent deeds to the same property.

According to Quinter Lucas's great-nephew, Daniel Worth Lucas, a wooden race carried water from a natural fall to the overshot wheel. The mill had stopped grinding before 1940, but the waterfall was a popular spot in hot weather for some time afterward. Quinter Lucas's sons Lee and Lex helped run the mill.[41]

Deeds

Floyd County Deed Book 31, page 568, 2 March 1903. Priscilla Lucas to W. Q. Lucas, 41⅛ acres, South Fork Roanoke River.

D. B. 40, page 586, 23 May 1915. D. S. Lucas to W. Q. Lucas, 31 acres, South Fork Roanoke River.

D. B. 67, page 103, 6 October 1947. William Quinter Lucas to Lee Lucas, two tracts land on waters of Roanoke River. Tract number 1, 31 acres, tract number 2, 41⅛ acres and all personal property, household goods, etc.

109. Moses Poff Mill

Mill Run (Goose Creek)

LOCATION: Route 659 (Griffith Creek Road NE) .25 mile northeast of Route 660 (Goose Creek Run NE); POWER: Overshot wheel; STONES: 1 set

History. Marion Hall (born c. 1867), a maker of coffins and caskets, bought this property from his father, Charles M. Hall, in 1891. Within the next six years he built a sawmill and woodworking shop, presumably driven by the same water power as the grist mill. Neither of the pertinent deeds mentions a mill of any kind or water power, leaving it unclear whether Hall bought the land with the grist mill in place, or built it himself. The mill's last owner, Samuel "Moses" Poff (1872–1947), tore the grist mill down around 1912–14. Quinter Lucas bought the stones and related machinery and installed them in his own mill (II, 108) on Dashing Run, a Goose Creek tributary.

Deeds

Floyd County Deed Book "V," page 123, 21 March 1891. Charles M. Hall and wife, Eliza, to T. Marion Hall, 20 acres on Mill Run (Goose Creek)

D. B. "Y," page 467, 14 September 1897. Marion T. Hall and Nanie, wife, and Jas. M. Conner and Melvinie, his wife, to Moses Poff, 20 acres plus or minus on the waters of Mill Run.

110. James McNeil Mill/ Charles Will Vest Mill/Roscoe Willis Mill

Flat Run (Tributary of Goose Creek)

LOCATION: Below intersection of Routes 659 (Griffith Creek Road NE) and 660 (Goose Creek Run NE) a short distance above the Roscoe Willis store building; POWER: 8-foot overshot wheel, later gasoline engine; STONES: 2 sets and sawmill; VISIBLE REMAINS: Willis's store still standing

History. The deeds to lands purchased by James Ireson McNeil (or McNiel) (1822–1904) on Flat Run and Goose Creek make no mention of a mill or mill property. This supports the belief that he was the builder, as well as the owner and operator, of "a combination grist mill [and] sawmill adjacent to Goose Creek."[42] The statement, in the previously cited source, that the mill was built "sometime after 1850" seems at first misleading, since one of two tracts James McNeil bought from his father in 1874 measured 40 acres, exactly the same acreage as the tract McNeil signed over along with his mills to David Link in an 1881 deed of trust. In other words, McNeil may not have owned the land on which his mill was built until 1874.[43] Ordinarily, this would be the earliest date that could be safely assigned to the mill based on deed book records. However, in *Floyd County, Virginia Heritage*, J. David McNeil writes that "James Ireson McNeil cleared part of his father's land and built a log house circa 1848 for his bride."[44] This was not at all unusual, but it is remarkable that twenty-six years later he and his family were still living on land that legally belonged to his father. A man who felt comfortable with this arrangement might also have built a mill and blacksmith shop on his father's property without worrying that some future circumstance or disagreement would threaten his family's immediate security, or his personal investment and future livelihood. Without additional documentation, it can be said with certainly only that this mill first ground cornmeal for James McNeil's family and neighbors sometime between 1850 and 1881, but a date early within that range is likely, based on family history. J. David McNeil notes that the mill was intended to produce both flour and cornmeal, but once the mill was complete its builders discovered that the 8-foot diameter wooden wheel and available water supply produced only enough power to grind corn. He estimated that the 2½ story building was 25 feet long and 15 feet wide.[45]

Although it never ground flour due to insufficient power, the McNeil/Vest/Willis Mill was of the two-story design typical of mid–19th century flour mills. The water wheel and race were gone when this was taken around 1970. (Frank Webb collection)

In 1883 James Ireson McNeil sold his mills and some land to James Sisson. Twelve years later Sisson sold the mills and 4½ acres to C. W. (Charlie Will) Vest. Millers who ran the mill for Vest included his son-in-law C. O. Carr and Joe Light. In 1923 Vest rented the mill and store to Roscoe Willis, whose wife, Bonnie Vest Willis, was Charlie Will Vest's granddaughter and James I. McNeil's great-granddaughter.[46] After the dam washed out in the 1920s, Willis adapted the mill to be driven by a gasoline engine. Like its counterparts in other sections of the county, the Willis mill probably experienced a fairly steady demand during the 1930s, followed by a steepening decline during and after World War II as more and more farm families stopped growing heirloom varieties of bread corn and came to depend instead on local, regional, or national brand names sold in stores and at larger mills. The mill was closed in the early 1960s and the building dismantled in the 1970s as a precaution against the inevitable collapse of the tall, century-old wooden structure.[47] Roscoe Willis's son Howard owned the stones and some other equipment in 1992.

Floyd County, Virginia Heritage includes more information on the mill and on the McNeil and Willis families, along with a photograph of James I. McNeil, a photograph of Roscoe and Bonnie Vest Willis, and two similar views of the Roscoe Willis mill and store.[48]

Deeds

Floyd County Deed Book "N," page 457, 11 April 1874. Jacob McNeil to James I. McNeil, two tracts, "a 40 acre tract on which the said James I McNeil resides" and an adjoining tract "supposed to contain 35 acres ... on the waters of the south fork of the Roanoke River."

D. B. "Q," page 268, 12 December 1881. James I. McNeil to David Link, Trustee, Deed of Trust, "40 acres including his mill property" on waters of Flat Run, a branch of the Roanoke River. *The debt secured by this deed of trust was satisfied and no transfer took place. It is included because this is the first deed book reference to the mill.*

D. B. "Q," page 687, 13 January 1883. James I. McNiel and wife, Lydia, to James M. Sisson, "4½acres of land including his mill property both his saw and grist mill and dwelling house" on the water of Flat Run.

D. B. "X," page 345, 5 July 1895. James M. Sisson to C. W. Vest, 4½ acres of land, Flat Run. Boundaries identical to preceding deed.

D. B. 64, page 584, 27 August 1946. C. W. Vest to Roscoe W. Willis. Four tracts, including the same 4½ acres as was conveyed to Vest by James M. Sisson in D. B. X, page 345.

111. W. O. Carr Mill

Flat Run (Tributary of Goose Creek)

LOCATION: Route 653 (Shawsville Pike NE) just below intersection with Route 660 (Goose Creek Run NE); POWER: Overshot wheel; VISIBLE REMAINS: None

History. The peculiar wording of Deed Book "X," page 465 (quoted below) leaves unclear whether W. O. Carr built this mill, or bought it along with the land on which it stood. The question remains unanswered, since no deed book record was found of a transfer of land from David Link to W. O. Carr. A smaller question arises from the fact that Joseph Samuel Carr, William O. Carr's oldest child by his second wife, purchased full interest in the mill and 4¾ acres in 1892 when the property was sold to settle the chancery cause between his father, W. O. Carr et al., and J. P. Conner et al. Twenty-four years later the elder Carr deeded to J. S. Carr ½ interest in the property the younger man presumably already owned. The 1910 census includes J. S. Carr as a 48-year-old miller, with his widowed father living as member of his household. No mention was made of the mill when in 1922 a tract of 4.75 acres was conveyed to Alfred Vest to a settle the chancery cause of *J. S. Carr vs. Mary L. Conner, et al.*

Frank Webb thought that James Ireson McNeil, who built what was later known as the Roscoe Willis Mill (II, 110) not far away on Goose Creek, might have been connected to the W. O. Carr mill, perhaps through David Link, whose name appears in deeds pertinent to both mills. But no definite connection was found. W. O. Carr also owned a grist mill and sawmill (I, 6) on the headwaters of Little River between 1875 and 1882.

Deeds

Floyd County Deed Book "X," page 465, 25 May 1892. J. L. Thompkins, Commissioner, to J. S. Carr in chancery cause of *J. P. Conner, et al. vs. W. O. Carr*, et al., 4¾ acres, "the land bought of David Link by W. O. Carr upon which there is a mill."

D. B. 41, page 539, 31 July 1916. Wm. O. Carr to J. S. Carr, "for and [in] consideration of the care and support of W. O. Carr by J. S. Carr, the said W. O. Carr deeds to J. S. Carr one half of 4½acres of land [on] which a saw and grist mill now stands and on which J. S. Carr's dwelling house stands."

D. B. 45, page 418, 21 August 1922. J. S. Carr to Alfred W. Vest. Two tracts, one 4¾ acres, another 114 1/4 acres. Land is "on the waters of the Roanoke River." Refers back to DB "X," p. 465, for 4¾ acres. Land sold to settle chancery case, *J. S. Carr vs. Mary L. Conner*, et al.

112. James R. "Jim" Hall Mill

Big Run Creek

LOCATION: Approximately 1 mile upstream from Route 221; POWER: Overshot wheel; STONES: 1 set and sash type (up and down) sawmill

History. An 1873 order of the Circuit Court of Floyd County grants James R. Hall (b. 1847) permission "to build a dam across Big Run Creek 18 foot high where his water saw and grist mill is established," but the exact meaning of "established" is open to speculation. It seems unlikely that Hall would have invested time and money erecting a dam and building a mill before obtaining, as required by state law, the Circuit Court's permission to block the stream with a dam and maintain a pond behind it. Approval was to be given only after a court-appointed jury of twelve examined the site of the proposed dam and made a recommendation, for or against, based on potential impediments to navigation and the passage of fish, potential threats to public health, and the extent of the damage that the construction and filling of the dam and millpond would cause to the real property of neighboring land owners.

But it was in fact common practice for would-be mill owners in the 19th century to raise dams on their own land without first obtaining permission. On the small, steep streams of Floyd County, there was no navigation to speak of, no commercial fishing, and little threat from pathogens found in stagnant water. The only real concern was damage to a neighbor's real estate. Since a mill of any kind was such an asset to a community, it was a pretty safe assumption that if a neighbor's land was affected, he would be compelled by the court to accept the compensation recommended by the same panel of jurors when they submitted the findings of their inquest. If no other person's property would be affected, there was virtually no chance that an application would be rejected.

Hall could have begun constructing his dam and mill as early as January 1867, following his marriage to Susanah Catherine Simpson (1831–1899), who had inherited the land on which the mill was built from the Wallace family.[49] His age was given as 20 in the marriage record, and hers as 36, but the birth date record indicates that he was only 19. Her father, James Simpson, and brother, William Allen Simpson, built and ran the oldest mill (II, 113) still standing in Floyd County.

Nothing more is known about this mill, except that the sawmill on the site was of the "up and down," which was typical of that time, rather than the circular type. The unusual height of Hall's proposed dam suggests that Hall overestimated his needs, or that the dam was to be built at the mill site or only a short distance upstream; water raised by a dam to a height of 18 feet above the stream bed could have supplied an overshot water wheel of at least sixteen feet in diameter working only a very short distance downstream. Few wooden wheels in Floyd County ever exceeded 16 feet in diameter and no turbine penstocks of that height are known to have existed. An alternative possibility is that Hall's saw and grist mills were built several feet above stream level, and perhaps some distance from the stream, as a precaution against flooding or because the terrain demanded it. An 1890 deed conveying the lands of James R. Hall on Big Run Creek to his son John William Hall does not mention a water power site or a mill, but one or both mills may still have been in operation.

Deeds and Other Public Records

Floyd County Court Order Book 1870–1877, page 324, 14 October 1873. On the writ of James R. Hall. He is authorized to build a dam across Big Run Creek 18 feet high where his water saw and grist mill is established. Application for this dam was made 15 July 1873, page 313, C. O. Book 1870–1877.

D. B. "V," page 446, 11 March 1890. James R. Hall to John William Hall, 77 acres on Big Run Creek.

113. William Allen Simpson Mill

Pine Branch

LOCATION: Route 654 (Goldfield Road NE) .75 mile west of Route 221 (Floyd Highway North); POWER: Overshot; steam engine during low water; STONES: 1 set (1870); 2 sets (1880); sash type sawmill (saw found in mill) VISIBLE REMAINS: The mill is in fair condition and contains some of the original wooden gearing. Water wheel is gone. Dug race clearly visible

History. After her father's stroke in 1925, fifteen-year-old Effie King (Brown) assumed the regular duty of riding horseback to the local mill with shelled corn to be ground into cornmeal, and mixed corn, rye, and oats to be ground into chop for cattle. At age 89, William Allen "Billy" Simpson (1836–1928) was still able to help her down and to lift the sacks of grain from the horse's back. Brown remembered her uncle as a kind man who allowed people to work out the price of corn meal or flour on one of his farms.[50] A devout Primitive Baptist, he lived his belief in working hard for six days then resting on Sunday, often advising friends and family to "work like you will live always, and pray as if you may die tomorrow."[51]

Miller and farmer William Allen Simpson with grandson Allen Cannaday, wife Geraldine "Girlie" King Simpson, and daughter Ethel Simpson Cannaday Barlow. Approaching 85 years of age in this c. 1920 photograph, Simpson was still running his mill six days a week. (Courtesy Effie King Brown)

When he retired in 1926 at age ninety, Simpson had been grinding wheat and corn for his neighbors for sixty years, having erected his mill a year or two after the close of the Civil War, during which he served four years in Jeb Stuart's Horse Artillery. There is no written record of a sawmill, but the blade and a forged tensioner of a sash type sawmill were inside the mill until the 1990s. The mill was built on land purchased in 1835 by "Billy" Simpson's father, James, whose occupation is listed as "miller" in the 1870 census. The elder Simpson died in 1874, leaving the mill and homeplace to his only son, who hoped no doubt to make a similar bequest at some future time. Billy Simpson was married at age 48 on Christmas Eve 1884 to 20-year-old Elizabeth Geraldine "Girlie" King. Their only son, James, died in 1919 of pulmonary tuberculosis at Western State Hospital in Staunton, where he was being treated for mental illness. James Allen Cannaday, the eldest son of Simpson's only daughter, and on whom Simpson may next have pinned his hopes, was killed in a truck accident at age 16. Nevertheless, William Allen Simpson's mill and the house built by James Simpson remain in the hands of Simpson descendants.

William Allen Simpson's mill has proven to be as durable as its builder and is one of only about ten Floyd County mills still standing in 2010. The Simpson mill is one of two or three water-powered mills in the county known to have employed a steam boiler and engine during times of low water, although a few mills were driven exclusively by steam at some point in their working lives.

The mill was closed to the general public in

The William Allen Simpson Mill water wheel was on this side or the lower gable end. A door on the far side opened onto a platform from which wagons could be loaded. Grain could be offloaded to elevators in the basement through a small door on the upper gable end. (Ricky Cox photograph, January 1993)

1926, but Simpson's son-in-law, William "Dee" Barlow, continued to grind for himself and close friends and family until his own death around 1942. The water wheel was sold to a buyer from North Carolina. Ethel Simpson Cannaday Barlow, wife of William Barlow and the only daughter of William Allen Simpson and Girlie King Simpson, was baptized in the mill pond.

Deeds and Other Public Records

Floyd County Deed Book "A," page 356, 26 August 1835. William Walters to James Simpson, 100 acres plus or minus land on Pine Branch waters of Little River.

Will Book "E," page 237, 1 June 1876. James Simpson wills to his son, W. A. Simpson, the home tract of land with mill and other appurtenances thereto.

W. B. 4, page 114, 5 April 1928. W. Simpson wills entire estate to wife, E. G. Simpson. (The text of the will is dated by Wm. A. Simpson, 24 February 1919. The will was admitted to probate on 3 April 1928.)

See Table B, *1870 Products of Industry.* Name of corporation or individual: Simpson, James

See Table C, *1880 Products of Industry.* Name of corporation or individual: Wm. A. Simpson

114. Joseph Johnson Poff Mill

Meadow Run Creek

LOCATION: North side of Route 221 (Floyd Highway N) in the village of Check, VA; POWER: 12' by 5' Fitz steel overshot wheel; STONES: 2 sets, feed mills, sawmill, wood shop, blacksmith shop; VISIBLE REMAINS: Steel overshot wheel and concrete wheel pit

History. This mill is very likely one of the "two grist mills" mentioned in the first deed below. In a transaction recorded in October 1891, 28-year-old Joseph Johnson Poff conveyed to his brother William J. Poff ½ interest in 2 acres of land with a "Store house[,] Granary[,] two Water Grist Mills[,] One Sawmill[,] Still House[,] Storage House with the stills[,] tubs[,] and every thing now belonging to said Distillery." Already a man of means at age 30, the younger Poff bought most or all of this property back two years later. The listing of only one mill in the 1893 deed suggests that W. J. Poff had retained his interest in that one enterprise or, more likely, that the second mill was no longer running. The mill dam constructed by Johnson Poff in 1911 may be the earthen dam described in his grandson's memoir, below.[52]

Joseph Johnson Poff's combining of several other water-powered manufacturing and fabricating processes with his two grist mills (he bought the Josie Custer mill [I, 137] in 1919) epitomized the ambition and progressive inclination shared by many millers in pre-automobile Floyd County. His ice house, store, woodworking machinery, blacksmith shop, and pre-prohibition distillery — and a nearby post office — provided most of the necessities and a few luxuries of rural life for patrons of his businesses. Basil "Poff" Simpson, Sr., described the facilities at his grandfather Poff's "upper mill" in an unpublished memoir:

Mill owner and entrepreneur Johnson J. Poff with current and future relatives, about 1905. The woman standing at far left is unidentified. Beside her are Miriam Kropff Poff (wife of J. J. Poff) and Alice Kropff (Miriam's sister-in-law). In the wagon, from left to right, are Lizzie Simpson, Pearl Poff (dark coat and hat), Ora Poff, Iva Poff (small girl), Bernice Poff (dark clothes and hat), Garnett Poff (little girl), Minnie Clower (white blouse), and Byron Poff (small boy, dark hat). Ardry Poff, whose hands are visible over Byron's left shoulder, didn't like the picture and scratched out her image. Horace Poff is astride the steer. Standing men, left to right, are Bill Poff, Herman Poff, Billy Kropff (Miriam's brother), Ross Simpson (brandishing tree limb), J. Johnson Poff, and Robert Kropff (Miriam's father). Ora, Bernice, Horace, Iva, Garnet, Byron, Herman, and Ardry were children of J. J. and Miriam. (Courtesy Bethina Simpson Huddleston)

Grandpa Poff was a miller and Blacksmith by trade and also had a sizable farm operation. He built an earthen dam on Meadow Run, one-third mile above Check. From there he sluiced the water through a race on the side of a hill to a flow-bay, 4' × 4' and 125' long, built of wood on wood piling made from whole trees. [*Flow-bay* may be a variant or a corruption of *fore bay.*] This bay carried the water to an over-shot water wheel 12' in diameter and 5' wide. When this wheel turned it produced tremendous power. The wheel was in the middle of the mill. Gears could be slipped so that the wheel would power either side. On one side was the buck-wheat flour, corn meal, and feed grinders. On the other side was the saw mill, planers, rip saw, shingle saws, turn lathes and Custer's cabinet room. The blacksmith shop was in a separate building adjacent to the mill. When the mill was in operation, the water that overflowed the wheel was picked up by another race below the mill and carried down stream another quarter mile and through another flow-bay to another overshot wheel where a similar mill was operated by the same water. At this mill a steam-operated dry-kiln cured lumber by using the shavings from the planers to fire the steam boilers. So you can see old Grandpa J. J. was quite an innovator. If he wanted a bit for his molding machine or a planer head, he simply went to his blacksmith shop and with forge and hammer and anvil made what he wanted. He was a shifty man and he had to be in order to clothe, feed and entertain that family he was rearing. They were used to more than the average in the area, and as the family grew so did the demands. They evidently lived a good life. I have heard my dad's sister, Aunt Mary, say that when she was a young girl, she wished many times that she could have been Johnson Poff's dog.[53]

Simpson noted that Johnson Poff was "one of the founders and a member of the Board of Directors of the State Bank of Check. He was also engaged in the mercantile business with partners who did the actual operation of the business. He owned the property. He was a busy and thrifty man."[54]

Advertisements for a mainstay of the mill's woodworking shop, coffins, appeared in the *Floyd Press* on June 15, 1922, and July 16, 1925, and most likely on many other occasions. A "lot of coffins, caskets and accessories" was advertised for sale with J. J. Poff's personal property at a public auction held on 9 March 1933. Among the farming equipment offered for sale by the administrator of the Poff estate were a mower, a hay rake, a grain drill, a grain binder, a threshing machine, and a Fordson tractor and plow. Together with an iron turning lathe, a cement mixer, and a shingle saw mill, these possessions show Poff to have been a wealthy man. Just how well off he was in 1933, compared to his neighbors, is suggested by the willingness of the sellers to accept "notes on Six months time, with approved personal security" for the purchase price of any item that sold for more than $5.00.[55]

Horace Poff, the fourth of Poff's fifteen children, ran the mill after his father's death. Berkley Rierson, Alvin Trigg Board, and Charles Wilson worked at both mills. Joseph M. "Josie" Custer was employed by Johnson Poff as a miller and a blacksmith after selling his own mill to Poff in 1919. "Custer's cabinet room" named in Simpson's memoir may have been the domain of Josie Custer, who eventually moved to Oklahoma.

Poff's mill in the village of Check, his "upper mill," ran until the late 1940s, as did the lower, known as the Josie Custer mill, a quarter mile down Meadow Run Creek on present day Meadow Run Road SE. The last of several buildings at the mill site were torn down in the 1960s.

Beth Simpson Huddleston, Johnson Poff's great-granddaughter, composed this poem about playing in the abandoned mill.

Papa Johnson's Mill

Though half-erect, a few of its
fattened rats living still on age-old remains—
the Mill told us stories
even our elders had lost in the
banks of their memories.

The wheels of commerce—
once spinning relentlessly from
the first sparks of dawn to the
dark blanket of night—
now served as the creative
toys of childhood.

We grabbed the spokes and
spun each other—
on challenge—
into dizzy oblivion.

We tossed chunks of crumbling
mortar through the whirling
Timbers. We fantasized
our own park of
inventive amusements.
But deep within,
we sensed the

importance of life,
avocation, and commitment from
a time before
us.

Deeds

Floyd County Deed Book "V," page 287, 8 October 1891. Joseph J. Poff to W. J. Poff, ½ interest in 2 acres on which are a store house, granary, two water grist mills, sawmill, still house, storage house with the stills tubs and everything belonging to said distillery, $1,000.00.

D. B. "Y," page 107, 12 September 1893. W. J. Poff and Minnie L. Poff to Joseph J. Poff, their ½ interest in saw mill, granary, still house, grist mill and store house, $1,000.00.

D. B. 32, page 571, 3 October 1906. W. E. Poff to J. J. Poff, 19 acres and store house on the Salem Turnpike and Daniel's Run Road.

115. Henry Wilkes (or Wiltz) Carr Mill

Lick Fork Creek (South Fork of Roanoke River)

LOCATION: Route 644 (Forest Road NE), approximately .5 mile north of Route 645 (Twin Falls Road NE); POWER: Overshot wheel; STONES: 2 sets, wool carding machine and sawmill

History. The H. W. Carr mill stood on land adjoining the site of Phillip Huff's "old mill" (II, 116). There was already a sawmill on this site when Carr, Peter DeWeese, and Andrew DeWeese bought it from Henry and James Huff in 1847. Henry W. Carr's first wife, Mary "Polly" DeWeese Carr, was killed in an accident involving a wool carding machine on June 18, 1851. In 1853 Carr sold to the DeWeeses a grist mill and 15.5 acres adjoining the Huff mill tract, but it is unclear whether he had built the mill on his own or in partnership with the DeWeeses. He apparently retained his one-third interest in the sawmill and twelve acres.

By 1868 Henry Carr (1809–1876) was the sole owner of both mills, having acquired from each of the DeWeeses a one-half interest in the grist mill and a one third interest in the sawmill. T. A. Webb was working for Carr as a miller in 1870. The circular saw noted in the *Products of Industry* report compiled that same year was among the first in the county, there being but one other on the list of 17 sawmills (whose products were valued at over $500) countywide. The *1870 Products of Industry* notes also that H.W. Carr produced 5,000 pounds of carded wool valued at $2,500.

The H. W. Carr mill, or a successor to it, was for a while known as the "Fowler Mill property," having been owned between 1886 and 1891 by N. T. Fowler (1842–1921) and his first wife, Maggie, who bought the mill from H. W. Carr's son, John. A self-described "builder, general contractor, and dealer in lumber and a mechanic"[56] as well as a millwright, Fowler was undoubtedly connected to the "Fowler and Huff" mill on the South Fork of the Roanoke River, a Floyd County mill that remains unidentified, despite having been described in the *Products of Industry* report that accompanied the 1880 United States census (see Table C). The H. W. Carr Mill, or at least the site on which it was built, was owned by Huffs and later by a Fowler, but the years separating those points preclude the suggestion that the H. W. Carr Mill and the Fowler and Huff Mill were one and the same.

Fowler's successor, J. H. Woodward, ran the mill for 22 years before selling it to J. W. Carr in 1913. Fifteen years later Carr sold it to Jasper Poff, who, according to Ralph Stump, kept it going until the early 1940s. Jim Craighead was his miller at some point. The mill stood vacant after Poff died in 1946, and collapsed sometime around 1970. Stump sold most of the equipment, including a steel Fitz water wheel, for scrap. Three millstones were sold to someone from Bedford County.[57]

Deeds and Other Public Records

Floyd County Deed Book "E," page 87, 7 August 1847. Henry Huff and James Huff to Peter DeWeese, Andrew DeWeese, and H. W. Carr, land "embracing the saw mill." Bounds mention Phillip Huff's old mill place and a mill pond.

D. B. "G," page 122, 21 May 1853. H. W. Carr to Andrew and Peter DeWeese, "all of a certain tract of land, 15½acres lying in the south fork of the Roanoke River ... adjoining Huff old mill place."

D. B. "L," page 239, 18 March 1865. Peter DeWeese to H. W. Carr, 2 tracts land, one known as grist mill tract and one known as sawmill tract, one-half interest in grist mill and one-third interest in sawmill.

D. B. "M," page 225, 21 November 1868. Andrew DeWeese to H. W. Carr, 2 tracts land, his one-half interest in grist mill tract of 15 acres known as the old Mill tract of Peter and Andrew DeWeese and his one-third interest in 12 acres known as the old sawmill tract.

D. B. "S," page 131, 18 November 1883. John Mirrett, Commissioner for heirs of H. W. Carr, to John Carr, mill tract containing 77½acres.

D. B. "S," page 389, 8 February 1886. John Carr to N. T. Fowler, 15 acres and mill.

D. B. "U," page 443. 1 March 1890. Deed of Trust, N. T. Fowler to W. L. Howard, trustee. "About fifteen acres of land upon which is situated a grist and saw mill and is known as the Carr Mill property." *The debt secured by this deed of trust was paid off.*

D. B. "V," page 181, 2 June 1891. N. T. Fowler and Maggie Fowler to A. A. Cannaday. About 15 acres "known as the Fowler mill property."

D. B. "V," page 241, 1 July 1891. J. H. Woodward from A.A. and Della Cannaday and N.T. and Maggie Fowler. Land "known as the Fowler Mill property" and "supposed to contain about fifteen acres ... lying on the headwaters of the South Fork of Roanoke River."

D. B. 39, page 253, 7 June 1913. J. H. Woodward to J. W. Carr, 10 acres and mill. One boundary line to run "nearly parallel with the race."

D. B. 48, page 51, 30 January 1928. J. W. Carr to Jasper Poff, "6 acres, more or less, gross sale, with mill on it." "J.W. Carr agrees for Jasper Poff to have water right Twenty (20) feet wide and also sufficient room for dam at head of race, also to turn water out at a low place specified in race about half distance of the old race at the old dam. J. W. Carr reserves the right to bridge across the race."

See Table B, *1870 Products of Industry.* Name of corporation or individual: Carr, H. W.

Carr is listed in separate entries as the operator of a grist mill, a wool carding mill, and a sawmill (entry not included in Table B).

116. Phillip Huff Mill

Lick Fork Creek

LOCATION: Intersection of Route 796 (Copper Hill Road NE) and Route 645 (Twin Falls Road NE); VISIBLE REMAINS: None are described in Frank Webb's notes, but at this location short, manmade ditches that appear to be two parallel races, one 5 or 6 feet above the other, are visible across Lick Fork from Route 645, just upstream from the intersection. Water could easily have been diverted into the lower one by a dam at the narrowest point of this valley, but no sign of such a dam is visible.

History. Phillip Huff was the first in a long line of Floyd County millers. His son Henry and grandsons Byrd and Robert ran a mill on Little River (I, 1). Irvin Huff (I, 2) was Phillip Huff's great-great-grandson.

In 1792 the Montgomery County court approved the building of a road from Phillip Huff's mill to Samuel Howard's, but neither mill is located by stream in the summarized court order.[58] The only Floyd County deed book references to this mill property appear in the first two deeds to an adjoining tract on which the H. W. Carr grist mill (II, 115) was built after the Phillip Huff mill had ceased to operate. Phillip Huff's mill tract passed out of the family's hands in 1871 when Henry Huff sold to H. W. Carr "said mill tract containing about 12 acres more or less it being all the unsold land of Phillip Huffs old mill place."[59]

Deeds

Neither of these deeds conveys the mill property.

Floyd County Deed Book "E," page 87, 7 August 1847. Henry Huff and James Huff to H. W. Carr, Peter DeWeese, Andrew DeWeese, parcel or tract of land "embracing the saw mill and bounded as follows ... ridge of rocks a corner of Philip Huff's old mill place."

D.B. "G," page 122, 21 May 1853. Henry W. Carr to Peter DeWeese and Andrew DeWeese, "all of a certain tract of land, 15½acres lying in the south fork of the Roanoke River ... adjoining Huff old mill place ... bounded as follows ... corner to Huff's old Mill place."

117. Peter Siner Mill/J. M. Perdue Mill

Lick Fork Creek

LOCATION: Route 645 (Twin Falls Road NE), about .1 mile west of Route 800 (Siner Road NE); POWER: Overshot wheel; STONES: 2 sets and sawmill; VISIBLE REMAINS: Part of dam still visible (1992)

History. The Peter Siner/Jim Perdue Mill was built before 1870, probably by or for Peter Siner himself, who came to Floyd County from Roanoke County after buying land here in 1846. It was dismantled in 1974, more than a century after it was built, by Algier Holt and Stanley Church, two years after Church bought the 6½ acre mill tract from James M. Perdue. The property had by then been in the Perdue family for almost fifty years.

The Siner/Perdue Mill, about 1950. The tiny flue poking through the wood-shingle roof marks this back corner as the "warming room," a refuge for millers and customers during the coldest weather. Millers feared fire almost as much as flood. Explosive dust could be ignited by open flames, overheated bearings, or sparks generated when a piece of iron or steel was accidentally fed between a pair of millstones. The concussion would jar loose more dust, resulting in what amounted to a chain reaction. (Courtesy Linda Casey Wimmer)

Peter Siner's son Phillip installed a steel overshot wheel around 1900 and the Perdues added a concrete mill dam in the 1940s or very early '50s. The dam, which had taken most of the Perdues' savings to build, was washed out not long after it was completed. Afterward, the mill ran for a short time powered by a gasoline engine. The buckwheat mill was sold at some point to Oliver Manning. In the early 1990s Algier Holt had part of the steel wheel and Donald Stoneman had at least one set of mill stones and a large steel drive gear from the water wheel.

In addition to Peter and Phillip Siner and Jim Perdue, millers over the years included (in no particular order): Jordan J. Shilling, Thomas DeWitt Perdue, Willie Callahan Doss, Porter "Snake" Perdue, Branard "Peanut" Perdue, Ballard Perdue, Leslie Perdue, and Mike Perdue.

Sharon Shaver Wood, whose father attended the Siner school, recalled that "when my dad [E. C. Shaver] was a boy, he used to take the corn to be ground. He couldn't lift the sacks himself, but he could get on the mule or horse by himself. Once he was seated, they'd put a sack in front of him and a sack behind him and off he'd go. At the mill, Mr. Perdue would unload the horse, grind the meal and then load the horse again, so that Dad could get the ground meal home."[60]

Deeds

Floyd County Deed Book "V," page 472, 28 March 1892. Peter Siner to Phillip M. Siner, 37 acres land and mill.

D. B. 39, page 494, 29 March 1914. Phillip M. Siner to W. S. Tice, 6½ acres land and mill.

D. B. 43, page 365, 24 December 1918. W. S. Tice to Jasper Poff, 6½ acres land and mill.

D. B. 44, page 583, 24 February 1922. Jasper Poff to J. W. Maxey, 6½ acres land and mill.

D. B. 45, page 295, 20 March 1923. J. W. Maxey to James M. Perdue, 6½ acres land and mill.

D. B. 106, page 471, 21 November 1972. J. M. Perdue to Stanley Church, 36½ acres land, including 6½ acre mill tract.

See Table B, *1870 Products of Industry.* Name of corporation or individual: Siner, Peter

See Table C, *1880 Products of Industry.* Name of corporation or individual: Siner, Peter

118. Ridinger and Goodson Mill

Laurel Creek

LOCATION: On Route 705 (Sowers Road NE), .3 mile West of Route 615 (Christiansburg Pike NE); POWER: Turbine; STONES: 2 sets and roller mill

History. The "Ridinger Mill Dam" is mentioned in a deed to land conveyed to Amos Ridinger by Simeon Alley on June 12, 1877, and recorded in D. B. "O," page 470. The mill had two sets of stones and was equipped with a roller mill, all turbine powered.[61] The roller mill and possibly the turbine as well were likely added not long before the mill was "destroyed by fire on the night of March 30 [1904]. There had been no fire in the mill for two days before the fire. It must have been the work of an incendiary. This is a heavy loss to the owners, as the insurance will not cover but about one half of the loss."[62]

Along with the financial loss, the advanced age of Thomas Goodson's partner and father-in-law, Amos Ridinger (born c. 1822), may have been a factor in their decision not to rebuild, despite the fact that the 25-year heyday of roller mill construction was just getting underway in 1904. Instead, Thomas Goodson (1854–1932) ground corn and feed at another building at the intersection of Routes 705 and 615, about a quarter mile away. The gas engine he is supposed to have employed would have been among the first in the county, as was his abandonment of water in favor of a more compact, convenient, and versatile source of motive power. There was also a blacksmith shop, operated by Grover Light, near the second mill.[63]

Melvin Light recalled that his grandfather Grover Light later owned the gasoline powered mill, but occupied himself mostly with blacksmith work, leaving the milling to his brother-in-law Curtis Iddings. The mill ran up until the early 1950s, after which time Grover Light tore it down and used the wood in a tobacco shed that was still standing in 2007.[64] Whether the "one set of corn rocks" sold along with Thomas Goodson's other personal property on November 11, 1933, came from the burned mill or was a smaller set purchased expressly to be driven by the gasoline engine is unknown.[65]

Deeds

Floyd County Deed Book "T," page 550, 2 June 1888. Amos Ridinger to Thomas Goodson, ⅓ interest in "land and mill property containing by survey 3⅓ acres, sawmill on the same excepted."

D. B "V," page 258, 19 December 1889. Emaline Alley to Amos Ridinger, 3⅓ acres land on Laurel Creek. (*Not the mill tract. See next deed.*)

D. B. 27, page 483, 1 April 1901. Amos Ridinger to S. R. Brame, Deed of Trust, ⅔ interest in mill tract. *This deed states that this 3⅓ acre tract is the same as conveyed to Ridinger by Emaline Alley as recorded in D. B. "V," page 258. The bounds make clear, however, that these are two different pieces of land, although of exactly the same size. Also, Ridinger had sold ⅓ interest in the mill property to his son-in-law Thomas Goodson a year and a half before the Alley tract was conveyed to him.*

D. B. 27, page 532, 24 May 1901. Amos Ridinger to James E. Ridinger, ⅔ interest in mill tract.

D. B. 33, page 428, 25 March 1907. Thomas Goodson to C. W. Stuart, mill, 3¾ acres tract. This land lies "on the North side of the public road" and is not the 3⅓ acre tract on which the Ridinger and Goodson mill stood, yet is identified in the deed as "the Mill tract." *The reference could be to an earlier mill, or the mill tract might once have been larger than 3⅓ acres.*

D. B. 36, page 279, 28 March 1907. James E. Ridinger to Thomas Goodson, his ⅔ interest in mill tract.

D. B. 53, page 269, 14 January 1936. C. R. Goodson, et al., heirs of Thomas Goodson, to G. A. Light, 3⅓ acres, "the same being referred to as the mill property."

119. Hosea Wimmer Mill and Huffville Roller Mill/Ezra Wimmer Mill

Brush Creek

LOCATION: Route 672 (Laurel Creek Road NE) 1 mile north of Route 612/660 (Daniel's Run Road NE); POWER: Overshot wheel, originally wooden, later Fitz steel wheel STONES: 2 sets, later Fitz roller flour mill, finally Meadows Mill; VISIBLE REMAINS: Mill

A mill was at work at the site of the Ezra Wimmer Mill no later than 1867, but this unusual (for a mill) shed-roofed structure was probably erected between 1906 and 1915. The rocky trench on the lower right is the remnant of the overshot wheel pit and tail race. (Ricky Cox photograph, 1992)

standing in fair condition. Wheel pit and dug race clearly visible.

History. The Hosea Wimmer mill is one of several in the eastern end of the county whose histories involve the Huff family. Esom Huff probably built the original mill or had it built, since it first appeared in a deed book record when he sold it and the six to eight acre tract on which it stood to Charles Huff in January 1867. The mill came into the Wimmer family for the first time less than a year later. Hosea Wimmer bought the mill and an unspecified number of acres in September 1868, and by the time he sold it to Harvey Jackson Ridinger in 1906, it was well known as the Hosea Wimmer mill property.

Marks left by a circular saw on the joists, studs, and siding of the existing mill indicate that the original pre–1868 mill was taken down and a new one built in its place, probably by Ridinger. This was done in or before 1915, as evidenced by a painted inscription on a concrete wall in the mill basement: "Harvey Jackson Ridinger, Merchant Miller, 1915." A few of the hand-hewn sills appear to have been salvaged from an earlier timber framed building, perhaps the original Hosea Wimmer mill. The new mill was probably built specifically to accommodate a roller type flour mill, in this case manufactured by the same people who made Fitz water wheels. Between 1900 and 1920 several other Floyd County millers opted to take down 19th century buildings and start from scratch rather than refit an undersized and aging building.

In 1920, the mill came back into the hands of the Wimmer family when Ezra Parlee Wimmer bought thirteen acres and a mill on Brush Creek from J.D. and Amanda Ekiss, who had purchased it from Harvey Jackson Ridinger in 1916. In addition to the Fitz Roller Flour Mill, which was probably a late nineteenth or early twentieth century addition, the Wimmer Mill operated a single run of stones. Sometime in the 1940s the machinery, including the Fitz roller flour mill, was sold for scrap. At about the same time the Fitz steel water wheel was sold to P. L. Shelor, who in turn sold it to Homer Roberson. It was put into service at the Roberson Mill (I,

27) in 1956 by Homer Roberson and his son Harry, the mill's current owner. The wheel ceased operating when the wooden race collapsed about 1970 but is still in place.

In 1949 the building was bought by Winston Wimmer. Ezra Wimmer ground corn there with a tractor-driven Meadows Mill for several years afterward. Winston Wimmer still owned the mill and land in 1996.

Deeds

D. B. "L," page 533, 10 January 1867. Esom Huff to Charles H. Huff, 4 tracts on Brush Creek. One tract of "6-8 acres embracing a Grist Mill."

D. B. "L," page 584, 12 September 1868. C. H. Huff to Hosea Wimmer, unstated acreage on Brush Creek, "with its appertenances imbracing a grist mill."

D. B. 33, page 84, 8 September 1906. Caroline Wimmer and Hosey Wimmer to H. J. Ridinger, no acreage given, parcel of land containing mill property.

D. B. 42, page 140, 1 November 1916. Harvey Jackson Ridinger to Amanda M. Ekis, 13 acres, "this being known as the Hosea Wimmer mill property." The inclusion of "a new line" in the bounds suggests that the acreage included with the mill expanded at this point.

D. B. 44, page 242, 20 August 1920. J. D. Ekiss and Amanda Ekiss to E. P. (Ezra P.) Wimmer, 13 acres and mill on Brush Creek. "This is known as the Hosea Wimmer Mill Property."

D. B. 70, page 212, 22 August 1949. Ezra P. Wimmer and Launah, his wife, to Winston Wimmer and Garnett Irene Wimmer, his wife, 30.3 acres.

See Table B, *1870 Products of Industry.* Name of corporation or individual: Wimmer, H.

See Table C, *1880 Products of Industry.* Name of corporation or individual: Wimmer, H., Jr.

120. J. P. Conner Mill

Goose Creek

LOCATION: Route 653 (Shawsville Pike NE), 3.25 miles north of Simpson, Virginia; POWER: Overshot wheel; STONES: 1 set; VISIBLE REMAINS: Dug race clearly visible from Route 653

History. The *Products of Industry* report that accompanied the 1870 United States census states that J. P. Conner's mill on Goose Creek could grind 28 bushels a day with its single run or set of stones. Most of the output was probably corn meal. The enumerator reported that the mill was open for only 6 months in a year, but does not say whether these months were consecutive or scattered throughout the year. The U. S. Post Office Department lists a Conner's Mill Post Office at 3.25 miles north of the Simpson, Virginia, Post Office.

Jacob Conner (b. 1813) bought tracts of land in this area in 1842 and 1859, both from Peter Geurrant, Jr. The stream name on the first of these deeds is illegible, but does not resemble any of the names now seen on maps of this area. The second tract of land could have been the future site of Conner's mill, if what now appears on maps as Goose Creek was known as Mill Run in 1859. Several other Floyd County streams now known by more distinctive names are referred to as Mill Run or Mill Creek in 19th century deeds. But if this was already Mill Creek when Conner arrived, what now forgotten miller preceded him on these waters? With its high head and low volume, this stream would certainly have appealed to pioneers with little time to plan or carry out elaborate construction projects, and limited access to the tools and materials required.

Conner bought land from John Light, Sr., his son Nathan's father-in-law, in 1871 and 1874, but these tracts do not adjoin the mill property unless the name Flat Run, which now refers to an upstream branch of Goose Creek, was used farther downstream at that time. Whatever the case, calling any stream in that part of Floyd County "Flat Run" shows as much appreciation for irony as the choice of "Mill Run" shows for practicality.

The mill was not mentioned when Conner's son William distributed one fourth of their father's real estate to his sister Liona Conner Link in 1900, but may still have been there.

Deeds

Floyd County Deed Book "C," page 146, 19 September 1842. Peter Geurrant, Jr., to Jacob. P. Conner, 60 acres on waters of ____ Run waters of South fork of Roan ___.

Deed Book "K," page 198, 7 May 1859. Peter Geurrant, Jr., to Jacob P. Conner, 160 acres on waters of Mill Run, waters of South Fork of the Roanoke River.

D. B. 31, page 414, 20 March 1900. William N. Conner and wife Elizabeth to Liona Link, "release and confirm" ¼ interest in the real estate of Jacob P. Conner, deceased, on the waters of Roanoke River.

See Table B, *1870 Products of Industry.* Name of corporation or individual: J. P. Conner

122. Scott Roller Mill/Floyd Milling Company/O.A. Webb and Son Mill

Dodd Creek

LOCATION: Route 8 (Webbs Mill Road N), 2 miles north of Town of Floyd; POWER: 2 turbines; STONES: Corn rock, roller mill, feed mill; VISIBLE REMAINS: In poor condition: windows broken, roof coming off, leaning toward Dodd Creek.

History. The original grist mill (II, 91) and a sawmill on this site were built by William Morricle (b. about 1777), who was granted permission to build a dam across Dodd Creek in 1835. The second owner, Matthew Scott (1816–1896), rebuilt and raised the dam around 1874 and may have built a new mill as well. In the 1896 settlement of Scott's estate, the mill and land were sold as directed to his son, Winfield "Field" Scott, who built a new mill after the building constructed by William Morricle or Matthew Scott burned to the ground around 1903. Scott's new mill caught fire soon after it was built but was repaired and opened for business in 1904.

Scott's was one of only a few local businesses of that era operating as a chartered corporation, with stockholders, directors, and officers. Scotts Roller Mill Company was organized in January 1902 with Scott himself serving as treasurer and business manager, as well as a director. In May 1909 the board of directors "passed a resolution declaring that it was advisable to change the name of the corporation and to effect this, that the company's charter should be amended so as to change the name of SCOTT ROLLER MILL COMPANY to that of FLOYD MILLING COMPANY INCORPORATED."[66]

That the stockholders approved the change is no surprise, considering that Scott is supposed to have disappeared sometime in 1908 with the funds of the Peoples Bank of Floyd.[67] A chancery suit filed against the Floyd Milling Company, Inc., et al., resulted in a public auction at the courthouse of Floyd County on 27 August 1910. Newspaper notice of the sale described a "modern six-stand Roller Mill of sixty barrels [per day] capacity, with corn, buckwheat, and Attrition Feed Mills attached. This is one of the best roller mills in the county ... and enjoys the largest custom of any mill in the county."[68]

In 1925 the mill was bought by partners O.A. (Oscar Ames) Webb and J. M. Harman. The job of painting their newly acquired business, let out to Horace Bowlin in 1926, required 60 gallons of orange paint. Harman sold his half of the business to Webb in 1934. Webb's son Luther worked alongside him in the mill from the beginning and became the owner in 1956. The appearance of O. A. Webb and his wife as grantors on subsequent deeds suggests that they had retained a part interest in one or more of the adjoining tracts, but not in the mill or mill tract. In 1967 Webb, his wife, and mother sold the mill property and he began to pursue the small engine business already taken up as a sideline at the mill. By the time it closed, Webb's Lawn Mower Service just southwest of the Town of Floyd on Route 221 had become as much a fixture of the community as the mill had been under the management of the Webbs, father and son. Men who worked in the mill for and with one or both the Webbs included Ercil Austin, George Via, Isaac Harman, George Willie Conner, and Roger Agnew.

During their 33 combined years of ownership, the Webbs devoted time and energy to the maintenance of the dam and turbines as well as the complex machinery inside the mill, no doubt dealing many times with situations like the washing away of one fourth of the dam by an overnight cloud burst in August 1942.[69] It was only a matter of time until safer, more powerful, and more reliable electric motors and diesel engines were installed in mills that lasted into the second half of the 20th century. The wooden center sections of the dams, often already in need of repair, were left to rot in place, or in pieces deposited along downstream banks by occasional floods.

To have remained in business for so long, Oscar and Luther Webb must have been both astute and accommodating businessmen, as well as good millers and millwrights. And in a way that few merchants would understand today, they had also to be tactful hosts, since going to mill was an all-day event, as much social as com-

Employees, neighbors, and customers gathered around, inside, and on top of Scott Roller Mill about 1912. Standing on the ground (from left to right) are Ollie, Dexter, Minnie, and Ursula Dickerson (children of Julius Dickerson), Griff Reed, Annie Sutphin, Roy Sutphin, Ike Sutphin (father of Sutphin children), Mrs. Ike Sutphin (holding infant), Harmon Sutphin, Pearlie Sutphin, Josie Dickerson Moses, Hiley Moses (miller and husband of Josie Moses), and Charlie Pugh. The two young men in caps seated on the front of their wagon are unidentified, and may have been from Patrick County. In the next wagon are Roy Moses (reclining, hatless) and his father Charlie Moses. Standing on the horse is Burks Moses (Roy Moses' son). Julius Dickerson, on horseback, is at the far right. On the porch roof are Claud Dickerson (son of Julius), Homer Dickerson, unidentified, Asa Dickerson, Posey Dickerson, and unidentified. (Courtesy Jessie Peterman Memorial Library, Floyd, VA)

mercial, for some customers. In a 1994 interview Luther recalled that selling bottled drinks and keeping a sack of roasted peanuts for customers to snack on lowered the stress levels of the millers "[Customers] wouldn't say nothing about being in a hurry if I had the peanuts out there."[70] To deal proactively with tobacco chewers and snuff dippers, the Webbs are supposed to have posted this message in a prominent location: "If you spit on your floors at home, feel free to spit on ours." One customer claimed that this conditional invitation concluded with the caution, "We'll use your sacks to clean it up." Either way, the point was made without confronting or embarrassing individuals.

Like many other millers, the Webbs were careful to notify the public of plans to close the mill. A customer who loaded a wagon, hitched a team, and pulled the load to the mill only to find it closed might become a former customer or at least a resentful one, if there were no other mills close by. Such notices from this and other mills typically used only a couple of sentences to say what they had to say, but O. A. Webb and son occasionally explained why they were closing: The mill was closed on July 16, 1940 so that Mr. Webb could attend his mother's sale[71]; "Inexperienced labor," resulting in the need for "necessary repairs," and a general cleanup closed the mill from Christmas Eve 1947 until January 5, 1948.[72]

They also went beyond the obligatory season's greeting on at least two Christmases, whose significance few of us now can fully appreciate. On December 21, 1944, the composer of the Webbs' holiday notice wrote, "Wishing you a

According to a 1910 newspaper advertisement, Scott Roller Mill, viewed in this undated photograph from the east side of Dodd Creek, was capable of producing over 11,000 pounds of flour in a single day. Both wide and long, and with a full three stories of work and storage space plus a basement, it is the largest mill ever built in Floyd County. The turbine house alone, perched atop the wooden dam, is larger than many of the county's corn mills. Frank Webb, who helped out here as a boy during summer vacations, believed that the dark mound at the base of the turbine penstock is made of buckwheat hulls. Floyd County was among the top two producers of buckwheat among Virginia counties in the 1920s. Although buckwheat was also fed to cattle, it follows that Floyd Countians were among the top consumers of buckwheat pancakes. (Courtesy Jessie Peterman Memorial Library, Floyd, VA)

Merry Christmas and Happy New Year and to those of you who have husbands, sons and daughters in the service of our country, may peace on earth soon reign and where possible, may homes be reunited once more. Respectfully, O. A. Webb and Son." On December 13, 1945, the news of their seven-day closing was followed by a message of gratitude and hope: "To our many friends may this Christmas be one of two-fold rejoicing. With the rebirth of freedom to all mankind, let us pay tribute, in prayer to those millions who have sacrificed on the far flung fields of battle that people of all faiths and all races might live in lasting peace and harmony. And pray God that their sacrifice has not been in vain. Yours for service, O. A. Webb and Son."

By the time Luther Webb sold the mill in 1967, the flour milling equipment had been removed and the mill equipped to operate as a "Grinding and Mixing Milling Business." Aside from a "4-foot Brush Mountain Rock" for cornmeal, all the equipment listed for sale with the mill in the auction brochure was for the production of animal feeds. The sale brochure mentions that feed franchises, presumably for regional or national brands, were available and that Webb's Mill was the local distributor for Big Spring Mill, Inc., in Elliston, VA, which is still producing much of the flour and meal sold in Floyd County forty years later. The mill's official name for at least part of its last years was Dodd Creek Mill. Joe Harman, Ike Harman, and Ercil Austin were all owners or millers at some point between 1968 and 1978, at which point Joe Harman sold the mill to Lynwood Belcher, who was the last to run it. It is supposed to have run at least until 1985[73] but had certainly ceased operation by the time it was bought by Kenneth Crawford in 1988.

Deeds

D. B. "Z," page 236, 6 June, 1896. Samuel Scott, executor for Matthew Scott, sold to Winfield Scott for $1,800 as directed in will. Land on the south fork of the Little River, being a part of the William Morricle tract on which the mill is situated, and it contains 20 acres and an adjoining upstream tract.

D.B. 29, page 84. 12 November 1902. Winfield Scott to Scott Roller Mill Company. This conveys all his right, title and interest in mill and land "upon which Roller Mill is now situated containing 5 acres more or less."

D. B. 37, page 188. 2 December 1910. Chancery Cause, Agnew and Proffit, Commissioners. Floyd Milling Co. to Proffitt, Proffitt, Lawson and Smith, mill bought at public auction.

D.B. 38, page 207. 26 February 1912. Chancery Cause, Agnew and Burwell, Commissioners, to Floyd Milling Company. Partners organized as the Floyd Milling Co. bought the mill at public auction. Proffit, Proffit, Lawson and Smith evidently sold the mill after December 1910.

10 December 1919. Floyd Roller Mill (H. L. Lawson, S. G. Proffitt, and S. E. Proffit) to Chank Weeks, land and mill known as Scott's Roller Mill.

D. B. 45, page 281, 1 June 1923. Chank Weeks to Kyle Weeks, Trustee, Deed of Trust, Scott Roller Mill and two other tracts to secure loan of $3,928.74.

D. B. 46, page 110, 6 April 1925. Kyle Weeks, Trustee, to J. M. Harman and Oscar Ames Webb. Scott Roller Mill on 13 1/20 acres and ½ acre tract "across the Riner road from the mill house above mentioned. Chank weeks defaulted on his deed of trust to Kyle Weeks. S. G. Proffit bought the mill at public auction and sold it to John B. Hylton, but no deed was recorded for either sale. All now agree to sell mill property to Harman and Webb. The tract across the road belonged to Peter Dickerson and wife. Along with the mill, Harman and Webb received from John B. Hylton "all of the corn in said mill" and "one little red six-year old cow in the possession of O. A. Webb."

D. B. 51, page 597, 5 November 1934. J. M. Harman to O. A. Webb, his ½ interest in mill.

D. B. 78, page 297. 15 August 1956. O.A. Webb and Lyona F. Webb to Luther H. Webb, "all that certain tract or parcel of real estate known as O. A. Webb and Son Mill." Waters of Dodd creek. Being part of the same real estate which was conveyed to J. M. Harman and Oscar Webb from Kyle M. Weeks, trustee, et al., deed, D.B. 46, page 110, April 6, 1925. Includes grain room extension reserved in deed to Luther H. Webb, D.B. 75, page 104, 2 February 1954.

D.B. 83, page 560. 31 December 1960. Luther H. Webb and Hattie D. Webb to Luther H. Webb and Hattie D. Webb. To "create a tenancy by the entirety with right of survivorship," 4 tracts. Fourth tract is Webb Mill House and lot from D.B. 78, page 297, August 15, 1956.

D. B. 91, page 96, 8 October 1965. O. A. Webb and Lyona F. Webb to Luther Hayden Webb and Hattie D. Webb, 1/6 acre. It is described and refers back to a deed in DB 46, page 110, dated April 6, 1925.

D. B. 94, page 320, 28 November 1967. Luther H. Webb, Hattie D. Webb and Lyona F. Webb, widow, to Emmet Raines, 6 tracts. Tract 4 is Webb Mill House. From D.B. 83, page 560.

D. B. 94, page 544, 24 February 1968. Emmet M. Raines to Joseph E. Harman and Norma A. Harmon, 3 tracts of land. Tract 1: 40 feet by 34 feet, known as shop lot. Tract 2: Webb mill house and lot, Tract 4 in DB 94, page 320, November 28, 1967. Tract 3: 1/6 acre.

D. B. 130, page 729, 14 September 1979, Joseph Harman to Lynwood L. Belcher, 3 tracts. Same as in DB 94, page 544 above.

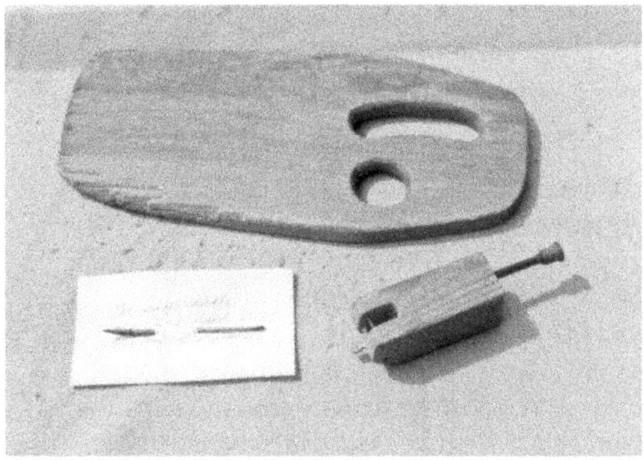

Tools and toys from the O. A. Webb and Son Mill. The miller's paddle scoops freshly ground cornmeal into the mouth of the customer's sack, which is held open by the miller's left hand and two nail points embedded in the upper edge of the meal box. The sack needle was an essential item before the arrival of hand-held electric sack stitchers. The third item is for playing practical jokes. When the miller demonstrates this lung power "tester" by blowing into the metal pipe, a wooden propeller spins. On the pretense of wiping the mouthpiece, the miller moves the pipe to a different hole before handing it to the unsuspecting contestant, who ends up with a flour coated face. (Ricky Cox photograph, August 1996)

133. John Stuart Mill

Laurel Creek

LOCATION: Route 705 (Sowers Road NE), .2 mile east of Route 616 (Otey Road NE). Location is marked by passage of power lines directly overhead. POWER: Tub wheel;

STONES: 1 set and sash type sawmill; VISIBLE REMAINS: Large stones stacked among natural stone formation that was the basis of the dam

History. Definitive written records for this mill are limited to the *1870 Products of Industry* report referred to below and *Chataigne's Virginia Gazetteer* (see Appendix A) which lists Stuart as the owner of a mill and sawmill in the vicinity of the Camp Creek, Virginia, Post Office in 1880, 1885, and 1890. The absence of a reference to a mill in the deeds to the three tracts of land Stuart acquired before 1870 suggests that he built the mill or had it built, but cannot be taken as proof positive since numerous deeds fail to mention mills known to have been standing upon the tracts being conveyed to a new owner. The single deed making specific reference to the "John Stuart Mill pond" almost certainly does not convey the tract of land upon which the mill stood. Only after a close study of the boundaries, one of which crosses the mill pond, could even the last deed be safely said to convey the actual mill property. Following that determination, careful comparison of boundary lines might identify which of the original tracts was then or afterward developed into a water power site.

The site of John Stuart's mill dam is among the most beautiful and naturally suited to the construction of a dam in the county. The use of a tub wheel suggests that the mill stood close by the dam, since the extra fall gained by building the mill downstream and channeling water to it with a race would have been of more benefit if applied to an overshot vertical wheel.

Deeds

Floyd County Deed Book "F," page 84, 7 January 1850. A. J. Kirby, Trustee for Abijah and Rila Booth, to John Stuart, 80 acres, Laurel Creek. Land sold to settle case of *George Booth vs. Abijah Booth.*

D. B. "L," page 405, 1 April 1867. Samuel T. Evans to John Stuart, 3 tracts of land, 1 of which is 60 acres on Laurel Creek and deeded to Abijah Booth from Asa L. Howard.

D. B. "L," page 744, 6 February 1869. J. L. Thompkins, Commissioner, to John Stuart, 230 acres land on Laurel Creek. Land sold to settle a dispute or debt between J.L. Howard and another party named Howard or acting for a Howard.

D. B. "Q," page 511, 28 March 1882. T.W. Spindle and Lavina C., and R. Grayham and M.E. Grayham, his wife, to John Stuart, 8 acre tract, "begins on North side of the bank of Laurel Creek at head of John Stuart Mill pond" with one line running "to a branch that runs in the mill pond opposite John Stuart house."

D. B. "U," page 167, 10 November 1886. S.B.B. Stuart and James E. Lawrence, Trustees for John Stuart heirs, to Jessie Edward, 47 3/4 acres land "on the waters of Laurel Creek." Boundary descriptions include a line "crossing Mill pond."

See Table B, *1870 Products of Industry.* Name of corporation or individual: Stuart, John

---

140. Kinsey Mill

Brush Creek

LOCATION: Between Flemon Wimmer Mill (II, 105) and Hosea Wimmer Mill (II, 119) POWER: Overshot wheel; STONES: 1 set, also sawmill

History. Armed with plotted maps of the two Wimmer Mill tracts (II, 105, and II, 119) and some familiarity with the old churches and homesteads along this part of Brush Creek, a patient researcher might either locate the Kinsey mill precisely or determine that it never existed. As noted in the history of the Flemon Wimmer Mill (II, 105), the deeds summarized below were removed from the records of two mills whose existence is unquestioned, because the deeds don't fit smoothly into the ownership history of either mill. Considered together, they appear to document a mill whose name and location have passed out of the living memory of that community.

The boundaries of the eighteen-acre tract conveyed by Burwell and Phoebe Wimmer and Mrs. John (Mary) Weaver to William R. Kinsey in 1874 refer to an adjoining property known as the "Kinser [probably Kinsey] Mill tract." This is almost certainly the O. T. Kinsey Mill on Brush Creek described in the *Products of Industry* report completed along with the 1870 census. O. T. (Otey T.) Kinsey was a blacksmith, and apparently a busy one. Two young blacksmiths, possibly apprentices, were part of his household in 1860. He is listed as a blacksmith in Montgomery County, Virginia, in the 1850 census, and the same in Floyd County in 1860 and 1870, and probably hired a miller to run the mill for him. If he died c. 1873, the chancery suit referred to in Deed Book "Q," page 467, may have been filed by his administrators against John Weaver, et al., as creditors or part owners of the mill.

Partners Bird Huff and William R. Kinsey, son of O. T. Kinsey, bought the mill at a court-ordered public auction on 21 March 1874 , but their ownership was not confirmed in a deed until May 1877. What happened to the mill after Kinsey acquired Bird Huff's one-half interest in 1881 remains a mystery. William R. Kinsey was a house carpenter living with his wife Emma, and son Walter, in Auburn, Montgomery County, when the population census of 1880 was taken.

Deeds

Floyd County Deed Book "D," page 356, 16 September 1846. John P. Lawrence to John Weaver, Jr., and Jacob Weaver, Jr., 5 acres on Brush Creek for purpose of a mill seat.

D. B. "K," page 494, 14 August 1857. John and Jacob Weaver to Esom Huff, 8 or 10 acres plus or minus, embracing a saw and grist mill on Brush Creek.

D. B. "O," page 467, 25 May 1877. J. L. Thompkins, Commissioner in Suit, *Otey Kinsey vs. John Weaver*, to William R. Kinsey and Bird J. Huff, 8 or 10 acres and mill property, "being the land and mill property conveyed by John Weaver and others to Esom Huff by deed dated 14 August 1857." Kinsey and Huff bought the mill at public auction in March 1874, but details of the chancery suit that forced the sale are not stated.

D. B. "Q," page 410, 17 March 1881. Bird J. Huff to William R. Kinsey, his 1/2 interest in mill and property "known as the Kinsey Mill," same as sold by Tompkins.

See Table B, *1870 Products of Industry.* Name of corporation or individual: O.T. Kinser (assumed to be O. T. Kinsey)

Section III. Mills West of Route 8 and North of Route 221

62. Presley Buckner Mill

Deep Water Creek

Location: Route 611 (Macks Mountain Road NW) just above (NE of) Carroll County line; Power: Overshot wheel; Stones: 1 set; Visible Remains: No trace, 2 stones lay between the road and stream until after 1993.

History. A pair of millstones, one with a large tree root grown through its center, lay between Deep Water Creek and Route 611 until around 1993. A local resident believes that a mill was operated at or near this site by Presley Buckner, who bought 144 acres on Deep Water Creek in 1925. If the mill ran even for a short time after 1925, it is strange that it was not better remembered by older residents of the neighborhood. Possible explanations are that Buckner ran the mill for someone else, at some earlier time, before buying land in Floyd County. It is possible also that the mill tract was on land purchased in an earlier transfer recorded in Carroll County instead of Floyd.

Presley Buckner is known to have operated a steam-powered sawmill on the Ed Turpin farm just over the hill on Berry's Creek. Turpin operated a small gristmill at the same spot, one of the few ever driven by steam power in Floyd County. According to J.R. (Roosevelt) Turpin, the corn and sawmills were operated only one day a week, probably Friday or Saturday, as was often the case with small-scale grist mills producing only corn meal or animal feeds. The owner and operator of such a mill was usually a full-time farmer as well, and ran the grist mill as both a service to neighbors and a means of acquiring extra corn for himself. The miller's toll, which could not exceed one-eighth of the corn brought to the mill for grinding, was calculated by volume rather than weight, and was commonly measured in a wooden box or "toll dish" that held exactly one-half of one gallon. This is one-eighth of the typical half-bushel (about 28 pounds) "turn" of bread corn carried by patrons coming on foot or horseback to the mill semi-weekly, or even weekly for very large families.

Deeds

D. B. 47, page 407, 13 May 1925. Lisco Hylton to Presley Buckner, 144 acres on Deep Water Creek.

63. Randolph Phillips Mill

Ran's Creek

Location: Route 619 (Phillips Road NW) .25 mile west of Route 756 (Hollandsworth Road NW); Power: Overshot Wheel; Stones: One

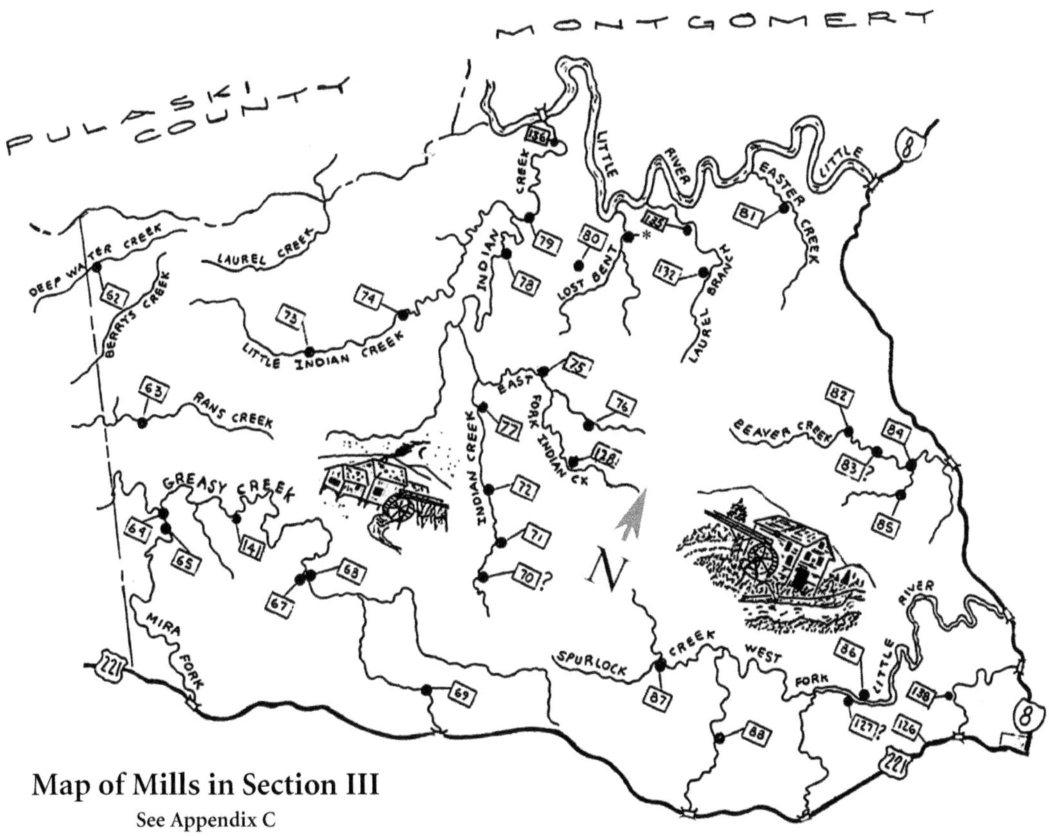

Map of Mills in Section III

See Appendix C

set; Visible Remains: Stacked stone foundation is clearly visible. A pair of millstones lay near the mill foundation into the 1990s. They may have been covered when the road was widened and paved c. 2000. Traces of the dug race remain at its lower end.

History. Randolph Phillips, Sr. (1797 or 1799–1870), was "listed as one of the county's large land owners" when the Floyd County was formed from Montgomery County in 1831. Six of his eight or nine sons served in the Civil War and at least four of them — Randolph, Jr., Enoch, Noah, and Henderson — were afterward involved with building or running water-powered grist mills.[1] Henderson and Enoch were both connected to the Enoch Phillips Mill (III, 74) on Little Indian Creek and in 1874 Henderson sold to Randolph, Jr., the site of the mill under discussion. After Randolph built a mill on the property, Henderson bought it back in 1883, in partnership with Noah.

The "Randal" Phillips listed as the owner of a mill on Ran's Creek in the *Products of Industry* report for 1870 could have been Randolph Phillips, Sr., who "died not long after the census was taken" in 1870.[2] If so, his mill may have been on the same two-acre site now under discussion. The deed written when Henderson Phillips sold this property to Randolph, Jr., in 1874 refers to a mill site (rather than a mill) and mentions the race several times in describing the boundaries of the property. Knowledge of when and how Henderson acquired this tract might settle this question, but the fact that the whole family bought and sold land regularly makes it difficult to trace a single small tract.

The mill identified in 1870 might also have belonged to Randolph Phillips, Jr., who is, like his father, listed as a miller in the 1870 population census. In this case, the *Products of Industry* record may have been describing a completely different mill, upstream from the site now under discussion, and not otherwise treated in this book. The reports from 1870 and 1880 differ in some particulars, but not enough to confirm that they refer to different mills. Also, points of comparison are limited by the lack of detail in the

1870 report. Support for the notion that these are different mills is found in Elza B. Cox's genealogy of the Phillips family. Cox wrote that Raney Phillips, for whom Ran's Creek was named, "lived at, or what was to become known as, Turman, VA, which was named after the Elijah Turman (1790–1868) family, early settlers, operating a Mill, General Mercantile Store and Post Office in the Turman Community." [3] The home of a younger Elijah Turman and Emazetta Phillips Turman (1854–1949), Raney Phillips's only child, still stands behind the Truman and Lala Spence house at what was once Turman, VA. The presence of a second Phillips Mill on what would be a northern fork of Ran's Creek is consistent with James "Arthur" Bond's recollection that at one time Raney Phillips had a mill farther up the creek (from this mill), near a little waterfall, below the Arthur White house still standing at 940 Hollandsworth Road NW. Bond (b. 1898) did not remember the mill running.[4] Whether or not it happened this way, Phillips's abandonment of an upper mill after 1870 and his purchase of the downstream site in 1874 would have made sense in light of the fact that even the lower mill, which stood below the fork in Ran's Creek, was eventually closed because of insufficient water.[5]

The mill named in the 1880 *Products of Industry* report most likely was at the site now under discussion and Randolph "Raney" or "Ranney" Phillips, Jr. (1834–1910), was almost certainly* its owner. Tom Lester was a miller here, as was Alex White, who ran the mill for Crockett "DeWitt" Phillips, Bennet "Graydon" Phillips, and Elkanah Phillips, all sons of Henderson C. Phillips. White is listed as a grist miller in the population census of 1900. Arthur Bond was old enough to walk the two miles from his home on Greasy Creek by himself to inspect the new wooden wheel built by Ben White, Alex's father. Bond's description of himself as "grown" when the mill was closed due to low water a few years later suggests a date between 1910 and 1917, when he would have been between 12 and 19 years old.

Vinton Gallimore (b. 1922) dreaded having to walk past the mill in the dark on the way home from his grandmother Liona "Sis" Montgomery Gallimore's house at the mouth of Ran's creek. Voices and a light inside the mill encouraged him to believe that the building was, as rumored, haunted, until he learned that it was a meeting place for card players, having been abandoned as a mill years before.[6] The small, box-like building was sided with vertical boards, like a farm building, and was not recognizable to her as a mill when Irene Hollandsworth, Vinton Gallimore's future wife, passed by in the late 1930s.[7]

Deeds and Other Public Records

Floyd County Deed Book "N," page 362, 26 November 1874. Henderson C. Phillips to Randolph Phillips, Jr., one mill seat and race, 2 acres.

D. B. "Y," page 116, 24 May 1883. Randolph Phillips to Henderson C. and Noah Phillips, one grist mill and 2 acres of land on "waters of Greasy Creek."

Floyd County Will Book 1, page 203, 19 August 1897. Henderson Phillips leaves to his sons Ellcany, Crockett Dewitt and Bennett Graydon the home tract of 196¼ acres. "Mill is on same tract of land."

D. B. "Y," page 548, 20 November 1897. J. L. Thompkins, Commissioner in suit of *Henderson Phillips vs. Noah Phillips*, sells unto Elkanah Phillips, Crockett DeWitt Phillips, and Bennett Graydon Phillips, all sons of Henderson Phillips, 2 acres and mill.

See Table B, *1870 Products of Industry.* Name of corporation or individual: Phillips, Randal

See Table C, *1880 Products of Industry.* Name of corporation or individual: Randolph Phillips

64. Phillippine Mill/Phillips Mill/Vaughn's Mill

Mira Fork Creek

LOCATION: Route 759 (Vaughn's Mill Road NW) .1 mile from Route 622 (Indian Valley Post Office Road NW); POWER: Turbine, then

**Local men hired to "enumerate" the industrial output of a given district sometimes filled in the blank labeled "name of corporation or individual" with the name of a popular or long-time miller or owner who was not the owner at the census date and in some cases had been dead for several years. Whether these individuals hadn't time to confirm ownership or simply misinterpreted the instructions, the presence of such errors in some records casts a shadow of doubt over all the information reported by that collector.*

A traveling photographer assembled this group at the Phillippine Mill around 1915. Left to right: Susan Marshall, Agnes Phillips, Lena or Mattie Phillips, Mary Cox, ? Cox (Mary's sister), Garfield Slaughter, Lynch Slaughter, unidentified, Bob Phillips (seated in doorway), Jim Mitchell (driver of the horses and wagon), Landon Harris, Early Phillips, Otis Phillips, Golden Harris (seated on ox), Reed Phillips, owner (in vest and black hat), George Slaughter, Wise Quesenberry, Garmon Phillips, and Samuel Jett Phillips (on the second story platform). Agnes, Mattie, Lena, Early, and Jett were the children of Reed Phillips, who also owned a store and the Phillips, VA, post office at the intersection of what are now Vaughn's Mill Road and Indian Valley Post Office Road. Bob Phillips, who worked at the mill, was married to Reed Phillips's oldest daughter, Josie. Agnes's future husband, Ephrem Vaughn, would own the mill for over forty years. (Ricky Cox collection)

22' by 3' Fitz overshot steel wheel, then diesel and electric; STONES: Sawmill (very early), Salem roller mills, buckwheat mill, corn rock; VISIBLE REMAINS: Mill still making feed and retailing meal, flour, and animal feeds in 2010. Natural stone abutments and remnants of the wooden dam remain c. 300 yards upstream. Dug race clearly visible along tree line on west side of stream. At the site of the dam, chiseled in the streambed, is a circular hole (perhaps to house a tub wheel) that is believed to mark the site of the Tobias D. Phillips (III, 65) mill or an earlier mill.

History. Located on the western edge of Floyd County, between Indian Valley and Dugspur (Carroll County), Vaughn's Mill is the only commercial mill still operating in Floyd County in 2010. The two and one-half story frame structure was built about 1900 by brothers Fleming "Reed" Phillips (1866–1934) and Bryant Phillips (1861–1941) and their brother-in-law Will Quesenberry. Their father and father-in-law, Tobias D. Phillips (1826–1885), owned a grist mill (III, 65) and probably a sawmill at about the same place or, more likely, upstream near where the Vaughn's Mill dam was located. The mill was officially named "Phillippine Mills," perhaps as a result of the recently ended Spanish American War, but was probably referred to as the Phillips Mill, since it stood just out the road from the Phillips Post Office and store, also owned by Reed Phillips. Bryant Phillips also had at least one sideline, supplying hand-sawn soapstone for building foundations, chimneys, hearths, spring bowls, and grave markers from a quarry near his home, one of several along a vein of soapstone that surfaces at several places in Floyd County.

The conspicuous absence of a roller mill in the first deed book reference to the mill, in which a one-half interest was sold to the skilled miller and millwright J. A. "Hiley" Moses and William Quesenberry, suggests that Moses was brought into the business specifically to oversee the outfitting of the new mill, including installation of the 3-stand roller mill system manufactured by the Salem Machine Works in Salem, Virginia. A mechanics lien filed on January 15, 1902, valued one eighteen-inch (turbine) water wheel, three roller mills, a receiving separator, a rotating bolter, a bran duster, a scourer and polisher and various other machinery, plus the cost of installation, at $1,600. The lien was for the unpaid balance of $1,100.[8] Subsequent deeds mention a buckwheat mill and corn rock as well as the roller mill. Feed grinders were no doubt included among the miscellaneous machinery. The sawmill is mentioned only in the first deed, and may have been a holdover from the original Tobias Phillips Mill. Arthur Bond, himself a sawmiller for much of his life, recalled that the sawmill was of the circular type, as opposed to the more primitive "up-and-down" sash saw. The teeth were cut into the rim of the saw plate,* requiring much more filing and "gumming" than modern circle saws, whose inserted teeth can be replaced individually when badly worn or damaged. According to Bond, Reed Phillips would set the log carriage in motion and sit down to read the newspaper while the cut was completed. The unedged boards were used, among other things, as sheeting beneath wooden shingle roofing.[9]

The Phillippine roller mill was first powered by a turbine wheel, as were most water-powered roller mills, primarily because of the turbine's higher shaft speed. But most roller mills were built on streams larger than Mira Fork (referred to as "Miry" Fork in some 19th century deeds) which joins the much larger Greasy Creek about a quarter-mile downstream. A turbine's immense consumption of water, compared to that of an overshot wheel, was not well understood by millers of that era, or they were anxious to believe salesmen or literature that emphasized the turbine's many legitimate advantages over a water wheel. These included lower weight, smaller size, smaller initial cost, greater horse power, higher shaft speeds, and greater mechanical efficiency. What they weren't told or did not fully comprehend was that while both the power and efficiency of a turbine could exceed those of a vertical wheel at maximum flow, both diminished rapidly as the volume of available water dropped below "full gate." According to S. Jett Phillips, who owned and ran the mill from 1931 to 1932, his predecessor, James "Odie" Marshall, replaced the penstock and turbine with an overshot steel wheel salvaged from the Willis Mill (III, 69), which was torn down around 1930.[10]

Ephrem Vaughn, who owned the mill from 1933 to 1975, added a fourth stand (with two pairs of rollers) to the roller mill and a middlings grinder to boost the quality and volume of flour. The Salem swinging sifter was replaced with a newer model which vibrated less in operation because of a movable counterweight, but had to be held when starting until the spring-loaded weight was moved into position by the motion of the sifter. A reel-type sifter was added as the last step in the flour making process, in case of a torn silk in the swinging sifter. The millwright hired to oversee these changes, a Mr. Cooper, was so severely arthritic that he had to be carried up and down the stairs in a chair. A local carpenter, Peyton Quesenberry, supplied the additional spouting required by the new machines. Around 1950 Vaughn replaced the cast iron ring gear on the overshot wheel with one from a wheel of identical height, bought from the owners of the Early Marshall Mill in the northeast corner of Carroll County, VA. They had disassembled their mill and moved it away from the creek to a safer and more convenient location, where it was powered by some combination of gasoline engine and electric motors.†

The two water wheels were essentially the same except for the configuration of the segment gear whose teeth faced inward on the original wheel and outward on the new one, an arrangement that was less likely to break gear teeth.[11] The segment gear is a large ring gear, made of separately cast identical segments, each representing some equal fraction of the total 360

**A saw of this type is on display at Mabry Mill (IV, 60).*
†This mill still stands on Route 611 (Deepwater Road) in good condition but far from any stream large enough to drive it.

Vaughn's Mill in an idle moment, around 1950. (Courtesy Gene Vaughn)

degrees. The segment gear sections are bolted to the spokes of the wheel and mesh with and drive a pinion gear affixed to the end of the main drive shaft. Placing the segment gear as near as possible to the periphery of the wheel drives the pinion and main shaft at much higher RPM than would be the case if power were taken from the slowly turning axle, as it usually was with wooden wheels. This adaptation countered one of the main advantages, higher initial shaft speed, that the turbine enjoyed over a vertical wheel, but required a more highly engineered all-steel wheel, and more precise manufacture and installation of the component gears, bearings, and shafting than was the case with more primitive drive trains.

A more visible and substantial change had been made to the water power system before 1943, by which time the pond behind the 50-year-old dam was filling with silt. Plans were made to replace the dam and possibly the wooden "box" race with new construction, made at least partly of concrete. But Ephrem Vaughn decided instead to extend the dug race upstream to a point near the old dam. This required hand drilling and blasting in the solid rock that formed the western abutment of the dam. Sammie Phillips held the drill and cleaned out the rock dust while Kinser Quesenberry hammered. Cutting a channel through the rock had been avoided originally by building a wooden race that angled from the center portion of the dam toward the bank and downstream. This allowed the wooden and dug races to join at a point below the bed rock that formed the western end of the dam.[12] When the blasting and digging were complete, the stream was simply blocked with a low dam of loose stone which, except for leaks in the rough dam, diverted the entire stream into the race. The full volume of the stream was now available at all times to drive the wheel, but at no time could flow exceed that amount since the mill pond and its capacity to store water for use during periods of low water had been eliminated.[13] When the wooden race that carried water from the dug race to the wheel collapsed in the late 1940s, the wooden support poles and cross braces were replaced by the cinderblock piers still in place in 2010. Two sections of steel pipe, standing one on either side of the wheel, supported the portion of the race that extended out over the wheel.

Eliminating the mill pond was not as shortsighted as it might first appear. The cost of building and maintaining a new dam and connecting race had been saved, and the constant threat posed to the mill and other downstream property, in the event of a washout, eliminated. The head or vertical fall provided by the height of the dam had been regained by extending the race upstream. The only feature of the mill pond that had been sacrificed, its capacity for storing water, had not been as useful as would have been the case on a larger stream.

Replacing the thirsty turbine with a water-saving overshot wheel twenty years before had only temporarily solved the problem of insufficient water. As the business grew to include the production of livestock feeds, so did the demand for horsepower. The flow of tiny Mira Fork had proven inadequate during droughts, even when the mill pond was intact. At the best of times, the two main operations of the mill, making flour and grinding feed, could not be carried on simultaneously. Switching from one process to the other required the shifting of a heavy power transmission belt from one pulley to another. Badly needed supplemental power was at first supplied by placing a Fordson tractor on a second-story platform between the mill and the adjacent hillside and connecting the tractor by flat belt to a line shaft above the flour sifter. Later, the same Fordson tractor, followed by a larger and more powerful Oliver tractor, was installed in the basement of the mill, and connected by belt to one of the main drive shafts of the mill when the flour mill was in operation. To accomplish this, the tractor had to be loaded onto a flat bed truck which was driven into the creek and backed up under the mill. Eventually a stationary Case power unit took over this duty.

The only way to extract more horsepower from the water wheel for an extended period of time would have been to build the new dam even higher than the old one to increase the fall and expand storage capacity. A wider and/or taller wheel would have been required to take advantage of the increased volume and greater head. A taller dam would have flooded more land and increased the danger of a catastrophic washout, and would only have delayed the inevitable conversion to more flexible and reliable sources of motive power. Investing in a new dam and significant upgrades to the water power system would only have extended Vaughn's commitment to a technology that was already near its useful limit, while offering only a short term solution to a chronic shortage of horsepower.

A large diesel engine was already driving the hammer mill used for grinding feeds when water power was completely abandoned in favor of electric motors. A *Galax Gazette* article from July 1969 states that the changeover occurred "ten years ago."[14] Similar transitions were happening at other mills as the demand for finely ground products like flour and meal diminished and the market for coarse, bulky feeds for dairy cattle and livestock grew. Ephrem Vaughn's son Gene, who joined the business after completing military service, also ran a small engine dealership and repair shop on the same premises from 1960 to 1993, selling and servicing Snapper lawnmowers and Homelite chainsaws. In 1983 Sammie Phillips sold his half of the business to Gene Vaughn and moved to the Boonville Flour and Feed Mill in Boonville, North Carolina, which his son, daughter, and grandchildren are still operating, with his help, in 2010. Phillips (born 1924) went to work in his Uncle Ephrem's mill at age 14 and, aside from service in the South Pacific with the U.S. Marines during World War II, has been a miller ever since.

Gene Vaughn shut down the flour mill in 1991 and since then has focused on grinding and retailing animal feeds to a dwindling number of local farmers. Once a mainstay, the dairy feed business has all but disappeared. Two of his sons, the great-grandsons of Reed Phillips, have managed the mill with part-time help from family and former employees.

As with so many other Floyd County mills, kinship by both blood and marriage figures heavily in the ownership record of this mill. Briefly, brothers Reed and Bryant Phillips sold shares in the mill to their Uncle Randolph Phillips and brother-in-law William Quesenberry. N. P. (Noah Preston) Phillips was a brother to R. T. "Bob" Phillips, husband of Reed Phillips' oldest daughter Josie. Early Quesenberry was married to Reed Phillips' niece, Susan Marshall Quesenberry. Samuel Jett Phillips (Sammie Phillips' father) was Reed Phillips' son. Ephrem Vaughn, the father of Gene Vaughn, was married to Reed Phillips's youngest daughter, Agnes, hence a brother-in-law to Jett Phillips. Connections to the Quesenberrys have not been explored, but two of Reed Phillips' sisters married into this large family.

The very last in a very long line of commercial mills in Floyd County, Virginia, August 2010. The original structure is visible only at the peak of the gable. The warehouse space to the left was added c. 1953, spanning Mira Fork Creek. By 2010 regional and national brands were supplying all the flour and cornmeal purchased in Floyd County. Bulk deliveries to the few remaining dairy farms were easily handled by large regional mills in adjoining counties, using larger versions of the truck parked in the foreground. All that remained was low-volume walk-in trade from horse owners, hobby farmers, and the handful of locals raising a few hogs or keeping chickens. (Ricky Cox photograph)

Deeds

Floyd County Deed Book 28, page 406, 3 February 1902. Alcy Phillips, F. R. Phillips, and Mary, his wife, and B. P. Phillips and C. L., his wife, to J. A. Moses and William L. Quesenberry, "½ interest in a mill house and 1/2 one half acre and the water power sufficient to run the mill and when the parties of the 2nd part is not using the water to grind with the parties of the 1st part is to have the waters to saw with."

D. B. 31, page 182, 30 December 1904. J. A. Moses and William L. Quesenberry to Sarah and Randolph Phillips, his ½ interest in roller mill, machinery, and water power.

D. B. 34, page 228, 30 November 1907. Fleming Reed Phillips and wife to Randolph Phillips, "one fourth of a Roller Mill and machinery corn mill buckwheat mill machinery water power" and one fourth of land and mill house.

D. B. 34, page 445, 9 March 1908. Randolph Phillips and Sarah, his wife, and to Mary O. Quesenberry (resident of Carroll County, Virginia), "¾ of one undivided interest in roller mill buildings, machinery and water containing 4 acres and 103 poles."

D. B. 36, page 160, 22 November 1909. Bryant P. Phillips and wife to Mary O. Quesenberry, one fourth interest in roller mill machinery, buildings, and water power, 4 acres and 103 poles.

D. B. 38, page 220, 23 February 1912. M. O. Quesenberry and M. Quesenberry, her husband, to R. L. (should be R. T.) Phillips and Noah P. Phillips (brothers), her entire interest in "certain piece or parcel of land ... together with the buildings machinery and water power containing 4 acres and 103 poles of land upon which the buildings are situated."

D. B. 39, page 204, 5 July 1915. N. P. Phillips, and Nina, his wife, to Early C. Quesenberry, their half interest in 4 acres and 103 poles, "with building, machinery and water power."

D. B. 40, page 458, 14 January 1915. R. T. (Robert Terry) Phillips and Jocie C. Phillips to G. E. Horton, "one half interest in Roller Mill Building Machinery and water power containing 4 acres and 103 poles of land."

D. B. 43, page 90, 4 October 1917. L. E. Horton to E. C. and G. L. Quessenberry, "one half interest in Roller Mill Building Machinery and water power containing 4 acres and 103 poles of land."

D. B. 44, page 4, 5 September 1919. E. C. Quesenberry and Susan E., wife, to Albert E. Quesenberry, his ½ interest in roller mill, building, machinery, water power and land (4 acres, 103 poles) upon which the buildings are situated."

D. B. 44, page 514, 29 March 1921. A. E. Quessenberry and Ocie Quessenberrry, his wife, and G. L. Quessenberry and Caroline Quessenberry, his wife, to Everett

Phillips, 4 acres 103 poles, one roller mill, buildings, machinery, water power and dam.

D.B. 45, page 12, 15 February 1922. Everett Phillips and Lucy E. Phillips, wife, to Richard A. (identified as "A. R. Quesenberry" in the deed) Quessenberry and J. O. Marshall, "certain 2 tracts or parcels of land, roller mill, building, machinery, water power and dam ... on the waters of Mira Fork." First is 4 acres, 103 poles. Second is 24 square poles. Refers back to DB 44, p. 514.

D. B. 49, page 548, 3 January 1931. James O. Marshall, and Ada, his wife, to S. J. Phillips, his ½ interest in roller mill and 2 tracts of land, 4 acres and 103 poles and 24 square poles. Mill is on first tract.

D. B. 50, page 327, 28 December 1932. S. J. Phillips and Freda, his wife, to Ephram Vaughn, his ½ interest in rolling mill and 4 acres and 103 poles.

D. B. 50, page 476, 8 May 1933. Emmazetta Quesinberry and Cassie I Quesinberry (widow) to Ocie M. Cox, ½ interest in mill and property, "with all water rights and privileges thereunto." Richard A. Quesinberry's property, including his ½ interest in the mill, was divided equally between his mother and his wife. Will recorded in Carroll County, VA.

D. B. 50, page 510, 28 June 1933. Ocie M. Cox and Tilden Cox, her husband, to Ephram Vaughn. "Parties of the 1st part are hereby conveying their ½ interest in the above described land and mill property as conveyed to Ocie Cox by Emmazetta Quesinberry."

D. B. 116, page 200, 23 December 1975. Ephriam B. Vaughn and Agnes Vaughn, husband and wife, convey " a one-half undivided interest" to Gene B. Vaughn and Winnie Q. Vaughn, husband and wife, and "the other one-half undivided interest" to Sammie E. Phillips and Ruby A. Phillips, husband and wife, 3 tracts, 3rd tract is mill.

D. B. 149, page 188, 30 January 1984. Sammie E. Phillips, and Ruby, his wife, to Gene Vaughn, and Winnie, his wife, "all their one-half undivided interest" in mill and land, 3 tracts, 3rd tract is mill.

65. *Tobias D. Phillips Mill*

Mira (or Mirey) Fork Creek

Power: Tub wheel; Stones: One pair, sawmill; Location: Route 759 (Vaughn's Mill Road NW), possibly near the site of the Vaughn's Mill dam, about 300 yards above present Vaughn's Mill. Visible Remains: Possible hole in rock

History. Tobias Phillips first purchased land on Mira Fork (Mirey Fork) in 1853, but no written reference to a mill appears in deeds involving Phillips until 1877. The boundaries to a tract he bought from Reuben Thompson and Owen Quesenberry and their wives mention "Mill Road" (probably present day Route 759, Vaughns's Mill Road NW). Later deeds refer to a mill pond, mill road, or Mill Creek. The first record that refers without question to the Tobias Phillips Mill is the 1880 *Products of Industry* report. Like the better-known decennial census of the population, the *Products of Industry* report was compiled every ten years by the federal government using (at that time) local men as its agents and enumerators. Phillips's mill was apparently small and fairly primitive in design, but at 10 horsepower, its wooden tub wheel was relatively powerful thanks to the eighteen feet of fall. The mill was large and powerful enough, at least, to produce goods whose value totaled more than the $500 minimum required for inclusion in the report.

In his 1883 will, Phillips left part of the home tract of land, including the mill, to his wife. Upon her death, a one-half interest in the mill and one acre of land was to go to each of his sons, Fleming "Reed" Phillips and Bryant Phillips. Alsey (or Alice) Phillips lived until 1902, but her sons were obviously in charge of the mill in the meantime and their shared interest in it continued with the construction of a much larger, modern roller mill (III, 64) c. 1901.

Who ran the Tobias Phillips Mill between 1880 and its closing (presumably near the time the new mill was completed) is not known with certainty, but F. R. Phillips is listed as a miller at Phillips, VA, in a business directory published in 1894 and 1897 (see Appendix A). The sawmill referred to in 1902 and an 1898 reference to a previously unmentioned mill race raise questions about the location of the original mill, since tub wheel mills, like the turbine powered mills that often succeeded them, were often built adjacent to a dam and needed no race.[15] Further, the eighteen feet of fall reported in 1880 is quite a lot to be achieved without a race of some kind between the dam and the mill. Whether and by how long the sawmill predated the new mill and what connection it bore to the race are also interesting but unanswerable questions.

As with so many other mill histories, this one offers enough clues to invite speculation, but not enough to provide answers that once could have been had for the asking.

Deeds

Floyd County Deed Book "H," page 191, 24 September 1853. John P. Smith to Tobias and John Phillips, 131 acres on Mira Fork Creek.

D. B. "Q," page 107, 11 September 1877. Reuben Thompson and Elizabeth, his wife and Owen Quesenberry and Luvenia, his wife, to Tobias Phillips, Sr., 38 acres on waters of Greasy Creek and the "right to way and water to a certain spring on the west side of saw mill road." Boundaries mention mill road.

Floyd County Will Book "E," page 461, 12 November 1883. Last will and testament of Tobias D. Phillips: "To my wife Alicey, a portion of the home tract.... At her death this is to go to son Fleming Reed Phillips except ½ of mill property consisting of one acre and mill which Bryant P. Phillips is to have." Rest of home tract to daughter Rosella Turman.

D. B. "Z," page 518, 14 February 1898. F. R. (Fleming "Reed") Phillips, and Mary, his wife, to B. P. Phillips, 2 ½ acres on Mira Fork Creek. "Beginning at a wild cherry by a mill race" and "to the mill dam...."

D. B. 28, page 406, 3 February 1902. Fleming Reed Phillips and Bryant P. Phillips and Alicey Beth Phillips to Johial A. "Hiley" Moses and William L. Quisenberry, "one half interest in mill house and ½ acre of land ... and water power sufficient to run the mill and when parties of the second part is not using the water to grind with the parties of the 1st part is to have the water to saw with." *The mill house referred to in this deed is the Phillippine Mill (III, 64) but the sawmill is of interest, as noted above.*

See Table C, *1880 Products of Industry.* Name of corporation or individual: Tobias D. Phillips

67. Greasy Creek Mill/Harris Mill

Greasy Creek

LOCATION: Route 787 (Indian Valley Road NW), about 3 miles north of Route 221 (Floyd Highway South); POWER: Overshot wheel, replaced by turbine; STONES: 1 set and Salem Foundry roller mill, hammer mill; VISIBLE REMAINS: Building in fair to good condition. Machinery removed. The wooden sills and part of the framework of the last dam remain in bed of stream, between sturdy concrete abutments. Foundation timbers of at least two earlier dams are visible in the stream bed upstream from the last dam.

History. The Greasy Creek Mill was the last water-powered mill built in Section III and is one of only two still standing in this northwest quarter of the county. As with virtually every water-powered roller mill in the county, it was built at a proven water power site.* Usually wider and longer, and invariably taller than the old buhr-stone flour mills they replaced, the new mills were designed specifically to house the extra machinery, shafts, elevators, and spouts required for the roller mill system for making flour. The new buildings typically occupied the spots vacated when their predecessors accidentally burned or washed away, or were deliberately razed. In this context, the Greasy Creek Mill is unique in having been built on the opposite side of the creek from the old mill. The new location required reconfiguration of the races but saved customers the trouble and perhaps danger of crossing the creek to get to the mill and also placed a protecting shoulder of land between the millpond and the mill. In spite of its distinctly different size and location, the new mill was known to some patrons by the name of its predecessor, "The Quarter Mill" (III, 68), for years afterward.

Based on inconclusive evidence, the best guess is that the Greasy Creek or Harris Mill was built for John Wiley Quesenberry under the supervision of master miller and millwright J. H. "Hiley" Moses, between May 1923 and July 1925, when the first reference to a roller mill appears in a deed. How long the old Quarter Mill stood across the creek after it was abandoned is unknown.

**Among these, all of which were built between 1895 and 1925, only the Duncan Mill on Indian Creek (III, 79) was not preceded by a corn or flour mill. But there had been a sawmill at the site.*

The past, horse-drawn and homegrown, meets the future, motorized and store-bought. Greasy Creek mill ran for about ten years after this picture was made in the mid 1950s. The team, Strawberry and Lady, are driven by Roy Turpin, eleven years old, waiting his turn on the mill porch. (Courtesy Gertrude Harris family)

In an oral history compiled in 1976, Stefan Klosowski wrote that Hiley Moses and partner John Quesenberry had bought the old Quarter Mill from Henderson Hylton and built the present mill on the west bank in 1920.[16] However, deeds show that Quesenberry did not own the mill (for the second time) until 1923, when he bought it at public auction conducted on behalf of Hylton's heirs. Also, the public sale notice published in the *Floyd Press* on August 17, 1922, clearly refers to the "old Saunders Quarter Mill," indicating that it was still intact and operable at that point.

Frank Webb's notes say the new mill was built in 1923, but no source is cited. A *Floyd Press* news item dated July 12, 1923, indicates that the old Hylton Mill was then still in use by reporting that "J.B. Woods [owner of the Willis Mill (III, 69)] has sold Hiley Moses and J. W. Quesenberry his flour mill machinery. They will install same at the Hylton Mill now owned by Mr. Quesenberry." A 1928 deed also refers to the mill as the H. H. Hylton Mill. Ordinarily, a new mill would not be known by the name of a man who owned the mill that was destroyed to make room for the new one, but at least one precedent does exist in Jack's Mill on Little River. The 1923 build date would make more sense if Quesenberry decided in late 1923 to build a new mill rather than refit the old one with the Willis mill machinery. Several other mill owners did choose to build from the ground up rather than refit, but no evidence exists, in this instance, to support such a convenient supposition.

In any case, it is plausible that Hiley Moses did supervise the construction of the mill, since he was clearly responsible for its operation several years before he became a partner in 1928. *The Floyd Press* reported in 1923 that Moses had moved to Indian Valley to take charge of running the mill, presumably on behalf of J. W. Quesenberry who, along with his wife, Clara Kesler Quesenberry, lived in the large and somewhat elaborate frame house now known as the Roscoe

Uriah Turpin (shown here) or one of his sons made the trip to the Greasy Creek Mill with a team and wagon at least twice a month from around 1950 until the mill closed. Baled hay and ear corn were ground, mixed with supplements, and sent home in 100 pound sacks to feed his horses and other livestock. The low steel wagon wheels didn't roll over rough ground as easily as the taller wooden originals, but they made the wagon easier to load and had no wrought iron tires or wooden spokes and felloes to loosen during hot, dry weather. (Courtesy Lillian Van Lear)

and Effie Reece house. Unoccupied and in disrepair, the house stands across from the mill and just north of it on Indian Valley Road, at the center of a complex of farm buildings— barns, granary, chicken houses, cellar, ice house, spring house, wash house, etc.— that once numbered seventeen.[17]

In any event, Johial "Hiley" Moses, who came from a family of millers and millwrights, is a prominent figure in the mill's early history, as was his wife Josephine "Josie" Dickerson Moses. The daughter of William M. "Billy" Dickerson and the granddaughter of William Morricle, both Floyd County millers, Josie Moses filled in for her husband at Greasy Creek and at the Dodd Creek mill now known as Webb and Son Mill (II, 122) whenever Hiley Moses was losing ground in his battle with alcohol. Customers at the mill on Dodd Creek were told that Moses had gone to the Town of Floyd and might not return that day. Greasy Creek customers were told that he had gone on business to the village of Willis, about three miles away. Customers soon learned that Josie Moses was well able to grind their corn, and possibly to grind flour as well, although that skill has not been attributed to her. Grace Phillips, whose husband once worked for Moses at the Greasy Creek Mill, recalled having herself been compelled to grind cornmeal when Josie Moses was away and Hiley Moses and his assistant Mr. Phillips had both "gone to Willis."

Moses' business partner Rollie Phillips bought a store and became postmaster of Indian Valley in January of 1927, so Phillips probably left the milling to his partners, first Quesenberry

The original mill structure, as seen from the east side, needed paint but still had a good roof and a sound foundation in 2008. (Ricky Cox photograph)

and then Moses. After becoming sole owner upon the death of 61-year-old Hiley Moses in 1931, Phillips hired George Boyd to run the mill and added a general store a few feet from the mill. Mack Turman and sons were hired around this time to construct a new dam, closer to the mill and taller and stronger than the one it replaced. The center portion of the dam, with its sloping plank face supported by interlocking timber framework, had dozens of counterparts throughout the county, but the massive L-shaped concrete abutments are unique, and appear sturdy enough to serve again if needed.

The turbine may have been installed at this point, as well. A spring loaded relief valve between the dam and the turbine prevented the buildup of excessive pressure in the enclosed pipeline that fed water to the turbine in place of the traditional vertical wooden penstock. This innovation was used at only one other Floyd County mill, the much smaller Chris McAlexander Mill (IV, 53) on Taylor's Creek, but doubtless would have been adopted by other mills had any of the larger turbine powered mills continued under water power long enough to warrant a major overhaul of the dam and penstock. Though it was not used after the turbine was installed, the overshot water wheel on the east end of the mill was still in place in the early 1950s.

George and Glenn Boyd bought the mill in 1936, but in 1941 let it go back to Phillips, who sold it again within three months, this time to (Mr.) Joyce Phillips. The last operators of the mill, brothers Harless and Noah Harris, bought it in 1943. A third brother, Arby Harris, ran the mill for them until around 1966. A diesel powered hammer mill for grinding the coarse and bulky feeds was added in the 1940s, when this mill, like most others that survived past mid-century, shifted its focus to animal feed, and away from flour and cornmeal. According to Klosowski, flour production had virtually ended by 1956[18] as fewer local farms grew wheat and more and more consumers bought regional and national brands of flour in local stores and supermarkets. The mill closed around 1966. Most of the milling equipment was removed and sold by Howard Mumford, but several pieces of machinery remained in 1995. These include a Salem rotary sifter, Eureka wheat scourer, wheat and bran bins, a flour mixer, various cleaners, and some shafting and belts. A great deal of concrete was used in the original or reworked foundation, and in a basement molasses tank. From this perspective, this may be the best-preserved non-running mill in the county. If the roof is maintained, it may last well into the 21st century.

The Greasy Creek Mill pond was, like every other, a year-round attraction to local children. Foster Quesenberry (b. 1929) remembered as a very small boy seeing the overshot wheel turning from an upstream hillside. As teenagers, he and his friends looked for turtles in the mud flats exposed when the pond had been drawn down by a long day of grinding. Saplings, cut off at an angle with an axe and preserved from decay by the covering water, were a danger to bare feet for years after the pond was filled for the first time.[19]

For earlier history of this mill site, see the Quarter Mill (III, 68).

Deeds

The first two deeds may or may not pertain to this mill, depending upon when it replaced the earlier mill.

Floyd County Deed Book 38, page 77, 16 December 1911. J. W. Quesenberry to Henderson Hylton, 2 acres and 55 sq. rods, being a "certain piece or parcel of land including gress [grist] mill and machinery thereto and water power to same mill known as Saunders Quarter Mill on Greasy Creek."

D. B. 45, page 267, 14 May 1923. J. E. Proffitt, Commissioner, *Joseph M. Hylton vs. Paul B. Hylton* (settling estate of H. H. Hylton, deceased), to J. W. Quesenberry, who bought at public auction a "parcel of land on the waters of Greasy Creek ... known as "Saunders Quarter Mill," containing 2 acres 55 sq rods and including all water rights which passed under said last named deed (D. B. 38, page 77).

D. B. 47, page 303, 28 July 1925. J. W. Quesenberry and Clara, wife, to Rollie N. Phillips, "one half interest in a certain tract ... land including a roller mill and machinery thereto and water power to same on Greasy Creek known as the quarter mill," 3 acres and 5 poles.

D. B. 46, page 269, 12 August 1926. J. W. Quesenberry and wife, Clara, to J. A. Moses and Rollie N. Phillips, 5 ¼ acres adjoining mill property.

D. B. 46, page 269, 12 October 1926. Margaret Hylton to Rollie N. Phillips and J. A. Moses, 2⅛ acres.

D. B. 48, page 227, 21 December 1928. J. W. Quesenberry and wife Clara to J. A. (Hiley) Moses, "½ interest in a certain piece or parcel of land including griss [grist] mill and machinery there to and water power to same mill known as H. H. Hylton Mill on Greasy Creek." H. H. Hylton grist mill and water rights.

D. B. 49, page 177, 11 June 1928. W. R. Reece and E. L. Reece to J. A. Moses and Rollie N. Phillips, 11 acres. Boundaries refer to "chestnut tree near mill pond."

D. B. 49, page 451, 27 February 1931. B. G. Howard, Commissioner, to Rollie N. Phillips, highest bidder at public auction in chancery cause of *Rollie N. Phillips vs. Josie C. Moses* (widow of J. A. "Hiley" Moses, who died in 1929). Four tracts of land "locally known as the property of the Greasy Creek Milling Company."

D. B. 54, page 500, 30 October 1936, Rollie N. Phillips and Zula Phillips, wife, to George L. Boyd and Glenn E. Boyd, "certain property known as the Greasy Creek Mill," 21⅜ acres and water rights. Also, "all and singular the personal property located in said mill and belonging to male party of the first part, including all grain, feed, machinery, book accounts and in short all items of personal property belonging to the male part of the first part and used in and appertaining to the business of the said mill." Buyers required to maintain insurance.

D. B. 59, page 276, 28 February 1941. George L. and Glenn E. Boyd to Rollie N. Phillips mill, land and rights.

D. B. 59, page 524, 14 May 1941. Rollie N. Phillips to Joyce Fitzhugh Phillips, mill, land and water rights. Buyer required to maintain insurance.

D. B. 61, page 204, 7 April 1943. Joyce Fitzhugh Phillips and Eula Phillips, wife, to Noah L. Harris and Harless E. Harris, mill and property locally known as "Greasy Creek Milling Company."

D. B. 65, page 410 and 411, 4 April 1947. Noah L. Harris and Ethel, wife, and Harless E. Harris, and Cleda Harris, wife, Deed of Trust to J. E. Proffitt and Kyle Weeks, 21 acres, mill and property. This contract says that the following equipment bought at Boone's Mill, Virginia, is to be moved to the mill and installed by N. L. and H. E. Harris: 4 ball bearing stands (rolls) 7 by 18 feet; 1 self-balancing sifter; 1 receiving generator No. 2; 1 scouring and polishing machine; 1 two barrel self rising mixer; two 30 by 8 foot reels; 1 flour packer; 1 2000 pound grain scales; 1 copper cylinder; 1 gas bleacher; 1 tempering conveyer; 300 foot 24 inch steel flume; one 16 foot by 40 inch overshot wheel; 2 pairs balancing scales; and miscellaneous roller bearing shafting, belts, etc. *Most of this equipment was later sold to Vaughn Mill per Sammie Phillips.*

D. B. 105, page 686, 18 September 1972. Roger Dale Harris (son of Harless) to Maury Mumford, mill and rights.

D. B. 127, page 227, 28 October 1978. Maury Mumford to David E. Holms, mill and etc.

D. B. 156, page 803, 14 November 1986, David E. Holms to Harry Roberson. Two tracts of 21⅜ acres and 19/20 acre with buildings, mill machinery and water rights.

68. Saunders' Quarter Mill/Quarter Mill

Greasy Creek

LOCATION: Route 787 (Indian Valley Road NW), approximately 3 miles North of Route 221 (Floyd Highway South); POWER: Tub wheel (1880), last used overshot wheel.

STONES: 1 set; VISIBLE REMAINS: Remains of dam framework in stream bed, upstream from Greasy Creek Mill dam.

History. "Saunders' Quarter Mill" was supposed by some to have been built and operated by slaves and was so named, it is said, because of its proximity to the slave quarters. Even the Greasy Creek Mill that replaced it in the 1920s was known to some as the "Quarter Mill," years after the original mill was gone. References to the "mill path" in the earliest deeds connected to this mill confirm its existence in 1849 on land owned by at least two generations of the Saunders family, who for a short time owned a smaller tract of Floyd County land and the Joshua Howell Mill on the South Fork of the Little River (III, 138). Since all are named as residents of Campbell County, VA, in the Greasy Creek property deeds, the family had by 1850 left Floyd County, or had never established permanent residences on the land referred to in 1875 as the "Quarter Farm."

By 1884 much of the Saunders farm had been acquired by Abraham D. Burnett, who had lost an arm at the battle of Frayser's Farm in June 1862 while serving as a private in Company A of the 24th Virginia Infantry. Burnett's fine Victorian frame house and even more impressive timber-framed barn still stand across from each other on Route 787 (Indian Valley Road), a little north and just out of sight of the mill site. At or following a public auction in March 1908 billed as "The Largest Land Sale Ever Known in Floyd County," Burnett's heirs sold the mill to J. W. Quesenberry, who ran it for two years before selling the mill and drastically reduced acreage to Henderson H. Hylton.[20] The *Floyd Press* reported this sale on April 1, 1909, but the deed was not recorded until December 1911. Quesenberry also conveyed water rights, which to this point had been unneeded, since the mill tract had included all land adjoining or affected by the mill pond and dam. At about the same time Quesenberry and his wife Clara built the house now known as the W. Roscoe and Effie Thompson Reece house just north of the mill on the opposite side of Indian Valley Road. One of the remaining outbuildings on the Reece farm is an ice house with 18-inch-thick, sawdust-filled walls. Its pit has been floored over to convert it to a storage building. Ice was cut from the millpond and kept until summer layered in sawdust, possibly from a sawmill driven by the same water power.

H. H. Hylton had contracted with the Saunders family to buy a one-half interest in the mill in 1877, nearly 35 years earlier. No deed for this conveyance was found, but the *Products of Industry* report of the 1880 census describes a mill owned by what appears to be "Saunders and Hylton" (part of the first name was cut off by the microfilm photographer) on Greasy Creek in the Indian Valley district. No other possible match is known, which suggests that Henderson Hylton was in some way connected to this mill in 1880, perhaps as the miller, whether or not he was a part owner.

In 1923 J. W. Quesenberry bought the mill for a second time at a public auction following Hylton's death. It is unclear whether Quesenberry was buying a newly constructed roller mill that Hylton had hired millwright and miller J. H. "Hiley" Moses to build on the west bank of the creek, or whether Quesenberry himself hired Moses to build the roller mill building still known as the Greasy Creek Mill. Whether the old mill was torn down or merely abandoned when the new one was ready to go to work is not known. For further history of this mill site, see the Greasy Creek Mill (III, 67).

Deeds

Floyd County Deed Book "F," page 316, 1 January 1849. Samuel Saunders and Mary Saunders, his wife, to Fleming Saunders, Sr., 1,430 acres on "Greasy Creek and its waters and the waters of Indian Creek." Deed doesn't mention mill but refers to "mill path."

D. B. "L," page 586. 12 October 1855. Fleming Saunders, Sr., to Robert C. and Fleming, Jr., all his "lands ... on Greasy Creek, its tributaries and the waters of Little River."

D. B. "K," page 59, 20 November 1856. Fleming Saunders, Sr., to Robert Saunders and Fleming Saunders, Jr. Land on Greasy Creek. Description of boundaries includes "two white oaks on the mill path."

D. B. "O," page 132, 31 December 1875. Robert C. Saunders and Fleming Saunders to A. D. Burnett, 632 acres, Greasy Creek. Refers to property as "Quarter Farm." Boundaries mention "corner to mill tract," but mill must not have been included, considering contract referred to in next deed.

D. B. "O," page 530, 23 October 1877. Robert C. Saunders and wife to W. W. H. Harris, Trustee. All his property to dispose of as trustee, including one-half interest in "900 acres of land in Floyd County owned in common with Fleming Saunders." Schedule A refers to an existing contract made by Saunders to sell to Henderson H. Hylton ½ interest in mill and 65 acres land on Greasy Creek. Hylton had yet to pay the purchase price and the sale was not completed.

D. B. "S," page 707, 23 September 1884. Fleming Saunders and wife and W. W. H. Harris, Trustee for Robert Saunders, to A. D. Burnett, part of mill tract, 34 acres plus or minus.

D. B. 36, page 238, 1 December 1909. H. P. Burnett, Special Commissioner for A. D. Burnett, deceased, to J. W. Quesenberry, "containing 111¾ acres and the said land being commonly known as the mill tract of A. D. Burnette land."

D. B. 38, page 77, 16 December 1911. J. W. Quesenberry to Henderson Hylton, 2 acres, 55 square rods, being a "certain piece or parcel of land including gress [grist] mill and machinery thereto and water power to same mill known as Saunders Quarter Mill on Greasy Creek."

D. B. 45, page 267, 14 May 1923. J. E. Proffitt, Commissioner, *Joseph M. Hylton vs. Paul B. Hylton* (settling estate of H. H. Hylton, deceased), to J. W. Quesenberry, who bought at public auction "parcel of land on the waters of Greasy Creek ... known as "Saunders Quarter Mill," containing 2 acres 55 sq rods and including all water rights which passed under said last named deed (D. B. 38, page 77).

See Table C, *1880 Products of Industry.* Saunders and Hylton.

69. H. P. Hylton Mill/S. P. and G. A. Willis Mill

Greasy Creek

LOCATION: Route 769 (Reed lane NW) .2 miles west of Route 768 (Firehouse Road NW); POWER: Overshot (1880), steel overshot wheel installed 1913–14; STONES: Two pairs, later added roller mill, planing mill; VISIBLE REMAINS: Dug race runs along west side of Greasy Creek; concrete foundations of mill on north side of Route 769.

History. Hardin P. Hylton owned numerous tracts of land around what is now the Willis community. In 1859 he bought two adjoining fifty-acre tracts "on the waters of Greasy Creek." The mill pond referred to in the boundaries to this land may have been part of Hylton's plans for a new grist mill, or could have served a now forgotten sawmill or grist mill near the same site. The *Products of Industry* report completed along with the 1870 census lists a mill on Greasy Creek in which H. P. Hylton was a partner. Hylton sold one-half interest in his mill, 52 acres, and the necessary water rights to George W. Spangler in February 1883. Two years later Spangler sold out to Peter Hylton who, in turn, sold the one-half interest to brothers Simon Peter (S.P.) and G.A. Willis in 1889. They bought the remaining half-interest from Hardin Hylton in 1890, and apparently continued as partners until S.P. died in 1915. The J. A. Moses running the mill for the Willis brothers in 1906 must be one of the five sons of John B. Moses who became millers.[21] Three of their first names begin with "J," but none are known to have the middle initial "A."

The settlement of the S.P. Willis estate in 1918 left the mill in the hands of G.A. Willis, William J. Phillips, and J.B. "Beth" Woods. Dissatisfaction with the partnership or poor business conditions led to the posting of the following announcement in the *Floyd Press* of November 25, 1920: "Willis Roller Mill for Sale: On Saturday, December 4 at 2 o'clock, in front of S.G. Cunduff and Co's store, we will offer for sale at public auction, to the highest bidder, the Willis Roller Mill and Planing Mill combined. This is good property and it will pay anyone interested in a roller mill to attend this sale. Terms made known on day of sale. Willis Roller Mill Co. Willis, Virginia."

A 1921 deed for land adjacent to the mill property refers to Woods as the sole owner of the mill, but he evidently was not successful, at least not at producing flour. The July 12, 1923, edition of *The Floyd Press* reported that "J.B. Woods has sold Hiley Moses and J. W. Quesenberry his flour mill machinery. They will install same at the Hylton Mill now owned by Mr. Quesenberry" (III, 67).

The steel overshot wheel from the Willis Mill was sold to the owners of Vaughn's Mill (III, 64) on Mira Fork Creek. It was installed there c. 1930 in place of the original turbine, which under the right conditions had several advantages over the cumbersome overshot wheel, but simply could not operate efficiently for very long at a time with the low volume of water supplied by a small stream like Mira Fork. The steel overshot wheel is still in place on the back side of Vaughn's Mill, Floyd County's only operating commercial mill, but was itself replaced around 1960 as a source of power by more convenient and more reliable diesel and electric motors.

In November 1928, Beth Woods deeded the land, building, and remaining works to Blue Ridge Bank, which sold the land, mill, and water rights for same to John and Mary Bishop in June 1929. A deed recorded two years later, in June

The Willis Mill, undated. The absence of a wooden race above the overshot wheel and the presence of the penstock-like structure to its left suggest that a conversion from turbine to overshot wheel power is underway. This change was made at three other mills due to insufficient water, but the scant history of this mill neither confirms nor denies that it was done here. (Frank Webb collection)

1931, between the Bishops and L. W. Page indicates that a mill was formerly located on the land being sold. A brief history of Willis, written in 1928 by 15-year-old Una Hylton (Slaughter) and reprinted in *Floyd County, Virginia Heritage* states that the mill built by Hardon Hylton "about fifty years ago ... was burned down a few years ago." [22] Assuming this account is accurate, either it refers to another, as yet unidentified mill built near Willis by Hardin Hylton, or this mill was rebuilt after the fire.

J.B. "Beth" Woods was helped in the mill by his father, Fred "Cricket" Woods, a well known old-time miller who had also worked at the Ed Strong Mill on Indian Creek. Cricket Woods is supposed to have told of being assigned as a Confederate soldier to a firing squad for a deserter. He did not want to participate and was saved from killing the man by a volunteer who took his place and fired, so Woods believed, the fatal shot.[23] The relatively flat terrain of that area and the width of the stream valley resulted in the use of a low dam and a dug race approximately ¼ mile long to raise the water to the height required by the overshot wheel. A wooden race suspended over Route 769 carried the water from the end of the dug race to the mill, and no doubt was a popular spot on hot summer days. The dug race is still clearly visible, and part of the foundation of the mill remains by the creek on the north side of Route 769.

Deeds

Floyd County Deed Book "K, page 604, 17 September 1859. Josiah Martin and Rebecca, his wife, to Hardin P. Hylton. Two 50 acre tracts combined, one boundary "crossing the creek near the head of the mill pond to 3 white oaks thence with H. P. Hylton's lines."

D. B. "R," page 24, 24 February 1883. H. P. Hylton to George W. Spangler, one-half interest in mill and 52 acres (water rights).

D. B. "S," page 69, 1 August 1885. George W. Spangler to Peter N. Hylton, his one-half interest in mill and 52 acres, also water rights.

D. B. "U," page 253, 29 August 1889. P. N. Hylton to S. P. and G. A. Willis, one-half interest in mill and land.

D. B. "U," page 695, 1 December 1890. H. P. Hylton to Simon Peter and G. A. Willis, his one-half interest in mill and land.

D. B. 43, page 269, 4 May 1918. B. G. Howard, Commissioner in cause of *G. A. Willis vs. V. Willis*, widow of S. P. Willis, to J. B. Woods, William J. Phillips and G. A. Willis, mill, land and water rights to same, 10⅛ acres.

D. B. 44, page 469, 30 July 1921. G. A. Willis and wife to S. C. Slusher, lot containing 17 acres. Both parties are reminded that the roller mill of J. B. Woods is built on part of this land and they have possession of that part as long as the mill is maintained.

D. B. 45, page 407, 5 December 1923. Deed of Trust, J. B. Woods to R. F. Thompkins, Trustee, 10⅛ acres land and roller mill thereon, satisfied 3 August 1944.

D. B. 48, page 209, 28 November 1928. J. B. Woods to Blue Ridge Bank, two parcels of land on Greasy Creek, one of 10⅛ acres and mill property. Second lot contains ¼ acre.

D. B. 48, page 566, 1 June 1929. Blue Ridge Bank to John and Mary Bishop, land, mill and water rights for same.

D. B. 49, page 406, 2 June 1931. John and Mary Bishop to L. W. Page. Mill was formerly located on this land.

See Table B, *1870 Products of Industry.* Name of corporation or individual: H.P. and J. Hylton.

See Table C, *1880 Products of Industry.* Name of corporation or individual: H.P. Hylton.

70. Eden Spangler Mill

Headwaters of Big Indian Creek

What Frank Webb believed to be a separate Eden Spangler (1845–1880) Mill on the headwaters of Indian Creek, perhaps near the Carvie Goad place on Route 765 (Firehouse Road, NW) and to which he assigned the number 70, now seems likely to have been the Henry Duncan Mill (III, 72). Spangler's parents, Henry Hylton Spangler and Elizabeth Shelor Spangler, were full or part owners of this mill between 1860 and the early 1870s. This mill is known to have housed wool carding machinery, which is unusual for a grain mill. Eden Spangler is listed as a wool carder in the 1870 census and the equipment and output of a wool carding operation run by E. Spangler were enumerated in the *Products of Industry* report completed in that same year. Spangler died of a fever in 1880, having reported his occupation as farmer when the U.S. census was conducted earlier that year.

See Table B, *1870 Products of Industry.* Name of corporation or individual: E. Spangler.

71. Fleming M. Duncan Mill/ William M. "Billy" Dickerson Mill

Big Indian Creek

LOCATION: Between Route 766 (Easy Drive NW) and Route 765 (Firehouse Road NW) just south of their intersection. POWER: 10 HP tub wheel (1870); STONES: 1 set; VISIBLE REMAINS: An abrupt drop below a flat area, just above where Route 766 now crosses the creek, could mark the location of the dam. If so, the flat expanse above was formed by topsoil settling behind the dam as storm runoff slowed upon entering the mill pond. The intersection of the two roads has been moved upstream considerably from its original location and may now cover the site of the mill.

History. Family connections, by blood or by marriage, and the opportunities and obligations that come with those connections, were a powerful factor in determining who built, owned, and operated Floyd County grain mills. Many mill owners and millers were doubtless led or pushed down that path by something besides the prospect of earning a better living than could be had by farming alone, or by working at some other essential skilled trade, such as blacksmithing or carpentry.

An ambitious couple might decide to build a mill at least partly because of their connection to a miller or millwright who could help with planning, building, or financing such a complex project. A son or son-in-law of a miller would have been encouraged to take up that trade by the possibility of inheriting part or full interest in a mill. Sons and grandsons might also be drawn by the familiarity of having grown up in or around a mill. Less easily discerned in deed records are opportunities to purchase part or full interest in a mill at a below-market price or under terms extended only to a close relative. In a less common scenario, older, well established relatives might find themselves in partnership with an energetic and persuasive kinsman who, lacking the capital to go it alone, had turned for help to relatives who had the means to help and could not refuse without risking social consequences. Kinship connections between owners whose tenures were usually consecutive but

William Dickerson and his sister, Emeline Dickerson Hylton, about 1921. (Courtesy Floyd County Historical Society)

sometimes separated by a decade or more emerge regularly when genealogy records are examined alongside the string of title deeds for a particular mill. The F.M. Duncan/Billy Dickerson Mill provides an outstanding but by no means unique example of this in the kinship ties of its two primary owners. Neither their surnames nor their wives' maiden names indicate family ties, and their tenures as owners were separated by four years of ownership by the Lesters, a family that produced many millers, but whose connections to either of the other families have not been explored. But a look at the family trees of Fleming Duncan and Billy Dickerson reveals very strong mill-related connections between the two families, as well as within each family.

Fleming M. Duncan's mother, Elizabeth "Betsy" Morricle Duncan, was the daughter of William Morricle, builder of a c. 1835 sawmill and later a grain mill (II, 91) on Dodd Creek. Duncan's second marriage, to Catherine Spangler, daughter of Henry Spangler, linked him into another early and prominent Floyd area milling family, that of Daniel Spangler, whose family ran the Spangler Mill (II, 99) more or less continuously for over 130 years. The Duncans were themselves probably the most prolific millers in the Indian Valley district, having had some part in the histories of at least ten different mills, most of which stood on one or another branch of Indian Creek.

William M. "Billy" Dickerson's mother, Naomi, was also a daughter of William Morricle (Billy Dickerson's middle initial stands for "Morricle"), hence a sister to Fleming Duncan's mother, Betsy Morricle Duncan. The two men were therefore first cousins, although not exact contemporaries, Duncan having lived from 1826 to 1885 and Dickerson (a Civil War Veteran) from 1837 until 1929. Dickerson's daughter Josephine (1868–1958) was married to Johial/Johile "Hiley" Moses, who is probably the best known of Floyd County's millwrights and "for hire" millers, and was part owner of several mills. Moses was the resident miller at Scott's Roller Mill when a widely circulated photo, including Hiley and Josie Moses, was made there about 1912. Later known as the O. A. Webb and Son Mill, Scott's Roller Mill stood on or near the site of the mill built by William Morricle, grandfather of Fleming Duncan and Billy Dickerson, nearly 80 years earlier.

Fleming Duncan gave his occupation as "sawyer" in the 1860 census, but whether or not he had by then built a sawmill on the Big Indian Creek land purchased in 1855 is not known. He was grinding corn there by 1870, but probably not wheat, since the *Products of Industry* enumerator noted only one set of stones, driven by a tub wheel supposed to be generating 10 horsepower. If the type and power of the wheel were reported accurately, this was an impressive output for a homemade wooden wheel on a small mountain stream.

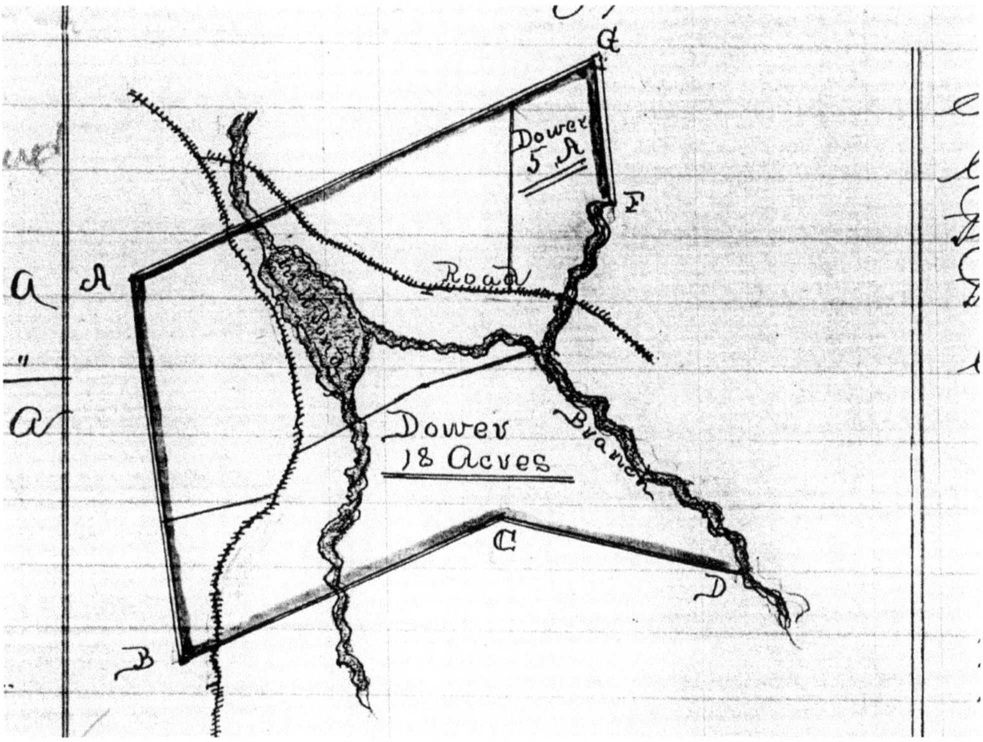

Mill ponds are frequently referred to in written boundary descriptions, but are rarely depicted in the few surveyor drawings included with deeds in the 1800s. The surveyor's clients may have commissioned this drawing of the Fleming Duncan Mill pond to establish that neither of the dower tracts assigned to Duncan's widow for her lifetime includes the mill pond or the mill. (Floyd County Deed Book "U," page 99, August 24, 1889)

Following his death in 1885, the mill and 47 acres were conveyed by a county court commissioner first to F. M. Duncan's widow, Catherine, then to Malachi Duncan, F. M. Duncan's son by his first wife, who sold it to George Lester. The legal tangle caused when this trade failed to go through was resolved when the court conveyed ownership to W. C. and Posey Lester in 1889. Catherine Spangler Duncan sold her remaining interest, dower rights to 23 acres, to the Lesters in 1894.

William M. "Billy" Dickerson was apparently the last to run this mill, which he bought in 1898 at age 61 and deeded to an unmarried daughter, Minnie, in 1915. Dickerson is listed as a miller in only one census, that of 1900; Minnie is listed as an unmarried daughter, 38 years of age, still living at home in 1910. The mill is not mentioned in deeds made in 1917 and 1921 between Minnie L. Dickerson and subsequent owners of three tracts of land, one of which must have included the mill site.

Deeds

Floyd County Deed Book "H," page 438, 3 August 1855. Peter Duncan, David A. Howell and Charlotte, his wife, to Fleming M. Duncan, 45 acres land on Indian Creek.

D. B. "S," page 528, 3 May 1886. J. L. Thompkins, Commissioner for F. M. Duncan, to Catherine Duncan, widow of F. M. Duncan, dower lands of 47 acres and mill thereon.

D.B. "S," page 581, 13 December 1886. W.D. Tompkins, Comr., cause of *Malachi Duncan vs. W.R. Duncan et al.*, to Malachi Duncan, 47 acres, on the waters of Indian Creek. Catherine, widow of F. M. Duncan, to have dower rights to 18 acres.

D. B. "U," page 99, 24 August 1889, J. L. Thompkins, Commissioner in cause of *Malachi J. Duncan vs. W. R. Duncan* and *M. V. Duncan vs. G. W. Lester and W. C. and Posey Lester*, to Posey Lester and W. C. Lester, a tract of land 47 acres and mill formerly belonging to F. M. Duncan, which was bought at first sale by M. Duncan, and sold to George Lester (no deed found) and is the same land conveyed to M. Duncan by W. L. Thompkins, 23 acres assigned as dower to Catherine Duncan.

D. B. "X," page 3, 14 June 1894. Catherine Duncan to P. G. and W. C. Lester, her interest in F. M. Duncan mill tract. "For a consideration $150 cash in hand the receipt of which is hereby acknowledged by Catherine Duncan, widow of F.M. Duncan do hereby sell, grant, and convey to W.C. Lester and P.C. Lester all my interest in a certain tract or parcel of land ... in the waters of Indian Creek it being the mill tract formerly owned by my late husband and conveyed to the said W.C. Lester and Posey G. Lester ... in D.B. "U," page 99 (Catherine Duncan held dower rights to 23 of the 47 acres).

D. B. 32, page 286, 8 July 1898, W. G. Lester and L.T. Lester, his wife, and P. G. Lester and Emmette Lester, his wife, to William M. Dickerson. 1 tract (mill tract): 47 acres, Indian Creek. Second tract: 12 acres.

D. B. 41, page 55, 18 May 1915. Wm. M. Dickerson to Minnie L. Dickerson, 2 tracts of land, one of 47 acres land (F. M. Duncan mill tract) and one of 12 acres land.

D. B. 43, page 103, 6 April 1917. Minnie L. Dickerson to H. C. Reed, 6 acres land on Indian Creek.

D. B. 44, page 470, 27 July 1921. M. L. Dickerson to Burdine Reed, two tracts of land, one of 41.5 acres and one of 12 acres.

D. B. 65, page 450, 28 April 1947. Burdine Reed to Edna F. Reed, Orvilla R. Cockram, Clarence E. Reed, Harvey L. Reed, and James E. Reed, and children of William H. Reed, two tracts of land, 41½ acres and 12 acres.

D. B. 77, page 59, 17 September 1955. W. E. Profitt, Commissioner for Edna F. Reed, deceased, to Clarence E. Reed, two tracts of land, one 41½ acres and one of 12 acres.

See Table B, *1870 Products of Industry*. Name of corporation or individual: F. M. Duncan.

72. Henry Duncan Mill and E. F. "Ed" Strong Mill

Big Indian Creek

LOCATION: East bank, estimated .4 miles northwest of intersection of Route 730 (Duncan's Chapel Road NW) and Route 766 (Easy Drive NW). When the mill was in operation, Route 650 (Mystic Lane NW) connected to Route 730 at both ends, passing directly by the mill on the east side. POWER: Overshot (Duncan Mill, 1880); steel overshot 20 by 2 or 3 feet (Strong Mill); STONES: 2 sets and Monarch Roller Mill (Strong Mill); VISIBLE REMAINS: Concrete abutment on east end of main dam and wooden fragments of it in the streambed. Dug race, between abandoned road and creek. Stacked stones that supported the framework of the wooden race at its lower end and formed a retaining wall between it and the road above. Concrete piers that supported the wheel. Both concrete abutments of upper reservoir dam on the south side of Route 730. Cast iron fragments at mill site.

History. After a grist mill and sawmill were established at this site by Henry Duncan between 1834 and 1844, the mill property went through a long and bewildering series of owners—including a Harman, a Howell, a Shelor, two Reeds, three Spanglers, and various Duncans—before Edwin F. Strong acquired full ownership of the mill property in 1902. He is said soon after to have hired Jessie Starr to build a three and one-half story roller mill to replace the old, perhaps original, mill which had housed wool carding machines as well as grain milling machinery. Carlysle Phillips recalled that the swinging flour sifter on the top floor of Strong's mill caused the whole building to shake so badly that, expecting its collapse, he would crawl on hands and knees across the floor of the upper story when visiting his father, Curtis Phillips, at the mill. The elder Phillips, a carpenter by trade, ran the mill between 1928 and 1932 and filled in at other times between regular millers.[24] One of these was his son Martin, who was running the mill around 1940. Other millers employed here were Cal/Kyle Slusher (1920s) and Clarence Wood (late 1930s). Hired millers lived in a small frame house on the hill above the mill.

In addition to cornmeal and wheat flour, the Ed Strong mill, like the mill it replaced, ground buckwheat. This could account for the second set of stones, which would ordinarily be unnecessary in a roller mill. Buckwheat mills were relatively few and far between because of the specialized equipment needed and the lower demand for buckwheat flour, even at the peak of Floyd County's farm population, which occurred about the time this mill was built. Aaron Altizer (born 1915) remembered riding in a wagon with his father from the White Rock community on Route 740 to the Strong Mill to have buckwheat flour ground.[25]

Strong's mill is thought to be unique among Floyd County mills in having had two mill ponds on the same stream. (Several mills had ponds or at least low diversion dams on two different streams.) The original, primary mill pond was necessarily nearest the mill, north (downstream) of Route 730; it fed water directly into the dug race. From there, water was led to the steel overshot wheel by a double-tiered wooden race, running diagonally between the lower end of the dug race and the mill. The second dam was built in the early thirties on the south or upstream side of Route 730, alongside Route 766 (Easy Drive NW). Its water was held in reserve until the lower pond was depleted and was also used, it is said, to produce electricity, but no description of the generating apparatus was given. The Strong home, an ornate structure with upstairs and downstairs porches, stood east of Big Indian Creek on the southeast corner of the intersection formed by Routes 730 and 766. A large store building built by Strong, but last known as Mangus's Store, stood across Route 730 from the house, also on the east side of Indian Creek.

Top: The model of the automobile and the supposition that the man on the porch is miller Clarence Wood date this photograph in the late 1930s. It is nearly impossible, today, to imagine a road this wide or level, or a building this large at the site of the Ed Strong Mill. (Courtesy Margie Moles). _Bottom:_ Martin Phillips, the miller at the Ed Strong Mill, snapped this photograph from the far side of Big Indian Creek about 1940. His wife, Vernie, stands in the door overlooking the overshot wheel, holding their daughter Ollie May. An older daughter, Erma, stands beside her mother. (Courtesy Erma Phillips Reed)

Ed Strong died in 1939. The last owner and miller, Strong's son-in-law Everett Mangus, closed the mill around 1950. The main dam had washed out sometime in the 1940s and the repairs were not completely successful. Mangus, who was known as "BumBum" to his neighbors, continued to run the store with his wife, Mary "Effie" Strong Mangus, until about 1960. The mill was dismantled around 1950 by a Mr. Bowman from Pulaski, VA. Ray Mitchell and Elvin Wade, Jr., each bought one mill stone. Uriah Turpin, the owner of a neighboring farm, bought a wooden grain bin. It was too large to fit through the door of any of his outbuildings, so he built a new granary around it.[26]

A cast iron bracket that once suspended one corner of the swinging sifter, the cast iron gate that controlled the flow of water onto the wheel, and the sheet steel apron that introduced water to the wheel were still at the site in early 2009.

Deeds

Floyd County Deed Book "A," page 267, 13 February 1834. George Duncan, John Duncan, Abel W. Duncan, and Spencer Duncan to Henry Duncan, 40 acres, being part of land of Blanch Duncan.

D. B. "B," page 13, 26 September 1836. Peter Duncan to Henry Duncan, 13 acres on Indian Creek.

D. B. "C," page 391, 25 January 1844. Henry Duncan, Deed of Trust, to A. Bishop, Trustee, two tracts of land, one of 13 acres, another of 40 plus acres, on Indian Creek.

D. B. "C," page 394, 6 February 1844. Trust deed, Henry Duncan to Harvey Deskins, 40 acres of land where Henry Duncan now lives, a sawmill and grist mill. Also 13 acres and half of undivided 195 acre tract.

D. B. "D," page 3, 24 February 1844. Henry Duncan and wife Elizabeth to Jacob Bishop, two tracts of land, one of 40 acres, on the fork of Indian Creek, and another of 13 acres.

D. B. "D," page 226, 17 March 1847. Henry Duncan to Thomas McCabe and William Thomas, 2 tracts of land on Big Indian Creek, one 41 acres and one 13 acres, deed of trust.

D. B. "K," page 376, 3 July 1860. David A. Howell and Luke Duncan to H. H. Spangler and Rufus O. Shelor, 41 acres on Indian Creek known as the Henry Duncan Mill.

No deed was found conveying the Henry Duncan mill to David Howell or Luke Duncan. Henry Duncan may have failed to pay a debt secured by a deed of trust and lost the mill through a commissioner's sale. He appears to have signed over the mill tract three times in the first two months of 1844, but still owned the mill in 1847. The ownership of the mill continues to be legally complicated for the rest of its history.

D. B. "L," page 21, 22 March 1862. Rufus O. Shelor to H. H. Spangler, his one-half interest in mill and 41 acres of land.

D. B. "L," page 724, 20 September 1869. H. H. Spangler to Asa Bishop, Trustee, "my mill property including four carding machines, mill and 43 acres of land to secure Samuel Spangler and others sureties on said H. H. Spangler bond as deputy sheriff."

D. B. "M," page 387, 2 May 1872. Deed of Trust, Solomon and Samuel Spangler to Jacob Mangus, 43 acres and mill on Indian Creek, also 136¾ acres where Solomon Spangler now resides.

D. B. "M," page 526, 3 January 1873. Asa Bishop, Trustee for Henry H. Spangler, to Solomon and Samuel Spangler, 43 acres and mill.

D. B. "P," page 224, 13 June 1879. Solomon and Samuel Spangler, Deed of Trust, to Eli Dickerson, 43 acres and mill and 136¾ acres land, Indian Creek, to secure Henry Correll.

D. B. "Q," page 508, 7 June 1882. Z. T. Dobbins, Trustee in cause of *Obediah Paris vs. Samuel Spangler's Administrators* to Solomon Spangler, his one-half interest in mill and 43 acres.

D. B. "R," page 450, 23 April 1893. Solomon Spangler to Lafayette Spangler, one-half interest in mill and 43 acres.

D. B. "V," page 63, 11 March 1891. Deed of Trust, Solomon Spangler to J. L. Thompkins, 2 tracts of land, one of 136¾ acres, one of 43 acres and grist mill.

D. B. "V," page 550, 30 March 1891. J. L. Thompkins, Trustee, to David Harman, 43 acres and grist mill and another parcel of land containing 136¾ acres.

D. B. "V," page 580, 2 July 1892, David Harman to Giles M. Reed, 43 acres and grist mill, known as Solomon Spangler mill property.

D. B. "W," page 47, 14 July 1892, G. M. Reed to W. D. Reed, ⅔ interest in mill and 43 acres.

D. B. "X," page 366, 2 July 1892, J. L. Thompkins, Trustee for David Spangler, to A. J. Duncan and W. D. Reed, 43 acres and mill.

D. B. "Y," page 227, 29 December 1896, William D. Reed to Landon S. Dickerson, ⅓ interest in mill and 43 acres.

D. B. 26, page 210, 10 October 1899, L. S. Dickerson to Edwin F. Strong, ⅓ interest in mill and 43 acres.

D. B. 28, page 239, 2 January 1902, A. J. Duncan to Edwin F. Strong, ⅔ interest in mill and 43 acres of land known as the old Spangler mill place.

D. B. 58, page 276 and 278, 11 January 1939, Ethel Strong Salmons and husband to Mary E. (Strong) Mangus, all lands of E. F. Strong, deceased, in Floyd County including "mill and store house."

See Table C, *1880 Products of Industry.* Name of corporation or individual: Solomon Spangler

73. Preston Spence Mill

Little Indian Creek

LOCATION: On Route 753 (Indian Creek Road NW) around curve from intersection with Route 751 (Indian Creek Road NW); POWER: Overshot Wheel; STONES: 1 set; VISIBLE REMAINS: Fragments of dug race visible along Route 753

History. References to Preston Spence's mill in the 1884–85 edition of *Chataigne's Virginia Gazetteer and Classified Business Directory* indicate that Lewis Preston "Press" Spence (1851–1923) constructed his mill on Little Indian Creek shortly after he bought a tract of 21 plus acres from Asa Harris in 1883. It is also possible that the mill was already there, even though none is mentioned in the deed.

Spence's subsequent use of his mill and land as security in a deed of trust was a common practice, but the 1898 deed by which Francina Spence relinquished her interest in her husband's real estate as part of a divorce settlement is the only such document recorded in this book. L. P. Spence later owned the Powhatan Williams Mill on Pine Creek for a short time in partnership with his son Amos.

Tobias C. "Little Tobe" and Mary Elizabeth Quesenberrry bought the mill and mill tract in 1900 and Phillips gave his occupation as "miller" in the 1910 census. Truman Quesenberry, (born 1901) remembered coming as a boy to this mill with his father, Jasper Quesenberry. Tobias Phillips' son Robert Terry "Bob" Phillips came there with his family to care for his mother in 1918 and may have run the mill before or after his father died in 1916. His wife, Josie Phillips Phillips, was a daughter of Fleming Reed Phillips, one of the builders of Vaughn's Mill (III, 64). The mill was gone by the mid–1920s and the race has since been all but eradicated by widening and maintenance of Route 753. What is assumed to be the top or runner stone from the Press Spence Mill, cut from what appears to be an inferior local stone, was recovered from the creek bed in the 1970s. The bed stone is supposed still to be there.

Deed

Floyd County Deed Book "R," page 156, 19 January 1883. Asa Harris and Lucy, his wife, and Henry Harris to Lewis P. Spence, 21 acres, 29 poles. Waters of Little Indian Creek.

D. B."V," page 328, 1 December 1891. Asa Harris and Lucy, his wife, and Henry Harris to Lewis P. Spence, 1½ acres, "on waters of Little Indian Creek ... adjoining lands of Sam Scott."

D.B. "Z," page 13, 7 February 1898. Lewis P. Spence and Francina Spence, his wife, to J. L. Tompkins, trustee, 2 tracts or parcels of land, "including residence and Mill seat first tract containing 21 acres and 29 poles."

D. B. 26, page 402, 5 May 1898. Francina Spence to Lewis P. Spence, her husband, her interest in land, 72 acres on Indian Creek. (No transfer.) "In consideration of the separation from bed and board and mutual agreement [parties] are to divorce." Francina Spence gave up her entire interest in 3 tracts totaling 47½ acres on Indian Creek, first tract, 21 acres, 29 poles.

D. B. 26, page 406, 27 January 1900. Lewis P. Spence, G. W. Dalton and Cordilla F. Dalton, his wife, to Mary E. Phillips, 22½ acres and 29 poles on Little Indian Creek in two tracts, one being 21 acres and 29 poles. (George Washington "Wash" Dalton, a local merchant, is a party of the first part as a result of the trust deed recorded in Deed Book "Z," page 13.)

D.B. 44, page 433, 17 August 1918, Mary E. Phillips, A.C. Phillips, G.L. Phillips, Laura E. Cox, Ruth J. Phillips, Nannie J. Phillips and Noah Phillips to R. T. Phillips, 24 acres.

74. Enoch Phillips Mill

Little Indian Creek

LOCATION: Near Route 631 (Duncan Valley Road NW) .2 mile upstream from ford of Little Indian Creek, near the mouth of Panther Branch; POWER: Overshot wheel; STONES: 2 sets and sawmill; VISIBLE REMAINS: Dug race is visible from Route 631 on south side of creek

History. Enoch Phillips (1840–1918) was a younger brother of Randolph Phillips, Jr., who owned a grist mill on Ran's Creek (III, 63), about four miles to the southwest. The "H. C. Phillips" referred to in the first deed was their youngest brother, Henderson, who owned 128 acres along this section of Little Indian Creek when he died in 1897.

Enoch Phillips acquired three acres for the mill and a lumber yard and an additional acre for the dam and race in 1890, but still had not carried out his plans for the mill when he sold both tracts in 1894 to A. E. and W. E. Phillips, probably his sons Albert "Early" and Walker "Emmett" Phillips, who were then 24 and 22 years old. The grist mill, at least, was operating five years later when the Phillipses sold out to W. A. Gray. The fate of the one-half interest in the mill site retained by H. C. Phillips in 1890 and continued in the 1894 deed is not known. The 1899 deed made to Gray implies that he bought the entire undivided interest in the mill, but he sold only a one-half interest to E. W. Boyd in 1902. Whether A. E. and W. E. Phillips had acquired their uncle's 50 percent, perhaps from his heirs, before selling to Gray in 1899, or the outstanding portion was gathered up by S. R. Brame before he sold the mill and three tracts of land to Emma S. Smith in 1909, is unknown.

The writing and recording of deeds for this mill may have generated as much income as the mill itself, which was bought and sold 15 times in 35 years. After E. H. Spence bought the mill in 1914, it changed hands almost yearly until settling with Walker Cox from 1920 until 1925. The Enoch Phillips Mill may have closed by 1926 when Perry Cox bought one of the two small tracts that had always been included in the sale of mill, probably because the race or millpond crossed or covered it.

Vinton Marshall thought that it may have been running or at least operable as late as 1930 and recalled that this mill ground only corn, though it was reported by another source to have had 2 sets of stones. It was torn down in the early 1930s by Perry Cox, who is supposed to have bought one of the mill stones for $1.00.[27]

Notable among this mill's many owners are land dealer S. R. Brame, who bought and sold many mill properties though he was not himself a miller, and "Blind Walker" Cox, who was assisted in the mill by his wife, and also supported himself and his family as a broom maker. Cox, who was blinded in a hunting accident, and his wife had also owned the Jim Dulaney Mill (III, 78) before buying this mill in 1920.

Deeds

Floyd County Deed Book. "V," page 183, 9 December 1890. H. C. Phillips and wife, Mary M. Phillips, to Enoch Phillips, 3 acres more or less, "taking in land enough for a Mill site and lumber yard." Boundary runs "with a new mill race and meanders taking land enough below said race to make a good wagon road.... Parties of the first part reserve one half interest in said Mill site for the purpose of erecting a grist mill, said party of the second part have the privilege of erecting a saw mill for his own use."

D. B "V," page 183, 11 December 1890. John M. Phillips and wife, Lucy, to Enoch Phillips, ___ acres land. Dam and race to be built on this land. Parties of the first part bind themselves not to claim damages by mill pond to land not conveyed. *A blank precedes the word "acres," but this has to be the 1 acre tract conveyed in the next and subsequent deeds.*

D. B. "X," page 113, 17 November 1894. Enoch Phillips to A.E and W. E. Phillips, 2 tracts. First tract, 2 acres for grist mill and lumber yard. One boundary running with new mill race. Second tract, 1 acre. One-half interest in grist mill retained for benefit of H. C. Phillips.

D. B. 30, page 152–153, 16 August 1899. A. E. and W. E. Phillips to W. A. Gray, 1 tract of 3 acres, 1 tract of 1 acre, and 1 tract of ½ acre (total 4½ acres), right of way reserved for John M. Phillips to mill.

D. B. 30, page 130, 3 November 1902. W. A. Gray to E. W. Boyd, ½ interest in mill site of 4½ acres.

D. B. 35, page 284, 22 January 1909. V. M. Sowder, Commissioner in Chancery Court, *S. R. Brame vs. E. W. Boyd*, 4½ acres land and grist mill. No transfer.

D. B. 35, page 283, 23 January 1909. S. R. Brame to Emma S. Smith, 4½ acres and grist mill.

D. B. 37, page 45, 4 October 1910. E. S. Smith to E. H. Spence, 4½ acres and grist mill.

D. B. 41, page 34, 6 April 1914. E. H. Spence to J. L. Bond (no transfer), 3 lots: number 1, 51⅝ poles and all water rights; number 2, 2 acres; number 3, 1 acre and 84 poles and all water rights for said mill.

D. B. 41, page 35, 18 June 1915. J. L. Bond to E. H. Spence, 3 lots: number 1, 51⅝ poles and all water rights; number 2, 2 acres; number 3, 1 acre and 84 poles and all water rights for said mill.

D. B. 42, page 78, 19 September 1916. E. H. Spence to Asa W. Bond, 3 lots: number 1, 51⅝ poles and all water rights; number 2, 2 acres; number 3, 1 acre and 84 poles and all water rights for said mill.

D. B. 42, page 281, 23 January 1917. Asa W. Bond to B. C. Phillips, 3 acres, 15½ poles and mill, and water rights. John Akers, K. M. Akers, J. Phillips and Asa Bond to have right of way to mill and road.

D. B. 43, page 114, 27 July 1917. B. C. Phillips to W. H. Bryan, 3 parcels of land and mill property.

D. B. 43, page 219, 19 February 1918. W. H. Bryan and Delilah Bryan, his wife, to B. C. Phillips, 3 parcels of land and mill property and "all the water right that they have for the benefit of the said mill together with all its appurtenances."

D. B. 45, page 278, 6 August 1920. B. C. Phillips to Walker Cox, 3 parcels of land, mill property, and all water rights. Mill on 2nd lot.

D. B. 45, page 289, 12 June 1925. Walker Cox to Virginia Catherine Duncan and son Charlie Duncan, 3 parcels of land and mill property.

D. B. 47, page 353, 13 June 1925. Virginia Catherine Duncan and Charlie Duncan, her son, to Wyatt M. Marshall, 3 parcels of land and mill property with all water rights. First parcel is 51⅝ poles. Second parcel described as lot number 2, "contains Lots 2 and 3 as in deed to parties of the first part from Walker Cox."

D. B. 46, page 199, 8 March 1926. Wyatt M. Marshall to Perry Cox, 51⅝ poles.

75. Noah Wade Mill

Headwaters of Big Indian Creek (North/East Fork)

LOCATION: On Route 814 (Dulaney Road NW) approximately 1 mile southwest from Route 740 (White Rock Road NW) just above ford of creek. Route 814 formerly connected to Route 803 (Akers Road NW) but the middle section, including the creek crossing, is no longer maintained. POWER: Overshot wheel; STONES: 1 set

History. This mill washed away before 1900 according to Alpha Dulaney, who retrieved one of the stones and displayed it by his driveway for many years. There was a sawmill here, as well. Carlysle Phillips believed that this was the old Andy Duncan Mill.[28] Solomon Duncan, the husband of the Sarah Duncan named in the first deed, did have a brother named Andrew, and their father, Blanch Duncan, has been mentioned in connection with several early mills. But no direct link between Andy Duncan and the Noah Wade Mill or this site was found.

Wade was unable to redeem the land after signing it over in a deed of trust to finance a Geiser sawmill. W. K. Fagg, trustee, took possession in 1901. The destruction of the mill by flooding around that time could have played some part in Wade's failure to cover the note. Noah Wade owned another mill (III, 76) on the headwaters of Big Indian Creek from 1883 to 1888. It too was given up to a creditor when the debt it secured could not be paid. The inclusion of Martha Wade's name, but not her husband's, on the first deed (below) to this, their second mill property, may have been a precaution against outstanding claims against property in his name.

The Wades ceded ownership of their other mill property (III, 76) on May 14, 1888, but were to retain possession of the house and 15 acres until November 1, 1888. Solomon Duncan actually sold the place now under discussion to Martha J. Wade, Randolph Phillips (her cousin), and Robert T. Wade (her son) on May 10, 1888, but Duncan's death on January 8, 1890, evidently resulted in legal snags that weren't sorted out until February 1891. In the meantime, Robert Wade had sold his one-third interest in this property to Randolph Phillips.

Deeds

Floyd County Deed Book "W," page 258, 12 February 1891. Sarah Duncan and Robert Wade to Randolph Phillips and Martha J. Wade, 40 acres.

D. B. 27, page 522, 11 April 1901. Noah and Martha Wade to W. K. Fagg, Trustee. Trust Deed, Geiser saw mill and 40 acres land on Indian Creek adjoining lands of Beauford Akers and John Duncan. "In trust to secure the Geiser Mfg Co of Waynesboro, Franklin County PA," $254.00. On headwaters of Indian Creek.

76. Isaac Reed Mill/Noah Wade Mill/Reed Mill

Headwaters of Big Indian Creek

LOCATION: On Route 750 (Alum Ridge Road NW) .5 mile west of Route 740 (Posey Road NW); POWER: Overshot wheel

History. In 1883 Isaac Reed sold to Noah Wade a tract of a little over an acre with a mill already on it. Subsequent deeds refer to a 15-acre tract bought around the same time and on which the dam, pond, and race for the mill were situated. The seller, Peter Reed, could have been Isaac Reed's son Peter A. Reed, who would have been 30 years old in 1883 and is listed as miller in the 1890 census, or Noah Peter Reed (b. 1829), who was working as a miller in Indian Valley Township in 1870 and in Alum Ridge District, where this property is located, in 1880.

Wade began work on rebuilding or replacing the mill even before the deed from Isaac Reed was recorded, and between October 1881 and July 1883 had accumulated debts for labor and materials totaling nearly $400. He was never able to pay his creditors and remove the mechanics liens affixed to his property by millwrights R. F. and D. L. Darnell and various other suppliers. The particulars of the mechanics lien follow the deed records, below. The mill was bought at a forced public sale by J. P. Proffitt, who sold it to D. L. Sowers, the first in a succession of short-term owners. The mill was last referred to in a deed book record when Noah Reed bought the mill and machinery known as Reed's Mill in 1897. General "Lee" Reed accidentally drowned in the Reed Mill pond in 1896.

Noah Wade ran another small mill (III, 75) on the headwaters of Big Indian Creek after relinquishing his claim on this one.

Deeds and Other Public Records

Floyd County Deed Book "R," page 35, 6 April 1883. Isaac Reed to Noah Wade, 1 rood 10 poles, on headwaters of Indian Creek near Hylton graded road, "together with the mill thereon and all water rights" and "right of ingress and egress for water purposes and right to repair water lane ... good roadway from said lot of land to the Hylton graded road."

D. B. "S," page 536, 13 December 1886. J. L. Thompkins, Trustee and Commissioner in cause of *R. F. Darnell vs. Noah Wade and Others*, to Joseph P. Proffit, 3 tracts land. First, 11 or 12 acres. Second, 15 acres Noah Wade bought of Peter Reed, Indian Creek. This is the "land on which Noah Wade resides and upon which the dam race, water power of Noah Wade mill is situated." Third, 1 acre plus or minus and is land on which Noah Wade mill is built and now stands. And which was conveyed to him by Isaac Reed together with all buildings, mill, water power, and water privileges and other improvements. *Proffit bought this property at a public auction forced by R. F. Darnell et al., who held a mechanics lien on the second two tracts of land, for work and materials supplied to Noah Wade for a mill.*

D. B. "T," page 366, 14 May 1888. Noah Wade to Joseph P. Proffitt, release of land in suit of *R. F. Darnell vs. Noah Wade.* Deed refers to suit in DB "S," refers to tract 2. Allows Noah Wade to occupy the dwelling house and 15 acres until November 1, 1888.

D. B. "W," page 460, 16 February 1891. J. P. Proffitt to D. L. Sowers, 2 parcels land on Indian Creek. One is the 15 acres Noah Wade bought from Peter Reed. One contains 1 acre Noah Wade bought of Issac Reed with mill and all water rights.

D. B. "W," page 480, 20 May 1893. D. L. Sowers to George T. Dulaney, 15 acres, it being the land bought from J. P. Profitt by said D.L. Sowers and also one-half interest in one acre of land, including "the mill together with one half of all the machinery, waterway and everything pertaining and belong to said mill."

D. B. "X," page 231, 21 January 1895. D. L. Sowers, and Virginia, his wife, and Geo. T. Dulaney, and Mary B., his wife, and Noah Reed and Nancy, his wife, to Wyatt and George W. Reed, mill, machinery and water power to run same. One rood and 10 poles adjoining Elder Isaac Reed and Isaac A. Reed.

D. B. "Z," page 289, 11 March 1897. George W. and Wyatt Reed, et al., to Noah Reed, "one Mill Known as the Reed's Mill, also one half acre of land more or less" and a second tract containing ⅛ acre more or less.

Record; Mechanic's Lien Book; Floyd County

D. L. Darnell, millwright and carpenter, Mechanics lien vs. Noah Wade, 15 February 1883, 80½ days work, total $60.37, paid $34.31, due $26.06 against 2 tracts land on Indian Creek, one purchased from Peter Reed, the other from Issac Reed. p. 3.

R. F. Darnell, millwright, mechanic's lien vs. Noah Wade, 15 February 1833, 115 days work, Total $230.00, $139.25 paid, $90.75 due. pp. 3–4.

John C. Wade, General Contractor, vs. Noah Wade, 2 April 1883, 90½ days work at $1.00 per day, due $90.50. pp. 5–6.

C. S. Dobyns of F. O. and C. S. Dobyns, Merchants, mechanics lien vs. Noah Wade, 4 April 1883, total due $27.11 for material for mill house, iron, nails, bolts, etc. pp. 6–7.

C. L. Darnell, millwright and carpenter, mechanics lien vs. Noah Wade, 17½ days work, total $16.87, paid $0.78, due $16.09, 5 April 1883. pp. 10–11.

Beauford Duncan, supplier, vs. Noah Wade, 24 April 1883, total $107.07, lumber for mill, beginning October 1881 through July 1883. pp. 12–13.

R. F. Darnell, millwright and carpenter, mechanics lien vs. Noah Wade, 22 days work, total $40.60, received credit of $3.40 in March and April 1883 by 1 bushel wheat, $1.00; ½ bushel buckwheat, 50 cents; 1 bushel wheat, $1.00; 1 bushel wheat ground for 44 pounds flour, 1.10; total, $3.60. pp. 14–15.

77. Crockett Duncan Mill

Big Indian Creek

LOCATION: About 400 yards SE of 1182 Sawmill Hill Road NW (Route 631). A public road that once ran between the Sam and Mozella Radford place on Route 631 and the intersection of Route 730 (Parker Lane NW) and Ambrose Park Road NW crossed Big Indian Creek less than 100 yards downstream from this mill. POWER: Horizontal wheel (tub or turbine); STONES: 1 set and a sawmill; VISIBLE REMAINS: Stacked stone foundation from this or an earlier mill on west bank of stream

History. Neither the 1908 deed recording Crockett Duncan's purchase of the land on which his mill stood, nor the 1915 deed conveying the same 22⅜ acres to his son, includes any sort of reference to a mill. The second deed is consistent with the supposition that the mill burned about 1914 and was not rebuilt, but the first is not so easily explained. The natural assumption, that Duncan built the mill himself or had it built after 1908, is undermined by three facts. None of them is by itself convincing, but the three together suggest strongly that Duncan bought a working mill, or at least a recently occupied mill site, from his wife's sister and brother-in-law in 1908.

First, deeds to numerous tracts on which mills were known to exist make no mention of them, especially if they were not in use. Even when they were still operating, small mills that ground mostly cornmeal on a single set of stones were subject to be overlooked, perhaps because they often represented relatively little cash outlay, regardless of the time and labor invested by the builders. Second, a 1904 deed to a 78-acre tract that included the land later known as the Crockett Duncan mill property refers to the entire 78 acres as "the mill tract," confirming that some type of water-powered saw or grist mill had once stood, or may still have been standing, along that stretch of Big Indian Creek. And third, Beauford Duncan, a previous owner of the tract, is listed as the owner of a corn or flour mill near the Beauford Post Office (just over the hill from this mill site) in 1894 and 1897 (see Appendix A).

Whether Crockett Duncan built a new mill on an old site or occupied an existing one built by Beauford Duncan or another previous owner of the land will probably never be known, but a careful search of previous deeds might provide clues as to when and by whom (if not Beauford Duncan) the earlier mill was built. The task will require some patience considering that Asa Bishop, a previous owner of the 78-acre "mill tract," purchased at least 47 other pieces of Floyd County land, much of it likely to be in the same general area.

A notation on another version of this photograph dates it at 1914. Yet another copy indicates that the photograph was taken after "the fire," although there is no clearly visible evidence of a fire. One source of identification names one Harris, six Smiths, and twelve Duncans in the correct male-female ratio of 8 to 10, but offers no help with matching the names, which include those of Crockett Duncan and his immediate family, with their owners. Another source states that Will Reed is standing on top of the mill and that U. T. Wade is the second of the three men at the lower left. (Courtesy Ethelene D. Hollandsworth)

Like those of his mill, Crockett Duncan's own origins are a little uncertain. Two accounts of his childhood agree that he stayed or was left behind when his father and mother or stepmother left for the Midwest, but differ significantly in the details. As a boy Carlyse Phillips was told that Crockett Duncan's parents, whose surname may have been Nester, were drowned while crossing Little River as part of a wagon train headed west. The infant Crockett, too small for an older brother to care for, was left to be raised by a John Duncan family, whose last name he took as his own.[29] Details offered in Barbara Reininger's online compilation of U.S. census data for Floyd County tell a similarly sad but less dramatic story. Her sources say that Crockett Duncan was born between 1854 and 1860 (as calculated from various census entries) to David William Duncan and Sarah *Nester* Duncan, who died of a fever in 1861. David Duncan remarried in 1863 but at some point Crockett was taken in by the Benjamin Duncan family and was listed as a member of that household in the 1870 and 1880 censuses. When his father, step-mother, and siblings left for Indiana about 1875, Crockett, who was by then at least 15 years old, stayed behind.[30]

Crockett Duncan's wife, Mary Ellen Duncan, was a daughter of Buford and Sarah Duncan. She inherited land on Big Indian Creek from her parents when the 78-acre tract referred to above was partitioned among her, her widowed mother, and two brothers in 1904. The tract on which the mill is thought to have been built was the portion of the 78 acres afterwards allotted to her sister M. "Mozella" Duncan Radford in 1906 and sold by Mozella and Sam Radford to Crockett Duncan in 1908.

Crockett Duncan's mill is supposed to have burned about 1914. A note on one version of the single known photograph of the mill says it was taken after the fire, but there is no clear evidence of a fire in the picture. If the fire did predate the picture, the water power, at least, seems to have sustained little damage. Regardless of exactly when the mill burned or how badly it was damaged, it was evidently not rebuilt, although even this is contradicted by one source, who reported that the mill ran approximately 1906 to 1923.[31] Crockett Duncan's house was across the creek from the mill.

Deeds

Floyd County Deed Book "R," page 529, 11 September 1884. Asa Bishop and Elizabeth, his wife, to Buford Duncan, 98 acres land "on the banks of Big Indian Creek."

D. B. 30, page 281, 11 February 1904. Buford and Sarah Duncan to Joseph G. Duncan, William A. Duncan and Mary E. Duncan, wife of Crockett Duncan, "sons and daughter of the said Beauford and Sarah E. Duncan," "78 acres more or less and known as the Mill tract or Bishop land."

D. B. 31, page 86, 92, 129, 135, 18 October 1904. Division of 78 acre tract per D. B. 30, page 281. William A.

Duncan, 17¼ acres, Joseph G. Duncan, 18 acres, Mary E. Duncan, 17 acres and widow's dower of 17 acres.

D.B. 32, page 481, 2 September 1906. William A. Duncan and Dara A. Duncan, his wife, Joseph G. Duncan and Nellie Duncan, his wife, Mary E. Duncan and Crockett Duncan, her husband, and Rebecca Cox and D.J. Cox, her husband, to Maud M. Radford, 22⅜ acres, waters of Big Indian Creek.

D. B. 34, page 318, 21 January 1908. Maud M. Radford and S. J. Radford, her husband, to Crockett Duncan. (Same tract as Deed Book 32, page 481. Believed to be the Crockett Duncan mill tract.)

D. B. 41, page 202, 7 August 1915. Crockett Duncan and Mary E. Duncan to H. J. (may be J. Harley) Duncan, their son, 22⅜(?) acres, "This being the land allotted to Maud M. Radford by Commissioners Jacob Smith, Eli Wade and Emery Alley and deed [ed?] from her to Crockett Duncan."

78. George R. Duncan Mill and Jim Dulaney Mill

Big Indian Creek

LOCATION: Off Route 752 (Simpkins Road NW) downstream about .5 mile from end of state maintenance. Current maps now show Route 752 turning south (upstream) as a private road when it reaches Big Indian Creek and state maintenance ends. Until sometime in the 1930s the road turned downstream at that point, crossed Big Indian Creek just below the mill, and came back upstream past the mill. Just above the mill the road turned steeply uphill and east toward Route 740, now White Rock Road. Its original eastern end remains as Route 752 (Mill Road NW), which now officially ends less than a mile west of Route 740. POWER: Overshot wheel (1880); Dulaney Mill overshot wheel and turbine, order uncertain. STONES: 2 sets, possibly a roller mill at one time. VISIBLE REMAINS: A fragment of stacked stone foundation or retaining wall

History. The Jim Dulaney Mill was probably built on or near the site of the George R. Duncan Mill described in the United States Census *Products of Industry* reports for 1870 and 1880. A surveyor's drawing on page 299 of Floyd County Deed Book "W," dated 1890, shows the "W. B. Dulaney Road" intersecting the "George R. Duncan Mill Road," and places both near the W. E. Teel place, which adjoined and would eventually include the Dulaney Mill tract. Also, when the 1.5-acre tract on which the Jim Dulaney Mill stood, or would later stand, was sold by W. B. Dulaney to Noah Reed in 1903, it was referred to as the Duncan Mill Tract and its bounds referred to a "mill pond corner."

That the building captured in a 1920s era photo was not the first mill on this site, regardless of a connection to the Duncan mill, also remains unproven, but this assumption is even better supported by various bits of evidence. First, the $100 paid by Reed to Dulaney in 1903 suggests that the 1.5 acre tract was more valuable than an ordinary piece of land of that size because of its potential for development as a mill seat, but the price is still a good deal less than the value of a working or workable mill. Second, the absence of any reference to a mill building or machinery in the description of the property suggests that, while the dam and pond were still there, the mill that had been driven by the water they stored was gone or beyond repair. Third, the belief that Noah Reed put up a new mill on the land bought from Dulaney is supported by an item in the "From Carthage" column on page 1 of the *Floyd Press*, February 18, 1904, which reported that "Rev. Noah Reed and Burdine Reed have completed their Roller mill, and are doing good business." The phrase "their Roller mill" could refer to the entire building or merely the flour-making machinery, leaving open the possibility that they had adapted an existing mill building to accommodate a modern roller mill. But considering the trouble and expense involved in modifying an existing building that was by then at least 30 years old, erecting a new one would have made more sense for a mill of this small size. Faced with the same decision, most Floyd County millers of that era chose to put up new buildings.

The most compelling support for the 1903 build date is that it is consistent with the report

The Jim Dulaney Mill, probably 1920s. The public road, which is no longer open to traffic or passable to an automobile, ran between the mill and dwelling house. (Ricky Cox collection)

from a local source that the Jim Dulaney Mill was built by Malachi Duncan (1855–1916), a millwright who lived less than a mile away at what is now 1184 Indian Valley Road. The 1903 date is equally consistent with the competing belief that the mill was built by Emory "Em" Duncan (b. 1856), a well known local carpenter who was almost certainly the builder of Burdine Reed's second mill, a gasoline-engine-powered flour mill at Carthage, about a mile to the east, sometime after 1910. Neither man could have built the George R. Duncan Mill, or any other mill that was in operation in 1870.

In 1908 a chancery suit against mill owner Burdine Reed resulted in a public auction at which the mill property was bought by S. R. Brame, one of a handful of Floyd County real estate speculators who bought mills all over the county, often under similar circumstances, and resold them at a profit anywhere from a few days to a couple of years later. Meanwhile, Burdine Reed was making plans to build a roller mill (III, 80) to be powered by a gasoline engine, about a mile east in the community of Carthage (now White Rock) on Route 740.

In April 1910 Brame sold the Dulaney mill and 3 acres to Vic Light, who is supposed to have rigged up an elevator inside the mill using the gear set from a horse-drawn mowing machine. In 1914 Light sold the mill to H. P. "Harve" Hall and afterward ran the country store later operated by his son-in-law Hayden Wade. Wade later moved from Alum Ridge to Christiansburg and built the first of a chain of Wade's Supermarkets, still operating in 2010.

The ⅔ and ⅓ interests held by Harvey Hall and Charles Hall, respectively, in the larger and newer Duncan-Hall roller mill a mile downstream explain a clause in the 1915 deed conveying the Dulaney mill to John and Susan Bond. The new owners were required to "agree and bind themselves that there will be no flour mill erected on land herein conveyed." [32] The question raised by the fact that the Reeds were supposedly running a roller mill (flour mill) here more than ten years earlier has not been pursued, but its solution may have to do with capacity rather than basic capability, since the mill could not have produced nearly as much flour as the new Duncan-Hall mill downstream.

The mill was next owned by Lacona Cox and her husband, "Blind" Walker Cox, who had also supported his family as a broom maker and afterward owned the Enoch Phillips Mill (III, 74) on Little Indian Creek. Although Cox sold this mill in March 1920 and bought the Phillips Mill only five months later, he gave his occupation as "broom maker" in the 1920 census. The mill passed from the Coxes to LeGrand Simpkins, a blacksmith as well as a notable maker of straight-backed chairs, and from Simpkins in 1927 to Jim Dulaney, by whose name the mill is now remembered.

Aaron Altizer (born 1915) accompanied his father to the mill and recalled that the wooden overshot water wheel barely cleared a large rock and that ice would form between the two in extremely cold weather. The ice had to be melted with pieces of heated iron each morning and during the day if the wheel was stopped for any length of time. He remembered seeing customers lined up and waiting to "get a little meal ground" while the miller worked to get the wheel turning again. Altizer believed that the mill may have been run at the very last by Martin Duncan, and that he may have installed a turbine water wheel at the bottom of a penstock constructed of stacked 2 by 4 lumber.[33] Elgin Bolt applied the same idea at his turbine-driven mill (IV, 49) on Burks Fork Creek.

Dale Teel (born 1928), whose father, Sidney Teel, bought the mill site in 1934, remembered seeing the remnants of the broken mill dam, and the miller's house, but had no recollection of the mill itself. A man from West Virginia paid Teel's father for the stones, but never returned to get them. They eventually crumbled and fell apart.[34] This fate suggests that they were made up of fitted pieces, backed by plaster of Paris and held together with shrink-fitted steel bands and were not of the more common, solid, one-piece type quarried and finished at one of several quarries on Brushy Mountain near Blacksburg, VA.

Deeds

Floyd County Deed Book 30, page 404, 15 December 1902, W.B. Dulaney and Mary E. Dulaney, his wife, to Noah Reed, 1½ acres, Big Indian Creek. Certain tract or parcel of land "containing 1½ acres ... known as Duncan Mill property lying on Big Indian Creek." Bounded as follows, "Beginning at a chestnut stump on east side of Mill pond corner to W.E. Teel and Martha Dulaney."

D.B. 30, page 403, 7 March 1903, Allen Dulaney and Martha A. Dulaney, his daughter, to Noah Reed, 1½ acres on banks of Indian Creek. (This is assumed to be tract 1 of the two 1½ acre tracts that make up the 3 acre mill property in subsequent deeds. Tract 2 is the mill tract.)

D.B. 30, page 409, 13 January 1904, Noah Reed and Nancy Reed, wife, to B. D. Reed, "½ interest in land and mill property know as Duncan Mill property," containing 3 acres.

D. B. 36, page 466, 1 April 1910, J. E. Burwell, Comr., Chancery cause of *Mary B. Dulaney vs. Burdine Reed, et al.*, and S. R. Brame and Darcy A. Brame, his wife, to V. O. Light, 2 tracts of 1½ acres each, Indian Creek. (On 26 September 1908, S. R. Brame had bought the property at a public auction held to settle the chancery suit between Mary Dulaney and Burdine Reed.)

D.B. 39, page 478, 9 February 1914, V. O. Light and Roxie E. Light, his wife, to H. P. Hall, 3 acres on Big Indian Creek, "a certain tract or parcel of land and mill property and known as the Duncan Mill property containing 3 acres" in 2 tracts of 1½ acres.

D. B. 40, page 502, 28 May 1914. H. P. Hall to C. G. Hall, 1/3 interest in 2 tracts of land, 1½ acres each, known as the Duncan Mill. In this same deed Harvey Hall also conveyed to his brother Charlie a ⅙ interest in the Duncan Roller Mill (III, 79) a mile or so downstream.

D.B. 41, page 20, 18 June 1915. H.P. Hall, C. G. Hall, and Lillie M. Hall, his wife, to John Bond and Susan Bond, "certain tract or parcel of land and mill property containing 3 acres" on the waters of Big Indian Creek. Two tracts, each tract 1½ acres. "The parties of the 2nd part hereby agree and bind themselves that there will be no flour mill erected on land herein conveyed."

D.B. 42, page 55, 10 October 1916. John Bond to Susan Bond, 3 acres, land and mill property.

D.B. 42, page 235, 21 November 1916. Susan Bond and John Bond, her husband, to Walker Cox, 3 acres in two 1½ acre tracts.

D.B. 44, page 132, 29 March 1920. Walker Cox and Lacona, his wife, and H.C. Cox, and Pearl Cox, his wife, to L.S. Simpkins, 2 tracts on Big Indian Creek, 3 acres more or less. Adjoining lands of W.E. Teel, W. B. Dulaney and Jane A. Simpkins.

D. B. 48, page 16, 23 July 1927. L. S. Simpkins and Sarah, his wife, to James M. Dulaney, 2 tracts land on Indian Creek containing 3 plus or minus acres (mentions mill pond).

D. B. 52, page 11, 4 December 1934. J. M. Dulaney and Lizena, his wife, and Mollie L. Duncan to Sidney Teel, 3¼ acres land. "The intention of this deed is to convey to Sidney Teel all land owned by J.M. Dulaney on Big Indian Creek and known as the mill tracts including a small strip of land bought from Mollie L. Duncan together with all its appurtenances." Bounds of second tract mentions a mill pond, so the pond at least was still there.

79. Duncan Roller Mill/ Hall Roller Mill/Hurt Roller Mill

Indian Creek

LOCATION: Route 787 (Indian Valley Road NW), in a curve .35 mile south of intersection with Route 740 (White Rock Road NW); POWER: Turbine; STONES: 2 sets, 4 stand roller mill, and buckwheat mill; VISIBLE REMAINS: Ends of timbers, possibly the base of the turbine penstock, visible in streambed when water is low, loose stones from mill foundation. The wooden dam was about 10 yards upstream from the mill, connected to the penstock by a wooden race or forebay.

History. To the first owners of the Duncan Roller Mill and like-minded men all around the county, the early 20th century seemed like a good time to build a mill. In 1900 Floyd County was home to over 15,000 biscuit and cornbread consumers. Most of them still lived on farms, and most of those farms were growing soft winter wheat to sell and to grind for flour. Most of those farms were also growing some variety of

The Duncan Mill on Big Indian Creek was architecturally unusual in several ways. The cantilevered front portion of the mill added extra,space to a small parking lot. The bits of framing protruding through the roof near the peak are unexplained and unknown in any other mill photograph. They may support the line shaft for the elevator heads. The shed-roofed extension at the far end has the look of a turbine house, but no others were known to be two stories high. (Courtesy Paul Duncan)

open-pollinated white "dent" corn to be ground into cornmeal for cornbread and cornmeal mush. A significant amount of meal was being cooked into mash and distilled into corn whiskey at several legal distilleries (more than forty in 1894, per Appendix A) and an unknown number of illegal ones operating in pre-Prohibition Floyd County.

On top of that, the buhrstone flour that had sufficed since the dawn of agriculture was being rejected by farm wives in favor of the extra-white, extra-light refined flour produced by the roller mill process. Regardless of their feelings about the density and color of the breakfast biscuits, thrifty husbands were happy to get more flour and less bran from a bushel of wheat by taking it to a roller mill. Farmers were also learning that grains and roughage, traditionally fed in whole (or, at most, shelled) form to livestock, were more appealing, more digestible, and more nutritious if first crushed, cracked, chopped, ground, or pulverized, and thoroughly mixed. Hogs grew fatter, faster, on the cream-colored wheat "middlings" extracted in the process of producing fine white roller mill flour; chickens laid more eggs when fed enriched laying mash. Following these progressive trends required the kind of equipment and horsepower that only the big roller-system flour and feed mills could offer.

Only in hindsight is it clear that the milling trade was already approaching a downward slope, one that would level off briefly during the 1930s as the industrialized national economy slowed, then turn more and more steeply downhill in the 1940s and the decades following. While the various owners of the Duncan/Hall/Hurt Roller Mill were responsible for its day-to-day management, the changing times were even more powerful in shaping its fate. The entire history of this mill — its conception, erection, modifications, and dismantling — occurred within a span of less than forty years. Its story fits neatly within a half century that began with boundless optimism in the milling business and ended with the dawning realization that if milling had any future in Floyd County, it would depend on the production of livestock feeds and be driven by sources of power that were stronger, more versatile, and more convenient than an overshot water wheel or a turbine.

Like its lifespan, the ownership history of this mill is relatively short, but it involves a fairly large number of transactions with many family connections between and among the various buyers and sellers. The mill was typical in this sense, but was unique in having been the only water-powered roller mill not built on the site of an earlier mill, though there had once been a sawmill a short distance upstream on the same tract of land. One previous owner of the site had at least thought of some other commercial potential since he retained ½ interest in a sulfur spring on the property, but — so far as is known — no effort was ever made to develop it. The retention of those rights may have been connected to the 1900s hope of a railroad coming into the county. One of several proposed routes was to have followed Little River and Indian Creek into the county and on toward Willis. Aaron DeHart, who was planning a smaller mill (III, 136) on a proven site only two miles downstream, sold a railroad right-of-way across his land in 1905.

Plans to construct a mill at the Duncan site

were put into writing in a 1900 deed made to N. J. (Noah Jackson "Jack") Duncan, who sold the property to his brother John Benjamin "Johnny Ben" Duncan. N. J. then bought it back, built the mill, and installed the roller mill in early 1912 in partnership with Johnny Ben Duncan and C. G. (Charlie) Hall, each of whom had purchased a ⅙ interest. Duncan's sale of two ⅙ interests is notable for two reasons: few other mill properties were divided for investment into such small fractions and no other deeds record the promise of one man's future labor in exchange for a cash investment. Noah Duncan probably needed a large chunk of cash to purchase and install the roller mill system, which — unlike simpler, more traditional milling equipment—could not be pieced together by resourceful millwrights, blacksmiths, and woodworkers using locally available materials. Unwilling to give up an equal or controlling interest in the business, he "indentured" himself to his investors in exchange for the needed cash.

Early in 1914 the Duncans sold their collective ⅚ interest to Charlie Hall's brother H. P. ("Harve") Hall. In May of that year Harve sold a second ⅙ interest in the Duncan Roller Mill to Charlie, along with ⅓ interest in an existing mill (III, 78) less than a mile, as the crow flies, upstream. Also known as the Duncan Mill (but distinguishable by the lack of the word "Roller" in its name), this mill was doubly handicapped as a potential competitor. It stood by a road much less travelled and it lay upstream from the confluence of Big Indian Creek and its main tributary, Little Indian, so had access to less water and less power than the Duncan Roller Mill. But the Hall brothers had too much at stake to take any chances; when they sold the old Duncan mill the following year, a clause in the deed required the new owners to "agree and bind themselves that there will be no flour mill erected on land herein conveyed."[35] A deed to the Newton Hylton/Walter Boyd Mill (IV, 35) property on Howell Creek places a similar restriction on the new owner.

In March 1915, C. G. Hall sold his ⅓ interest to William Ira "Columbus" Duncan and H. P. sold his ⅔ interest to John Duncan, a nephew of Noah and Johnny Ben. John Duncan would be a full or part owner until 1930, making his involvement in the mill longer by far than any other owner's. His partners were, at various times, Asa Carner (a distant relation), Wyatt "Lynch" Duncan (his brother), and Taz Hurt. According to John Duncan's son Paul, John Duncan and millwright Asa Carner together modified some part of the mill's machinery, but whether that work coincided with Duncan's addition of the turbine house at the rear of the mill is not known. Taz Hurt ran a small store that Duncan, who had no regular hired help in the mill, built to minimize interruptions of his work as a miller. Initially his wife, Maude Young Duncan, sold meal, feed, and flour in the store. Grocery items and a gasoline pump were added at the request of customers wishing to combine errands. The store was on the same side of the road as the mill, close by on the upstream side and very near the dam and millpond.[36] Paul Duncan remembered as a small boy pulling a cardboard box around the parking lot by a string. A horse kicked at the box and put its foot through the radiator of one of the few automobiles in the community, a Ford Model T car. The miller's house sat across the road and "up the bank" from the store and mill, about where the widened and paved roadway is now.[37]

In addition to supplying local demand, John Duncan sold flour and meal under the name "Big Indian Mill" (or Mills) to wholesalers like Gordon Dalton in Radford, Virginia, and occasionally to commercial buyers from as far away as Roanoke, Virginia. "Hucksters" taking local produce to peddle door to door in West Virginia mining towns might also stop in for locally branded mill products they knew would appeal to homesick former Floyd Countians living in company housing. Duncan also took advantage of the ideal conditions presented by mill ownership to raise 20 to 30 hogs each year, feeding them middlings, "chop," and bran from the flour mill along with surplus, old, or spoiled corn. Duncan's ownership of the "Hurt place," the 80-acre farm that joined the mill tract on the downstream side, provided a convenient spot for the hogs.[38] Keeping the hogs out of the immediate vicinity of the mill was not merely a concession to the sensibilities of customers; it also complied with a state law mandating that mill owners wishing to raise hogs must keep them on a separate tract of at least 50 acres to insure sanitary conditions at the mill. The success of this measure must have depended somewhat upon which portion of the 50 acres was allotted to the hogs.

After Lynch Duncan — John's brother and former partner — gave the mill up in 1936, it passed through several hands, some familiar and some new. In 1946 Gervase Duncan, a son of one of the original partners, Johnny Ben Duncan, bought the mill and by 1950 had dismantled it and reused parts of it and its stone foundation in and around a general store that still stands in Graysontown, Pulaski County, Virginia. Rufus Duncan and his brother Martin are supposed to have run the mill and store for Charlie Hall.

Exactly when it ceased grinding is unknown, but Charlie Hall's son, Henry (b. 1934), remembered being persuaded along with his playmates that a Japanese soldier was hiding out in the mill during the Second World War. Although willing to overlook some rather large geographic difficulties in order to accept this story, they probably would have been more skeptical if customers had been going in and out of the mill.

Deeds

Floyd County Deed Book 27, page 281, 24 December 1900. M. T. Hurt and wife and W. C. Duncan and wife, Julia, and N. C. Duncan to N. J. Duncan, 6¾ acres land on which mill is to be built and water rights for dam. Two tracts of 6 acres and ¾ acres. Bounds of said land to begin "above old Saw mill corner ... thence down the creek as it meanders, extending all the way down far away enough from the creek to cover and take all land that may be used in putting in a dam across said creek and all land that water from dam may cover or damage to a maple thence crossing the creek to the beginning." Hurt retained a right of way to lands not conveyed and "one half interest ... in the sulphur spring on land conveyed and full access to same."

D. B. 28, page 566, 5 August 1902. N. J. Duncan and wife Rosa to J. B. Duncan, six acres on the waters of Big Indian Creek, "Known as the old Hambrick Sawmill Property." *The "Hambrick Sawmill Property" referred to is one of two on Big Indian Creek. The other was at the site of the Aaron Dehart Mill (III, 136).*

D. B. 38, page 242, 20 October 1905, M. T. Hurt and R. B. Hurt to N. J. Duncan, 4 acres more or less on Indian Creek.

D. B. 38, page 241, 16 November 1911. J. B. Duncan and wife to N. J. Duncan and wife. The same "mill seat" tract conveyed from N. J. Duncan to J. B. Duncan, "6 acres more or less with buildings and machinery and all appurtenances."

D. B. 38, page 388, 16 November 1911. N. J. Duncan to J. B. Duncan, ⅙ interest. N. J. Duncan to put in roller mill first of February 1912. N. J. Duncan to operate mill at no cost to J. B. Duncan for three years, also to insure property against fire for three years. Half of $1,000 policy to be paid to J. B. Duncan in the event of fire.

D. B. 39, page 427, 16 November 1911. N. J. Duncan to C. G. Hall, ⅙ interest "in all appurtenances consisting in water power buildings and roller mills and bur mills and all other machinery connected with said mill." Party of the first part to "put in roller mill machinery in by the first of February 1912 and to operate same without any cost to party of the second part for his labor for the term of three years."

D. B. 39, page 479, 3 February 1914. N. J. and J. B. Duncan to H. P. Hall, N. J. Duncan's ⅔ interest and J. B. Duncan's ⅙ interest in the Duncan and Hall Roller Mills and 10¾ acres of land.

D. B. 40, page 502, 28 May 1914. H. P. Hall to C. G. Hall, ⅙ interest in N. J. Duncan Roller Mills, also ⅓ interest in 2 tracts of land, 1½ acres each, known as the Duncan Mill property (Dulaney Mill, III, 78)

D. B. 40, page 515, 12 March 1915, C. G. Hall to William I. C. Duncan, his interest in N. J. Duncan Rolling Mills, 10¾ acres land and mill property,

D. B. 40, page 522, 23 March 1915, H. P. Hall to John W. Duncan and Maude Duncan, ⅔ interest in 10¾ acres land and mill property, N. J. Duncan Roller Mills.

D. B. 41, page 153, 25 September 1915, William I. C. Duncan and Mattie, his wife, to Asa L. Carner, ⅓ interest in 10¾ acres land and mill property and all water rights.

D. B. 43, page 565, 1 January 1919. Asa L. Carner to John W. Duncan, ⅓ interest in mill known as the N. J. Duncan Roller Mills.

D. B. 45, page 41, 22 September 1921. John W. Duncan to Wyatt L. Duncan, ½ interest in mill and 10¾ acres land.

D. B. 47, page 118, 2 February 1925. Wyatt L. Duncan to M. T. ("Taz") Hurt, ½ interest in mill.

D. B. 49, page 508, 25 October, 1929. M. T. Hurt to W. L. Duncan, ½ interest in mill.

D. B. 49, page 508, 26 September 1930. J. W. Duncan and Maude Duncan to Wyatt L. Duncan, ½ interest in mill.

D. B. 49, page 509, 9 September 1931. Wyatt L. Duncan to W. S. Roop, Trustee, 10¾ acres land and mill property which W. L. Duncan acquired a ½ interest in from M. T. Hurt by deed 25 October 1929, "not yet recorded but to be recorded contemporaneously with this deed and the other ½ undivided interest being acquired by the said Wyatt L. Duncan by deed dated 26 September 1930 from J. W. Duncan and Maude Duncan (both deeds, see D. B. 49, page 508). *Wyatt "Lynch" Duncan apparently paid the debt secured by this deed of trust. It is included to explain when and from whom he acquired the mill property.*

D. B. 53, page 499, 5 September 1936. W. L. Duncan to M. T. Hurt, 10¾ acres and mill known as the Duncan and Copper Valley Mill property. Also "1 Peerless steam engine, 1 American saw mill and saw and drive belt, the same being the complete saw mill and equipment owned by the said W. L. Duncan." Property conveyed to Hurt to settle $2,050 debt.

D. B. 58, page 400, 4 March 1939. M. T. Hurt to C. G. (Charlie) Hall, 10¾ acres and mill. Settlement of chancery cause *Peoples Bank vs. Mrs. R. B. Hurt et al.* Hall bought three other tracts at the same time.

D. B. 58, page 561, 12 August 1939. C. G. (Charlie) Hall to A. L. O'Dell, 10¾ acres. Mentions mill pond but not the mill.

D. B. 64, page 553, 31 May 1946. A. L. O'Dell to G. E. Duncan. Mentions mill pond in boundaries, but not the mill.

80. Burdine D. Reed Mill

LOCATION: Route 740 (White Rock Road NW) at intersection with Route 636 (Carthage Road NW)
POWER: 12 HP, 1-cylinder gasoline engine; STONES: 2 meadows mills, roller mill and store; VISIBLE REMAINS: None

History. In 1910 Burdine Reed (1865–1956) bought 49 square poles (less than one third of an acre) in Carthage, Virginia, from Asa Burnett. He almost certainly intended to build a mill there, but exactly when that happened is unknown. Reed had owned a one-half interest in the Jim Dulaney Mill (III, 78) on Big Indian Creek between 1904 and 1908 and must have known of plans being made as early as 1900 for the Duncan Roller Mill (III, 79) no more than a mile downstream and north of his mill. The new roller mill, only a few feet off Indian Valley Road (Route 787), would be much easier to get to, and also larger and more powerful than Reed's own mill, since the confluence of Big and Little Indian Creeks lay between the two mill sites. Although the property changed hands several times in the meantime, the new roller mill was in operation by the end of 1912 and undoubtedly took some business from the smaller, less accessible upstream mill, although the latter ran for another 20 years. After it was closed, the road on which it stood was so seldom travelled that it was abandoned by the state highway department at the request of Sidney Teel, a subsequent owner of the Dulaney Mill site.

The site of Burdine Reed's new mill in the community of Carthage (now called White Rock) is, like the Dulaney Mill, only about a mile from the John Duncan Mill site, but it lies to the southeast, with no road running directly between the two. The site chosen for Reed's new mill also lay on the western edge of a large and well populated section of the county bounded by Little River on the north, Routes 740 and 730 on the west and south, and Route 8 on the east. Reed could have expected to draw trade from at least half this area of over 20 square miles, since no other roller mill existed, then or later, within those bounds.

The mill was built for Reed by a man known as "Em" Duncan, who was almost certainly Emory M. Duncan (b. 1856), a carpenter, wagon builder, and woodworker well known in that community. Between 1920 and 1950 numerous small, part-time corn and feed mills were built expressly to be driven by steam or gasoline engines or electric motors, but Reed's was probably the only full-scale Floyd County flour mill ever built to be driven by a gasoline engine. It was also the only one built on a ridge top, since its gasoline engine, unlike a water wheel or a steam engine, needed water only for cooling.

Progressive thinking was also reflected in Reed's choice of self-contained, semi-portable mills manufactured by the Meadows Mill Company in North Wilkesboro, North Carolina, for grinding corn and chop. The stones inside were smaller and lighter than those of a conventional "corn rock," but the runner stone was balanced to run at several hundred RPM and was fixed to a shaft turning on heavy-duty babbit metal bearings. While not simple, Meadows Mills required less skill to set up, operate, and maintain than their predecessors. Starting and maintaining the 12 HP gasoline engine would have required some mechanical know-how and probably a good amount of physical strength, but compared to the time involved in building, digging, and installing, and afterward maintaining a dam, race, and water wheel, the hours needed to place a gasoline engine, set it up, and belt it to the main shaft were insignificant.

Despite its strategic location and incorporation of the latest technology, the Burdine Reed Mill ran for a much shorter time than most of its water-powered counterparts. It had certainly ceased to operate by the time Charlie Dulaney sold the land and building to Columbus Reed in 1927. The flour mill may even have been removed before Dulaney bought the property in 1925. An undated *Floyd Press* advertisement offers for sale "one flour mill, twenty-five bbl. capacity. A good bargain for quick sale. Will sell property and machinery together, or separately.... Burdine Reed, Carthage, Virginia." Though unrecorded, the date of the advertisement had to precede the 1925 sale of the land and building to Charlie Dulaney. An August 26, 1926, *Floyd Press* advertisement for a "Deputy Sheriff Sale of Personal Property" held on August 30, 1926, to settle debts owed by Dulaney to the Floyd County Bank and J. D. Ekiss doesn't mention a flour mill, but Dulaney had also run a steam- powered sawmill and planer and ground cornmeal with a Meadows type mill at another location now known as the Vada Altizer place near Carthage (White Rock), so the absence of the flour mill in the sale bill is inconclusive.

After buying the mill tract and building in 1927 from Noah Alley, who had bought it at auction forced by a chancery suit against Dulaney, Columbus Reed ran a store in two rooms of the mill for a short time. He and his wife were still living in part of the building when a son was killed in World War II. Like many other millers, Burdine Reed owned a store near his mill. This was deeded, along with a house and 91 acres, to J. Dan Ekiss and his son-in-law Sherman Ridinger in 1923, in exchange for a store and a mill in Christiansburg, Virginia, near where the road still known as the Mud Pike joins Route 8. Burdine Reed also owned a leather tanning yard located at what is now known as the Hubert Reed place on Route 739 (Reed Hill Rd. NW) and run by Tom Sowers, who made saddles and harnesses from the leather. Aaron Altizer recalled that the series of hand-dug pits through which the animal hides were gradually moved resembled rows of graves in size and arrangement.[39]

Deeds

Floyd County Deed Book 40, Page 524, 30 April 1910. Asa Burnett to B. D. Reed, 49 poles at Carthage, Virginia.

D. B. 47, page 525, 2 December 1925. B. D. Reed to C. R. (Charlie) Dulaney, tract of land of 49 poles and B.D. Reed Mill property, Carthage, Virginia, adjoining Roy R. Dulaney store property.

D. B. 46, page 351, 30 April 1927. C. R. Duncan to W. C. (Columbus) Reed, his mill property of 49 poles. Settlement of chancery cause "*J. W. Brammer, et al., vs. C.R. Dulaney, et al.* and N.W. Alley and M.S. Aley, his wife, parties of the first part, and W.C. Reed, party of the 2nd part." Sold to A.W. Alley at public auction on December 18, 1926. N.W. Alley "has directed in writing that said land be conveyed to W.C. Reed." At Carthage, 49 poles refers back to DB 47, page 525.

---

81. James G. Lester Mill

Easter Creek

LOCATION: .25 mile upstream from Route 738 (Easter Creek Road NW); POWER: Overshot wheel; STONES: 1 set; VISIBLE REMAINS: None. Nearby waterfall suggests favorable location.

History. In 1834 Francis Gardner, then serving in the U. S. Navy, deeded his interest in the lands of his late father, John Gardner, to his sister Mary Gardner Lester and his step-mother, Ann Gardner. There is no mention of a mill, but the property description is minimal. At least two mills, probably a grist mill and a sawmill, were on this site when in 1854 Ann Gardner and Mary Lester sold George Shelor 199 acres of land "with mills excepted." Three years later James G. Lester along with his wife and his mother deeded the mill to Christina, Samuel, Thomas, and Abram Lester, the wife and minor sons of Leftwich Lester. Leftwich Lester gave his occupation as "miller" to census takers in 1860 and 1870.

The entry in the *1870 Products of Industry Report* cited below doesn't neatly match this mill for two reasons: The name of the stream was left off the transcribed record and James Lester, et al., had sold this mill to Leftwich Lester (deed recorded in the names of his wife and three of his children) in 1857. On the other hand, *Products of Industry* enumerators sometimes named as the owners of mills former employees or owners. The most compelling evidence that these do refer to the same mill is the fact that James G. Lester didn't buy the land for his next mill (II, 95) until 1871.

Deeds

Deed Book "A," page 264, 17 March 1834. Francis R. Gardner of the United States Navy to Ann Gardner and Polly Lester, "his stepmother and ... his sister." All his interest in lands of his father, James Gardner, deceased.

D. B. "G," page 467, 22 July 1854. Ann Gardner and Mary Lester to George Shelor, 199.5 acres land with mills excepted.

D. B. "J," page 186, 15 May 1857. Mary G. Lester, James G. Lester and Mary Ann, his wife, to Christina Lester, Samuel Lester, Thomas Lester and Abram Lester. "A certain mill seat and mills which is now on it ... on the waters of Easter Creek in the midst of the land belonging to George Shelor the quantity of land being what the said mills now sits upon and what the water covers from the dam not exceeding 10 feet high."

See Table B, *1870 Products of Industry.* Name of corporation or individual: Lester, James

82. George H. Booth Mill/ George B. Booth Mill/Noah Booth Mill

Beaver Creek

LOCATION: Upstream from Route 705 (Beaver Creek Road NW) and Dobbins Store, approximately 1 mile; POWER: Originally overshot wheel (1880), then 2 turbines and 2 penstocks. Mill powered by gasoline engine for a few years after penstocks and forebay failed. STONES: 2 sets; VISIBLE REMAINS: Dug race

History. There could have been a mill on this site as early as 1812 when Abijah Booth was granted permission by the Montgomery County Court to build a mill dam on Beaver Creek.[40] But no mention of the mill was found in a deed record until 1889, when George B. Booth's land was divided after he was killed by a falling tree. But as noted in the overview of Floyd County mill history, this is by no means a reason to conclude that the mill was not on one of the tracts conveyed by the earlier deeds. Writers of deeds regularly failed to mention active mills as well as abandoned ones, especially when large tracts of land were involved.

Proof of the mill's presence in 1870 and 1880 is included in the *Products of Industry* reports for those years which name as owners of the mill George H. Booth (1795–1872) and George B. Booth (1835–1889), son and grandson of Abijah Booth. George H. Booth was probably not the owner in 1870, having conveyed large tracts of land to George B. in 1860 and 1866, but this, again, is not surprising. *Products of Industry* enumerators sometimes listed as owners recent, long-time, or popular millers, without determining who actually held legal title to the property.

The commissioner's sale that followed the court-ordered division of George B. Booth's land put the mill in the hands of Samuel Dulaney. In general, public auctions resulting from chancery suits or the division of an estate among multiple heirs provided the only opportunities for people who weren't close blood relatives or in-laws to acquire an existing mill.

The mill came back into the hands of the Booth family when Noah Booth bought it and thirteen acres in 1906. Although his name has continued to be associated with it, Noah Booth kept the mill only until 1914, when John W. Hall paid $900.00 for the mill and 27 acres of land. Hall hired Monroe Akers to run the mill for him. Ten years later Hall sold the property, for the same price he paid, to William H. Booth. In 1927 William Booth sold the twenty-seven acres to Giles G. Dobbins, with whose name the property is still associated since he owned and ran a well-known store near present day Route 705.

Noah Booth's nephew Leonard Booth, interviewed by Kathleen Ingoldsby on February 17, 1993, said his first memory of the Noah Booth mill was as a very young boy when he would ride to the mill behind his father on horseback. Leonard was born on June 22, 1902, and he places this early memory about 1905. He said "uncle" Noah ran the mill (Leonard's father and Noah were brothers) and at this time or earlier, it had a mill dam and a dug raceway that connected to a wooden race with a 15-foot fall to an overshot water wheel. About that time the mill was converted to turbine power. "It had two pin [pen?] stock, full of water," he recalled.

Before Noah Booth, George Hurt ran the mill. Hurt was Cassie Mannon Booth's grandfather (Leonard's wife). Monroe Akers was the last miller. The mill had two stone burrs, one for flour, one for corn. In early times it had two stones for rye flour, and it ground buckwheat. When roller mills came into use, rock ground couldn't compete with the rollers. Some of the stones are now in a house "on top of the mountain," as seen from Route 8 and the Blue Ridge Parkway, south of the parkway.

One of the turbines powered the sawmill, the other the grist mill. Both were fed by the still visible dug race, approximately 250 yards long, and a wooden forebay that is believed to have carried the water about 150 feet across a meadow from the race to the mill. Some part of the mill machinery, most likely the set of stones used to grind cornmeal, was driven by a gasoline engine for a few years after the wooden forebay and penstocks became unusable.

Deeds

Floyd County Deed Book "B," page 208, 11 September 1838. Abijah Booth to George Booth, 305 acres on Beaver Creek.

D. B. "K," page 449, 17 November 1860. George H. Booth to George B. Booth, 359 acres on Beaver Creek.

D. B. "M," page 357, 13 February 1866. George H. Booth to George B. Booth, 80 acres on Beaver Creek adjoining above tract at Gap Branch.

D. B. "U," page 185, May 1889. Division of land of George B. Booth. Surveyor's map shows mill in Lot No. 1.

D. B. "U," page 299, 23 November 1889. Z. T. Dobyns, Commissioner in Cause of J. M. Booth vs. Austen M. Booth, to Samuel W. Dulaney, Lot No. 1 of 49 acres and mill in division of land of George B. Booth.

D. B. "U," page 711, 14 July 1890. Samuel W. Dulaney, and Malinda E. Dulaney, his wife, to Wm. Dulaney and A. J. Reed, Jr., 49 acres, Lot No. 1.

D. B. 29, page 335, 30 March 1903. A. J. Reed, Jr., and Sarah Reed, his wife, to William Dulaney, their ½ interest in parcel of land of 49 acres on Beaver Creek known as the mill tract.

D. B. 29, page 598, 1 April 1903. William Dulaney to George C. Hurt, ½ interest in 13 plus acres and ½ interest in one grist mill and sawmill and machinery thereon, and water power to run same.

D. B. 35, page 513, 24 July 1906. Wm. A. Dulaney and George Hurt to Noah Booth, 13 plus acres on Beaver Creek.

D. B. 39, page 497, 11 February 1914. Noah Booth to John W. Hall, 27 acres and mill for $900.00. Monroe Akers ran mill for J. W. Hall.

D. B. 47, page 95, 21 August 1924. John W. Hall, and Mildred, his wife, to Wm. Boothe, 27 acres for $900.00.

D. B. 54, page 261, 12 March 1937. Wm. Booth and wife to Giles G. Dobbins, 27 acres. Boundaries mention mill pond.

See Table B, *1870 Products of Industry*. Name of corporation or individual: Booth, George H.

See Table C, *1880 Products of Industry*. Name of corporation or individual: Boothe, George B.

83. George Booth Mill

Beaver Creek

The George Booth Mill was first thought to be distinct from the Noah Booth Mill (III, 82) but they are one and the same. George H. Booth and his son George B. Booth were long-time owners of the mill or at least a mill site that probably dates back to the early 1800s. The site assigned to mill number 83 on Frank Webb's map is near what was once Dobbins' Store. An undocumented mill called the Jim Boothe Mill, said to have washed away before the 1920s, was also supposed to have been near this spot, but no connection has been found between them.

84. Bonsack-Eller Wool Carding and Oil Mill

Beaver Creek

LOCATION: Route 705 (Beaver Creek Road NW) about .5 mile west of Route 8 (Webb's Mill Road N); POWER: Overshot wheel; STONES: 2 carding machines, 6 pestles

History. Because of the specialized machinery required, there were only a handful of wool carding mills in Floyd County at any point (see Appendix A). The last of them, owned and operated by J. T. Dunn when it closed in the 1940s, survived that long only because it could produce finished blankets and coverlets on its power-driven looms. As long as wool growers were trading raw wool for finished woolen goods or retrieving at least part of it for use at home after it had been cleaned, washed, and carded, and perhaps spun into yarn, a local woolen mill had a powerful advantage—convenience—over more distant competitors. Sheep farming and shearing continued, albeit on a smaller scale, past World War II, but once shipping the wool became a one-way process, farmers ceased caring how far they had to send their wool to get the best price.

Among the woolen mills that did not produce woven goods, the Bonsack-Eller mill is the best documented. George Eller is named as the operator (he did not own the mill until 1886) of a water-powered wool-carding mill in the 1860 *Products of Industry* report and reportedly produced 5,000 pounds of wool rolls valued at $2,000 during the six months that the mill was active. There were other wool carding mills at work in the county in that year, but presumably none of them produced the minimum $500 worth of goods required for inclusion in the report. Eller appears in the *Products of Industry* report again in 1870 and had one of two woolen mills listed in *Chataigne's Virginia Gazetteer and Classified Business Directory* in 1885 (see Appendix A), one of three in 1889, and one of five named in 1894 and 1897.

The Bonsack-Eller operation is sometimes referred to as "carding machine," which seems to have referred to the entire building, not just the machines inside. Dr. Amos Wood indicates that the entire property was once known as "Bonsack's Machine" and also as the "Oil Mill."[41] The latter name corresponds with the 1870 *Products of Industry* report's listing of Eller as the operator of an "Oil Mill," which crushed linseed or flax seed using "six pestles." The crushed seeds were then boiled to extract the valuable oil inside.

On August 10, 1905 the *Floyd Press* reported, "Mr. George Eller, of Posey, died Friday night after a long illness.... He was born and raised in Botetout County and came to this county to work in a carding mill. He soon bought the property and ran a carding machine up until a few years ago. He was a good and influential citizen and well thought of by all who knew him." Dr. Amos Wood also mentions Eller's move to Floyd County, but his placement of Eller's arrival in "about 1868"[42] is contradicted by Eller's inclusion as a wool mill operator in the 1860 *Products of Industry* report for Floyd County.

Deeds

Floyd County Deed Book "H," page 82, 7 December 1854. George H. Booth to John Bonsack, 27 acres land on Beaver Creek. *John Bonsack is presumed to have built this mill.*

D. B. "O," page 51, 18 February 1886. Jacob Bonsack, executor for John Bonsack, deceased, to George Eller, 163 acres land and mill on Beaver Creek.

D. B. 27, page 456, 4 April 1901. George Eller to his son, D. L. Eller, 146 acres land on Beaver Creek, where he now lives. Mill no longer running.

See Table B, *1870 Products of Industry.* Name of corporation or individual: George Eller

Eller is listed in separate entries as operator of wool carding mill and producer of linseed oil.

85. Henry H. Earles Mill

Earls Branch (of Beaver Creek)

LOCATION: One-half mile south of Route 705 (Beaver Creek Road NW) on Earle's Branch; STONES: One set; VISIBLE REMAINS: A few stones from the mill dam

History. Like many millers, Henry Harrison Earles* (1842–1923) was a man of many talents and enterprises. He was a brick mason and a maker of bricks and operated a legal distillery. Leonard Booth recalled that his was "'the biggest apple brandy operation around'.... When Leonard's father-in-law, N. S. Mannon, bought the Earles' farm in 1925 'all the hills were covered with apple trees.'"[43] India Earles Dickerson, in a 1982 apple variety survey conducted by Kathleen Ingoldsby, remembered the family farm as having many varieties of apples: Paradise Winter Sweet, Old Ben Davis, Black Twig, Early Ripe, Rusty Sweet/Coat, Fallawater, Newtown Pippen, Winter Pearmain, Harmon, Red June, Yellow Transparent, Wolf River, and Smokehouse, among others. Earle's grist mill ground corn on one set of stones.[44] Little else is known about the mill beyond the location, which is marked by a few remaining stones from the dam. Earles purchased the 20⅛ acres on which his mill stood in 1875, the same year he married his first wife, Susannah Lester, daughter of Fleming Lester. The property was deeded, this time in a tract of 27¾ acres, to Henry Earles' son Charles E. Earles in 1905. A flyer advertising the sale of the Earles farm in April 1925 makes no mention of a mill.

Deeds

Floyd County Deed Book "N," page 438, 29 March 1875. J. L. Thompkins, Commissioner in suit of *Wm. B. Shelor vs. Fleming W. Lester, et al.*, to Henry H. Earls 20⅛ acres of lands of F. W. Lester and wife on Beaver Creek. Boundaries include "a hickory stump near a carding machine" and cross "corner of machine to mouth of a small branch."

D. B. 32, page 188, 1 December 1905. H. H. Earls to Charles E. Earls, 27¾ acres on waters of Beaver Creek.

**The family name is consistently spelled "Earles" in census records. The name of the stream, which may or may not be named after this family, appears as "Earls" in the two deeds summarized for this mill's history.*

86. Charles Turman Mill/ F. M. Slusher Mill/W. C. Lester Mill

West Fork of Little River

LOCATION: 1.25 miles southwest of Route 719 (Laurel Branch Road NW); POWER: Turbine; STONES: 1 set and sawmill

History. Charles Turman, who is thought to have built the first mill on this site, left the executors of his estate specific instructions regarding the fates of his slaves. Following the death of his widow, they were to be hired out and their wages set aside until the youngest turned 21. At that point all were to be freed and removed to "some free and non slaveholding state or country" and their transportation paid for with the money they had earned. Any money that remained was to be used to purchase land for them in "the state or country to which they may be sent." Considering that his widow, who was 84 years old in 1850, could have lived on for several more years, and that the youngest of his slaves could have been an infant in 1847 (or at the time of his wife's death, depending upon how the will was interpreted), Turman's slaves may not have benefited from his intentions for many years, if ever. But considering that neither the Civil War nor its consequences were a foregone conclusion in 1847, his plan was, in the context of its time, commendable, in spite of the built-in delays.

The last man to run this mill, Winfield C. Lester, may have been of similarly compassionate disposition, having been the superintendent of a poor house, and his wife the matron, before they bought the mill in 1881. Lester is said to have been killed in an accident at the mill in 1921, after which time it was closed.

The history of this mill during the forty years that passed between the tenures of these two owners is not fully or convincingly documented. The supposition that Turman ran a grist mill is based on oral sources. Some related deeds dated between 1847 and 1887 mention the mill dam, a couple the mill itself, and some neither. The *Products of Industry* report for 1860 names Jacob Slusher as the owner of a sash type sawmill, but not a grist mill. His will, dated August 9, 1871, provides for the disposition of his sawmill, but does not mention a grist mill. A couple of deeds that are included because of a boundary reference to the mill pond or race may actually convey adjacent property and not the mill site itself.

Deeds and Other Public Records

Floyd County Will Book "B," page 93, 1 April 1847. Charles Turman willed to wife and son Mathew the home tract (presumed to include the mill if he did in fact build or own it). The other half of his estate is left to two daughters. *The will includes his plan for the eventual emancipation of his slaves.*

Floyd County Deed Book "J," page 431, 7 May 1858. Asa Howard and, Ellen, his wife, to Mahlon Turman. "Division of a tract of land." $1.00. Lying in the waters of West Fork, Little River, 47 acres. "Adjoining the lands of Col. Jacob Slusher and others." "Beginning at a large rock corner to Col. Jacob Slusher." This was referred to in DB "T," p. 361.

D. B. "K," page 212, 15 September 1859. Mathew Turman and others to Jacob Slusher, two tracts of land, Lot No. 3 containing 55 acres and Lot No. 4 containing 19 acres on West Fork Little River. Both were "part of the old Charles Turman tract of land." Boundaries refer to "a rock at the abutment of the old mill dam."

D. B. "K," page 401, 27 August 1860. Jacob Slusher and Telitha, his wife, to Francis M. Slusher, 154½ acres on West Fork Little River and 19 acres on West Fork Little River.

D. B. "K" page 402, 27 August 1860. Jacob Slusher and Telitha, his wife, to Oliver Slusher, 156¾ acres on the waters of West Fork Little River. Boundaries mention the abutment of the mill dam and the mill pond. *This land may adjoin the mill tract.*

D. B. "P," page 198, 22 August 1878. John Earls, Commissioner, to Burdine Bishop, 136 acres, waters of West Fork Little River. "Land herein conveyed formerly belonged to Charles Turman deceased and there to Mathew Turman and his heirs" is described in Deed Book "K," page 212. *This deed, too, may refer to a tract adjoining the mill tract.*

D. B. "Q," page 241, 21 October 1881. J. L. Thompkins, Commissioner in Cause of *A. J. Kirby and C. M. Stigleman vs. F. M. Slusher and others*, to Winfield C. Lester, 146½ acres known as the F. M. Slusher home and mill place on West Fork Little River. Land conveyed to F.M. Slusher by Jacob Slusher and wife by deed dated 27 August 1860, recorded D.B. "K," p. 401, except that part of it conveyed to Asa Bishop by F.M. Slusher and wife on 18 February 1871 in D.B. "M," p. 83. Land sold to Kirby and Stigleman, who sold it to W.C. Lester.

D. B. 52, page 235, 25 April 1935. Libby Garretson, heir in law of W. C. Lester, to H. Learn Hubbard, 55 acres of land on West Fork of Little River, part of 300 plus acres owned by W. C. Lester at the time of his death. *This land is thought to be the mill tract.*

87. Eli Wade Mill

Spurlock Creek

LOCATION: Route 740 (Ferny Creek Road NW) .25 mile north of Route 746 (Graham Road NW); POWER: Overshot wheel; STONES: 2 sets

History. Eli Wade (1861–1946) inherited the land for his mill when the estate of his grandfather Henry Wade was settled in 1879. Eli and his siblings Mary and John W. Wade divided the one- sixth interest of eighty acres that would have gone to their father, also named John W. Wade, who died in the Civil War. *Chataigne's Virginia Gazeteer and Classified Business Directory* lists Eli Wade as the owner of a grist mill and a sawmill near the Abraham post office in Floyd County in 1897.[45] By the end of that year he had sold one-half interest in the mills and land to G. W. Reed and the other half to John M. Wade.

The two one-half interests in the mill circulated separately for a while but had been recombined by the time D. F. Hylton bought the Eli Wade Mill property in February 1909. On February 18 Hylton posted this notice in the *Floyd Press*: "My mill known as the Eli Wade mill has now been put in first-class condition by Thos. Shelor and operated by M. F. Hylton as miller. Thanking my customers for past favors, I ask a continuance of same in future. D. W. Hylton, Big Sandy, W. Va."[46]

Whether his customers withheld the continuance of their past favors or he got tired of managing a mill long-distance, Hylton sold the Eli Wade Mill to James Goad eight months later. Goad's will, dated December 2, 1935, refers to the tract bought from Hylton in 1909 but does not mention the mill.

Deeds

Floyd County Deed Book "P," page 347 plus, 7 May 1879. Estate of Henry Wade. Elizabeth Wade, widow of Henry Wade, to 5 living children, ⅙ interest (Lot 1, 80 acres) to children of son, John W. Wade, deceased, Eli Wade, Mary Wade, and John Wade.

D. B. "Y," page 536, 29 December 1897. Eli Wade to G. W. Reed, ½ interest in saw and grist mill and 16 plus acres land.

D. B. "Y," page 572, 19 June 1897. Eli Wade to John M. Wade, ½ interest in saw and grist mill, water rights and 16 plus acres land.

D. B. "Z," page 329, 7 May 1898. George W. Reed to M. T. Akers, "½ interest in one grist mill, one sawmill, and one dwelling house" and 16 plus acres land.

D. B. 26, page 268, 30 November 1899. M. T. Akers to W. K. Fagg, Deed of Trust for mill and 18 plus acres on Spurlock Creek, grist mill known as the "Eli Wade Mill," and Peerless steam engine and sawmill.

D. B. 27, page 416, 30 May 1900. John M. Wade to B. O. Shelor, his ½ interest of ¼ acre and storehouse on Spurlock Creek.

D. B. 28, page 66, 11 April 1901. B. O. Shelor to G. W. Reed, ½ interest in land and store house on Spurlock Creek.

The two previous deeds do not convey the mill property, but do show that the owners were running a store near the mill.

D. B. 31, page 381, 23 February 1905. John M. Wade to Wm. A. Hylton, ½ interest in grist and sawmill and 16 acres on Spurlock Creek known as the Eli T. Wade mill. Mill and land sold to Martin Quisenberry and by said M. Quisenberry to Joseph Helms and by J. Helms to Wm. a. Hylton, but no deeds were recorded. All parties consented to this deed.

D. B. 35, page 297, 2 February 1909. H. W. Simmons, Commissioner in Chancery Cause of *J. B. and D. F. Hylton vs. D. F. Hylton*, at public sale. D. F. Hylton is highest bidder for Eli Wade mill property and 17 acres land.

D. B. 36, page 89, 7 October 1909. D. F. Hylton to James A. Goad, 17 acres, Eli Wade Mill and property, Spurlock Creek.

88. Solomon Harman Mill/John Harman Mill/ Caleb "C. W." Harman Mill

West Fork of Little River

LOCATION: East side upstream from Route 729 (Laurel Branch Road NW) 1.5 miles North of Route 221 (Floyd Highway S); POWER: Tub wheel(s), then turbine; STONES: 1 set (1870), 2 sets (1880), and roller mill (1910)

The remodeled Harman Mill. One large print of this photograph is dated 1914. The various shades of the roof shingles on the right end suggest that it was made within a year or two after this part was added or reroofed. Harman's choice to add to his old mill, rather than replace it entirely, resulted in a unique roofline. (Frank Webb collection)

History. In *Harman Genealogy (Southern Branch): With Biographical Sketches, 1700–1924*, John Newton Harman notes that "Solomon [Harman] had a little corn mill, at the old homestead in Floyd County, Virginia, known as the 'Harman Mill.'"[47] Writing around 1932, seven years after Harman's book was published, Amos Wood was only a little more specific in his claim that "Solomon Harman and his wife Elizabeth Slusher, settled on West Fork seven miles west of the town of Floyd and here they built one of the first mills of Floyd County." [48] The most recent and most specific claim concerning the mill's origins appeared in an item headed "Ancient Mill is Washed Out" on page 3 of the *Roanoke Times* on Thursday, August 12, 1943, two days after the Harman Mill was destroyed by a flash flood on the night or late evening of Tuesday, August 10.

> Floyd (Special)—A cloudburst accompanied by the worst electrical storm, lasting around two hours, that old residents can recall, late yesterday afternoon practically destroyed a century and a half old grist mill in the Topeco section on Rural Route 2, a few miles west of here, and did considerable damage to bridges, fences and lowland crops. The mill was owned and operated by Giles Epperly, formerly of Roanoke, and known as the old Harmon mill.
>
> The structure was in two sections, one ancient, the other newer. The older part was washed out and the newer section upset. A bridge 56 feet long over West Fork creek at the bottom of a hill below the mill was washed out, and the creek valley was inundated and damaged over a distance of five miles.
>
> Meadows were covered with water, a great deal of corn was blown down by the accompanying high wind, and a bridge at the George Slusher farm two miles west of here, washed out.
>
> While there is not any authentic estimate of the damage to the mill property, it was said by some that it would exceed $2,000. Extent of the other damage could not be estimated. A lot of grain stored in the mill was destroyed.
>
> The mill was built in 1800 by Solomon Harmon and has been in continual use since.

Solomon Harman (1775/1779–1842) was 21 or 25 years old in 1800 and could conceivably have built a mill in that year, but not on this site, since he didn't buy the land from Jacob Goodykuntz until 1809. Nor did he build the first mill here. A mill of some kind was already on the property in 1809, as it had been when Jacob Goodykuntz bought the same land from Guy Smith in 1803. So the questions of exactly who built the first mill at this site, and when, remain unanswered, but it may be safely said that Solomon Harman was grinding corn for his family and neighbors at this site early in the 19th century.

Considering the long life of this mill, the record of its ownership is remarkably brief. Solomon Harman's son John apparently bought the mill from his siblings in the process of settling their father's estate. It is not certain whether this decision was made by the Harman heirs after their father died in 1842 or mandated by their father's will, which was not found, in spite of written references to it. Following the death of John Harman (1814–1893), the mill and 44 acres along with one smaller tract were assigned by the Floyd County Circuit Court to his widow, Mary, in May 1894. A year later, a chancery suit filed by John and Mary Harman's daughter, Margaret Harman Hylton, resulted in the sale of the mill and 89 acres to Margaret's husband, Jacob Hylton. By 1905 Hylton had died

North side of the mill, between 1910 and 1918. On the high platform (from left to right) Shirley Kingrea consults his watch while Edgar Weddle, Otho Weddle, and owner Caleb W. Harman look on. Others are unidentified. (Frank Webb collection)

or had deeded the property to his wife, who in that same year conveyed it to their daughter Alice, wife of Caleb W. Harman.

Caleb Harman (1882–1918) owned the mill until his death in 1918. In 1920 his widow and his mother-in-law, who had retained some part interest in the mill, deeded the mill and land to Minnie Hylton Harman, daughter and sister, respectively, of Margaret Harman Hylton and Alice Hylton Harman, as well as the wife of Caleb Harman's older brother, Arthur (1876–1941). At a 1934 sale forced by bankruptcy, Arthur and Alice Harman bought back their own mill with Norman Weddle as Arthur's new business partner. Four years later the three of them sold out to Giles Epperly, a son of miller and millwright John W. Epperly.

Giles Epperly was still running the mill when it was more or less destroyed in August 1943. On page 17 of the morning edition on Sunday, August 15, the *Roanoke Times* ran another short news item about the flood damage and a picture of the washed out mill. The caption stated that a cloudburst "late Tuesday afternoon practically destroyed a century and a half old grist mill in the Topeco section of rural route 2 in Floyd county." The accompanying photograph "shows the remaining part of the Harmon mill after the older of the two sections had washed away. Parts of the mill could be seen all along the stream for a mile."

One piece of the mill floor washed more than five miles downstream and lodged on the west bank of the creek just beyond what was then the Conner farm north of where modern day Valley Drive NW approaches West Fork. It lay there rotting well into the 1950s.[49]

Whether or not he exercised this option, Giles Epperly had acquired the right of raising the dam an additional two feet when he bought the mill from Harman and Weddle in 1938. If he did raise the dam to its full permitted height, the pond behind it would have been both wider and longer, as well as deeper than before, and would have held an additional 43,560 cubic feet of water, nearly 326,000 gallons, for every one-half acre of pond surface. Add to this an unusually hard rain and the fact that the turbine and penstock, as seen in the *Roanoke Times* photo, remained in place when the short section of dam between them and the east bank washed out, and it is easy to see how the section of the mill nearest the creek was undercut and carried away, which then caused the newer portion to upset. The brief article on the same page of the *Roanoke Times* stated that "two mill rocks weighing two tons each were washed down the creek from the mill house 300 feet."[50]

The mill was notable in having been owned by Harmans for more than 100 of its 125 plus years. Jacob Hylton, who owned the mill for about ten years, was the son-in-law of John Harman and the father-in-law of Caleb W. Harman. The last two Harman owners, brothers Caleb W. and Arthur T. Harman, were related to Solomon Harman but not direct descendants.

The 1870 *Products of Industry* report noted a single run or set of mill stones and a tub wheel at the John Harman Mill on West Fork. The notation of two run of stones ten years later is consistent with John Newton Harman's statement that Solomon's son "John put in Burrs in 1872,"[51] assuming that he means an *additional* pair of burr (buhr) stones intended for grinding flour. Assuming further that the mill had ground mostly corn before this time, the addition of a set of stones and other machinery for making flour would have more than doubled the number of machines and required additional drive train components inside the building. Although there is no evidence to support such a supposition, this would have been a good time to build a new, larger mill from the ground up.

Jacob Hylton rebuilt the dam[52] during his tenure as miller but the basic set up described in 1880 apparently served until the advent of the roller mill compelled flour millers region wide either to give up flour milling or install a roller mill. Most who chose to make this leap tore down the existing mill and started from scratch, but C. W. Harman hired John Epperly to add onto the existing building and install the roller mill machinery.[53] His choice to enlarge and remodel rather than tear down his old mill, which hints at the possibility that the original building had been replaced since the early 19th century, resulted in a structure unlike any other mill in the county. The tall center section, whose purpose was probably to house the elevators that ran from basement to attic in every roller mill, gave the mill a curiously modern appearance, with its clerestory windows and irregular roof line.

Harman's refusal to sacrifice the existing building is mirrored in what may have been a similar reluctance to abandon the centuries-old process of grinding flour between two stones. A mid-January 1910 item in the *Floyd Press* noted that "C. W. Harman has one of the best buhr mills in the county and wants to prove he can make you the very best flour. See his ad." On January 29, 1910, Harman was extolling the virtues of the traditional buhr mill in a *Floyd Press* advertisement:

> Notice of New Buhr Mill
>
> I have just completed one of the finest buhr mills that can be built and can make the very best of buhr flour and I have the best process for making buckwheat flour. Can crush our corn. Toll will only be one gallon to the bushel. Your patronage solicited. Give me a trial.
>
> C. W. Harman

(Although Harman states that his toll for grinding corn is only "one gallon to the bushel," most of his readers probably knew that the charge for this service was limited by state law to one-eighth portion of the corn to be ground, i.e., "one gallon to the bushel.")

Harman might already have been planning to modernize and was merely trying to hold on to his flour customers until the mill could be remodeled. But considering the time and expense of installing even a single run of stones, let alone additional machinery, it is also possible that he himself suddenly saw the writing on the wall only after his flour business continued to dwindle and had then scrambled to heed its warning. By the end of May 1910, only four months later, both Harman and his mill had been converted to the more complex and more efficient roller system of producing flour:

> New Roller Mill Started
>
> I have completed my new roller mill and have commenced grinding. I have one of the most up-to-date mills in the country, fitted with the latest improved machinery. Your patronage is solicited. All I ask is to give me a trial.
>
> Respt.
>
> C. W. Harman

In addition to the owners, men who worked at this mill included, in no particular order, Oscar Ames Webb, Dalton Wade, Emmett Davis, Shirley Kingery, and Hobart Kingery.

Deeds

Montgomery County Deed Book "C," page 632, 14 May 1803. Guy Smith to Jacob Goodykuntz, mill and land.

D. B. "D," page 666, 6 November 1809. Jacob Goodykuntz to Solomon Harman, mill and land. *Mills and land willed to John Harman by Solomon Harman.*

Floyd County Deed Book "X," page 179, May Court

Winter, 1914–15. The men at right are Arthur Harman, John Hylton, Lee Jennings, and owner Caleb Harman. In the 1943 flood that destroyed the mill, the dam gave way between the turbine penstock (the roofed structure at top center) and the eastern (mill side) bank of the stream. The mill was undercut and part of it collapsed into West Fork. (Courtesy Margaret Harman Hubbard)

1894. J. J. Leak, J. P., assigns to Mary Harman, widow of John Harman, two tracts, one of 44 acres including mill and mansion house, another of 12½ acres on West Fork, Little River.

D. B. "Y," page 221, 8 January 1895. J. T. Conduff, Commissioner in Chancery suit of *Margaret Hylton vs. Mary Harman*, sells to Jacob Hylton 89 acres. Land formerly belonging to John Harman, Jr.

D. B. 31, page 403, 27 March 1905. Margaret Hylton to Alice Hylton, mill and land on West Fork Little River, known as the John Harman Mill place with all rights.

D. B. 44, page 84, 23 February 1920. Margaret Hylton and Alice Harman, widow of C. W. Harman, to Arthur T. Harman, mill and land known as Harman's Mill, West Fork of Little River.

D. B. 51, page 527, 3 July 1934. Joseph Sowers, Trustee in bankruptcy of A. T. Harman, to A. T. Harman and Norman E. Weddle, one-half interest each in mill and land.

D. B. 57, page 433, 17 December 1938. Alice Harman, A. T. Harman, and Norman E. Weddle to Giles M. Epperly, tract of land known as Harman's Mill, including mill, mill house, dam, water race and other rights, one to raise mill dam two feet higher than present height.

D. B. 58, page 405, 19 August 1939. A. T. Harman to G. M. Epperly, ¾ acre tract adjoining mill tract.

D. B. 62, page 553, 24 February 1945. G. M. Epperly to Marvin W. Trail, two parcels of land, one ¾ acre and mill tract known as Harman's Mill and another of ¾ acre. (Mill gone.)

See Table B, *1870 Products of Industry.* Name of corporation or individual: Harman, John

See Table C, *1880 Products of Industry.* Name of corporation or individual: Harman, John

127. Beauford Howell Mill

West Fork of Little River

There is some evidence that a Beauford Howell woolen mill once stood on the West Fork of Little River (West Fork Creek), distinct from the Dodd Creek woolen and grist mill owned by his father and grandfather, and eventually by Beauford Howell himself. This now seems unlikely, for reasons explained in the history of the Joshua Howell/David Howell/Beaufort Howell Grist and Woolen Mills (III, 138).

128. Giles Wade Mill

Headwaters of Big Indian Creek (North/East Fork)

LOCATION: Route 730 (Ridgeview Road NW), approximately .5 mile east of Route 750 (Alum Ridge Road NW); POWER: Overshot wheel; STONES: 2 sets

History. The last man to run this mill, Giles Wade, left Floyd County before 1920 and died in Callaway County, Missouri, in 1937. Both the log mill house and the dam were in disrepair in the early 1900s, according to John Poff, son of Albert Poff. This is the only log mill house on record, but there were almost certainly others, especially among the earliest small mills in the county. Deeds to this property offer no clue as to when the mill was built, since none of them mention the mill at all. Although conspicuous, this absence would not disprove the mill's existence, even if there were no credible oral history surrounding it. Deeds to large tracts of land frequently fail to mention mills known to have been standing when the deed was written.

Charles Gallimore removed three millstones from the ruins around 1960.

Deeds

Floyd County Deed book "J," page 149, 13 February 1857. John R. Duncan and Anna, his wife, and Nancy Duncan, wife of George Duncan, to Isaac Wade, 140 acres on waters of Little River.

D. B. "K," page 537, 19 March 1861. Andrew A. Reed and Nancy, his wife, to Isaac Wade, 112 acres on waters of Indian Creek.

D. B. "U," page 369, 28 February 1890. Trust Deed, Z. T. Dobyns, Commissioner in cause of *Martha Wade vs. Harvey S. Wade and Others* to Landon S. Dickerson, 240 Acres plus or minus. Martha Wade was the widow of Isaac Wade.

D. B. 26, page 520, 22 January 1900. Landon S. Dickerson and wife to Giles Wade, 171 acres land on waters of Indian Creek. The same land conveyed to L.S. Dickerson in D.B. "U," p. 369.

D. B. 37, page 264, 1 February 1911. Giles Wade and Melvina E. Wade, his wife, to Albert J. Poff, 210 acres on tributary waters of Indian Creek.

---

132. James T. Lester Mill

Laurel Branch

LOCATION: Upstream from Route 601 (River Hill Road NW). Now a dead-end road, Route 601 at one time forded Little River and continued into Montgomery County. Local wagon traffic was still using the ford in the 1930s.

History. In 1834 James T. Lester, Sr. (1779–1860), was granted permission to build a mill and dam on a large tract of land he had bought in two parts in 1831. It would seem foolish in today's rule-bound system to invest much time and money in a business before acquiring the necessary permits. But Lester, like other Floyd County millers of that era, had already erected his mill and dam and filled the millpond before complying with a state law requiring him to petition the county court for permission to do what he had already done. Since no other person's land would be affected and a convenient mill was an asset to a community and the county, there was little danger that any of the jurymen chosen to evaluate his request would object.

The site of Lester's mill, now among the most obscure and inaccessible in the county, probably seemed a likely enough spot in a time when trails and settlements lay along the valleys of streams, whose presence guaranteed at least a narrow passage through the hills. Along larger streams like Little River, roads and travelers converged at places where the bank was low on both sides and where the stream was shallow enough and its bed solid enough to ford. In the year ending on June 1, 1850, Lester drew enough trade from Montgomery and Floyd County customers to report having ground 2,000 bushels of corn and 500 bushels of wheat.[54]

Few could have predicted which of those early fording or ferrying places would later be crossed by bridges and main highways, and which, like this one, would wind up off the beaten path, at or beyond the ends of dead-end roads. By 1907 deeds to the property included the dam as a landmark in boundary descriptions, but neither the mill nor the pond is mentioned, which suggests that the mill was inoperable (if there at all) and the pond had been drained. Even though it was possible into the 1930s to ford Little River with a horse-drawn wagon on Route 601, giving patrons from both sides of the river access to this mill site, traffic was already being channeled along roads (like modern day Routes 8 and 787), whose gentler grades and more direct routes to trade centers like Christiansburg and Radford would ensure their future improvement.

Deeds to the two tracts Lester bought in 1831 mistakenly refer to Laurel Creek instead of Laurel Branch. Lester's March 1834 request for the county court's permission to build his dam and mill repeats that error, but the ruling issued the following month refers to the stream as Laurel Fork. Taking the original as their model, writers of 19th century deeds repeated the substitution of "Creek" for "Branch," apparently without bothering anyone. Confusion was likely in any case, since there were two other Laurel Creeks and one Laurel Fork, and probably other Laurel Branches elsewhere in the county.

Fully ten years after James Lester, Sr., died in 1860, administrators of his estate sold at auction 112 acres, including his mill property. The winning bidder, Crockett Reed, quickly sold the land and mill to Madison Reed and had the deed made to him. Two years later Reed bought the mill in partnership with Harrison D. Reed and became sole owner in 1876. He sold the mill and 372 acres in 1878 but may have run the mill for Fanny Burnett since he appears as a 69-year-old miller in the 1900 census. Linwood Altizer (b. 1917) knew that Crockett Reed ran this mill but couldn't recall a successor. He also recalled that Jehu Duncan was supposed to have built a house on a flat created by a mill pond. Deeds continue to mention the mill dam as a landmark until 1907.

Deeds and Other Public Records

Floyd County Deed Book "A," page 15, 11 June 1831. William Clifton to James Lester, 104 acres on the head of Laurel Creek, waters of Little River.

D. B. "A," page 86, 21 November 1831. Charles Simmons to James Lester, 400 acres. Land formerly in Montgomery County, now Floyd. Mentions Laurel Creek.

Floyd County Court Order Book 1, page 127, March, 1834. On the application of James Lester to build a water grist mill on Laurel Creek, and it appearing to the Court that he is the owner the land on both side of the said stream. It is ordered that a writ of Ad quod damnum issue (?) to the Sheriff of the county to be executed the 4th day of April next to inquire into damages which any person or persons above or below may sustain by the erection of the said Mill.

Floyd County Court Order Book 1, page 129, April, 1834. On the petition of James Lester for leave to build a water Grist Mill dam across a creek called the Laurel Fork—This day came the said James Lester and the Sheriff returned an inquest taken pursuant to a Writ of *ad quod damnum* to him issued which inquest is in these words to wit: "Floyd County to wit: We the jurors (to wit) Samuel Dulaney, James Graham, Richard Dulaney, John Sowers, Robert Graham, William Sowers, Humphrey Reed, Joseph Howard, Thomas H. Steger have this day met upon the land of James Lester in conformity to a writ of Ad quod damnum which issued from the County Court of Floyd, and after having first been sworn and empanelled have proceeded to examine the land above and below and do make the following report, that is to say the mill and dam having been already erected and the water overflowing no other land but the land of the said James Lester, there is no damage sustained. Given under our hand and seals this 4th day of April 1834." Whereas upon consideration of the said inquisition, and after hearing the said petition and the circumstances being weighed It is ordered that the said James Lester have leave to build the said mill and dam.

D.B. "M," page 37, 1 November 1870. F. A. Winston, Commissioner for James Lester, Sr., to Madison D. Reed, 112 acres "known as the James Lester, Sr. mill place." *This deed summarizes more than one transaction, but in essence conveys ownership of the James Lester, Sr., Mill from Lester's heirs through Crockett D. Reed to Madison D. Reed.*

D. B. "N," page 43, 31 December 1872, John R. Duncan, Annias Duncan and Eveline Duncan, his wife, and Allen T. Duncan and Rebecca, his wife, and Wyat Sumner and Rachel, his wife, to Crockett D. Reed and Harrison D. Reed, 35¾ acres.

D. B. "O," page 294, 16 October 1876, Harrison D. Reed and Sarah, his wife, to Crockett D. Reed, 35¾ acres more or less on the waters of Laurel Creek "with mills and all appurtenances thereunto."

D. B. "P," page 138, 2 December 1878, C. D. Reed to Fanny Burnett 372 acres on "waters of Laurel Creek and Long Shole Branch waters of Little River" with one boundary line "beginning at two linns [a variant of "linden," or the American basswood tree] on the bank of the River" and running "to two chestnut sprouts near the mill on the bank of the creek."

D. B. "W," page 335, 6 September 1890, Fannie Burnett to Dosha Duncan, 2 acres on Laurel Creek "beginning at six spruce pines ... on the bank of the creek by an old mill Dam."

D. B. "Z," page 551, 28 December 1897, J. J. Duncan and Doshia C. (?) Duncan, his wife, to Stillman Reed, land on waters of Laurel Creek. Bounds mention "a white oak on the mill pond." No acreage given.

D. B. 38, page 428, 16 January 1907, C. D. Reed and wife, Abigail, to Stillman Reed, 25 acres on Laurel Branch. Mentions mill in description of the property boundaries. Reed accepted as a condition of the conveyance the duty to "support and maintain C.D. Reed and Abigail Reed during their natural life." *C. D. and*

Abigail Reed were the father and unmarried sister of Stillman Reed. Abigail died in a house fire after 1910.

See Table A, *1850 Products of Industry.* Name of corporation or individual: James Lester (and Son?)

135. Amon G. Lester Mill/John Harris Mill

Laurel Branch (Laurel Creek in some deeds)

LOCATION: One-quarter to .75 mile below crossing of Route 687 (Pugh Road NW) near a waterfall. The road to the John Harris place turned left off Route 687 a short distance before the road crossed Laurel Branch. At one time Route 687 connected with what is now Route 704 (Dickerson Lane NW), but the middle section, including the stream crossing, is no longer maintained as a public road; POWER: (probably) Overshot wooden wheel

History. Deeds to large tracts of land regularly fail to describe mills and mill property within their bounds, but the deed to a 2.5 acre parcel could hardly fail to note the presence of working sawmill and grist mill. It is all but certain therefore that between May 1901 and November 1907 Amon Lester built a sawmill and a grist mill on the land he purchased from Amos and Minnie Pugh in 1901, probably with this purpose in mind. Linwood Altizer (b. 1917) never saw the mill house but remembered that the mill pond was just upstream from a sudden drop (not a sheer waterfall) of approximately 20 feet. Placing the mill below the drop would have insured plenty of fall to partially compensate for the low volume of water.[55]

In 1907 John T. Harris deeded to Amon Lester two tracts of land of 26 and 2.5 acres, both farther down Laurel Creek, in exchange for the two mills on 2.5 acres of land and $600 cash. Neither these nor earlier descriptions of the tracts Harris traded to Lester say anything about a mill, but the deed by which Lester's land and mill property is conveyed to Harris states that Harris had given "$600 paid in hand and Mill Property" to Lester. The near identical phrasing in the two deeds (D. B. 34, page 389 and D. B. 38, page 3), the fact that they were written on the same day, and the absurdity of each man giving the other $600, make clear that this is merely an error made by a bored, distracted, or overworked justice of the peace. The failure of the deputy clerk who entered the deeds into the record to notice this irregularity is explained by the fact that the deeds, though written the same day, were recorded in March 1908 and December 1911, more than three years apart. This small error is worthy of mention because it is of so little consequence and yet it is the most serious clerical error that has come to light in the examination of many hundreds of deeds.

John T. Harris is listed as a "miller in his own mill" in the population census for 1910. The mill is not mentioned in a 1925 deed of trust in which Harris put his place up to secure a debt to a carpenter. Alice Reed Harris's brother Stillman "Still" Reed (1866–1930) may have operated this mill as well. Still Reed's wife, Margaret, is said to have been the sister of John Harris, making the two men brothers-in-law, two times over. John T. Harris also served the community as a "horse doctor," or self-taught veterinarian.

Deeds

Floyd County Deed Book 32, page 393. 22 May 1901. Amos and Minnie Pugh to Amon G. Lester, 2½ acres. No mention of sawmill, grist mill, or water power.

D. B. 38, page 3, 8 November 1907. Amon G. Lester to J. T. Harris, 2½ acres and sawmill and grist mill property.

D. B. 47, page 431, 15 October 1925. Deed of Trust, J. T. Harris to J. Houston Harris for labor and materials used on house. *The 5 acre tract signed over to secure this debt included within its bounds the 2½ acre tract on which the sawmill and grist mills had stood. Neither is mentioned.*

136. Jonathan Elswick Mill/ George Goodby (or Godbey) Mill/ George Hambrick Mill and Aaron DeHart Mill

Indian Creek

LOCATION: Route 787 (Indian Valley Road, NW) approximately .25 mile upstream from Little River and the Montgomery County line. POWER: 5 HP tub wheel (Elswick 1880); turbine (DeHart Mill); STONES: 1 set (1880); 2 sets (DeHart Mill)

History. The site of the Elswick and DeHart Mills near the mouth of Indian Creek is among the oldest and most interesting, in regional and national contexts, in the county. The Elswicks were among the earliest settlers in Montgomery County, VA. John Elswick died there about 1749, not long after arriving. His son Jonathan owned several tracts of land on Little River near present day Copper Valley and in 1797 was granted "leave to build a water grist mill on Indian Creek near the mouth of same."[56, 57] Although the mill is not specifically mentioned in the deed, it was sold with several other tracts of land to Elswick's son-in-law George Goodby or Godbey in 1817.

Elswick's acquisition of parts of his Little River holdings from the executors of land speculator James Patton links him, indirectly, to the earliest pioneers in the New River Valley. Patton was among those killed by Shawnee raiders at Draper's Meadow (now Blacksburg, VA) in July 1755. Mary Draper Ingles and two children were captured and taken to Ohio, beginning the ordeal chronicled in *The Long Way Home* outdoor drama in Radford, VA, written by Earl Hobson Smith, and in James Alexander Thom's novel *Follow the River*.

Although complicated by the inclusion of multiple tracts in various combinations, and one or two failed transactions, the story conveyed by the deeds connecting George Godbey's 1817 purchase to Aaron DeHart's in 1904 is relatively straightforward. Taken together, the deeds depict the mill site as having been the commercial and industrial center of the community for most of the 19th century, with stores, a stable, blacksmith and wood shops, and several private homes nearby. The U.S. Census *Products of Industry* report for 1870 notes that George Hambrick was employing water power, perhaps generated by the tub wheel at his grist mill, in his blacksmith shop as well.

One of the last deeds to the original grist mill and property links George Hambrick to the future site of another Indian Creek mill. In addition to three adjoining tracts at the old Goodby or Godbey place, the deed recorded in November 1888 in Floyd County Deed Book "T," page 498, refers to a second Hambrick sawmill tract some 2 miles above the mouth of Indian Creek. This tract surfaces again in 1900 as the future site of the Duncan/Hall Roller Mill (III, 79), but is not directly related to the history of the DeHart mill property, except perhaps as a factor in the Hambrick family's money troubles.

The story of the second mill is equally interesting, both as a mirror of changing times inside the county and for its connection to events, political and industrial, of regional and national significance. Aaron DeHart, who built the last mill on this site around 1905, was one of several Confederate veterans who built grist mills in Floyd County. The stories of most of his comrades' Civil War service and subsequent lives are unrecorded beyond what exists in military and family records, but a sketch of DeHart's brief, uneventful service and the long, productive life that followed is included in Jay S. Hoar's *The South's Last Boys in Gray*. The two-page profile includes a photograph of DeHart at age 98 and a verbatim copy of a detailed letter written in 1983 by his granddaughter Olga DeHart Harvill, who bought the DeHart homeplace and mill site in 1968.

Aaron DeHart was only 15 when he entered the service near the war's end. Besides being one of the last 16 CSA veterans in Virginia when he died at age 99 in 1947, DeHart was probably among a similarly small group of soldiers who came home with something of a hero's reputation, without ever having seen action.[58] According to Jay Hoar, Dehart "was not quite sixteen when he met some recruiters who had in tow one

A long wooden race carried water directly from the millpond to Aaron DeHart's turbine-driven mill. (Courtesy Olga Harvill)

Peter Cruise — a cripple, a family man (and father of the girl Aaron was to marry). On the spot, Aaron volunteered to take Peter's place in the Patrick County Home Guard, in which he was shortly promoted to corporal. Cpl. DeHart, in his final weeks of C.S.A. service, was assigned to Co. D, 51st Va. Inf., but he saw no fighting."[59]

DeHart married Dollie Cruise, began raising a family, and built and taught at a school in his home community of Vesta in Patrick County, VA.[60] As a middle-aged man, he pulled up stakes and left, and by the time the 1900 population census was taken was living in the Indian Valley District of Floyd County, where he gave his occupation as "bar keeping in a saloon." In 1904 he bought 36 acres and the mill seat portion of the old Elswick property owned since 1860 by father and son George and John Hambrick. The deed mentions the mill dam, but not the mill itself. J. C. Hambrick was in financial trouble and needed to sell the mill or the mill site/seat and the surrounding cluster of buildings and equipment that as recently as 1895 had included "a sawmill, griss[grist] mill, blacksmith shop and wood shop" and at least one dwelling house.

Whether the old mill was gone, beyond repair, or merely outdated, DeHart was planning a new turbine-driven mill in 1905. He also ran a country store and maintained the Copper Valley Post Office in his eight-room house where travelers could stay the night for 25 to 50 cents.[61] The difference in rates probably depended upon whether lodgers were walking, riding, or driving a wagon, since horses as well as travelers had to be fed and cared for. The DeHart house was among the first in that area to be illuminated by electric lights, thanks to a DC generator driven by the mill's turbine.[62]

A deed made by Aaron DeHart in May 1905 as he was planning his new mill is also of interest in the broader context of county and regional history. For the sum of $500 DeHart signed over to Archer Phlegar and J. A. Grayson the rights to "all of the minerals and ores in and upon" his land, along with the right to "mine and remove the minerals and ores over the said tract of land and to erect a washer on the land." DeHart did at least stipulate that "the same is not to be erected where it will interfere with the Mill and Mill race which the said Aaron W. DeHart proposes to erect on the place." As far as is known, no minerals worth extracting were ever discovered and DeHart got $500 at no expense to himself or his property. But thousands of people who signed similar "broad form" deeds in the coal counties of Appalachia lived to regret it since the surface of land and the rights of its owner are virtually worthless when the owner of mineral rights is unrestricted in his means of extracting his property.

In the same deed, DeHart granted a right of way for a railroad on the east side of Indian Creek, along one of several proposed railroad routes into Floyd County. None of these now seems very plausible but at that time local papers were filled with glowing reports of the railroad's imminent arrival and of the universal prosperity that would surely follow. The proposed Radford to Willis route following Little River and Big Indian Creek was one of the more practicable because of the relatively gentle slope. Older residents of Copper Valley say that grading was actually commenced and that evidence of it remains between Route 787 and Little River near where the two go their separate ways in Montgomery County. Whether Phlegar and Grayson were themselves industrialists or merely local speculators acting on some inside tip that a mining company or railroad were headed in that direction is unknown.

Until well into the first half of the 20th century the majority of millers were probably in some way connected to the liquor business. The most upright probably supplied corn meal to distillers, whether they knew it or not. A good number of millers ran or were knowingly linked to distilleries, both legal and illicit, and one or two were known to be distributors of untaxed whiskey. But DeHart was the only miller who operated a public bar room, having purchased a 3.5 acre tract known officially as the "Lawson bar room property," and informally as "the Stillhouse" in 1915. This is probably the "saloon" referred to in DeHart's entry in the 1900 census. The building in which DeHart sold locally-made "government" liquor, which had been tested, taxed, and stamped accordingly, stood on the west side of Route 787, a short distance south and uphill from what is now 278 Indian Valley Road.[63] Olga Harvill recalled that her father, George DeHart (born 1888), did not care for alcohol as an adult, having gotten more than enough as a boy, siphoning whiskey from barrels into the jugs from which it was dispensed to customers.[64]

Around 1926 DeHart tore down the bar room, which had presumably been closed since the advent of Prohibition in 1920, and built the Zion Primitive Baptist Church on the other side of the road.[65] Unused for many years, the church stood at least until the mid–1970s. The water-powered sawmill had also been abandoned by the mid-twenties and was rented for two winters (1926 and 1927) to Harrison Dulaney and Callahill Altizer, who named his son after Aaron DeHart. The partners were operating a threshing machine in the Copper Valley and Carthage (now White Rock) neighborhoods in the summers and needed something to keep their portable steam engine working between threshing seasons. The "bill" (the wooden framing, siding, rafters, and sheeting boards) for the church was among the lumber they sawed with the rented mill.

Garfield Mitchell, the last man to run the mill for the elderly DeHart, was still grinding in the early 1940s, but exactly when the mill closed or what happened to it is unknown.[66] DeHart broke up housekeeping in 1943 and stayed with one or another of his children until his death in 1947.[67]

The race and turbine remained but the mill had apparently been destroyed by accident or deliberately dismantled by the time the mill property was advertised for sale at auction in the *Floyd Press*, April 15, 1948: "At auction, A. W. DeHart estate Sat. April 17, 1948, 10:30 A. M. 200 acres in 4 tracts, Tract #1, 150 acres, 5 room house, spring and outbuildings. Tract #2, 2 acres, 9 room house, barn and outbuildings. Tract #3, 3 acres bottom land. Tract #4, 40 acres land with mill dam. Plenty of water to run mill. A good race and water wheel."

Probably no one in the auction crowd or anywhere within a hundred miles in April 1948 had any serious thoughts about building a water-powered mill. By then a decade had passed since the most recent and ultimately the last water-powered mills were built on any Floyd County stream. A more mobile, cash-based economy had been ushered in by World War II. Small, engine-driven "corn rocks" running a day or two each week would be popular for another ten or fifteen years, but the small scale water-driven mills were closing, and the large ones were cutting back on flour production as subsistence farms disappeared and the county's population fell. Millers with an eye to the future were instead increasing capacity to produce, store, and deliver large volumes of dairy feed using powerful, reliable diesel and electric motors.

Less than 20 miles away on the newly built Blue Ridge Parkway, Ed Mabry's old mill, which was built five years after DeHart's, had just been restored as a tourist attraction. It had been revived not as a continuation of the pre-industrial past, but as a symbol of it, as a producer of nostalgia rather than food. Had they known all this, the DeHart heirs might have saved themselves a few cents on the cost of the advertisement. The turbine water wheel mentioned there is said still to be buried in the mud between Indian Creek and the enlarged and brick-cased A. W. DeHart house.

Deeds and Other Public Records

Montgomery County Court Order Book 10, page 30, 8 March 1797. Jonathan Elswick has leave to build a water grist mill on Indian Creek near the mouth of same.

Montgomery County Deed Book "F," page 398. 18 December 1817, Jonathan Elswick to George Goodby.

Floyd County Deed Book "E," page 211, 17 September 1848. George Gobey to David P. Hall, 300 plus acres on Little River, including land and mill property on Indian Creek. *"[T]he poor house tract" mentioned in the description of the boundaries is presumably the same poor farm that once occupied the river bottom farm now owned by Charlie H. Hall. It begins about a mile downstream from the mouth of Indian Creek, near the point where Little River encounters the eastern end of Mack's Mountain and veers sharply northwest away from Floyd County.*

D. B. "H," page 208, 27 August 1853. David P. Hall to Charles W. Roop, 100 acres land on Indian Creek and Little River.

D. B. "H," page 350, [n.d.] February 1856. Charles W. Roop to Henry Showalter, James Haley, and John Carmicle, the third of three tracts described as 100 acres land known as the old Goodby mill place.

Henry Showalter to C. W. Roop, no transfer, mill and 12 acres. This undated note probably refers to a failed transaction that took place between February 1856 and May 1857.

D. B. "J," page 296, 15 May 1857. Charles W. Roop to William Smith, 1 tract of land 30 acres +- "lying at the mouth of Indian Creek upon which there is a grist and saw mill." One other tract in Floyd County and two in Montgomery, totaling 250 acres.

D. B. "K," page 407, 6 November 1857. Henry B. Showalter to Adam Garnand, 12 acres land. Boundaries mention mill pond. The terms of this deed explains a series of unrecorded transactions involving Showalter, Roop, Smith, and Garnand: "Whereas sometime ago Henry B. Showalter sold to Charles W. Roop a tract of land lying in Floyd County at the mouth of Indian Creek containing 12 acres more or less ... receipt where of is hearby acknowledged and for which there has never been a deed executed by the said Showalter to the said Roop on the 15 day May 1857 conveyed the same to Wm Smith and whereas said Wm Smith on 6th day Nov 1857 after complying with provisions ... sold the said tract of land at publick auction." Adam Garnard bought the 12 acres and mill property at the auction.

D. B. "K," page 575, 3 March 1860. Adam Garnand to George Hambrick, 12 acres and mill property at the mouth of Indian Creek.

D. B. "L" page 129, 26 May 1863. Samuel Lucas, Sr., and Samuel Lucas, Jr., to George Hambrick, 4¾ acres "adjoining said Hambrick's land. Mentions a blacksmith shop, storehouse, and stable.

D. B. "T," page 498, 12 November 1888. Deed of Trust, George Hambrick to John R. Johnson, Trustee, 4 parcels of land on Indian Creek, "one known as the old saw mill tract about two miles above the mouth of the creek, supposed to contain 6 acres which was purchased of HB Showalter ... and 3 adjoining tracts about ¼ of a mile above the mouth of the creek purchased one of Adam Garnard the others from Saml Lucas and Saml Lucas Jr. the three supposed to contain 22 acres or about that on which are Hambricks mill both saw and grist, store, house dwelling, black smith shop and c."

D. B. "X," page 176, 3 January 1895. G. C. Hambrick to J. C. Hambrick, tract "on which there is a sawmill, griss mill, blacksmith shop and wood shop." "Being the same of which Geo Hambrick gave a deed of trust ... and is known as the Geo Hambrick mill property on which besides the buildings already mentioned there is a storehouse and 4 dwelling houses and land containing 22½ acres." *Although John Hambrick did not own this mill until 1895, it is listed under his name in the 1880 Products of Industry report.*

D. B. 29, page 564, 17 August 1903, *J. Singleton Diggs, vs. J. C. Hambrick*, for the sum of $200.00, mill and 26 acres pending in chancery suit. "Suit in chancery now pending in Montgomery County against J.C. Hambrick for $200."

D. B. 30, page 504, 5 June 1904. J. C. Hambrick to Aaron W. DeHart, 36 acres on Indian Creek and Little River. Includes mill dam, but does not mention mill. DeHart reserves right of way through 3 acres still owned by J.C. Hamrick and G.C. Hamrick.

D. B. 31, page 430, 11 May 1905. Aaron DeHart to Archer Phlegar and J. A. Grayson, all minerals and ores on his land. Ore washer not to obstruct dam and race where A. W. DeHart proposes to erect grist mill. Right of way for railroad.

D. B. 43, page 113, 31 December 1915. A. D. Akers to A. W. DeHart, 3½ acres land known as the Lawson bar room property.

D. B. 44, page 517, 23 August 1921, J. C. Hambrick to A. W. DeHart, 3 acres land at the mouth of Indian Creek. This is probably the three acres referred to in DB 30, page 504.

See Table C, *1880 Products of Industry.* Name of corporation or individual: Jno. C. Hambrick.

138. Joshua Howell Mill/ David Howell Grist and Woolen Mill/ Beaufort Howell Grist and Woolen Mills

South Fork of Little River (Dodd Creek)

LOCATION: Route 696 (Penn Road NW) approximately 1 mile northwest of Floyd County Courthouse; STONES: At least one, and woolen mill machinery

History. Several mills, with or without the blessing of the county court, were probably at work before Joshua Howell's mill on Dodd Creek. But the approval of his mill dam by the Montgomery County court in 1784 is the earliest court order pertaining to a mill whose subsequent history and location are known.

In 1807 Joshua Howell sold his mill and 227 acres to R. C. Saunders and Co. A year later he moved from Floyd County to Champaign County, Ohio. He died in Muncie, Indiana, in 1836 at 61 years of age. Joshua's son David Howell bought what had been his father's land and mills from the Saunders company in 1811 and seems to have spent the rest of his life working in the mill(s). He is referred to as "David Howell (Miller)" in several deeds, probably to distinguish him from two other David Howells living near Floyd in 1840 and afterward. He is listed as an 83-year-old miller in the 1850 census. His son Beaufort (also spelled Beauford and Buford) Howell is described in that same census as a 35-year-old wool carder.

At age 90 a widowed David Howell was living with Beaufort and his wife and six children and was still able to work in the mill. In a January 1857 letter to his brother Daniel, who had moved with their parents to Ohio in 1808, David writes, "Some warm days I go to the mill and grind for the neighbors when Beaufort is busy though I am under no necessity laboring at all and in truth, if I do any work at all they scold me for it."[68] The letter includes local grain prices—wheat: $1.00 per bushel; corn: 60 cents per bushel; oats: 33 cents per bushel—along with a bitter denunciation, written by Beaufort, of the Democrat controlled state government of Virginia, which opinion may have had something to do with his plans to leave Virginia. Beaufort went on to write, "I have been engaged in a woolen factory for many years and having become tired of the business, I am selling out with a view toward moving to Kansas and farming and grassing [grazing?] stock. I can go to Kansas and buy land for all my children and pay myself and still have something left to go on and I am not able to locate them in Va. with such homes as I would like." David had written in the same letter that "Beaufort sold a part of my old plantation for $41.00 per acre and is going to move to Kansas after he can sell all his machines. He has ten wool-carding machines, 8 power looms, 1 spinning machine of 216 spindles yet to sell."[69] Two years later Beaufort Howell moved with his family to Tennessee. They moved again, to Indiana, in 1869, and finally made it to Kansas in 1879. Beaufort Howell died in Chanute, Kansas, in 1902.

The U.S. *Products of Industry* report, compiled in parallel with the population census in 1860, names B. Howell as the owner of a "Wool Carding and Coloring" business on West Fork, Little River. (The 1850 report included him as well, but the "schedule" filled out by the 1850 enumerators didn't include a space for a stream name.) That Howell is supposed to have moved away in 1859 presents no obvious contradiction, since the names of previous owners often clung to mills, even in documents as official as the federal government's enumeration of industrial facilities and output. The location of the mill on the West Fork of Little River rather than the South Fork (Dodd Creek) is more puzzling. At least one deed mistakenly places the David Howell mill property on West Fork, an error that is extremely unusual, but perhaps understandable in light of the fact that the two streams are less than a mile apart where Penn Road crosses Dodd Creek.

An explanation that is simpler but harder to support is that Beaufort Howell did at some point operate a woolen mill on West Fork where he owned land until 1857. Frank Webb speculated that "Beauford Howell sold his mill on the West Fork of Little River and moved temporarily to the David Howell Mill on South Fork, Little

River, and then sold out and moved to Kansas." However, no record was found of the "Little River Manufacturing Co." associated by Frank with the Beauford Howell Mill on West Fork, and none of the West Fork property deeds mention a mill. Finally, the very high number assigned by Frank to the Howell Mill on Dodd Creek and the fact that it does not appear on his map of Floyd County mills suggest that he was never fully confident of the existence of a Howell mill on West Fork.

Whatever the fate of the West Fork carding and dyeing mill, if there ever was one, the grist and woolen mills on Dodd Creek were sold in July 1856 to a group of seven investors. The day before this transaction, 73 acres (assumed to be the 73½ tract that contained the mills) was deeded to Howell by R. M. Farraday, who had bought it for $3,000 (the same price paid by Headen, et al., for the land and mills) only a few months before, but had never recorded a deed or made any payments. The ten one-tenth interests scattered among the investors were reassembled in 1871 when A. J. Kirby bought the six-tenths interest of Henry H. Spangler at a public auction forced by a chancery suit filed by Kirby against Spangler. The mill probably last ran under Kirby's direction and may have been dismantled then as well. The buildings and machines are mentioned in the 1871 deed, but the property is referred to only as "the Howell mill tract" when Kirby sold it and two smaller tracts to Martin Penn in 1883.

Deeds and Other Public Records

Montgomery County Court Order Book, Number ?, 13 April 1784. Joshua Howell (1745–1836) granted leave to build a water grist mill on his own land on the South Fork of Little River (Dodd Creek).

Montgomery County Deed Book "D," page 504, Montgomery County, 18 May 1807. Joshua Howell to R. C. Saunders and Co., 227 acres land and mills on Dodd Creek.

Montgomery County Deed Book "E," page 419, 3 June 1811. R. C. Saunders to David Howell, 227 acres land and mills on Dodd Creek.

D. B. "D," page 205, Deed of Trust, 14 March 1846. Beauford Howell, David Howell, and Alexander Howell to George Rosenbaum and Jackson Godbey, 3 tracts land. One: 90 acres on waters of West Fork, Little River. Two: tract of 48 acres adjoining Charles Turman and patented to Beauford Howell from Commonwealth 1845. Three: tract containing one acre, including dwelling house of Beauford Howell. Also the following personal property: two double-geared carding machines, two picking machines, one large dyeing kettle and all fixtures of Beauford Howell's dyeing establishment. David Howell sells unto said G. Rosenbaum and J. Goodby three certain tracts of land. One: 100 acres known as home tract. Two: "one other tract containing 225 acres known as the mill place." Three: 50 acres adjoining same. "These three tracts of land embrace the land on which the said David Howell lives [and] Beaufort Howell mills and dwelling house." *The $1,832.57 debt to Harvey Deskins secured by this deed of trust was evidently paid off and the property retained by the Howells, father and son. The deed is summarized here for the sake of the property descriptions.*

D. B. "G," page 94, 26 May 1853. David Howell to Beauford Howell, 73½ acres, waters of Little River. David Howell retains life estate, "reserving the right to himself to keep possession and the free use of the same during his life."

D. B. "J," page 18, 23 July 1856. R. M. Farraday to B. Howell, 73 acres South Fork, Little River. Farraday is returning to Howell land he contracted to buy on 7 April 1856 for $3,000. Had paid nothing on it. No deed found, Howell to Farraday.

D. B. "J," page 278. 24 July 1856. Beaufort Howell, Susanna, his wife, and David Howell to James B. Headen, David Kitterman, James W. Leseuer, William B. Shelor, Payton G. Richardson, Samuel Dobyns, and Henry Lane, 73½ acres, "being the same land upon which the said Beaufort Howell now resides," $3,000. David Howell releases, for $1.00, the life estate retained in this property when he sold it to Beaufort Howell.

D. B. "L," page 209, 6 October 1859. Henry Lane, J. W. Lesure, and J. B. Headen to James R. Williams and Luther Bowen, 6/10 interest in the Beauford Howell Mill Property, 73½ acres including land, machines, and fixtures. H. Lane, 1/10 interest, Lesure, 1/10 interest, and J. B. Headen, 4/10 interest.

D. B. "L," page 178, 7 March 1864. Jacob Bower, Admins. for Luther Bower, decd., to James R. Williams, his entire interest in a certain carding and wool ? factory and mill known as Bowers Mill.

D. B. "L," page 668, 20 March 1868. David Kitterman to Henry H. Spangler, his interest in mill, carding machine, and land of Beauford Howell mill.

D. B. "L," page 548, 14 April 1868. J. R. Williams to Henry H. Spangler, 6/10 interest in 73½ acres known as the Howell Mill property.

D. B. "M," page 243, 30 October 1871. J. L. Tomkins, Commissioner in suit of *A. J. Kirby vs. H. H. Spangler*, of the first part, and [to] Andrew J. Kirby of the second part. J. L. Tomkins grants and conveys to Andrew J. Kirby said land, buildings, mill and machine property on waters of South Fork, Little River, which formerly belonged to David Howell, then Beauford Howell, then J. B. Headen, Dobbins, Kitterman and others. *The 3 one-tenth interests originally owned by Dobyns, Richardson, and Shelor were still outstanding, but Kirby either held liens against them or paid the owners and so acquired sole ownership through this deed.*

Floyd County Deed Book "R," page 265, 31 December 1883. A. J. Kirby to Martin T. Penn, about 100 acres containing three tracts: the Howell Mill tract, the Tice tract, and small tract from Mrs. S. Dobyns.

D. B. 34, page 12, 16 July 1907. Martin T. Penn to H. W. Spencer, 3 parcels land containing 127½ acres, South Fork of Little River, one tract being A. J. Kirby's land of 102⅜ acres.

See Table A, *1850 Products of Industry.* Name of corporation or individual: B. Howell. (Business described as "Wood Carding and Coloring")

141. Elijah Turman Mill

Greasy Creek

LOCATION: East side of a loop of Greasy Creek, just north of the exact center of a line drawn between the mouth of Shady Branch and where Route 624 (Shady Grove Road NE) crosses Greasy Creek; POWER: Unknown; STONES: Unknown; VISIBLE REMAINS: None

History. This Elijah Turman Mill (there were two other Elijah Turmans and at least one other mill by that name) exists in living memory not as a mill but as "the mill seat." The label "mill seat" is typically applied to abandoned mill sites strongly marked by some natural feature, such as a water fall, or a rock formation on one or both sides of the creek, that is clearly visible long after any artifact or physical evidence of the mill itself has disappeared. Unlike a mill site, which is important only for what it once was, a mill seat continues to serve as a neighborhood landmark, equally useful in telling a story or giving directions.

One man who grew up in the neighborhood and had to cross Greasy Creek regularly remembered this as the only spot in several miles at which a young man could jump from one side of the creek to other.[70]

The indirect description of the mill's location results from the fact that the wagon road that must once have reached the spot was never developed into a public road and does not appear even on U.S. Geological Survey maps.

Section IV. Mills West of Route 8 and South of Route 221

28. Christopher Smith Mill/Luke Martin Mill

Dodd Creek (South Fork of Little River)

Location: Route 8 South (Parkway Lane S), approximately 1.5 miles from Floyd Court House; Power: Breast wheel; Stones: 2 sets and Sawmill

History. According to Amos Wood, "Christopher Smith ... built one of the first flour mills in Floyd County" on Dodd Creek.[1] In 1828 "Stuffle" Smith was granted leave to build a dam and erect a grist and saw mill by the Montgomery County court.[2] As required by state laws governing the building of mill dams, Smith paid his neighbor Henry Smith $23.00 to compensate for the lost use of lands affected by the dam or mill pond.

The *Products of Industry* report completed in 1850 states that Smith used as raw materials 4,000 bushels of wheat and 1,000 bushels of corn between June 1849 and June 1850. Smith's mill is listed again under his name in the *Products of Industry* report for 1860. He died in December of that year, leaving the home tract, including his mills, to his wife, Wilmoth Smith. Upon her death, the home tract and mills were to be sold and the proceeds divided as directed in the will. William R. Martin bought the mill in 1869 and is listed as its owner in the *Products of Industry Report* for 1880 and in a directory of Virginia businesses in 1881 and 1885 (see Appendix A). He also ran a legal distillery at this site. The 1880 description of the mill notes a breast wheel and two runs of stones. The 1870 *Products of Industry* report thought to describe this mill lists John Earles as the mill's owner or operator. John Earles never owned this mill nor is there written evidence of his having run it, but the circumstances do invite speculation. A John Earles was the son-in-law of Christopher and Wilmoth Smith twice over, having first married their daughter, Julina. Following her death from a fever at age 25, he married her sister Mary, his former sister-in-law. Earles could have run the mill for his widowed mother-in-law and possibly for Martin as well, thus associating his name with it.

The Christopher Smith mill was operated for many years by Hosea Correll, and was also known as the Luke Martin Mill, he being the son of Wm. R. Martin. Giles M. Whitlow is believed to have been the last person to run this mill. The "old mill house" was still standing when Amos Wood was composing his history of Floyd County around 1932.[3]

Deeds and Other Public Records

Floyd County Will Book "C," page 597, 20 January 1861. Christopher Smith to Sarah Smith Jenkins, 137 acres. C. Smith to Christina Smith Whitlock, 80 acres. C. Smith to Sarah Smith Jenkins, 27 acres. C. Smith to Mary Jane Smith Earls, wife of John Earls, 137 acres. Remainder of homeplace, including mills, to wife, Wilmoth Smith, for remainder of her natural life.

Floyd County Deed Book "M," page 217, 8 February 1869. Mason Jenkins, surviving administrator of

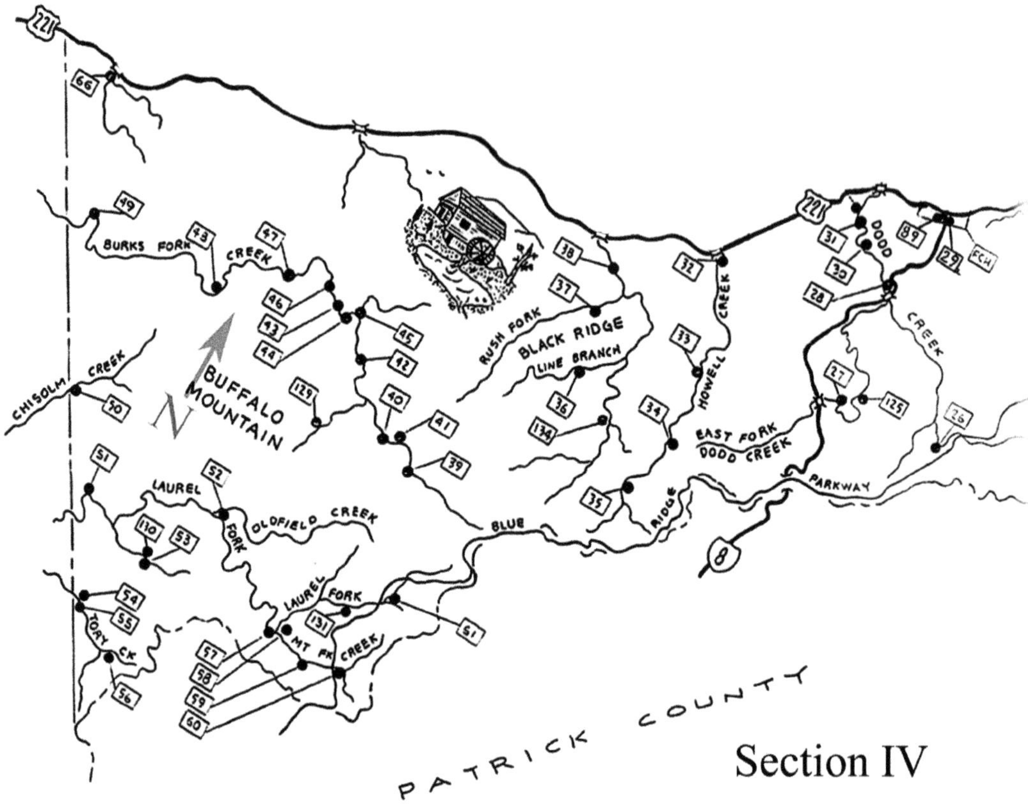

Section IV

Christopher Smith, deceased, to William R. Martin, 145¼ acres, including 37 acres land deeded to Sarah Smith Jenkins by Christopher Smith. Boundaries refer to "a cleft of a rock below the mill."

D. B. "V," page 544, 25 May 1892. J. L. Thompkins, Commissioner in cause of *Martha A. Martin vs. Joseph P. Martin and Others*, to J. A. Martin, 196 acres land and mill tract.

D. B. "X," page 319, 20 December 1894. J. A. Martin, et al. to Thomas H. Howard, 196 acres land and mill on South Fork, Little River.

D. B. 30, page 345, 16 March 1904. T. H. Howard to Giles M. Whitlow and children, 167 acres on South Fork, Little River (Dodd's Creek), part of what is known as the T. H. Howard-Martin Mill place.

See Table A, *1850 Products of Industry.* Name of corporation or individual: C. Smith

See Table B, *1870 Products of Industry.* Name of corporation or individual: Earles, John

See Table C, *1880 Products of Industry.* Name of corporation or individual: Martin, Wm.

30. Eli Williams Wool Carding Mill

South Fork of Little River (Dodd Creek)

LOCATION: On Route 720 (Epperly Mill Road SW), 1.2 miles south of Route 221 (Floyd Hwy. S); POWER: Unknown; STONES: Unknown, wool carding mill, sawmill

History. This mill is thought to have been built by John Gardner and later operated by Thomas Tousey and Eli Williams. It was a wool carding mill, but may have ground grain also. Ira Howard had a mill farther downstream on this property after Eli Williams sold out and moved to Craig County, Virginia. Eli Williams's sons Eli, Jr. (II, 101), and William J. Williams (I, 20) also ran mills.

Deeds

Montgomery County Deed Book "F," page 462, 28 February 1818. William Gillum to John Gardner, 236 acres on South Fork, Little River.

Montgomery County Deed Book "K," page 481, 17 September 1830. Robert Craig to Thomas Tousey, 2 tracts of land on South Fork Little River, one of 38 acres purchased from Nickolas Tice by John Gardner, the other tract of 138 acres, a part of 238 acres purchased from William Gillum, also 1 carding machine with all its apparatus. (This was known as Gardner's Old Machine).

Floyd County Deed Book "A," page 51, 3 June 1831. Thomas Tousey (of Philadelphia, PA) to Eli Williams, two tracts of land. One of 38 acres, "the same which was conveyed by Nicholas Tice and wife to John Gardner by deed bearing date the 3rd day of February 1817" and one of "138 acres, [which] together with the said tract of 38 acres [was] conveyed on the 3rd day of June 1824 by the said John Gardner to Robert Craig and William B. Charlton." Conveyed with the land were "the carding machine and cards, fulling apparatus, tools, implements and all buildings, water, water rights and privileges thereunto belonging or in any wise appertaining to him the said Eli Williams and his heirs."*

D. B. "A," page 431, 10 June 1831. Deed of Trust, Eli Williams to T. Tousey.

D. B. "A," page 355, 22 September 1835. Deed of Trust between Eli Williams, Harvey Deskins and William Thomas to M. Sandifur, a tract of 38 acres, a tract of 238 acres, and a tract of 170 acres. First two tracts purchased from T. Tousey, 3rd tract from J. Tice.

D. B. "A," page 363, 6 October 1835. Deed of Trust between E. Williams and George Shelor and M. Sandifur. E. Williams grants to M. Sandifer one Negro woman and her two children, James and John, seven head of cattle, 30 head of sheep, a carding machine and four head of horses.

D. B. "B," page 219–220, 16 April 1839. Deed of Trust, George Shelor and R. D. Montague, Trustees of Eli Williams to Thomas Tousey (of the State of New York), several tracts, 1 of 176 acres from J. Tice, one of 138 acres from T. Tousey, one 38 acres from T. Tousey, and one of 52 acres from Eli Phlegar. Refers to "appurtenances and improvements" but no specific mention of a mill.*

D. B. "C," page 378, 23 December 1843. Jobe Wells and David Goodykoontz, Trustees for E. Williams to Joseph L. and Asa Howard, 3 tracts land, one of 170 acres, one of 38 acres, and the other of 138 acres, with all appurtenances with same. Refers to "appurtenances and improvements" but no specific mention of a mill.*

31. Ira Howard Mill

Dodd Creek
(South Fork of Little River)

LOCATION: Route 720 (Epperly Mill Road SW) approximately 1.5 miles south of Route 221 (Floyd Hwy. S); POWER: Breast wheel (1880); STONES: 2 sets; VISIBLE REMAINS: Some features such as the dug race and parts of a dam associated with the John W. Epperly mill (IV, 126) may date back to the Ira Howard mill, but it is believed that the dam for the Howard mill was farther upstream than the one built and used by John Epperly.

History. The land that would become known as the Ira Howard (1797–1865) mill property came into the Howard family in 1843 when Joseph and Asa Howard bought three tracts of land totaling 343 acres on the South Fork of Little River (Dodd Creek). No mill was mentioned in the deed, but the Eli Williams wool carding mill (IV, 30) was on one of these tracts, upstream from this mill. In 1846 Joseph Howard sold his ½ interest in the property to Ira Howard, who in 1850 reported having produced 2,000 bushels of corn meal and 150 barrels of flour in the last twelve months (Table A). When Ira died in 1865, both a sawmill and a gristmill were at work at or near the present day site of the John Epperly mill. In 1874 Joseph L. Howard (1821–1889) bought a ⅓ interest in his father's mill and the next year acquired the two remaining ⅓ shares from his partners, Hardin P. Hylton and

**Contrary to his usual practice of reducing the text of a deed to its essential facts, Frank Webb included the complete verbatim texts of these three deeds in his notes on this mill. Possible reasons include the following: References to previous owners of the property and the mill (e.g. John Gardner); the unusual involvement of a partner, supplier, or financier from well outside the Floyd County area, in this case Thomas Tousey of Philadelphia (1831) and later New York (1839); the large amounts of money (for that time) involved; or the references (in the first of the three) to the machinery used in a wool carding mill. Readers are invited to examine the complete texts of these deeds in the Office of the Clerk of the Circuit Court in the Floyd County Courthouse. Also listed were two Montgomery County deeds (D.B. "H," page 147, and D. B. "I-J") which have no apparent bearing on this property, but do illuminate the very complex business dealings of some of the men involved with it.*

Elijah W. Hylton. The deed recording this sale spells out in unusual detail exactly what Howard was getting for his money. In addition to the land and building, the buyer was guaranteed "right and title to and interest in the saw and grist mill upon said land and all the tools fixtures appurtenances personal property and implements connected with said mills and each of them or used in connection with the working running or repairing of said Mills or for handling or storage of bags [?] grain meal and flour together with water rights and privileges connected with or appertaining to said Mill."

Although Ira Howard had by then been dead for 15 years, he is listed as the owner of a mill on Dodd Creek in the 1880 *Products of Industry* report. The breast wheel listed as his source of power is one of only a few in the county and, at seven feet, is the widest documented wheel. The diameter of the wheel was not recorded, but with ten feet of available fall, must have been at least ten feet tall to qualify as a breast wheel. Among possible configurations using a vertical water wheel, a wide breast wheel is best suited to a high-flow, low head site. The two turbines installed when John Epperly built a new mill on this site would have taken full advantage of these conditions.

When Joseph Howard's estate was settled in 1899, Pattie Howard inherited the mill and 11 acres of land and was given the option of honoring the contract of the current miller, or employing a new one. Sometime before 1899, Pattie Howard Tipton sold her eleven acres with the mill and all water rights and privileges to John W. Epperly, who built a new roller mill here between 1899 and 1905. Amos Wood claimed that the Howard Mill "has been burnt down and rebuilt two or three times"[4] but gave no dates. If the mill was damaged or destroyed by accident around that time, its destruction probably saved Epperly the trouble of razing it, since only one or two mills built to make flour with buhr stones were deemed worth the expense of remodeling and refitting with a roller mill. In every other case, the owners chose to knock down the old mill and build a bigger, more modern mill from the ground up.

Deeds and Other Public Records

Floyd County Deed Book "C," page 378, 23 December 1843. Jobe Wells and David Goodykuntz, trustees for E. Williams, to Joseph L. and Asa Howard, 346 acres on South Fork Little River (Dodd Creek) in three tracts: No. 1, 170 acres, No. 2, 38 acres, No. 3, 138 acres. This land was formerly owned by Eli Williams.

D. B. "D," page 271, 29 June 1846. Joseph L. Howard to Ira Howard, his ½ interest in 345¾ acres land on Dodd Creek.

Floyd County Will Book "D," page 125. 12 May 1865. Ira Howard leaves his mill and property and 343 acres to his wife, Dee Emaline Howard.

D. B. "N," page 285, 4 August 1874. John Merritt, Commissioner in suit of *Granville Montgomery vs. James B. Puckett and Others* to Hardin P. and Elijah Hylton and Joseph L. Howard, "property known as the Ira Howard Mill Property." The Hyltons bought the mill at auction and sold ⅓ interest to Howard before the deed was made.

D. B. "N," page 461, 5 May 1875. Harden P. Hylton and wife, Francis, and Elijah W. Hylton and wife, Hannah, to Joseph L. Howard. All their interest in 15¼ acres land on South Fork of Little River and known as the Howard Mill property formerly belonging to Ira Howard, saw and grist mill and water rights.

D. B. "P," page 125, 11 January 1879. This unusual deed simply confirms that Joseph L. Howard is the sole legal owner of the Ira Howard mill and as such has "all water rights and privileges that ever belonged to Ira Howard." It identifies and corrects deficiencies, including inadequate descriptions, in the deeds recorded in Deed Book "N," pages 285 and 461.

D. B. "U," page 482–489, November 1889. *Peter Howard vs. Joseph Howard*, division of property of Joseph Howard, deceased. Pattie L. Howard Lot 5, mill lot and 11 acres. May continue the present contract with the miller as or may employ a new miller. The disposition of the mill appears on page 489.

D. B. 26, page 244, 3 October 1899. Walter S. Tipton and Pattie L. [Howard] Tipton, Maud L. Howard and T. H. Howard, her husband, and Cassie Dobyns to J. W. Epperly, mill property, lands and water rights and privileges, known as the Ira Howard Mill property. (Maude L. Howard, T. H. Howard and Cassie A. Dobyns had bought the mill from the Tiptons and paid for it, but no deed had yet been recorded when Epperly bought the property, so the Tiptons are also listed as sellers in the deed.)

See Table A, *1850 Products of Industry*. Name of corporation or individual: Ira Howard

See Table C, *1880 Products of Industry*. Name of corporation or individual: Howard, J. L.

32. George W. Shelor Mill

Howell Creek

LOCATION: Intersection of Route 807 (Canning Factory Road SW) and Route 221 (Floyd Highway S); POWER: Turbine; STONES: 1 Meadows type mill and sawmill. VISIBLE

George Shelor's grist mill lived a longer and more widely known existence as the Huff Cannery. Howell creek passes beneath the row of buildings grafted to the mill. (Frank Webb photograph, 1992)

REMAINS: Building converted to commercial cannery, and later workshop and storage use. Dug race on east side of Route 807 (Canning Factory Road SW). Tail race filled or covered in the process of grading Routes 221 and 807, which intersects 221 just west of Howell Creek.

History. Although it stands beside one of Floyd County's two busiest highways at the corner of Route 221 (Floyd Highway South) and Route 807 (Canning Factory Road), few of the hundreds of people driving past it each day recognize the old Huff Cannery west of the Town of Floyd. Fewer still know that the part of it nearest to Route 221 was built by George W. Shelor to be a grist mill. Shelor had moved his wife and three youngest children to a farm near Oxford, Maryland, but was commuting back and forth and working on the mill when he died in 1922. The unusual (for a mill) shed-roofed structure was never completely outfitted as a mill, but may have ground for a short time for locals. The mill building was eventually absorbed by a series of expansions of the canning factory, which was initially a separate building behind the mill. The one-time mill has for many years been used as a farm shop and storage space.

The mill and mill race mentioned in 1892 on page 507 of deed book "V" were probably already there when Shelor bought the land. Proven water power sites were often reused to save the work of digging new races and acquiring water rights through adjoining property. How much use Shelor made of this water power site before building his own mill is not known. Copies of policies of the Mutual Benefit Association of Virginia for 1896 list corn and sawmill in good condition insured for $50.00. On 17 November 1905 the MBA listed mill and contents insured for $50.00. On 1 January 1910, there was listed "a mill and house, made of wood, roof of shingles in good condition," insured for $1,000.00. These suggest that the mill was at least in usable condition, whether or not it was in use. But an 1895 item in the *Weekly News* describing businesses and farms on Howell Creek mentions at this site only "G. W. Shelor's sawmill which is doing a good business.[5]

Adapting the old race and dam to his new turbine powered mill would have required only the reworking of the lower end to accommodate a penstock or other closed passage such as a pipe between the race and turbine, which was installed under the building at slightly above the level of the tailrace water. George W. Shelor's grandson Herbert Shelor recalled that instead of linking directly to the stream at the nearest point, the tail race may have run more northerly, paralleling the creek, toward Route 221, possibly piggy-backing this function onto an existing drainage ditch. Assuming that the drainage ditch or tail race was more gently sloped than the bed of the stream, this arrangement would have allowed for a lower placement of the turbine, thus providing a few more inches of fall in the penstock.

George W. Shelor was a blacksmith and a machinist and the owner of a legal distillery, as well as a would-be miller. Following his death in Oxford, Maryland, in 1922, his large land holdings were divided among his twelve children. The Shelor homeplace and mill house are still in the family. George W. Shelor also built the Shelor/Nolen/Pine Creek mill (I, 123) in 1913 and sold it to his son Pelham Shelor shortly after.

Deeds

Floyd County Deed Book "V," page 507, 30 April 1892. George W. Shelor homestead exemption: 1 acre land with mill, mill race, 1 acre land and dwelling house on Floyd-Carroll Turnpike, 1 shotgun, household and kitchen furniture, 1 yoke oxen, 3 old broken-down wagons and 1 shingle saw and table.

D. B. "V," page 538, 20 May 1892, G. W. Shelor homestead exemption: 1 barrel whiskey containing 46 gallons valued at $1.50 per gallon.

D. B. "Z," page 229, 12 February 1898. Carrie P. Dobyns to G. W. Shelor and O. W. Jenkins, two tracts land, of 37 acres and 42½ acres, on Howell Creek. *This may be the tract, south of Route 221, on which a steam powered sawmill is still situated in 2010. If so, it is not the Shelor grist mill tract.*

D. B. 27, page 323, 15 September 1888. F. O. Dobyns to G. W. Shelor, ½ acre land on Howell Creek and Hillsville Turnpike. *It is not clear that this is the mill tract but it was included in Frank Webb's notes. The selling price of $200 for only one-half acre suggests valuable improvements of some kind on the property, but only the bounds of the property are described in the deed. This may be the land on which the sawmill, referred to in the newspaper notice in 1895, was at work.*

33. Eli Smith Mill/Simon Smith Mill

Howell Creek

LOCATION: Route 716 (Cox Store Rd SW), .8 mile south of Route 717 (Kristi Lane SW); POWER: Overshot wheel; STONES: 1 set; VISIBLE REMAINS: Dug race partially visible

History. Eli Smith built this mill on land he bought from George Whitlock in February 1868. With only one set or "run" of millstones, Smith probably ground mostly cornmeal. In the later years of this mill's history, customers wanting wheat ground into flour may have traveled on up Howell Creek to a mill (IV, 34) that was probably built by L. D. Wickham, and run by several others after him.

Ownership of Eli Smith's mill and lands passed to his children upon his death in 1905. Eli Smith's son, Asa Simon, is believed to have run the mill after his father's death, hence the name. These lands, plus a third tract of 55 acres, were sold at public auction in 1929 but no mention of the mill appears in that deed.

Deeds and Other Public Records

Floyd County Deed Book "L," page 531. 29 February 1868, George Whitlock to Eli Smith, 3 acres on Howell Creek.

Floyd County Will Book 1, page 436, 1905. Mill willed to Eli Smith's children.

D. B. 48, page 434, 4 June 1929. J. E. Proffitt, Commissioner in cause of *Mary J. Weddle vs. Asa S. Smith and Ezra Wickham,* does sell at public auction three parcels of land on Howell Creek containing 55 acres, 17 acres and 3 acres, being the lands of the late Eli Smith.

See Table B, *1870 Products of Industry.* Name of corporation or individual: Smith, Eli

34. Andrew Weddle Mill/L. D. Wickham Mill/ Luther Webb Mill/Murray Agee Mill

Howell Creek

LOCATION: Route 724 (Old Furnace Road SW) .1 mile west of Route 720 (Fairview Church Road SW); POWER: Overshot, turbine, overshot; STONES: 2 sets, sawmill, and blacksmith shop. VISIBLE REMAINS: Dug race is clearly visible on both sides of Route 724.

History. Weddle family history maintains that Andrew Weddle ran a grist mill on this site, but will and deed records suggest that Weddle's mill was a sawmill, and that it was L. D. Wickham who built the grist mill on more or less the same spot after buying the "Saw Mill seat" from Andrew Weddle in 1910. However, while deed records rarely include incorrect information, they sometimes fail to mention important improvements such as existing mills. So it is possible that the Weddles were grinding corn as well as sawing lumber for themselves and their neighbors. In any case, water power was at work here no later than 1886, when a deed between Robert Weddle and T.H. Howard specified that Andrew Weddle was to have a mill race and water rights through Howard's newly acquired land, as long as the race was not over 100 yards long. Weddle retained his right to maintain a mill race, as long as it was not moved or extended, when Howard sold the same property to J. B. Belcher in 1908. A Civil War veteran, Weddle also owned and ran the Weddle, Virginia, post office from 1890 to 1905.

L. D. Wickham bought the one-third or one-half acre "Saw Mill seat" in 1910 with the express intention of building a mill on it. The mill, water rights, and 9⅞ acres of land were sold to Walter Boyd in 1916 for $500.* Two months after he bought Wickham's mill, Boyd sold the land on which he had previously run a mill (IV, 35) farther up Howell Creek. To protect his investment in the Wickham Mill, Boyd stipulated in the deed that the new owners could not erect a grist mill, or give up the water rights to any other builder without his (Boyd's) consent, unless he had by then sold the Wickham mill.

Walter and Nannie Boyd sold the Wickham mill and land, by then known as the Walter Boyd Mill, to Luther Hayden Webb in 1919 for $2,500. Webb operated a blacksmith shop at the site along with his mill, and at times rented both the mill and shop to other operators. Although his mill and all others with two sets of stones were designed to produce bolted "buhr stone" wheat flour using some version of the Oliver Evans process, at some point Webb gave up making flour and began selling flour from the C.W. Harman mill on West Fork Creek. The roller mill installed in the Harman mill in 1910 produced whiter, finer flour, and made more of it more quickly than did the old stone grinding process. The roller process also relieved the miller of the regular and laborious task of sharpening his stones with metal hammers called mill picks. County-wide, old time flour millers were facing the decision of whether to convert their old mills to the roller mill process, build roller mills from the ground up, or concede the business of flour milling to larger and newer competitors.

Webb's purchase of an adjoining tract in 1927 may have been an investment or an expansion of the mill property, in which case the size of the tract (50 acres) was coincidental. He might also have been planning to raise or fatten hogs. Abundant water along with access to mill byproducts and surplus or spoiled grains had made raising hogs such an attractive sideline for early Virginia millers that concerns had arisen over public health. The result was a state law requiring millers who wished to keep hogs to

**The purchase prices recorded in most real estate title deeds are not usually included in the mill histories in this book. They are included here to show that the value of mill properties and the incentive to build new mills generally mirrored the economy, peaking in this case with the war economy of the First World War, and dropping steeply as the Great Depression set in. The late but inevitable arrival of an industrial-based economy in Floyd County also had a powerful effect on the desirability and price of mill properties. As roads improved and the shift away from a small farm economy accelerated during and after the Second World War, the number of mills declined sharply. No water-powered mills are known to have been built after 1936; those that survived began shifting to electric and diesel power for at least some of their operations.*

The mill as it was when owned by Luther Webb, Frank Webb's father, around 1929. The bell to call the miller, who often farmed or pursued another trade when no customers were at the mill, and the turbine house at the upper end, are typical. Murray Agee replaced the water-hungry turbine with a more efficient albeit more primitive wooden water wheel and added a sawmill. (Frank Webb collection)

own a 50-acre tract that could adjoin the mill lot but must be separate from it. Wheat "middlings" too coarse or dark to be sold as flour was a popular hog feed. Ironically, from a nutritional standpoint, farmers would have been better off to eat the middlings themselves and feed the fluffy white breakfast biscuits to the hogs.

In 1929 L.H. and Nancy Webb sold their turbine powered mill, which they were no longer operating, to W.C. and Gladys Keith for $1,500. Two years later the Keiths sold the mill and land to M.V. Agee for only $650. Agee is believed to have hired A.N. (Newton) Hylton and Wooster Cockram to build and install an overshot wheel in place of the turbine, which must have been chronically short of water on such small a stream. The addition of a sawmill at the same time required cutting a hole in the front wall of the mill to accommodate the large wooden pulley and the flat belt that drove the sawmill. Agee's sawmill and one at the Flint Mill (II, 106) on Beaver Dam Creek were among the very last water-powered sawmills in the county. And they were among the few that employed circular saws, which had by 1930 completely replaced the older, slower "up-and-down" sash saw. Once the required belts and pulleys were in place, the power for these sawmills was virtually free. But it was limited at best and probably could not have driven both the grain and saw mills at the same time. It remains an interesting question whether either man would have gone in this direction in better economic times.

In 1939 Agee sold the 9⅞-acre mill tract and an adjoining fifty-acre tract (bought from L.H. Webb in 1931) to F.L. Harman, who ran the mill for only a short time afterward. The mill was torn down around 1950 or '51 by Jervis "Ted" Harman, who bought the same two tracts from F.L. Harman in 1944.

Luther Hayden Webb, the father of author Frank Webb, was running the Floyd Restaurant when he died in 1932 of heart failure at age 38.[6] He was the brother of Oscar Ames Webb, longtime owner of the O.A. Webb and Son Mill (II, 122), still standing along Route 8 (Webb's Mill Road) north of town. Oscar Webb's son, also named Luther H. Webb, was among the last and best known Floyd County millers.

Some notices from the *Floyd Press* reveal further history:

7 September 1922:

To my customers:

I am going away and have rented my mill for the coming year. I wish to ask all who owe me to come in and settle at once. Thanking you for past favors and beg for a continuance in my absence.

Respectfully,
Luther Webb
Route 6, Floyd, Va.

24 January 1924:

Notice:

I have rented the Luther Webb Mill, which is in good repair, and I solicit the patronage of the people. Am prepared to shoe your horses and do other shop work.

Maston Harter

20 September1928

Notice:

My mill will run only three days a week—Monday, Wednesday, and Friday from this date until further notice.

Luther Webb

Deeds and Other Public Records

Floyd County Will Book "E," page 225, 14 January 1879. John Weddle to Andrew Weddle, the land where he now lives. Acreage not stated (D. B. 39, page 304, says $119\frac{1}{2}$ acres).

Floyd County Deed Book "S," page 436, 18 August 1886. Robert Weddle to T. H. Howard, $122\frac{3}{4}$ acres land on Howell Creek. Andrew Weddle to have mill race and water rights. Race not to exceed 100 yards long through this property.

D. B. 35, page 148, 3 November 1908. T. H. Howard to J. B. Belcher, tract known as the Weddle Place, $122\frac{3}{4}$ acres. A. Weddle to have mill race on said land not to exceed 100 yards in length at the same place where it is now located.

D. B. 36, page 362, 19 February 1910. Andrew Weddle to L. D. Wickham, $\frac{1}{3}$ acre known as the "Saw Mill seat" and water rights granted to Andrew Weddle through J. B. Belcher land.

D. B. 39, page 304, 3 March 1913. B. G. Howard and V. M. Sowder, commissioners, and David E. Weddle (son of Andrew and Sarah) and Sarah Weddle to L. D. Wickham. Chancery causes of *F. M. Weddle, vs. G. M. Weddle, et al.* and *John D. Weddle vs. Bertie DeHart* resulted in sale of Andrew Weddle's land of $119\frac{1}{2}$ acres excluding tracts of $\frac{1}{2}$ acre (should be $\frac{1}{3}$) previously sold to L. Wickham and $13\frac{1}{10}$ acres sold to S. J. Weddle. Wickham to have water rights through $13\frac{1}{10}$ acre tract.

D.B. 41, page 345, 25 March 1916. L.D. Wickham to Walter Boyd, $9\frac{7}{8}$ acres land, mill, and water rights. $500.

D. B. 43, page 592, 2 December 1919. Walter and Nannie Boyd to F. L. Harman and Luther Hayden Webb, mill and $9\frac{7}{8}$ acres (F. L. Harman had bought mill and land but no deed made), and water rights to same. $2,500.

D. B. 46, page 276, 12 March 1927. S. P. Hylton to L. H. Webb, 50 acres land on Howell's Creek adjoining mill tract.

D. B. 49, page 2, 25 March 1929. L. H. Webb and Nancy, his wife, to W. C. and Gladys Keith, Walter Boyd mill and $9\frac{7}{8}$ acres land and water rights to same. $1,500.

D. B. 49, page 335, 20 March 1931. W. C. Keith to Murray V. Agee, Walter Boyd Mill, $9\frac{7}{8}$ acres, mill, and all water rights. $650.00.

D. B. 49, page 573, 13 November 1931. L. H. Webb to M. V. Agee, 50 acres.

D. B. 58, page 279, 28 December 1939. M. V. Agee to F. L. Harman, $59\frac{7}{8}$ acres and mill.

D. B. 63, page 87, 11 December 1944. F. L. Harman to Jervis "Ted" and Edna Harman, $59\frac{7}{8}$ acres and mill.

35. Newton Hylton Mill/Walter Boyd Mill

Howell Creek

LOCATION: Route 723 (Patrick Road SW) about .25 mile south of Route 720 (Fairview Church Road SW); POWER: Overshot wheel; STONES: 1 set; VISIBLE REMAINS: Dug race, breached earth and stone dam

History. Alonzo "Newton" Hylton was one of the great old-time craftsmen and millwrights of Floyd and Patrick counties. He worked mostly on mills of the smaller, more primitive type, similar to Mabry Mill (IV, 60) and characterized by hand-made wooden and wrought or cast iron components. A friend and favorite subject of photographer Earl Palmer, Hylton is featured in every photograph in a chapter titled "Eternal Mountaineer" in *The Appalachian Photographs of Earl Palmer.*[7] Newton Hylton built several wooden water wheels for other mills, including two for Mabry Mill, but this little mill on the upper reaches of Howell Creek is the only mill he is known to have owned and operated. The belief of Hylton family historians that Hylton built this mill is supported by the fact that no mill was mentioned in the deed when Hylton bought the land from Jessie Carter and wife in 1894. *Chataigne's Virginia Business*

The dam of earth and piled stones at the site of the Hylton/Boyd Mill is the largest example of its type anywhere in the county, and the only one that would have retained a sizable pond. Similar dams, now forgotten, may have been removed to recover the stone for some other purpose, or simply to clear and consolidate a field for some post-millpond use. (Ricky Cox photograph, 1992)

Directory and Gazetteer for 1911 lists A. N. Hylton running a grist mill on Route 6 in Floyd County.[8]

Walter Boyd, who bought the mill in 1912, is supposed to have done away with it when he bought what was then the L. D. Wickham flour mill (IV, 34), a short way down Howell Creek, in March 1916. Whatever the fate of the Hylton mill, when Walter Boyd sold his old mill tract to John Boyd two months after buying the Wickham mill, he took pains (see below) to prevent the reappearance of a competitor for his new business. A deed for the Dulaney mill (III, 78) on Big Indian Creek includes a similar limitation on the new owner. That there are only two such instances is surprising, considering how many men were involved in some way with more than one mill.

The Hylton/Boyd mill was built on a steep bank, and the pilings and gears were still visible there in the early 1930s.

Deeds

Floyd County Deed Book "X," page 130, 6 December 1894. Alonzo Newton Hylton from Jessie G. Carter and wife, 50 acres plus or minus on the headwaters of Howell's Creek.

D. B. 38, page 196, 21 March 1912. A. N. Hylton and Sarah, wife, to Walter and Nannie Boyd, 20¾ acres of land on headwaters of Howell's Creek, adjoining Joseph Hancock, John Boyd, and others.

D. B. 42, page 304, 23 May 1916. Walter Boyd and Nannie, wife, to John S. Boyd, 20¾ acres on Howell's Creek. The party of the second part agrees not to erect any grist mill on the above mentioned tract of land and will not allow any person water rights to erect grist mill without getting the consent of the parties of the first part, provided however that the above agreement in regard to the erection of a mill and water rights shall be null and void in case the L. D. Wickham mill tract of land is sold which now belongs to Walter Boyd.

The Asa Pratt mill as it looked on June 7, 1931. A serious amateur photographer, Kenneth Pratt recorded the shutter speed and lens opening used to take to this carefully composed photograph. (Courtesy Genevieve Starkey)

36. Asa Pratt Mill

Line Branch (Tributary of West Fork of Little River)

LOCATION: Route 726 (Black Ridge Road SW) approximately 2.5 miles south of Route 221 (Floyd Highway South); POWER: Wooden overshot wheel; RUN STONES: 1 set; VISIBLE REMAINS: Dug race, fragments of low stone dam

History. The *Floyd Press*, February 25, 1937, shows a front view of the mill and its overshot wooden wheel, which is unusually large relative to the size of the building. The caption states that the mill was built by LeRoy Davis some fifty years earlier (c. 1887). No deed book record connects Davis to this property, but he could have been hired to build the mill. As to when the mill was built, the fact that neither of two deeds recorded in 1901 (Hubbard, et al., to Whitlow and Whitlow to Pratt) mentions a mill suggests that Asa Pratt built the mill after he bought the land. However, over and over, deeds to property known to include mills, both abandoned and still running, fail to mention them, so the photo caption from 1937 may be correct.

George Shelor recalled that Walter Slusher lived in a frame house above where the mill must have stood. Presumably the Asa Pratts lived there as well, within sight and easy reach of the mill. The absence of any mention of the mill, race, or water rights in the deed between sellers Asa and Francis Pratt and buyer Walter Slusher, and the mill's condition when photographed by Kenneth Pratt in 1931, suggest that it was no longer in operable condition in 1929.

Deeds

Floyd County Deed Book 27, page 422, 14 January 1901. Rufus Hubbard, Maggie Hubbard and Mary E. Hubbard to W.H. Whitlow, 2 tracts of land on Line Branch. Tract 1: 27¾ acres. Tract 2: 27 acres, 12 poles.

D.B. 27, page 425, 25 January 1901. W. H. Whitlow and wife, Lucinda, to F. S. Pratt (wife of Asa L. Pratt), 2 tracts land on Line Branch Tract 1: 27¾ acres. Tract 2: 27 acres, 12 poles.

D. B. 48, page 496, 28 September 1929. Francis S. and Asa L. Pratt to Walter E. Slusher, two tracts land on Line Branch, waters of West Fork, Little River, Lot No. 2 contains 27 acres and 12 poles, Lot No. 1 contains 27¾ acres.

37. German Belcher Mill

Rush Fork Creek

LOCATION: Route 804 (Rush Fork Road SW) 1.6 miles west of the intersection with Route 726 (Black Ridge Road SW); POWER: Undershot turbine; VISIBLE REMAINS: Stacked stone dam abutments and part of mill foundation nearly intact. Mill pond silt along stream banks immediately above dam.

History. German Belcher came from the Slate Mountain section of Meadows of Dan and bought two tracts of land on Rush Fork around 1900. His mill, like several other smaller mills, was built, operated, and abandoned without leaving any known written record. Only the inclusion of one "10-inch turbine wheel" in a list of personal property to be sold at auction [9] offers a documented clue to the presence of his mill.

R.O. Slusher, Jr., and his sisters Ruth Ann Slusher and Neva Slusher Simmons recalled carrying sacks of corn across the rump of a horse or in a buggy to this mill as young children. After calling at the house to let Mr. Belcher know he had customers, they waited while he closed the dam and filled his mill pond, supplying enough water to grind for about an hour. As was the case with other small mills, Mrs. Belcher sometimes ran the mill when her husband was away or busy outside the mill. The mill's Dutch doors, which let in air and light but kept out dogs, cats, and chickens, were a common feature on Floyd County mills. Ruth Ann Slusher recalled that the Belchers invited the entire community to their home every March 12 to a birthday party for Mr. Belcher or his wife. Mrs. Belcher smoked a pipe, a fairly common practice among women of her generation.

The fine proportions of the building and the rocky, tree-lined valley around it made the German Belcher Mill one of the most attractive in the county. Kenneth Pratt may have snapped this photograph during the same June 1931 visit that produced the picture of the Asa Pratt mill (IV, 36) on Line Branch. Belcher's wood-shingled building and attached turbine house stood on a carefully laid stone foundation. (Frank Webb collection)

Deeds

Floyd County Deed Book 26, page 459. 13 February 1900. Walter F. Dillon to German Belcher, 153½ acres.

D.B. 29, page 290. 7 March 1903. Harvey and Margaret Weddle to German Belcher, 4 acres.

D.B. 58, page 312. 2 February 1940. German Belcher estate to heirs (listed in deed).

D.B. 58, page 365. 1 March 1940. John Q. Belcher, et al., two tracts sold at public auction, 102⅞ acres and 84⅝ acres.

38. Martin Weddle Mill and Cara Simmons Mill/Obediah Simmons Mill

West Fork of Little River

LOCATION: About 1 mile upstream from Route 221 on Route 824 (Mill Pond Road SW); POWER: Turbine, 3-foot diameter (possibly a wooden tub wheel); STONES: 2 sets; VISIBLE REMAINS: Stacked stone dam abutments on both sides of the creek and a "mud log" (the bottom log of the wooden dam) between them in the bed of the stream. (1992)

History. In May 1833 Cara Simmons (1793–1868) petitioned the court of Floyd County for the right to build a dam and gristmill on 144 acres bought from Jacob Goodykuntz in 1831. The land was sold to Goodykuntz by David Weddle in 1819 and before that had belonged to Benjamin Weddle, father of both David and Martin Weddle. The 1819 deed mentions an old mill house and the 1833 deed notes that the tract included an old mill dam formerly belonging to Benjamin and Daniel Weddle. This earlier mill may have been built by Martin Weddle, who was granted permission to erect a dam on the West Fork of Little River in 1805[10] and later ran a mill in Lee County, Virginia. Had Cara Simmons built his dam and mill at the same spot, and he may have, this would be among the earliest Floyd County mill seats whose exact location is known today. His 1833 application to the Circuit Court of Floyd County for a *writ of ad quod damnum* may mean only that the time allotted by law for the rebuilding of a previously approved dam had expired, or it may mean that the new dam, and perhaps the mill as well, were to be in new locations on the same tract of land.

Simmons reported having ground 2,000 bushels of corn and 1,000 bushels of wheat in the twelve-month period ending on June 1, 1850 (see Table A). Upon his death in 1868, all his property went to his wife Catherine, and upon her death, to his son, Obediah. Obediah died in 1893, leaving all lands and property to Elizabeth A. Simmons, his wife. The land was deeded to her daughter, Minnie Howard (Mrs. A.T. Howard) in 1901. The mill stood until the early 1900s, and the stacked stone abutments of the dam were still clearly visible in 1992.

Neither of the *Products of Industry* records of this mill indicates the presence of a sawmill, but this may mean only that the annual value of the sawn lumber was less than $500. *Chataigne's Virginia Gazetteer and Classified Business Directory* lists Obadiah Simmons as the owner of a sawmill in 1897 (see Appendix A). Simmons was by then dead, but had left the mill to his wife Elizabeth, who is surely the "Miss O. Simons" whose name also appears among the owners of sawmills, albeit at a different but nearby post office, in the aforementioned 1897 directory. R. O. Slusher recalled having seen, as a boy, the ruins of a sash type sawmill on West Fork between his home on Rush Fork and Route 221.

Deeds and Other public Records

Montgomery County Deed Book "G," page 254, 6 September 1819. David Weddle to Jacob Goodykuntz, mentions mill house on West Fork of Little River.

Floyd County Deed Book "A," page 30, 10 October 1831. Jacob Goodykuntz to Cara Simmons, tract of land containing 144 acres, containing old mill dam formerly belonging to Benjamin and David Weddle.

Floyd County Will Book 1, page 190, 9 January 1868. Cara Simmons willed to his wife, Catherine, all property during her lifetime and then it was to go to son Obediah.

W. B. "F," page 535, 7 October 1893. Obediah Simmons willed to his wife, Elizabeth A. Simmons, all of his property and real estate including the mill tract.

D. B. 26, page 164, 21 March 1899. Elizabeth A. Simmons agreement with daughter Minnie E. Howard (Mrs. A. T. Howard.)

D. B. 28, page 220, 12 December 1901. Elizabeth A. Simmons to Minnie E. Howard, all of her property both personal and real willed to her by Obediah Simmons.

See Table A, *1850 Products of Industry.* Name of corporation or individual: Cory Simmons

See Table C, *1880 Products of Industry.* Name of corporation or individual: Simmons, O.

39. Martha "Mattie" Weddle Mill

Burks Fork Creek

LOCATION: Route 799 (Conner Grove Road SW) about ¼ mile above the Eli Sowers/ Rufus Harris Mill (behind DeHart's Store); POWER: Tub wheel; STONES: 1 set; VISIBLE REMAINS: None

History. Local people around 1990 had heard of this mill, but remembered only that Martha "Mattie" Weddle (b. 1866), a small woman, ran the mill herself. Only one other mill, Nancy Shank's on Little River (no number, location unknown), was known by a woman's name. But it's impossible to say how many women like Josie Moses, Millie Jane Belcher (IV, 37) Ella Thomas (IV, 41), and Rachel Hundley (I, 6) regularly helped or filled in for their husbands.

Although Joseph Hylton had owned the land at least since 1889, his mill was probably built about 1907 when he traded George Vaughn a strip of road frontage in exchange for a deeded right of way for a dug mill race. It is possible, also, that the mill and mill race were already there and that this deed merely formalized an existing hand-shake agreement. The 1915 deed between Hylton and his daughter and son-in-law Martha and Martin P. Weddle (1867–1924) is unusual in that the Weddles each had a specific fractional interest in the land, rather than owning the whole tract in common. The 1924 deed conveying Martin Weddle's two-thirds interest to Mattie mentions the "old mill and mill site," suggesting that the mill was no longer in use. There is no further mention of the mill in deed records.

Deeds and Other Public Records

Floyd County Deed Book "L," page 703, 1 June 1866. Richard Thompson to Joseph Hylton, 206 plus or minus acres on Burk's Fork Creek.

D. B. "V," page 510, 21 May 1889. Paulina Hylton to Joseph Hylton, 22½ acres on Burk's Fork Creek.

D. B. 37, page 480, 28 December 1907. George Vaughn to Joseph Hylton, for a strip of land between the road and Vaughn's land, Vaughn conveys to Joseph Hylton a water way to dig a race to his mill from the mill pond.

D. B. 41, page 90, 15 March 1915. Joseph Hylton to Martin P. Weddle and Martha P. Weddle, two tracts of land on Burk's Fork Creek, one of 80 acres and one of 44 acres. Martin P. Weddle, ⅔ interest, Martha P. Weddle, ⅓ interest.

D. B. 45, page 581, 13 September 1924. Martin P. Weddle to his wife, Martha P. Weddle, his ⅔ interest "for life or so long as she remains his widow." Upon her death or remarriage, Martin Weddle's ⅔ interest would be divided among their children. Laura Virgie Weddle was to receive ⅓ of the land, it lying on the southwest side of the place containing the old mill and mill site. The remaining ⅓ (half of Martin Weddle's ⅔) to be divided among other heirs in equal proportions.

40. W. J. H. Barringer Mill

Burks Fork Creek

LOCATION: Route 799 (Conner Grove Road SW) .25 mile west of Route 805 (Emmanuel Road SW); POWER: Unknown; STONES: Unknown, roller mill added 1902.

History. William J. H. Barringer (b. 1854) was the son and grandson of millers Jacob and John A. Barringer but was himself more active as a mill owner than as a miller. Barringer's belief in keeping the public abreast of his intentions shows up even in the deed to the one-acre lot he bought in 1890 "for the purpose of building a Saw mill and griss mill." *Chataigne's Gazetteer and Virginia Business Directory* lists Barringer as mill owner or operator of a Floyd County mill near Pax in 1894 (see Appendix A). By 1895 he was running a "Grist-Mill, Saw-Mill and Box Factory"[11] and a general store. His advertisements, like those of W. H. Harman on Little River, are not burdened by modesty, false or otherwise. Although it is less impressive without the eight different typefaces used in the original, the text of his advertisement in the December 6, 1895, edition of the *Weekly News* merits reproducing in full:

New Goods! New Goods! W. J. H. Barringer, PAX, VA. I have just returned from the Northern markets where I purchased the largest and best stock of goods that I ever bought before. I struck a streak of good luck while North and bought 46 dozen pairs of boots and shoes cheaper than ever before and I can sell them very low. I also bought a large bill of Gent's Clothing consisting of fine Ready-Made Suits and Overcoats which will be sold cheaper than ever, also the finest Felt and Straw Hats for the ladies ever brought to this section. I also have a large stock of ladies' fine dress goods, and in fact I have bought a large stock of everything usually kept in a country store, and I can and will sell goods cheaper than ever before. The reason I can do so is just this. I bought my goods in large quantities and paid spot Cash for them, and I run my own teams and do my own clerking and have no partners, and make my support on my farms and by mills and factory. Everybody is respectfully invited to call and examine my goods before purchasing elsewhere. My GRIST-MILL, SAW-MILL AND BOX FACTORY at Pax is in good running order, and I have two good, accommodating millers, Mr. Joseph Vipperman and son. My mill on Laurel Fork is also in first-class condition, and Messrs. C. T. and Archibald Hylton, millers, are ready to serve the people.

I am truly thankful to my friends and customers for their past patronage and I hope to have a share of your trade in the future. W. J. H. Barringer.

In 1902 Barringer built a new roller mill, or—more likely considering that it was only about ten years old—adapted his existing mill to the new system, as was happening all over the county. According to an alert observer, "Thirteen wagons, loaded with machinery for Mr. W. J. H. Barringer's roller mill at Pax passed through town Wednesday. We understand Mr. Barringer will have a complete up-to-date mill when completed."[12] On August 28, Barringer published notice of his readiness to hire a miller: "I want a first-class miller to attend my roller mill in about 60 days from this date. I want one who understands it perfectly. Anyone wanting a job will please write me."[13] A notice on page 2 of the *Floyd Press* published on October 23, 1902, stated that "Mr. J. B. Moses has gone to Pax to assist W. J. H. Barringer install his machinery in his new mill. It is rumored he will take charge of the mill when completed. We compliment Mr. Barringer as Mr. Moses is a professional and successful mill-man." John Berry Moses (b. 1846) was himself the son of a miller, and the first five of his ten sons—Charles, Johial or Hiley, Jacob, George, and John—gave their occupations as "miller" in at least one census.

The *Floyd Press* continued to follow the story and on November 26, 1902, reported that "Mr. W. J. H. Barringer has completed his roller mill. He has corn, wheat, and buckwheat roller mills, all doing good work." If this report is accurate, Barringer's mill and millers were truly ahead of their time. As far as is known, all other Floyd County mills continued, until they closed, to grind cornmeal and buckwheat flour with a set or run of traditional mill stones.

The mill is said to have burned and was apparently gone in March 1907 when Barringer sold the one-acre lot on which it had stood to O. W. Slusher. W. J. H. Barringer was living in Blacksburg, VA, in December 1908.[14] W. J. H. Barringer also owned the J. W. "Pob" Boyd mill on Laurel Fork Creek (IV, 61) from 1894 to 1904. It was the "mill on Laurel Fork" referred to in his 1895 newspaper advertisement.

Deeds

Floyd County Deed Book "Q," page 189, 28 February 1881. Zebrum Keith to W. J. H. Barringer, 80 acres land, Burk's Fork Creek.

D. B. "V," page 590, 5 July 1890. Austin Hylton to W. J. H. Barringer, one acre land adjoining above land for the purpose of building a saw mill and grist mill.

D. B. 33, page 528, 25 March 1907. W. J. H. Barringer and M. E. Barringer to O. W. Slusher, four tracts, total of 48¼ acres; 4th tract, 1 acre. Same bounds as D.B. "V," page 590.

41. Eli Sowers Mill/Rufus Harris Mill

Burks Fork Creek

LOCATION: Route 799 (Conner Grove Road SW) about .1 mile west of Route 805 (Emmanuel Road SW). Mill stood on the south bank of Burks Fork immediately upstream from where it is crossed by Route 799; POWER: Wooden overshot wheel; STONES: 1 set; VISIBLE REMAINS: Ruins of a large wooden pulley and shaft visible through 1990s. Frag-

Ice accumulated on the wooden race and wheel illustrate why the owners of some small mills simply closed down during frigid weather. A nearly solid curtain of ice hangs from the race as it crosses over Burks Fork to the mill. The photograph has been manipulated to reduce the contrast between the brilliant ice falls and deep shadows beneath the race. (Courtesy Genevieve Starkey)

ments may remain. Dug race on north side of Burks Fork obscured by private road.

History. A story supposed to have appeared in the *Floyd Press* in December 1941 says that this mill was built by John Barringer and had as millers a Mr. Hookes, a Mr. Pratt, and Rufus Harris (1851–1936). There is no record of John Barringer (1829–76) ever owning the mill, but he could have built it for his brother-in-law Eli Sowers. It's possible also that he had built it for his father-in-law, William Sowers, on the land that Eli Sowers would later inherit or buy from the other heirs. The second scenario seems unlikely, since no mill is mentioned in the will and because the terms of the will suggest that Barringer and his father-in-law were not on very good terms. In dividing his property among his children, Sowers stipulated that the portion of his property allotted to his daughter Tripthena was to go to her and her heirs, but not to her husband, John A. Barringer.[15]

Brothers Ishmael (1832–8?) and Jessie Pratt (1840–1911) bought the mill in 1875. In 1888 Ishmael's one-half interest was sold to Rufus Harris to settle a chancery suit having to do with the settlement of Ishmael Pratt's estate. Pratt's oldest son and the administrator of his estate, Asa, later owned and ran a small corn mill on Line Branch (IV, 36). Rufus Harris and his wife Louvenia became full owners of the mill when she bought the remaining one-half interest from Jessie and Susan J. E. Pratt in 1894. As a "house carpenter," Rufus Harris worked away from home at least part of the time, leaving his wife to run the mill. The mill was rebuilt sometime in the early '30s with A. N. (Newton) Hylton building the overshot wooden water wheel.

After Rufus died in 1936, Louvenia deeded the mill to her daughter Ella and son-in-law Homer E. Thomas (1882–1954). Like her mother, Ella Harris Thomas (1890–1954) ran the mill by herself. Their son Waller was probably the last to run the mill. It is not mentioned in Homer Thomas's will or in the partitioning of his property after his death in 1954.

Among more than 140 Floyd County mills, only this and a couple of others are known to have been on the sides of the creeks opposite from the dug races that served them. Other examples are the A. W. Stanley Mill (IV, 130) and the Agee Mill (IV, 57). Wooden races occasionally had to cross small branches or even public roads to deliver water to overshot wheels, but only these few actually crossed over the creeks from which the water they carried had come. In this case, the south side of the valley was unsuitable for a dug race because of its distance from the creek and the obstacle presented by the road running between it and the creek. But placing the mill on the north side would have required customers traveling on Route 799 to cross the creek to get to the mill. Since the creek is neither wide nor deep at this point, and not far from the slope on the northern side of the valley, the builders may have reasoned that it would be only a little more trouble for them to build and maintain a race instead of a wooden bridge across the creek, and much safer and more convenient for customers.

The construction of the Rufus Harris Mill water wheel, rebuilt c. 1930, displays a feature seen on at least two other wheels built by A. Newton Hylton. The spokes are socketed into the oak axle or "spindle" but do not pass through it as in some wooden wheels. The socketed spokes are less than half as long as the mortised type and are not weakened by the notching required to fit the mortised type together where they intersect at the center of the axle. At the same time, the axle is not weakened by mortising, nor do the shallow sockets provide an avenue for damaging water to penetrate to the center of the axle. (Frank Webb collection, undated)

Deeds and Other Public Records

Floyd County Will Book "C," page 596, 15 February 1862. William Sowers, home tract to wife, Rhoda Sowers. Balance of estate to be sold and equally divided among children. "[I] desire that my heirs do make there right to Eli Sowers to his part of the land where he lives being one half of that tract."

Floyd County Deed Book "L," page 146. Andrew Keith and wife, John A. Barringer and wife, John Sowers and wife, and heirs of Wm. Sowers, dec'd (in compliance to will of Wm. Sowers, dec'd.) to Eli Sowers, 195 acres, waters Burks Forks. Eli Sowers bought an additional 83½ acres of his father's property.

D. B. "O," page 270, 27 November 1875. Eli Sowers and Mary Ann, his wife, to Ishmael Pratt and Jessie Pratt, 3 acres land and mill on Burks Fork Creek.

D. B. "V," page 58, 9 November 1888. Commissioner's Sale, W. L. Howard, Commissioner, chancery cause of *Asa D. Pratt Adm. of Ishmel Pratt, vs. James G. Harris of the One Part and Rufus Harris of the Other.* Sold to Rufus Harris, "one half interest in land and mill property in said cause mentioned known as the Eli Sowers mill property.

D. B. "W," page 472, 23 January 1894. Jessie and Susan J. E. Pratt to Louvenia Harris, ½ interest in mill property known as the Eli Sowers mill.

D. B. 53, page 297, 20 June 1936, Louvenia Harris to Ella. M. Harris (daughter of Louvenia Harris) and Homer. E. Thomas, her husband. Two tracts, one of 10½ acres, the homeplace. Second tract, 2 acres, "what is known as the mill tract," including the mill, mill machinery and all water rights.

Will Book 6, page 52, 14 September 1953. Homer E. Thomas, all property to children, Wallar E. Thomas and Ora T. Harman. Land was partitioned per Deed Book 76, pages 586–87.

42. Henry Slusher Mill and Buffalo Roller Mill and Andy Alderman Mill

Burks Fork Creek

LOCATION: Route 799 (Conner Grove Road SW) .3 mile north of Route 727 (Union School Road SW) across creek from Thistle Hill Road SW; POWER: Overshot wheel

Like the plans of its builder and major stockholder Winfield Scott, the Buffalo Roller Mill, shown about 1908, came to a sudden and unfortunate end. It was much larger and far more elaborate than either the Henry Slusher Mill, its predecessor, or the Andy Alderman Mill, its successor. The man and boy standing in the wagon on the left are Elza and Roy Weddle. (Courtesy Vernon Harris family)

(Slusher, 1870); turbine (last two mills); STONES: First mill, one set. Second, roller mill. Third, at least one set. VISIBLE REMAINS: Dug race along the east side of creek; mud log in stream bed marks location of dam

History. No mill was mentioned in the deed when Henry Slusher (1829–1898) bought 112 acres on Burks Fork from his maternal grandfather, Archeleus Hylton, in 1857, so it is probable that Slusher built the mill or had it built. But whether this was done before or after the Civil War is unknown. Henry Slusher's career as a mill owner, like those of Aaron Dehart, Abraham Burnett, Caleb Sowers, William Allen Simpson, and others, was preceded or interrupted by service in the Confederate army. Slusher was evidently already a leader among his friends and neighbors when, at age 32, he entered the Confederate army as an elected captain in Co. D, 54th Virginia, in October 1861.[16] Slusher "enlisted at Jacksonville [now Floyd, VA] ... but resigned a year later in Beaver Creek, Kentucky. Probably, like most men of the 54th, he thought he would be fighting in Virginia and they never were. He reenlisted as a private in Co. H, 51st Virginia Infantry and served the remainder of the war. He was captured in Waynesboro [VA] on 3/4/65 and sent to prison at Fort Delaware. He was released on oath on 5/5/65."[17]

One of Slusher's nearest competitors in the mill business was his half-brother, Jackson W. Harris (the younger), who owned and probably built the Martin Harman Mill (IV, 44) and who was also was involved with several other Burks Fork mills in the last quarter of the 19th century. In 1883 Slusher sold Zebram Keith "½ interest in 1 rod and 26 poles land and ½ of mill property on said land."[18] Both men were dead by the time partners Winfield Scott and David Slusher acquired the mill site and an adjoining seven and three-fourths acres, but exactly how the land passed through the hands of the Slusher and Keith heirs is unclear. At least part of the Slusher portion was owned for a while by his daughter Emazetta E. Webb and her husband, Thorburn Webb, the parents of O. A. Webb (II, 122) and Luther H. Webb (IV, 34).

The second mill on this site lasted only a very short time compared to the mills that came before and after it, but was in some ways more interesting than either. Both the mill itself and the plans of the men who built it were modern, ambitious, and destined to go up in smoke. In retrospect, the central figure in this story is Winfield Scott, who bought the Henry Slusher mill in partnership with David Slusher in January 1906. One year later the two sold the same property to Buffalo Roller Mills, Inc., a chartered, stock-issuing corporation, with officers and a board of directors.

The stated purposes of the new corporation were to "to do a general milling business, buy grain, manufacture flour, meal, foods, feeds and chop and sell said products and exchange for grain and grind for toll such as is selected to grind and to purchase mill products and sell same."[19] The business was to be capitalized at not more than $20,000 and not less than $10,000. Winfield Scott was one of three directors as well as the business manager and treasurer. The only other Floyd County mill owned by such an elaborate organization was the Scott Roller Mill Co., later the Floyd Milling Company. It, too, was formed to own and manage a modern flour mill built and then sold to a corporation soon after it was built by Winfield Scott. It, too, enjoyed the services of Scott as one of three directors and the treasurer and business manager of the corporation.

In the meantime, the old Henry Slusher mill had been destroyed by some accident, or more likely was razed to make way for a state-of-the-art roller mill, equally modern but much smaller than Scott's roller mill (II, 122) on Route 8. On July 18, 1907, the *Floyd Press* reported that "The Buffalo Roller Mill is now complete and grinding for the public, wheat, buckwheat, corn meal and feed mill attached." The future looked bright for the stockholders, the miller, and patrons, since this was the only roller mill for miles around, and light, white roller mill flour was proving itself the favorite over the old-fashioned stone-ground flour. Buhr mill flour, as it was sometimes called, contained less bran than modern "whole wheat" flour, but was nonetheless noticeably darker and coarser than highly refined roller mill flour.

The bright future dimmed considerably when Scott disappeared with the funds of the Farmers Bank of Floyd in 1908. His assets, including hundreds of acres of land as well as stock in the Buffalo Roller Mill and the Scott Roller Mill, were seized and sold to pay his debts.[20] Scott owned 1,000 of the 1,500 shares of Scott Roller Mill, Inc., stock and 126 of the 240 total shares of Buffalo Roller Mill, Inc., stock.[21] The advertised par value of the Buffalo Roller Mill stock was $50 a share, but the 126 shares brought less than $8.50 a share when "[Winfield] Scott's Buffalo Roller Mill stock was sold to George W. Helms for $1055.00."[22] When the dust settled the Scott Roller Mill had been reorganized as the Floyd Milling Company. The building it occupied survives today as the Webb/Dodd Creek Mill (II, 122) on Route 8 north of Floyd.

Whether the Buffalo Roller Mill Corporation would also have survived the scandal and the financial fallout of Scott's disappearance will never be known since the mill burned to the ground around 1910. In any case, the *Floyd Press* of September 8, 1910, carried this announcement:

> Stockholders to meet. You will please take notice that at a meeting of the Board of Directors of which meeting each member had personal notice as provided by law, a resolution was passed declaring it advisable and for the benefit of the corporation that the same be dissolved and that a meeting be held in the office of the company at G. W. Agnew's law office, in the Town of Floyd, Virginia, on the 8th day of October, 1910, at one o'clock P. M. to take action on said resolution.
>
> J. P. Hurd, President S. S. Hurd, Secretary

The chancery cause filed against Scott by the corporation resulted in the sale of the mill tract to Andy Alderman (1865–1947). The two turbine wheels that came with it are presumably the only pieces of equipment that survived the fire, placed as they were near the surface of the water in the tail race. In size, appearance, and function, Alderman's new mill building was probably more like Henry Slusher's mill than Winfield Scott's. It was, however, modern in its use of a turbine or turbines, with one driving the sawmill and the other the grist mill, most likely in alternation. Burks Fork at that upstream point would not be able to supply water to two turbines for very long at a time.

Adina Alderman was responsible for the mill, grinding corn meal and buckwheat flour, while her husband worked as a farrier and blacksmith, shoeing horses and making and repairing farm wagons. The mill stopped running sometime between 1938, when Andy Alderman deeded the mill to his wife, and his death in 1947.

Andy Alderman in front of his new mill, about 1913. The unstained siding and bright, flat roof shingles confirm that the mill, the third at this location, was nearly new when this panoramic view was taken. But the sawmill, which may have been used to saw the lumber for the mill, has already accumulated a sizeable pile of slabs, and sawn boards have been put to use on fences. The wide angle lens makes the valley between the mill and modern day Conner Grove Road appear to be wider than it is. (Frank Webb collection)

D. B. Goad bought the mill from Adina Alderman in 1948 and later tore it down. The Andy Alderman house, which stood very near the mill site, burned sometime around 2008.

Deeds

Floyd County Deed Book "J," page 282, 12 September 1857. Archeous (Archeleus) Hylton to Henry Slusher, 112 acres land on Burks Fork Creek (part of Thomas survey).

D. B. "R," page 254, 22 December 1883. Henry Slusher to Zebram Keith, ½ interest in 1 rod and 26 poles land and ½ of mill property on said land. *No record found of the conveyance of Keith's or Slusher's one-half interest in the mill tract.*

D.B. Z, page 358, 17 October 1898. Mary E. Harman, H. M. Knowles and J.T. Knowles, her husband, and Nancy Slusher, widow of Henry Slusher, to E.E. Webb, dividing up Henry Slusher lands, 2 tracts, 31¼ acres and 7¾ acres "beginning at a bunch of maples by a mill race." E.E. Webb is not to have possession of second tract of land until death of Nancy Slusher, it being part of her dower right.

D. B. 32, page 205, 17 January 1906, E. E. Webb and T. F. Webb and Nancy Slusher, widow of Henry Slusher, to Winfield Scott and D. Slusher, 7¾ acres, $250 to be paid at the death of Nancy Slusher. Annual interest to be paid to Nancy Slusher in the meantime. Mentions mill race or mill road.

D. B. 33, page 213, 1 January 1907, Winfield Scott and David Slusher to Buffalo Roller Mills Co., Inc., two tracts land on Burk's Fork Creek, one tract of 1 rod and 28 poles formerly belonging to H. Slusher and Zebram Keith, with all water rights and mill thereon. Second tract of 7¾ acres. $12,000 and $250 to be paid to Mrs. E. E. Webb at the death of Nancy Slusher.

D. B. 39, page 179, 11 April 1912. Andy J. Alderman from G. W. Agnew in Chancery cause of *W. J. Scott vs. Buffalo Roller Mills Company, Inc. and Others*, and Mr. J. P. Hurd and wife, L. C. Hurd and George W. Helms and wife, parties of the 1st part, and Andy J. Alderman, 2nd part (W. J. Stump, S. S. and J. Hurd acting as trustees). Sold on 14 July 1911, on premises at public auction, and Mrs. L. C. Hurd being the highest bidder. Bought said land of 1 rod and 20 poles with water rights, and a 7½ acres tract, and 2 turbine water wheels wherein she sold land above to G. W. Helms, and G. W. Helms conveyed same to A. J. Alderman.

D. B. 56, page 529, 16 April 1938. A. J. Alderman to Adina Alderman, his wife, above land.

D. B. 67, page 396, 27 March 1948. Same to D. B. Goad.

See Table C, *1880 Products of Industry.* Slusher and Keith Co.

43. Jacob Barringer Mill

Burks Fork Creek

LOCATION: Route 657 (Weddle Road SW) approximately .2 mile west of Route 799 (Conner Grove Road SW). State maintenance of the road ends very near this point (2010); POWER: Unknown; STONES: Unknown; VISIBLE REMAINS: A level plot, formed by sediment settling behind the dam, remains inside an oxbow bend in the creek.

History. Jacob Barringer is assumed to have built this mill within a few years of purchasing a single tract of about 500 acres from Thomas McCabe in 1845. One of two administrators for the estate of George Hylton, McCabe bought the property at a public sale and resold it to Barringer without recording a transfer from Hylton to himself. The deed was instead made directly from the Hylton estate to Barringer, who was at that time living in Montgomery County, VA. Whether Barringer built the mill, or had it built, or found it in place (it is not mentioned in the deed), it was the first of at least three mills constructed before 1900 on a narrow creekside strip of what had been Hylton's land. Because of overlap in their histories, readers are encouraged to examine all three (IV, 43; IV, 44; and IV, 45) to get the most complete picture of any one.

The mill appeared twice in deed book records during its working life, first as a landmark when Jacob Barringer sold 9 acres of land to Jackson Harris in 1854, and as an "appurtenance" to the transferred property when Barringer's son John bought from his parents the 125 acre tract of land on whose southwest corner the mill was located. Though apparently long gone by then, the mill resurfaced in deed book records in 1896 when J. W. (Jackson) Harris sold the "old mill pond tract" of 1¼ acres to John Sowers. When Sowers sold his one-half interest in the mill known at that time as the "Slusher, Harris, and Spangler Mill" (IV, 44) in 1908, he included in the deal the old Barringer Mill tract. Though completely irrelevant to the operation of the mill upstream, this tract was still attached to the mill property when Martin Harman, the last man to operate the mill, bought out Hezekiah Hylton's one-half interest and became sole owner in 1936.

The name of the Barringer Mill has disappeared from living memory, but the site has still attached to it the story of an Alderman family grandmother who, upon finding a hound-driven deer treading water in the mill pond, hurried home for a knife, swam or waded into the pond, and captured and killed the deer for her family.

Deeds and Other Public Records

Floyd County Deed Book "D," page 211, 16 December 1845. Thomas McCabe to Jacob Barringer, 500 acres, more or less, on Burks Fork.

D. B. "G," page 391, 1 April 1854. Jacob Barringer to Jackson W. Harris, 9½ acres land on Burks Fork Creek. *This deed does not convey the mill property, but the bounds described therein pinpoint the location of the Jacob Barringer Mill.*

D. B. "J," page 224, 23 July 1857. Jacob Barringer and wife, Araminta, to John A. Barringer (their son), 125½ acres and mill on Burks Fork Creek.

D. B. "Y," page 95, 15 February 1896. J. W. Harris and wife, Sarah, to John Sowers, 1¼ acres land on Burks Fork Creek, this being the old mill pond tract.

---

44. Jackson W. Harris Mill/ Martin Harman Mill

Burks Fork Creek

LOCATION: Route 657 (Weddle Road SE) approximately .2 mile west of Route 799 (Conner Grove Road SW); POWER: Overshot wheel (1880), turbine?, wooden overshot wheel, gasoline engine; STONES: 2 sets; VISIBLE REMAINS: Dug race on west/south side of Burks Fork

The Martin Harman Mill had been serving as a cattle shed for many years by 1992. The end of the dug race is visible in the center of the left side of the picture. (Frank Webb photograph, August 1991)

History. This mill is one of three (the others are IV, 43 and IV, 45) known to have stood within the bounds of a 500-acre tract bought by Jacob Barringer in 1845. Jackson W. Harris, the elder (1818–1864), bought the land on which this mill was built from Barringer and his wife, Araminta, in 1855, possibly specifically for that purpose. The deed conveyed a narrow, one and one-half acre strip on the north or east bank of Burks Fork and a fourteen-acre tract on the south or west side of the creek. Exactly where the tracts lay in relation to each other is not described in the deed, nor is their relationship clarified by comparison of plat drawings. Taken together, the two tracts may or may not have provided the land Harris needed on both sides of the creek for a dam and a mill pond a suitable distance upstream from the mill. If he did own the land on both banks, Harris may have erected a mill on this land before his death in 1864. At least one subsequent deed does seem to indicate that the mill was built on the one and one-half-acre tract bought in 1855.

The more likely scenario, based on comparison of the bounds of all deeds and the course of Burks Fork, is that J.W. Harris (Jackson Harris, the younger) built the mill on a small tract (described in one instance as two acres and thirty poles, and in others as one and one-half acres) carved out of the fourteen-acre tract and combined with a small square of land on the north bank of Burks Fork. This last would have been possible only after 1878, when Harris bought land on the north bank of Burks Fork that had belonged to his deceased father-in-law, John Barringer.

The earliest confirmation of this mill's presence is an 1882 deed recording the sale by Jackson W. Harris the younger (1849–1923) of a one-half interest in the mill and a lot of two acres and thirty poles to Archeleus Slusher. As described in the deed from Harris to Slusher, the lot seems to encompass part of the fourteen acres and a small square of a land on the opposite (north or west) bank of Burks Fork. This is probably the Jackson Harris Mill described in the *Products of Industry* report for 1880, but there is a slight possibility that that report refers to the nearby John Barringer Mill (IV, 45) which may still have been in operation, and which Harris would have acquired along with the John Barringer land in 1887.

In 1887 Harris and Slusher added a third partner, Catherine Spangler, after which the mill was known as the Harris, Slusher, and Spangler Mill. Harris and Slusher sold their respective interests in 1888 and 1891, leaving the mill in the hands of John Sowers (one-third interest) and Catherine and G. W. Spangler (two-thirds interest). Between 1891 and 1908 John Sowers or Mathias Hylton, or the two in partnership, acquired the Spanglers' two-thirds interest, but the deed or deeds showing exactly how this was accomplished were not found.

Through Sowers' connection to the mill, an additional downstream tract of one and one-fourth acres was added to the two south bank tracts that were typically sold along with the mill. Referred to as the "old mill pond tract," it is believed to be the site of the Jacob Barringer mill (IV, 43). The spot is marked by a massive natural rock formation in Burks Fork Creek, very near the point at which state maintenance of Route 657 (Weddle Road SW) ends (2010). A flat piece of land enclosed by a bend immediately upstream from the rock formation was formed by silt settling in the mill pond and was for many years used as a potato patch.

A column called "Burks Fork Items," submitted to the *Floyd Press* by "A. Pink," reported on April 9, 1903, that "quite a severe hail storm visited our section Friday night, blowing over H. Hylton's mill house." Among the Burks Fork Mills, none was owned by an "H. Hylton" in 1903, but Hezekiah Hylton (1869–1945), who was later an owner of this mill, may have been connected to it at that time through his father, Mathias Hylton (1836–1925). Hardin P. Hylton owned part or all of several Willis and Floyd area mills around 1900, as well.

The mill was last run in the late 1930s, possibly by Hezekiah "Kiar" Hylton, who sold his one-third interest to Martin Harman in 1936. Harman had bought the interest of the third partner, his brother Posie, in 1923. Harman's grandson, Ronald Weddle, recalled that a wooden overshot wheel and wooden race leading directly to it from the end of the dug race were still in place in the 1940s. The race was taken down after a horse found its way into the race and, unable to turn around, had walked all the way to the wheel before falling to its death. Weddle also recalled having unearthed and sold as scrap a device he had taken to be a turbine water wheel.[23] Oddly, a couple of 19th century deeds mention "an old turbine wheel" as part of the transferred property. Although historically older and technologically more primitive, overshot water wheels replaced turbines at several sites where the water supply was inadequate for a turbine, whose great power and efficiency declined rapidly at less than full flow. So if this was in fact a turbine, it could have been used at this mill, or merely salvaged from one of the two nearby Barringer mills that predated the Martin Harman Mill.

When videotaped in 1995 the mill building had been converted to use as a barn. The wooden floor of the lower story had been removed. One grain cleaner, some spouts, and the stairs from the second story to attic remained. Square holes in the floor of the second story marked the former locations of spouts and elevators.

Deeds

Floyd County Deed Book "H," page 335, 24 November 1855. Jacob Barringer and Araminta, his wife, to Jackson Harris, one tract of 1½ acres on Burks Fork and a second tract of 14 acres on Burk's Fork Creek. *A reference to "the mill pond branch" in the description of the second tract is believed to pertain to the Jacob Barringer Mill. At some point between the date of this deed and the next, ownership of this property had to pass from Jackson W. Harris, the elder, to his son, also Jackson W. Harris. The elder man died in 1864, and the younger, born in 1849, was only six years old in 1855.*

D. B. "O," page 699, 8 June 1878 (See IV, 45).

D. B. "Q," page 588, 25 August 1882. J. W. Harris to Arch Slusher, "a certain tract or parcel of land containing 2 acres and 30 poles of which one half the said Harris deeds to the said Slusher for the sums above [$600] mentioned with all water privalages for the mill."

D. B. "T," page 32, 17 January 1887. J. W. Harris and wife and Archelous Slusher to Catherine Spangler, one-third interest in land of 1½ acres plus or minus, "including grist mill thereon situated and old turbine water wheel together with all the machinery and fixtures thereto belonging." Also, J. W. Harris to A. Slusher and Catherine Spangler, one-third interest each in parcel of land containing 12½ acres on Burks Fork Creek.

D. B. "T," page 177, 19 March 1887. Weeks, Helms and Company to Arch Slusher, J. W. Harris and Catherine Spangler, three lots on Burks Fork Creek, Lot No. 3 containing 23 poles, Lot No. 4 containing ? poles, and Lot No. 5 containing 2 rods and 7¾ poles.

D. B. "T," page 324, 20 April 1888. J. W. Harris to John Sowers, his one-third interest in three lots deeded to A. Slusher, J. W. Harris and Catherine Spangler by Weeks, Helms and Company. Also the mill and mill lot known as the Slusher, Harris, and Spangler Mill, containing 14 acres plus or minus on Burks Fork Creek.

D. B. "W," page 199, 20 February 1891. Arch Slusher to Catherine and G. W. Spangler, his one-third interest in mill and two lots on Burks Fork Creek, containing 14 acres.

D. B. "Y," page 95, 15 February 1896. J. W. Harris and wife Sarah to John Sowers, 1¼ acres land on Burk's

Fork Creek, this being the old mill pond tract. *No deed found from G. W. and Catherine Spangler or to M. F. Hylton.*

D. B. 35, page 150, 19 August 1908. John Sowers and wife Celia, Jonas Harman and Lavinia Harman, and M. F. Hylton and wife Ellen to Hezekiah Hylton, Posey Harman, and Martin L. Harman, mill, land, and water rights, containing 14 acres on Burks Fork Creek known as the Slusher, Harris and Spangler Mill. Also, 1¼ acre below the mill, being the old mill pond land owned by John Sowers. "Jonas Harman and wife joins the deed for no other purpose than to make good the water way and 1/8 of an acre more or less of land and the said MF Hylton does not warrant the above named 1¼ acres of land it being John Sowers individually."

D. B. 45, page 245, 28 April 1923. Posie Harman and wife Lala to Martin L. Harman, his one-fourth interest in mill and 14 acres, and 1¼ acres and ⅛ acre land.

D. B. 53, page 303, 23 June 1936. Hezekiah Hylton to M. L. Harman, his ½ interest in land and mill on Burks Fork Creek.

See Table C, *1880 Products of Industry.* Name of corporation or individual: Jackson Harris

Though it is unlikely for several reasons, this information could pertain to the John Barringer/Burks Fork Union Mill (IV, 45). Harris had bought a half interest in John Barringer's mill in 1875 and acquired the remainder, with or without an operable mill on it, in 1878.

45. John Barringer Mill/Burks Fork Union Mill

Hungate Branch/Wade Branch (Tributary of Burks Fork)

LOCATION: Probably near the mouth of a Burks Fork tributary apparently once known as the Wade or Waid Branch or Hungate Branch, near where it is crossed by Route 657 (Weddle Road SW) a short distance from that road's intersection with Route 799 (Conner Grove Road SW); POWER: Overshot wheel (1870); STONES: 2 Sets (1870); VISIBLE REMAINS: None

History. John Barringers's "Burks Fork Union Mill" was the second of three mills built along a three-quarter-mile stretch of Burks Fork. The mills were almost in sight of each other, and were apparently constructed within a span of less than 40 years. All three stood on land that had been part of a 500-acre tract bought by John Barringer's father, Jacob, from Thomas McCabe in 1845. The original Barringer tract was crossed near its southern end by Burks Fork, in a northward-arcing east-west line. Most of the portion lying south of the creek was sold to Jackson Harris in three tracts, the first (9 acres) in 1854. Conveyed in 1855, the last two (1.2 and 14 acres) are important to the history of the J. W. Harris/Martin Harman Mill (IV, 44). The elongated shape of the smaller, upstream tract suggests that Harris had in mind all along to build a mill, but there is no record of a mill on the south bank until 1878, by which time Jackson Harris had died and the land passed to his son, also named Jackson W. Harris, but referred to, at least in deed book indexes, as J. W. Harris.

In 1857 John Barringer bought, in a single tract of 125.5 acres, that part of his father's original tract that lay immediately to the north of Burks Fork. It encompassed land on both sides of what is now Route 657 (Weddle Road) from its intersection with Route 799 (Conner Grove Road) westward to the point at which Route 657 turns sharply north and state maintenance of the road now ends. A mill presumably constructed by or for Jacob Barringer (IV, 43), but referred to only as "the mill," was standing near that point, between the road and Burks Fork, when the deed was written. Whether or for how long John Barringer operated his father's mill is unknown, but it was presumably dismantled or abandoned before 1868 when the younger Barringer is supposed to have completed a mill of his own on a large tributary that joined Burks Fork between the old mill and the Burks Fork Road. Modern maps give this stream no name; locals refer to it as Chestnut Branch. But a hand-drawn map (on a napkin) with information attributed to "Mrs. Hodges" names the stream "Hungate Branch" and identifies it as the site of a Barringer Mill.

The name "Wade's Branch" has also been applied to what is believed to be the same stream. On July 2, 1868, Henry Slusher, who himself had a mill (IV, 42) a short distance up Burks Fork, wrote to his brother Stephen in Jackson County, Kansas, that "John Barringer has built a grist mill. It is overshot and he brings the water from Waid Branch around the hill side from above his house. It is nearly ready for running."[24] Richard

Hungate and Owen Wade both owned land on Burks Fork before 1850, but no effort has been made to place the holdings of either man within the watershed of this stream. This information, or the pinpointing of the site of John Barringer's home, could confirm or demand the rethinking of the conclusion that Barringer's mill stood by this stream.

The John Barringer Mill is described in the *Products of Industry* report for 1870 as having an overshot water wheel and two runs of stones. By the time the next report was compiled in 1880, John Barringer was dead and his son-in-law Jackson (J.W.) Harris owned the mill property. Harris is named as the owner of a mill in the report compiled for 1880, but the enumerator's placement of Harris's mill on Burks Fork itself rather than a tributary, and, more importantly, the listing of only one set of stones instead of two, suggest that he was describing a different mill. Even if that is the case, the absence of Barringer's mill from the 1880 report does not, by itself, prove that the mill had ceased to operate, since, according to column headings in the *Products of Industry* reports for 1870 and 1880, only mills producing over $500 worth of products annually were supposed to be listed and described.

That grinding had ceased at the Burks Fork Union mill is more strongly implied by the mere fact that Jackson Harris almost certainly did by that time own another operating mill only a short distance down Burks Fork. He probably would not have operated two mills so close together. Had there been sufficient trade for two mills, it would have made more sense to make the new mill, across and a short distance down Burks Fork, larger, and save the cost of labor and upkeep for two separate mills. The assumption that the mill Harris owned on the south side of the creek had at this time just been built is based on the 1882 date of the first deed book reference to a mill on the land purchased by Harris's father in 1855. It may be significant also that it was not until 1878, 23 years later, that the land on both sides of the stream (essential for building a dam across Burks Fork and flooding the land above it on both sides of the creek) was finally in the hands of one owner. The unlikely discovery of a court order for a *writ of ad quod damnum* issued to either of the Harrises, father or son, between 1855 and 1882 would resolve several questions.

As regards the fate of John Barringer's Burks Fork Union Mill, as noted Harris acquired a one-half interest in the mill from Barringer, his father-in-law, in 1875. No water rights were included in the deed to the mill and a tract of "one rood and twenty poles" (a little over a quarter of an acre) on which it stood. No record of the disposition of the other one-half interest was found. Harris would naturally have acquired the remaining interest when he bought all the late John Barringer's 125-acre tract at auction in 1878, but the deed includes no mention of that fact, which leaves open the unlikely possibility of other explanations. A photograph of a building standing close by Route 799 north of Weddle Road, taken or collected by Frank Webb, is labeled as the "Burks Fork Union Mill," but the fact that he stated no location for a mill of that name or a John Barringer mill in his final notes makes that identification questionable. Locals have identified the building as Harris's store.

Deeds

Floyd County Deed Book "J," page 224, 23 July 1857. Jacob and Arminta Barringer to John Barringer, 125.5 acres and (Jacob Barringer) mill.

D. B. "N," page 545, 21 May 1875, John A. Barringer to Jackson W. Harris, one-half interest in "Burk's Fork Union Mill."

D. B. "O," page 699, 8 June 1878, J. A. Barringer, deceased, to Jackson W. Harris, 126 acres. Tripthena Barringer (John Barringer's widow) to have dower rights to 40 acres for her lifetime. Owner of tract containing the mill or the dower tract is to have use of spring near the boundary between the two.

See Table B, *1870 Products of Industry.* Name of corporation or individual: Barringer, John A.

---

46. William Alderman Mill/ Shepherd Alderman Mill

Burks Fork Creek

LOCATION: Route 657 (Weddle Road SW), past end of state maintenance. East side of Burks Fork Creek, between road and creek, near the Shepherd Alderman house. VISIBLE

REMAINS: Part of stone dam abutment on east side of creek, near the Shepherd Alderman house

History. One of the first owners of this mill was Hardin Hylton, who bought the sixteen and one-half acre tract on which it stood from brothers-in-law Jackson W. Harris and Abram Harris in 1878. Jackson W. Harris or his father, also Jackson Harris, probably built the Martin Harman Mill (IV, 44) not far upstream. Hylton himself had already built and still owned what would later be known as the Willis Mill on Greasy Creek (III, 69).

The 1878 deed to the Burks Fork acreage does not mention a mill building, but one of the upstream boundary lines crossed "the head of a mill pond," indicating that a mill dam and pond, at least, were in place on the property. References to "the head of the mill pond" in later deeds that do mention the mill may encourage the supposition that the mill as well as the dam and pond was already there when Hylton bought the land from the Harrises. On the other hand, the difference between the 1878 selling price of $100 and the $600 that William Alderman's wife, Sabra Ann Alderman, paid to Hylton in 1888 for "the mill tract" suggests that Hylton built the mill or had it built within that ten year period. A real estate boom (if there was one in that time period) could hardly account for six-fold increase in the value of a mill property.

Hylton may have replaced or refurbished a mill that had been destroyed or damaged, which would explain the presence of the dam and pond. A less likely but still plausible solution is that the Harrises had planned to build a mill on this land, built the dam and pond, then sold the land to Hylton before building the mill. As with so many other mills, there are enough facts to encourage all sorts of speculation, but too few to support firm conclusions. Alderman family history says that Joel Weddle (1843–1936), the owner of an adjoining piece of land, actually built the mill. A man born in 1843 certainly could have built a mill for Hylton after 1878 but probably not much earlier.

The mill was owned by Aldermans, with one exception, from 1888 until it closed. William Alderman (1830–1915) ran both the mill and a blacksmith shop but for some reason does not appear as an owner on any deed, except as a party to a chancery suit that resulted in a public auction of the mill property in 1889, the year after the Aldermans acquired it from Hylton. William's wife, Sabra Ann Barringer Alderman (1836–1910), the daughter of Burks Fork miller Jacob Barringer, bought the mill and land, again, at the commissioner's sale with her son Joseph as an equal partner and is consistently listed as the owner alone or in combination with her brother-in-law or one or the other of their sons, Joseph and Shepherd.

William's brother, Thomas, bought a half-interest in the mill and five-plus acres from Sabra and Joseph in 1890 and sold it to C. C. Weeks in 1892. Weeks sold it to Shepherd Alderman in 1906. A blacksmith and a musician as well as a miller, "Shep" Alderman lived in a house which still stands across Burks Fork from the mill site. He was the last to run the mill and his name and William's are the most firmly associated with it. Sabra Alderman deeded her portion of the one-half interest she and Joseph still owned together to Shep in 1906. Ten years later he became the sole owner after paying Joseph $100 for the remaining interest.

One account of the Aldermans and related families has it that in one severe winter, when most of the other mills along Burks Fork were disabled by ice, the Alderman Mill ran continuously for several days. William operated the mill in daylight hours and Shepherd ran it at night to keep the community supplied with corn meal.[25] A photograph of Shepherd Alderman and additional biographical information are included in the source just cited. Exactly how long before his death in 1950 Alderman stopped running his mill is unknown.

There was also a sawmill at this site. Though hard to picture now, both mills were on the east side of Burks Fork Creek, between the road and the stream. No longer maintained by the state or county, Route 657 once followed Burks Fork to where the creek is crossed by Route 786 (Boone Road SW), near the site of the William C. Potter Mill (IV, 47).

Deeds

Floyd County Deed Book "O," page 625, 26 March 1878. Jackson W. Harris and wife, Sarah, and Abram Harris and wife, Sarah, to Hardin P. Hylton for $100, 16½ acres more or less on Burk's Fork Creek. "Beginning at two chestnuts near a small branch ... crossing the head of a mill pond."

D. B. "X," page 55, 16 November 1888. H. P. Hylton and Francis, his wife, to Sabra Ann Alderman, 16½ acres on Burks Fork "known as the mill tract" and the mill thereon. "Adjoining the lands of Joel Weddle, Nathaniel Keith, and Jacob Alterman." $600.

D.B. "U," page 297, 30 October 1889. Z. T. Dobyns, Commissioner in cause of *John W. Hooker vs. Wm. Alderman et al.*, to Joseph Alterman and Sabra Alterman, 16½ acres known as the mill tract.

D. B. "U," page 643, 14 July 1890. Joseph A. Alterman and Sabra Ann Alterman to Thomas Alterman, ½ interest in 5 acres, 2 roods and 29 poles, "together with its water power and all its appurtenances."

D. B. "W," page 111, 2 December 1892. Thomas Alterman and Elizabeth Alterman, his wife, to C. C. Weeks, "our entire interest which is one half interest in a certain tract or parcel of land on Burks Fork containing five acres more or less with its apertainances the other half is now owned by Joseph A. Alterman and bounded by the land of Elizabeth Keeth, Catharine Alterman, J W Harris, B E Maxey and Joseph A. Alterman and Sabra Ann Alterman."

D. B. 33, page 78, 27 October 1906. C. C. Weeks to S. A. Alderman, his ½ interest in William Alderman mill place containing 5 acres plus or minus.

D. B. 33, page 104, 16 November 1906. Sabra Ann Alderman to S. A. Alderman (son), her entire ½ interest in all her lands. She is to control until her death. Joseph A. Alterman (son) owns other ½ interest.

D. B. 43, page 264, 27 September 1916. Joseph A. Alderman and Lottie Alderman, his wife, to S. A. Alderman, no acreage given. On waters of Burks Fork, for $100.00, "their entire interest" in a tract "beginning at two chestnuts corner to Joel Weddle's land ... crossing the creek at the head of the mill pond." Refers back to DB "U," page 297 30 October 1889.

47. William C. Potter Mill

Burks Fork Creek

LOCATION: Route 786 (Boone Road SW) just upstream from low water bridge; POWER: Unknown; STONES: Unknown; VISIBLE REMAINS: Dug race on north side of the creek, lower end obscured by farm road

History. When the United States census was taken in 1860, William Columbus Potter was a farmer and blacksmith living in Carroll County, Virginia. Ten years later, he was a physician living in Floyd County. Few people have improved their situations as much in only ten years.

Potter's mill was built on land he bought in 1875 and was probably erected soon after he obtained a deeded right of way for his race from Abram Harris in 1876. In 1894 William C. Potter's son Flournoy Bishop Potter and C. G. Vaughn bought the mill tract of 55 poles (about ⅓ acre) from W. C. Potter's widow and other heirs. Another son, Thomas Henry "Harry" Potter, bought his brother's one-half interest in 1903. C.G. Vaughn, whose half interest in the Potter Mill hasn't been accounted for, is probably Christopher Green Vaughn, a prominent figure in the history of the Vaughn-Dunn woolen mill, a short way down Burks Fork. Vaughn's and Potter's shared ownership of a tract too small for most purposes suggests that the site was still thought to have business potential, whether or not the mill was in place and operating. Two years later, T. H. Potter began selling off his father's lands. Of several deeds recorded in Floyd County Deed Books 32, 33, 35, 44, and 47 between T.H. Potter and various buyers, only the first, through which Potter conveyed 80 acres to Amos Hylton in October 1905, mentions a mill race. None of these deeds mention the mill itself, or the rights to the water required to run it. T. H. Potter was living in Botetourt County with his second wife and twelve children, six from each marriage, in 1910.

Deeds

Floyd County Deed Book "P," page 177, 20 February 1875. Frederick Johnston, administrator of Charles L. Snyder, deceased, to William C. Potter, land lying in Burks Fork, 270½ acres.

D. B. "W," page 191, 25 November 1876. Abram Harris to W. C. Potter, agreement for water race through land to W. C. Potter mill. Price to be agreed upon when race is complete. If not, "leave the damage to two disinterested men with privilege of calling on the third if they cannot agree."

D. B. "U," page 511–515, 30 December 1889. J. L. Thompkins to Julia A. Potter, wife of W.C. Potter, dower land, mill lot number 9, containing 55 poles.

D. B. "X," page 309, 16 June 1894. Julia A. Potter and heirs in law of W. C. Potter to F.B. Potter and C. G. Vaughn, 55 poles land and mill tract of W. C. Potter. *No deed found for C. G. Vaughn's ½ interest in the mill tract.*

D. B. 30, page 136, 13 March 1903. F. B. Potter to T. H. (Thomas Henry "Harry") Potter, ½ interest in 55 poles and mill tract of W. C. Potter and another tract of 63¾ acres.

D. B. 32, page 76, 28 October 1905. T. H. Potter and, Versie, his wife, to Amos? (Amon?) F. Hylton, $1,950, "beginning at a mill race by the creek."

48. Vaughn-Dunn Woolen Mill

Burks Fork Creek

LOCATION: Route 630 (Burks Fork Road SW), north of Route 732 (Buffalo Springs Lane SW); POWER: Turbine, then overshot water wheel; STONES: 1 set. Sawmill. Machinery for wool carding, spinning, and weaving. VISIBLE REMAINS: None

History. The Vaughn-Dunn mill is the best known and longest-lived of the Floyd County businesses that used water power to process locally grown wool. The earliest and smallest of these provided only the service of cleaning and carding wool for customers to spin, knit, and weave at home, or to use as quilt batting. More sophisticated woolen mills, like the Vaughn-Dunn mill and Beaufort Howell's mill (III, 138), could clean yarn more thoroughly using mechanical "picking" machines before it was carded, and were also equipped to spin the carded wool into yarn and to dye it. Only the Howell mill and the Vaughn-Dunn mill are known to have installed the power-driven looms that turned yarn into finished blankets and coverlets on the spot. Members of the Vaughn and Dunn families, as well as collectors in and outside the county, have blankets and coverlets made at this mill before it closed in the 1940s.

The seventy-five-year history of this business began when Wilson T. Vaughn bought 250 acres plus an additional one-acre tract from Bryant Hylton in 1861. "At the close of the War Between the States, Vaughn set up a roll carding machine and wool picker in a little building on the property. Power to run the machines was provided by a turbine wheel which in turn was motivated by water running through an ancient mill race previously used to run a grist mill."[26] In 1880 W. T. Vaughn sold 105 acres to Thomas A. Vaughn, but reserved for himself one-half acre, with a carding machine, mill, and water rights. Three years later, Wilson Vaughn sold to Thomas A. Vaughn, Columbus Penn ("Lum") Vaughn, and Christopher "Green" Vaughn a tract of eighty poles, including the rights he had retained in the 1880 deed between himself and T.A. Vaughn. The 1883 deed mentions an "old machine house," a saw mill, carding machine, mill, and a "new [machine] house." In 1890 T. A. deeded his interest in a "certain tract of land and wool factory, carding machine, wooler, saw mill and water power and all machines attached" to C.G. and C.P. Vaughn.

On February 15, 1902, a page 2 item in the *Floyd Press* noted that "Vaughn Bros. and Co., proprietors of Vaughn's Woolen Mill, of Vaughn, Va. ... is now in the weaving business and doing good work." In May of the same year the brothers installed a new water wheel and a governor to steady the motion of the machinery and minimize thread breakage. They had by then employed as foreman J. T. (Joseph Thomas) Dunn, who in 1903 bought from the Vaughns one-half interest in an enterprise then described as "mill, wool carding factory, machines, sawmill, and gristmill."[27]

According to J.T. Dunn's son, Angus, his father was brought in to set up new machinery and fell in love with Columbus Penn Vaughn's daughter, Allie. Dunn, a native of Prince William County, VA, was 30 and Allie Vaughn 19 when they married in 1902. Dunn purchased the remaining half interest in the mill, machinery, water rights, and half acre of land from the Vaughns in 1919. The mill continued to be known by many local people as Vaughn's Woolen Mill for as long as it ran, despite the change in ownership.

At the peak of its activity, probably in the 1920s, this processing and manufacturing plant employed as many as ten local women and girls plus Dunn, his sons and daughters, and occasionally his wife. In so doing, it provided the county's nearest approximation, and a very distant one at that, of the huge, water-powered textile mills that industrialized many of the river cities of New England and later the Piedmont towns of Southside Virginia and the Carolinas. The work was light enough for women and more like traditionally domestic occupations than were grain milling and most other early nonfarm occupations

But as with every industry that ever budded, bloomed, or died on the vine anywhere, geographic and economic realities had the last word despite the industry and ingenuity of the owners. By the beginning of the 20th century the biggest demand in textiles locally, nationally, and worldwide was for inexpensive, brightly-printed cotton. Floyd County's growing season is too short for cotton; lack of access to water or rail transportation and inadequate sources of power made it hopelessly impractical to import and process cotton either to be sold here or sent back out of the county. Demand for the wool, linen, and wool-mix fabrics relied upon by pioneer families, and suitable for production in limited amounts in Floyd County, was declining, thanks to cheaper and cheaper cotton goods. The demand for both wool and cotton would soon be further undermined by the development of synthetic fabrics.

J. T. Dunn and Buford Howell, who processed wool on Dodds Creek, were the only "millers" whose products, like this patterned woolen coverlet pictured here with J.T. Dunn, might still be found on the shelves of Floyd County homes. Overhead line shafts run the length of the building, transmitting power to machinery on both floors. Extra shuttles hang on the frame of the loom. (Frank Webb collection, undated)

As a result of these major shifts in the textile industry and the realities of geography, Floyd Countians were spared the discomforts and dangers as well as the economic confinement that typically accompanied single-industry company towns, payrolls, and stores. But had conditions been right for the textile industry to take root and thrive here, as did the garment industry for a brief few decades after mid-century, local people would no doubt have lined up for the same financial opportunities hundreds of other Floyd Countians sought in the mill towns, coal camps, and industrial centers within a few hundred miles in every direction.

To prevent wasted trips and hard feelings, the owners of the larger grain mills, whose customers might come from several miles away, regularly announced planned closings and new features and services in the local newspaper. The processing of wool was not as time sensitive as milling grain and grinding feed, and customers could probably not expect same day service in most cases. But since they might be coming from anywhere in the county or from adjoining counties and not get news, or at least not the correct version of it, by word of mouth, Dunn sometimes posted notices and advertisements in the *Floyd Press.* On August 17, 1922, he wrote to squelch a false rumor about his business and to explain delays in orders already placed.

WOOL GROWERS ATTENTION

It has been reported that the Vaughn Woolen Mill has been sold and is not running. I wish to say that the above statement is false. The Vaughn Woolen Mill is running and is receiving wool every day. Have about caught up with last year's work. I am doing the very best I can for the people. So send in your wool and I will return your work as promptly as possible.

Thanking you for your patronage, I am
Yours truly,
J. T. DUNN, Proprietor

Ten years later he was still at work and, hopefully, more or less caught up, when this advertisement appeared on page 4 on April 20, 1933:

WOOL

Send me your wool to be made up as follows:

10 lbs. of washed wool will make one pair of all wool blankets, weighing about 8 lbs. to the pair, charge $3.00 for making.

8 lbs. of washed wool will make one pair of cotton warp blankets, weighing about 6 lbs. to the pair, charge $2.00 for making

All blankets measure or size to order 72 × 84 inches.

10 lbs. of washed wool will make nine pounds of yarn, for either stockings or sweaters, charge for making 12cts per lb. for single yarn, and 18cts per

J. T Dunn's building and business both declined as Floyd Countians came to depend on stores and catalogs for fabrics and clothing that had once been manufactured, then worn and worn out, within a few miles of home. But the new roof on the old part of the mill means that there was still at least some carding, spinning, dyeing, and weaving to be done. (Frank Webb collection, taken after 1937)

lb. for 2 or more ply. Unwashed wool will lose from 25 to 60 per cent in weight in washing.

Will take wool at Market Prices for above charge.

TERMS: — Collect on Delivery. Customers to pay transportation charges.

Address all communications to
J. T. DUNN,
Willis, Va.

By 1942 Dunn was 70 years old, his sons were in service or had chosen other careers, and the booming war economy was bringing "store bought" consumer goods within the financial reach of people who had historically been compelled to be nearly self-sufficient. But he wasn't quite ready to give it up, and having waited until the spring shearing was over, posted this brief notice in the June 25, 1942, issue of the *Floyd Press*: "WOOL. Send me your WOOL, to make into BLANKETS, COVERLETS and knitting yarn. Prices Reasonable, J. T. DUNN, Willis, Va."

In 1948 Dunn sold everything to Elijah H. Vaughn, who tore down the mill and sold all the machinery for scrap except the steel overshot water wheel and the wool carding machinery. The wheel went to a restaurant on Route 52 between Hillsville and Cana, VA, and the wool carding machinery was bought by a mill in Mount Airy, North Carolina. When the grist mill and sawmill stopped running is unknown. Both were apparently sideline operations after Dunn took over. While still a partner in this mill in 1894, C. G. Vaughn bought a one-half interest in the W. C. Potter mill (IV, 47) just up Burks Fork, but how actively he was involved and when he sold that interest are unknown.

Short, unpublished histories of the Vaughn-Dunn woolen mill have been written by Catherine Vaughn Pauley and Oneida Dunn Banks. Pauley's account focuses on the early history and the role of the Vaughns; Banks explains how Tom Dunn came to be in Floyd County and outlines the processes of washing, carding, dying, spinning, and weaving the locally grown wool. On July 1, 1937, the *Floyd Press* featured a rather flowery history and description of the operation titled "Textile Plant Owned by J. T. Dunn; Coverlets Made to Please the Purchaser." One week later, on July 8, the *Press* printed pictures of the dyeing vat and wooden race, the mill, and a finished coverlet under the heading "An Ancient Textile Plant in this County."

Deeds

Floyd County Deed Book "L," page 510, 20 February, 1861. Bryant Hylton to W. T. Vaughn, 250 acres plus one acre tract.

D. B. "P," page 450, 10 March 1880. Wilson T. Vaughn to Thomas A. Vaughn, 105 acres, except one-half acre, including carding machine, mill, water rights.

D. B. "R," page 148, 15 March 1883, Wilson Thomas Vaughn to Columbus Penn Vaughn, Thomas Everett Vaughn, and Christopher Green Vaughn, 80 poles, includes water power. This land and water power was reserved by W. T. Vaughn in deed to T. A. Vaughn, "whereby except the old machine house ½ saw mill, carding machine and mill the right to use them in the new house as long as we want."

D. B. "V," page 194, 28 July 1890, T. A. (E?) Vaughn [and wife?] to C. G. and C. P. Vaughn, their entire interest in certain tract of land and wool factory, carding machine, wooler, sawmill and water power and all machines attached.

D. B. 30, page 73, 25 May 1903. C. P. Vaughn and C. G. Vaughn to J. T. Dunn, one-half interest in mill, wool carding factory, machines, sawmill and gristmill.

D. B. 43, page 493, 18 April 1919. C. P. Vaughn to J. T. Dunn, one-half interest in wool carding factory, saw and grist mill, one-half acre and all water rights.

D. B. 52, page 615, 13 March 1935. J. T. Dunn to Angus Dunn, two tracts, on a tract of one acre and one of 50¼ acres.

D. B. 54, page 3, 26 October 1936. Angus Dunn to J. T. Dunn, same two tracts.

D. B. 72, page 453, 29 May 1948. J. T. Dunn to Elijah H. Vaughn, one-half acre, locally known as the Vaughn Woolen Mill with all machinery, grist mill and water rights.

49. Alfred O'Neal Mill/Elgin Bolt Mill

Burks Fork Creek

LOCATION: Short distance upstream from Route 625 (Bolt Mill Road SW) approximately .5 mile west of Route 758 (Buffalo Mtn. Road SW); POWER: O'Neal Mill, tub wheel (1870, 1880); Bolt Mill, two turbines; STONES: O'Neal Mill, 1 set, sawmill, wool carding machine. Bolt Mill, 2 sets and Midget Marvel self-contained roller mill. VISIBLE REMAINS: One concrete abutment of Bolt Mill dam.

History. Alfred M. O'Neal (1809–1898) was living in Carroll County, Virginia, when he began buying land along Burks Fork on the western edge of Floyd County in 1843. He eventually acquired several hundred acres and by 1870 (possibly much earlier) was grinding corn, rye, wheat, and buckwheat, and carding wool using water power. O'Neal's mill was the farthest downstream of at least fourteen Floyd County mills on Burks Fork.

If the Carroll County portion of the stream is as well suited for mills, there must have been several between O'Neal's and the meeting of Burks Fork and Big Reed Island Creek about five miles, in a direct line, southwest of the Floyd/Carroll line. The distance is easily twice that following the course of the stream. Big Reed Island Creek drains much of western Carroll County before emptying into the New River about a mile southwest of the Pulaski County, Virginia, community of Allisonia.

None of the deeds made to O'Neal mention a mill, but as may be seen in this history and numerous others, this cannot be taken as proof that O'Neal was the first to build a mill here. Neither is it known when O'Neal began operating his mill. His son Isaac appears in the 1860 population census as a single 22-year-old miller, but a reference to "Greasy Creek" may place him at another mill on the western side of Floyd County. The 1870 *Products of Industry* report for Floyd County records that Alfred M. O'Neal owned one double wool carding machine as well as a grist mill; his son Isaac gave his occupation as "wool carder" in the 1870 population census.

The 1880 *Products of Industry* report noted a ten horsepower tub wheel and one run of stones at this site. Six years later, O'Neal, by then in his late 70s, sold the mill property and his dwelling house to Isaac. The size of the parcel and the terms of the sale, which include lifetime maintenance for the elder O'Neal and his wife and a future payment to a sister, are similar to those set forth in a deed made at about the same time to L. T. "Landy" O'Neal, who is assumed to be Leander O'Neal, another of Alfred O'Neal's sons. In the deed to I.W. (Isaac) O'Neal, A. M. O'Neal and L. T. (Leander "Landy") each retained a one-third interest in a sawmill on the site, but there is no mention of a grist or wool carding mill.

In 1899 Isaac O'Neal sold essentially the same property, with slightly reduced acreage, to his son and daughter-in-law, John K. and Nancy Bolt O'Neal. There is no mention of a mill in the deed, but John's younger brother, Louis, had described himself as a 19-year-old single miller when his marriage was recorded in 1896. Whether or not the grist mill was still in operation, it is likely that wool carding did cease around this time, given that the Vaughn Woolen Mill (IV, 48) the largest and most versatile woolen mill anywhere in the county, was growing steadily only a few miles up the creek and would soon be under the ambitious guidance of J. T. Dunn.

J. K. and Nancy O'Neal, who had married at the ages of sixteen and fourteen and were still in their mid-twenties in 1901, had owned the property for less than two years when creditors forced a sale, at which the mill property was bought by W. E. Weeks. Again, no mill was mentioned, nor was there reference to a mill in 1904 when C. F. Cain bought the same property from W. E. and Laura Weeks. Julius "Elgin" Bolt (1880–1968), whose paternal grandmother was

The O'Neal/Bolt Mill served customers in Floyd and Carroll Counties for at least eighty years and was the last of more than a dozen Floyd County mills on Burks Fork to grind grain. Expansion of the main structure resulted in uneven window spacing on the long side. The small flue in the shed roofed addition, as far as possible from the dusty milling floor, is typical. (Ricky Cox photograph, 1992)

Isaac O'Neal's sister, Martha, acquired the mill or at least the mill property in 1911 after C. F. Cain's financial woes led to another commissioner's sale. The highest bidder was one of Cain's creditors, C. K. Burnett, who sold the mill to Bolt in time to have the title assigned to him when the deed was made. The mill was again conspicuously absent from the deed, as it was when Bolt conveyed the title to his wife in 1931, in an apparent effort to fend off his own creditors. The mill is mentioned, finally, in a "Memorandum of Lis Pendens" filed with the Circuit Court of Floyd County by H. H. Nester in 1934. Nester sought to "set aside" the 1929 deed and force a sale in order to recover $1,058.77 owed to him since November 1931, plus the accrued interest. He was apparently unsuccessful, perhaps because Bolt had conveyed the mill property to his wife in August of 1931, and so did not legally own the mill when the debt to Nester was incurred. Nester's document is valuable, at least, for its description of property "upon which there is situated a dwelling house and large rolling mill, known as the Elgin Bolt rolling mill."

The lengthy absence of the mill from deed records could suggest that O'Neal's mill was at some point destroyed and replaced by a mill that was built or last owned by Elgin Bolt. But as has been observed multiple times, deeds frequently fail to mention mills, especially when they are situated on tracts as large as this one. The O'Neal mill was reported to have a single run or pair of stones in 1880, but the Bolt Mill had two sets, though neither was in place when it was videotaped in 1995. The second set, used to grind only buckwheat, was installed by O'Neal in a part of the mill that was added on to the original. The foundation and timbers of the 2½ story mill that last stood here showed plainly that a substantial addition had been put on the side facing what is now Bolt Mill Road. There was also an addition on the side opposite the creek. This cannot be confirmed, but it is likely that these additions were done to accommodate the extra machinery required by roller mills. Bolt is known to have added the Midget Marvel rolling mill and the accompanying cleaners, elevators, shafting, and spouts soon after he bought the mill, near the very peak of building or refitting mills for roller mill systems. The mill may have enjoyed its greatest prosperity during World War I, at which time it sometimes ran day and night to keep up with the demand for flour. Since demand within the county would, if anything, have fallen while servicemen were away, the implication is that local farmers or Bolt were selling surplus flour outside the county at inflated wartime prices.

Julius Elgin Bolt also did blacksmith work in a shop building that stood upstream from the mill, along what was the public road before a loop running beside the creek between the two Burks Fork bridges on Route 625 (Bolt Mill Road SW) was cut off. Before the easternmost bridge, just off Route 758 (Buffalo Mountain Road SW) was constructed, winter travelers sometimes drove wagons across the frozen head of the mill pond instead of fording the creek.

When the mill was videotaped in the summer of 1995 it was more or less fully equipped, minus the stones and turbines, but structurally unsound due to age and a bad roof. Nine elevators were in place, along with spouting, a receiving separator, and various cleaners. The top half-story (third story) was unfloored. The penstock, constructed of stacked 2 x 4 lumber, had once extended from the turbines to the dam thirty to forty feet upstream. The remains of an earlier dam, probably O'Neal's, could once be seen another fifteen to twenty feet upstream. The newer dam was probably built when the roller mill was installed, since millers regularly built new dams or raised existing ones to supply the extra water needed to power the cleaners, rollers, and sifters of the roller mill system. The wooden center portion of the dam washed out in the late 1940s. The mill ran part time, using a tractor for power, to grind for the family and for neighbors on weekends until sometime in the 1950s. Elgin Bolt's son Floyd bought his siblings' interest in the mill property in 1969, a little more than a year after their father's death. The mill was razed sometime after 1996. Sammie Phillips purchased some of the machinery from the mill, including one set of millstones.

In the 1980s, the current owner, Elgin Bolt's grandson, J. E. (Julius Elgin) Bolt, built a dam just below the second Route 625 bridge over Burks Fork Creek (downstream from the mill site), dug a race along the west side of the creek, and began producing hydroelectric power for his home, using first one of the turbines from the mill, and then a steel overshot wheel. The project, which is described in a 1981 *Roanoke Times and World News* article,[28] was abandoned and the dam destroyed after several years.

Another of Alfred O'Neal's sons, Benjamin Franklin O'Neal (1840–1931), ran a steam-powered corn mill in Indian Valley in the 1920s. Sherman Quesenberry moved the mill up the hill into a specially constructed building that still stands beside Route 755 (Slaughter Hill Road NW) and powered it with a salvaged automobile engine. Around 1945 Sherman's son Cecil moved the mill to a building on the northeast corner of Route 754 (Macks Mountain Road NW) and Route 619 (Horse Ridge Road NW). The stones, main shaft, and main pulley were salvaged when it was demolished in the 1980s.

Deeds and Other Public Records

Floyd County Deed Book "S," page 290, 6 April 1886. A. M. and Frances O'Neal to I. W. O'Neal, 124¾ acres. The sellers are "to have their maintenance from said land during the remainder of their lives" and I. W. O'Neal is to pay Rhoda Reynolds $50 if she is still alive when he comes into full possession of the land. "A. M. O'Neal and L. T. O'Neal to hold one third interest each in saw mill." Grist and carding mills not mentioned.

D. B. 26, page 154, 6 September 1899. Isaac O'Neal to J. K. and Nancy O'Neal, 114¾ acres land.

D. B. 28, page 6, 9 July 1901. J. K. and Nancy O'Neal (by Commissioners) to W. E. Weeks, 114¾ acres.

D. B. 31, page 362, 20 March 1904. W. E. Weeks to C. F. Cain, 114¾ acres.

D. B. 37, page 485, 17 March 1911, B. G. Howard and V. M. Sowder, Commissioners for *C. K. Burnett, et al., vs. C. F. Cain* to J. E. Bolt, 114¾ acres.

D. B. 49, page 472, 27 August 1931. J. Elgin Bolt to M. E. Bolt, his wife, 114¾ acres.

D. B. 51, page 224, 20 February 1934. *H. H. Nester vs. J. E. Bolt, M. E. Bolt, Gracie Alderman, Blue Ridge Bank, Inc., and Peoples Bank of Floyd*, memorandum of *lis pendens* asking court to "set aside" deed made by J. E. Bolt to wife, M. E. Bolt, D. B. 51, page 224, 114¾ acres land commonly known as the O'Neal homeplace, "upon which there is situated a dwelling house and a large rolling mill, known as the Elgin Bolt rolling mill." Seeking to recover $1,058.77 owed, with interest, from November 1, 1931.

D. B. 96, page 632, 25 April 1969. J. E. Bolt, deceased, to Floyd and Oka Bolt, heirs of J. E. Bolt, selling estate in two tracts. Tract 1, 94¾ acres. *Floyd Bolt bought the other 20 of the original 114¾ acres earlier, D.B. 92, page 519.*

See Table B, *1870 Products of Industry.* Name of corporation or individual: A. M. O'Neal & Co. Grist Mill. O'Neal is listed is a separate entry as the operator of a wool carding mill.

See Table C, *1880 Products of Industry.* Name of corporation or individual: O'Neal, Alford

50. George Bolt Mill

Chisolm Creek

LOCATION: Route 629 (Long Mountain Road SW) about 1.5 mile west of Route 758 (Buffalo Mountain Road SW); POWER: Turbine and later overshot wooden wheel. Steam engine. STONES: 1 set

History. The original mill had a turbine for power and was built by George Bolt, Noah Dodd, and Mac Moles. Moles and Dodd sold their respective one-third interests in the one-half-acre lot and mill back to Bolt in 1920. After the original dam broke it was replaced by a low dam, almost certainly farther upstream, and dug race. These supplied water to an overshot wheel built by Bolt and A. Newton Hylton. A steam engine that powered a sawmill at the same location was sometimes used for grinding.

Deeds

Floyd County Deed Book 47, page 144, 12 June 1909. H. L. Lawson to G. W. Bolt, 27½ acres on waters of Laurel Fork Creek.

D. B. 41, page 25, 5 June 1915. G. W. Bolt to Mack Moles, ⅓ of ½ acre land on Chisolm Creek and a grist mill thereon and said mill now belonging to N. M. Dodd, G. W. Bolt, and Mack Moles and water power for said mill.

D. B. 41, page 26, 5 June 1915. G. W. Bolt to N. M. Dodd, ⅓ of ½ acre land on Chisolm Creek and a grist mill situated thereon, said mill now belonging to N. M. Dodd, George W. Bolt, and Mack Moles, and also water power for said mill.

D. B. 44, page 188, 12 June 1920. N. M. Dodd and Mack Moles to G. W. Bolt, ½ acres land on Chisolm Creek and all interest on grist mill on same land.

D. B. 79, page 463, 28 December 1957. J. E. Proffitt, Commissioner in Cause of *Elmer Bolt vs. Junior Branscomb*, to S. G. Bolt, 36 acres plus or minus, being lands of George W. Bolt, deceased.

51. Logan's Mill

Laurel Fork Creek

LOCATION: Upstream from crossing of Route 603 (McAlexander Road SW) about 1.75 miles west of Route 632 (Shelor Road SW) VISIBLE REMAINS: None

History. Reporting from the vicinity of the Quoit, VA, post office, a correspondent calling him or herself "Who Next?" contributed the following brief item to the *Floyd Press*. It was published on November 27, 1902:

> LARGE HOGS!
>
> There are two hogs in this neighborhood vying for the prize. One of them owned by Mr. Zebram Keith, of near Union Church, which is certainly a very fine animal; the other one being fattened by Mr. Lloyd X. Sutphin, of Logan's Mill, on Laurel Fork, Floyd county. This last hog was raised by Mr. J. H. Hoback of the east end of the county near Stonewall. The grandparents, (Romeo and Juliette) of this hog were brought from Kentucky by Mr. Robert Logan, and were a registered Berkshire No. ___.

How active Robert Logan was in the management of a mill on his Floyd County property in 1902 is an open question since at one point he controlled more than 7,000 acres, mostly in Floyd County, and because he wasn't living in Floyd County, or Carroll or Patrick County (both of which join Floyd County near this mill) in 1890, 1900, or 1910. But his interest in hogs and his ownership of a mill may be more than coincidence, as may be the fact that Zebram Keith, whose prize hog was the other contender in this contest, was half-owner of the Henry Slusher Mill (IV, 42) on Burks Fork. Access to fresh, flowing water; spoiled or surplus grain; and inexpensive milling byproducts made the raising of hogs a logical and once popular sideline for millers. To minimize risks to the health of customers, not to mention insults to their noses, a state law was passed before 1800 requiring a mill owner who wished to raise hogs near his mill to employ for this purpose a tract of at least fifty acres separate from the mill tract.

Before the passage of this law, millers who kept hogs close around the mill may have thought along the same lines as the farmer who saved himself the work of building thirty feet of fence by using the back of his house as one side of his hog lot. He claimed to have done it so that his wife could accomplish one of her many chores, feeding the hogs, by merely stepping out onto the back porch. When his wife protested on the grounds that the smell would be unbearable, he allowed that since he had gotten used to it, the hogs probably could too.

Logan's Mill was also known as the Proffitt Mill.

Deeds

Floyd County Deed Book "O," page 258, 1 July 1876. C. W. Burwell to Martha D. (Burwell) Logan, 6,782 acres in Floyd County and 361 acres in Carroll County. This deed confirms the partitioning of the lands of C. W. Burwell, as stipulated in his will, among three heirs. A fourth heir, Nathaniel Burwell, was given 500 acres, which included a mill, where he was then living. Lands divided among the other three heirs lay in Floyd, Carroll, and Patrick Counties, and totaled, by survey, 19,208 acres. (Most of this land conveyed to C. W. Burwell by General Harry Lee, Amy H. Lee, C. C. Lee and R. E. Lee.)

D. B. "O," page 269, 3 October 1876. Martha D. Logan to Robert Logan, power of attorney over her lands,6,782 acres in Floyd County and 361 acres in Carroll County.

D. B. 33, page 139, 15 December 1906. J. Lewis Logan to H. L. Lawson, S. G. Proffitt, J. A. Hooker and Issac T. Hooker, 7,143 acres in Floyd and Carroll Counties, excepting lands sold since 1876.

D. B. 35, page 545, 14 January 1909. H. L. Lawson to S. G. Proffitt, ⅙ interest in land purchased of Robert Logan.

52. Charles Carter Lee Mill

Laurel Fork Creek

LOCATION: Route 758 (Buffalo Mountain Road SW) approximately 1.5 mile east of Route 629 (Long Mountain Road SW) just above junction of Oldfield Creek and Laurel Fork Creek

History. The site of the Charles Carter Lee mill was part of a huge tract deeded to Henry "Light Horse Harry" Lee III, father of Robert E. Lee, for his extraordinary service in the American Revolution. Henry Lee (1756–1818) willed a portion of this land to his son Charles Carter Lee, where he was to live and maintain a law office. This land was sold to the Burwell family and later to the Lawson, Proffitt and Hooker Company. Some of the land was still owned by the Lawson family in 1996, at which time the building in which Charles Carter Lee is supposed to have practiced law was still standing but in very poor condition. Many of Lee's fellow mill owners innocently or consciously ignored a state law requiring them to file a *writ of ad quod damnum* with the local circuit court, and then to wait for the court to appoint and hear the recommendations of a jury before building a dam. As an attorney, Lee was less likely to have been excused on either ground — that he didn't know the law, or that he had chosen to disregard it — so he dutifully applied for and was granted that permission in 1833. The mill is said to have been built and run by slaves of the Lee family. A story persists that a slave who was a miller here is buried on the property with a bag of silver coins he got as tips at the mill. The story makes little practical sense from any perspective, but, like all folk legends, may have filled some psychological need for the people who kept it alive.

Frank Webb believed that this mill and its race were separated when the land was divided among Nathaniel Burwell's heirs in 1876. The race, which was on the portion given to Nathaniel Burwell, Jr., is not mentioned when his (Nathaniel Burwell, Jr.'s) own lands were in turn divided among his heirs in 1900, which suggests that the mill was no longer running. While there is no record of such a transaction, it is possible, also, that the mill and mill race tracts were in the meantime rejoined under one owner. Other mills built on the portion of the Lee-Burwell tract that lay in Floyd County included the William DeHart Mill (IV, 131) and the Daniel Smith Mill (IV, 58). A backward trace of the deed records of others in the area would probably place them in this group as well. The husband or heirs of Martha Burwell Logan (Patsy Diggs Logan), daughter of Nathaniel Burwell, Sr., built Logan's Mill (IV, 51) some two miles farther down Laurel Fork Creek.

Deeds

Floyd County Deed Book "O," page 258, 1 July 1876. C. W. Burwell to Martha D. Logan (Patsy Diggs Logan)

and others, 15,000 acres, but by recent survey containing 19,208 acres in Floyd, Patrick, and Carroll Counties. C. W. Burwell to sell property and divide proceeds of sale to Martha D. Logan, Nancy Carter and C. W. Burwell. They entered into agreement to divide said land. In the division of the land, Nathaniel Burwell, Jr., to have 500 acres of land embracing the house and mill race where N. Burwell, Jr., now lives. (This is the mill race of the C. C. Lee mill on Laurel Fork Creek.)

D. B. 27, page 586, 26 November 1900. Division of Nathaniel Burwell's lands. In total 544¼ acres. Nine lots. His widow, Nancy, gets 102 acres, the rest to his children. Each tract drawn and described. No mention of a mill race or other feature of a mill.

53. Chris McAlexander Mill

Taylor's Creek

LOCATION: Built on the site of the A. W. Stanley Mill (IV, 130) on Route 603 (McAlexander Road SW) .2 mile west of Route 632 (Shelor Road SW). Mill was 20 to 30 feet below the dam on north side of the creek, within a foot of Route 603. POWER: Turbine; STONES: 1 set, feed grinder; VISIBLE REMAINS: In 2010 the collapsed remains of the building lie on the stream bank and partially in the creek. The breached stacked-stone and earthen dam is still largely intact with the culvert pipe that fed the turbine still in place.

History. Both the mill and the earth and stacked-stone dam were built around 1934 by master millwright and blacksmith A. Newton Hylton and Chris McAlexander, with help from McAlexander's father-in-law, James E. Thompson. The mill was built in an informal partnership with McAlexander's widowed sister, Dollie McDaniel, who had bought the old A. W. Stanley mill place after her husband died while working in a CCC Camp. While there was little cash money to be made in a mill grinding mostly corn and feed on an exchange basis, both partners would at least have plenty of corn meal and extra feed for their animals. To retain her interest in the mill, Dollie McDaniel specifically excepted the one-acre mill lot and water rights when Chris and Goldie McAlexander bought the adjoining 35-acre tract from her in 1937 (see A. W. Stanley Mill, IV, 130). The one-acre mill tract was transferred to Chris and Goldie McAlexander along with another tract in 1946.

Around that same time, and probably after the deed was recorded, the dam was raised about two feet and the wooden race and penstock, which had by then rotted, were replaced with galvanized culvert pipe inserted near the base of the dam. This eliminated the leaky wooden race and penstock, and, since it drew water from much lower down, allowed the miller to draw off much more of the mill pond water in the process of grinding. At the same time, a gated 2 by 3-foot concrete passageway was made through the dam to serve as both a flood gate and a cleanout passage. During periods of high water, the gate could be raised to relieve pressure on the dam and prevent erosion if the concrete spillway seemed unable to accommodate the runoff. The passage also allowed for the periodic removal of accumulated silt in the mill pond, which was nearly filled with sediment when the dam was reworked. Water rushing toward the opened gate would cut a passage through the silt, flushing it back down to the level of the gate, thereby increasing the capacity of the pond and extending the running time of the mill. In a discussion at the McAlexander mill site on July 19, 2010, Chris McAlexander's son Calvin recalled that with the gate only two-thirds opened, the pond would empty in about ten minutes, briefly flooding the downstream bottom with muddy water.

The task of raising the dam fell mostly to sixteen-year-old Calvin, who used a team of horses, a chain, and a pry bar to load field stones onto a sled. The team and sled were driven across the top of the dam to place the stones forming the face of the dam. Dirt was excavated from the road bank and piled on top of and behind the stones with a drag pan.

The mill pond, which must then have been only recently filled, was the scene of a double tragedy on June 21, 1934, when Martin Dalton drowned in the mill pond. Lonnie Webb, a powerful swimmer, was summoned to the scene from approximately a mile away. Having run all the way to the mill site, he plunged in and was partially paralyzed by the shock of the cold water. Dalton was pulled from the pond but

Chris McAlexander's Mill, with its wood-shingle roof and single set of hand-cut stones, would have blended easily into the landscape a full century before it was built about 1934. But its turbine power and the steel feed pipe that replaced the wooden race and penstock in 1946 represented state-of-the-art hydropower technology. (Ricky Cox photograph, August 1996)

could not be revived. Webb was disabled for the rest of his life.

Along with Flint Mill (II, 106), c. 1932, and Jack's Mill (I, 5), c. 1935, the Chris McAlexander Mill was one of the very last water-powered mills built in Floyd County. After two decades of active construction of new mills and refitting of older ones, mill development was in decline by 1920, and continued to slow throughout the decade despite a strong national economy. The 1930s saw even less activity thanks in part to the nationwide economic depression. Probably a factor, as well, was the perception, based on a future glimpsed in the 1910s and 20s, that an economy founded on small, multi-crop farms would eventually be replaced by a wage-based system in which fewer and fewer farmers fed more and more people. By the time the economy was revived by World War II, better roads and increasing dependence upon non-farm income in Floyd and surrounding counties had convinced local entrepreneurs to invest in businesses more attuned to an emerging industrial economy.

Although it took another sixty years for meal and flour production to disappear from Floyd County, few owners of existing mills thought it worthwhile to do more than maintain meal and flour making equipment or upgrade the power sources after the mid–1930s. From the '40s onward, the most significant investments in milling equipment were made in hammer mills, bin mixers, and bulk delivery trucks, the latter mostly to serve a dairy industry that would itself virtually disappear by the end of the 20th century. The only new mill built after 1940, the Farmers Exchange Mill (I, 139), was constructed in the 1950s to produce cattle feeds, exclusively, using electric-powered hammer mills and mixers.

Aside from the use of a turbine instead of an overshot wheel at the Chris McAlexander Mill, the last three water-powered mills built in Floyd County followed a decidedly primitive pattern. In all three, water power turned the runner stone and shook the gravity-fed corn from the wooden hopper into the mill. And in at least two of them, a belt turned the shaft of a small feed grinder, sometimes referred to as a "scrusher," which ground small volumes of animal feeds. Otherwise, all handling, measuring, and bagging were done manually. Grinding only corn and having only a single working floor (Jack's mill did have a second story) precluded running up and down stairs, but still every pound of corn had to be lifted and poured into the hopper by hand and then scooped up and bagged once it had been ground. So, although Chris McAlexander's Mill was chronologically among the newest of all Floyd County mills, it was, like Mabry Mill only a few miles to the east, technologically one of the oldest.

Calvin McAlexander ran the mill for a few years in the late 1940s while his father cut timber. The mill ceased grinding around 1950. When the building collapsed around 2005 it still had what appeared to be the original sawn-shingle roof. The Fitz-Burnham turbine and pair of stones had already been sold to an antique dealer in Meadows of Dan, Virginia, and went from there to an unknown buyer in Bristol, Virginia.

Deeds

Floyd County Deed Book 50, page 256, 21 September 1932. J. E. Burwell, Commissioner for W. J. Bowman, deceased, to G.W. McAlexander and assigned to Dollie V. McDaniel, 36 acres more or less.

D. B. 56, page 114, 15 December 1937. Dollie V. McDaniel to Chris McAlexander, 36 acres. "Exempted from and not conveyed by this deed about one acre of said tract of 36 acres ... upon which is situated the corn mill" and "all water rights belonging to and appertaining to the conduct and operation of said mill."

D. B. 63, page 523, 1 February 1946. Dollie V. McDaniel to Chris and Goldie J. McAlexander, two tracts of land, "Tract No. 1 containing 1 acre, more or less, and being the remainder of a 36 acre tract of land ... known as the corn mill lot."

54. William Jackson "Miller Will" Stanley Mill

Tory Creek

LOCATION: .1 mile upstream from Route 632 (Shelor Road SW) on north side of creek. Location indicated by end of dug race. POWER: 20- to-24 foot wooden "high breast" or "pitchback" wheel; STONES: 2 or 3 sets; VISIBLE REMAINS: Sections of the dug race can be seen in woods along the north side of Tory Creek

History. William Jackson Stanley (1868–1947) bought 57 acres on Tory Creek from C. Levi Culler in December 1902. Out of that parcel Culler retained a tract of less than an acre on which stood his mill. He also reserved the rights to the water needed to run the mill. Five years later Culler and his wife deeded the "3 rods and 5 poles of land" that had originally been withheld plus their mill and "water power for same" to W. J. Stanley. Stanley probably operated the Culler mill before hiring A. Newton Hylton to help him build a larger mill about 1918. The new 3½ story mill was 200–300 yards downstream from the Culler mill and on the opposite side of Tory Creek. Power was furnished by a 20- to 24-foot diameter high breast or pitchback wheel, almost certainly the tallest wooden wheel ever built or used in Floyd County. An earthen dam diverted water into a dug race about four feet wide and approximately one-quarter mile long. Coleman Dalton dug the race in exchange for a cow. Water was carried from the end of the dug race and over what was then the public road to the water wheel by a wooden race. Calvin McAlexander recalled being dripped on when he passed under the race in the back of his father's 1935 Chevrolet pickup.[29] The Stanley mill was usually closed during extremely cold weather since the wheel, like all wooden wheels, was more susceptible to freezing than metal wheels.

Stanley operated his mill until 1947, with his son Connie doing the milling during the last two years. The mill was torn down and the contents sold for scrap in the 1960s. Two sets of stones were sold to "Bill" Cox Weeks (Mrs. Kyle Weeks), who assembled a large collection of millstones, many of which were at one time displayed behind her home on West Main Street in the Town of Floyd.

While not the largest or best known of Floyd County mills, the Will Stanley mill is among the more interesting mechanically. It was one of the last flour mills built in the county, and

"Miller Will" Stanley poses by his pitchback water wheel. This type of wheel was rarely seen in the mountain South, but was designed to make the most of varying water levels in high head, low volume installations. (Courtesy Octavia Stanley Cox)

was almost certainly the last built with two or more sets of stones, one specifically for wheat flour. Its builders rejected the high-speed roller mill system that was installed in most other flour mills built after 1900, and for which several older mills were refitted between 1900 and 1920. Stanley chose instead to employ the much older Oliver Evans flour-making system using large, low-rpm mill stones and probably a reel-type sifter or dresser to remove the ground up bran. The Evans system was slower and did not produce as much or as many grades of flour and wheat by-products, but it required less machinery, lower shaft speeds, and less capital outlay.* Perhaps most important in this instance, it required less power, and therefore demanded less water.

Stanley and millwright Newton Hylton may have understood, as some of their contemporaries did not, that the high-speed turbine wheels typically installed to drive roller mills had one serious drawback. While they could more easily supply the high shaft speeds and the horsepower demanded by the newer roller mill system, turbines large enough to power a flour mill also required more water than a small mountain stream could reliably supply, even when a dam was used to build up a reserve of water. And they may have benefited from the several examples of unsatisfactory turbine installations that had occurred by the time Stanley was ready to abandon the old Levi Culler Mill (IV, 55) and build a new one. Stanley would certainly have known about the failure, due to an inadequate water supply, of the turbine-equipped Wooster Cockram mill (IV, 59) on Fork Mountain Creek, only a few miles east of him, between 1913 and 1918. His choice of a high breast or pitchback configuration water wheel and the decision to make such a large diameter wheel out of wood rather than purchase one made of steel also invite speculation. The choice of material was probably due to cost, but the choice of a pitchback set up reflects advanced or at least atypical thinking, since only a handful (see I, 123 and Tables A, B, and C) of more than 100 vertical wheel installations in Floyd County used the pitchback or breast configuration instead of the overshot.

But if he was part of a small minority county-wide, Stanley was in the majority nationally and internationally. In his exhaustive study of water power, *Stronger Than a Hundred Men: A History of the Vertical Water Wheel*, Terry S. Reynolds reports that by 1850 British and American hydraulic engineers and millwrights were convinced that the high breast configuration is generally superior to overshot water wheels, chiefly because it can take advantage of changes in the level of water supplied to the wheel, and because its motion is not impeded by tail water returning to the stream after being dumped from the wheel.[30] Overshot vertical water wheels are typically made as tall as possible to take full advantage of the available head; when stream levels were higher than normal, an over-

**The claim that this mill had three sets of stones instead of the two typically found in buhr-stone flour mills in Floyd County might suggest that Stanley employed the "New Process" system of making flour, which was briefly popular in the U.S. before the roller system virtually eliminated stone grinding of wheat flour. But no other New Process mill is known of in Floyd County, and too little is known about Stanley's mill to speculate further.*

shot wheel of optimum diameter could be slowed by the tail water moving in a direction opposite that of the wheel's rotation.

Stanley may have taken a scientific approach, or he may simply have been good at learning from the experiences of his fellow millers and millwrights. The original 16-foot diameter wheel at nearby Mabry Mill, built around 1910, was replaced with a smaller and less powerful 14-foot, 10-inch wheel because the original "wallowed" in the tail water. In other words, its motion was impeded by water flowing back to the creek, because the bottom of the wheel, turning in the opposite direction, was in effect trying to force the tail water uphill. Assuming that the bottom of Stanley's pitchback wheel stood only a few inches above the tail water at normal stream levels, it would have extracted the maximum power, over a wide range of stream conditions, from the force of gravity acting on the waters of Tory Creek.

Will Stanley also made furniture, including beds and chairs.

Deeds and Other Public Records

Floyd County Deed Book 31, page 50, 20 December 1902. C. L. Culler and Martha, his wife, J. P. Marshall and Lelia, his wife, and Floyd S. Dickerson and Almedia, his wife, party of the first part, all of Carroll County, Virginia, and William J. Stanley, of Floyd County, party of the second part, two tracts on Tory Creek and Danville Pike. First tract, 44 acres and 6 poles but subject to certain considerations. Second tract, 13 acres, 3 rods and 25 poles. The said Levi Culler and wife also grants to W. J. Stanley free access to nearest spring and land enough on which to build a springhouse but said Levi Culler retains a right to 3 roods and 5 poles of land on which the mill is situated. The said Cullers retain the right of water power to be had by means of dam, race, and flume and also the trees suitable for timber that lies above said mill (right ceases in ten years). This conveyance also involved transfers of cash among the parties of the first part and conveyance of land from Stanley to J. P. and Lelia Marshall (does not convey the mill or water power). Stanley to secure Marshalls against losses due to Lucinda Light's claim of dower rights.

D. B. 36, page 472, 25 December 1907. C. L. Culler (of Carroll County) and wife to William J. Stanley, land on both sides of Tory Creek containing 3 rods and 5 poles more or less and "full water power by dam, race and flume."

D. B. 67, page 371, 8 March 1948. Executor for W. J. Stanley, deceased, to Octavia Thompson, land of 45 acres more or less on Tory Creek and Danville Pike.

55. C. Levi "Lee" Culler Mill

Tory Creek

LOCATION: upstream from the Will Stanley Mill (IV, 54); POWER: Overshot wheel; STONES: 1 set; VISIBLE REMAINS: None

History. This small mill on Tory Creek was the site of a sash type ("up and down") sawmill as well as a grist mill. The sawmill and single run of stones were powered by an overshot water wheel. C. Levi "Lee" Culler (1853–1933), sold his mill to William Jackson "Miller Will" Stanley in 1907, having sold him 57 acres of adjoining land five years earlier. Stanley operated this mill before building a much larger mill in 1918, downstream and across the creek from the Culler Mill. The new mill (IV, 54) featured a wooden wheel of large diameter, set up in an unusual pitchback or high breast configuration.

Deeds and Other Public Records

Carroll County Deed Book 15, page 313, 6 January 1883, Hillary Brim and wife to Levi Culler, tract of land on Tory Creek in Floyd County and Carroll County, 104 acres on both sides of Danville Turnpike. *This may or may not be the tract on which the Culler Mill was built. The Danville Turnpike would seem to be modern day Route 58, which is some distance from the mill site.*

D. B. 36, page 472, 25 December 1907. C. L. Culler (of Carroll County) and wife to William J. Stanley, land on both sides of Tory Creek, containing 3 rods and 5 poles more or less and "full water power by dam, race and flume."

56. William McAlexander Mill

Tory Creek

LOCATION: Private road off Route 58 (Danville Pike SW) south of Tory Creek Church
POWER: Overshot wheel; STONES: 1 set

History. Family tradition says that William McAlexander had a small mill near his home on Tory Creek in the extreme southwest corner of Floyd County. Only one deed was found for William McAlexander. This land was still owned by his descendants in the mid–1990s. Deeds and census records for Carroll and Patrick Counties could include references to McAlexander and his mill.

Deeds

Floyd County Deed Book "K," page 629, March 1862. Edmond Marshall to William McAlexander, 125 acres land lying in Floyd, Patrick and Carroll Counties.

57. Agee Mill

Laurel Fork Creek and Mountain Fork Creek

LOCATION: East side of Laurel Fork Creek just below where it is joined by Mountain Fork Creek just south of intersection of Route 779 (DeHarts Store Road SW) and Route 603 (Fork Mountain Road SW); POWER: Wooden overshot wheel; STONES: 2 sets, later roller mill and sawmill

History. The mill is thought to have been built around 1912 by John T. Agee (1847–1926) and sons, and Alonzo "Newton" Hylton, a son-in-law. Since 1888 Agee had owned a smaller mill a short distance upstream. Earlier ownership history of both sites is included with the history of the Daniel Smith Mill (IV, 58).

This mill was originally powered by water from two streams, as were several other headwater mills in Floyd County, including nearby Mabry Mill (IV, 60). Around 1916 Agee hired a Mr. Nunn to install a Sprout Waldron Roller Mill system; the sale of a one-half interest in the mill to his son, John Walker Agee (1881–1947), in early 1917 was probably due both to advancing age and the need for cash to pay for the new machinery. After one of the wooden races or sluices collapsed the water came entirely from Mountain Fork Creek. The race had to cross over Laurel Fork, the larger of the two streams, to deliver its water to the mill. Other millers here included "Cap" Ayers, "Wooster" Cockram, Murray V. Agee (1893–), Alonzo "Newton" Hylton (1872–1957), and Leonard Agee, Murray's son.

A sawmill appears for the first time in deed records in 1947, when Murray Agee bought the property from Cap Ayers. This was no doubt of the circular saw type, similar to and perhaps the same one that Agee installed at the Weddle/Wickham/Webb/Agee Mill (IV, 34) on Howell Creek, which he owned from 1931 to 1939. A November 1947 news item from the *Floyd Press* notes, "The Agee Mill, located in Meadows of Dan section for more than 50 years, has been bought by Murray Agee from T. P. Ayers and who will operate it. This mill has an outstanding reputation in the manufacture and sale of Water Ground Buckwheat throughout the state."[31]

Even though a 1965 deed listed the mill, a sawmill, and a lumber planer, implying that all were valuable features of the property, the mill collapsed very soon after. Dan Hubbard heard the mill falling from his yard across the road and turned in time to see the tallest part of the mill collapse straight down, as if it were being deliberately demolished.[32]

Deeds

Floyd County Deed Book "U," page 667, 24 March 1888. J. H. Cruise to John T. Agee, 42½ acres on Laurel Fork Creek. *The relevance of this deed and the next to the history of the Agee mill depends upon which tract of land — the one conveyed in this deed or the four acres on which the Daniel Smith mill stood — the mill was built. Comparison of the boundaries of the two tracts may answer this question.*

The Agee Mill, photographed by Earl Palmer on June 22, 1952. Metal roofing was widely available when this mill and the attached sheds were constructed beginning around 1912, but wooden shingles, sawn or split from locally cut trees, were less expensive. (Courtesy Blue Ridge Institute and Museum of Ferrum College/Earl Palmer Collection)

D. B. 40, page 183, 31 March 1908. J. D. Blackard, N. G. Blackard, and J. F. Reynolds (and wives, all of Patrick County) to John T. Agee, 4 acres and mill and "about one fourth of one acre of land lying at the butment on the opposite side of the creek from the land conveyed in the above named writing" it being land conveyed to Daniel Smith by J. L. Thompkins and conveyed by D. Smith to Blackards and Reynolds.

D. B. 46, page 196, 16 January 1917. John T. Agee to J. W. Agee, ½ interest in roller mill and water rights.

D. B. 46, page 386. 3 February 1927. Heirs of J. T. Agee, deceased, to John Walker Agee, "one undivided half interest in a roller mill situated upon about ½ acre of land."

D. B. 65, page 402, 31 March 1947. J. W. Agee to Thomas Pinkney Ayers and Eva Ayers, "all his real estate containing about 80 acres."

D. B. 67, page 151, 10 November 1947. T. P. "Cap" Ayers to Murray Vivian Agee, 2 acres plus or minus land, grist mill, sawmill, planer, and all water rights.

D. B. 90, page 197, 21 May 1965. Minnie T. Agee, widow of M. V. Agee, and Henry Leonard Agee, son of M. V. Agee, to Hugh A. White, Sr., 4 tracts of land, including tract No. 4, containing 2 acres land and grist and saw mill, planer and all water rights.

D. B. 105, page 154, 7 July 1972. H. A. White, Sr., to James A. Keesee, 4 tracts land, does not mention mill.

58. Daniel Smith Mill

Laurel Fork Creek

LOCATION: Just above intersection of Route 779 (DeHarts Store Road SW) and Route 603 (Fork Mountain Road SW); POWER: Overshot wheel; STONES: 2 sets and sawmill; VISIBLE REMAINS: None

History. The land on which J. H. Cruise built his mill and sawmill was very likely part of the estate of Nathaniel Burwell, Sr., who "by his last Will and Testament of record in Roanoke County authorized and directed his Executor (Charles W. Burwell) to sell all his lands lying in the Counties of Floyd, Carroll, and Patrick, supposed to contain about 15,000 acres."[33] The survey showed that they had underestimated by more than 4,000 acres, which would in today's market be worth at least $20,000,000. The total of 19,208 acres did not include Burwell's holdings, if any, in Roanoke County, where he was living when his will was written.

The sawmill, which may have been built first to saw the lumber for the grist mill, appears to have been abandoned before 1900, by which time sash type "up and down" sawmills, often driven by inefficient "flutter wheels," were being replaced by more or less portable sawmill outfits with circular blades. These steam-driven outfits were more powerful, faster, and more efficient. Best of all, they could be moved to the woods, saving the cost of hauling heavy logs over rough roads to the nearest water-powered sawmill.

The grist mill was abandoned around 1912, about which time John and Walker Agee built a new mill (IV, 57) just downstream from this one. The Daniel Smith mill is said to have collapsed in the mid–1920s.

Deeds

Floyd County Deed Book "O," page 600, 13 November 1877. J. H. Cruise from C. W. Burwell, 96 acres land on Laurel Fork Creek.

D. B. "R," page 583, 18 October 1884. J. Harvey Cruise to Daniel Smith and John V. Cruise, sawmill and mill site, and 4 acres land.

D. B. "V," page 117, 12 January 1891. John Thomas Agee to Daniel Smith, his ½ interest in mill (sawmill) deeded to him by John V. Cruise, per D. B. "V," page 670, 24 May 1888.

D. B. "Y," page 3, 30 July 1895. Daniel Smith to J. D. Blackard, W. G. Blackard, and J. T. Reynolds (all of Patrick County, VA), 4 acres and grist mill on Laurel Fork Creek. Includes ¼ acre on opposite side, "lying at the Butment."

D. B. 40, page 183, 31 March 1908. J. D. Blackard, N. G. Blackard, and J. F. Reynolds (and wives, all of Patrick County) to John T. Agee, 4 acres and mill and "about one fourth of one acre of land lying at the butment on the opposite side of the creek from the land conveyed in the above named writing," it being land conveyed to Daniel Smith by J. L. Thompkins and conveyed by D. Smith to Blackards and Reynolds.

---

59. J. W. "Wooster" Cockram Mill

Mountain Fork Creek

LOCATION: .5 mile downstream from Mabry Mill on Route 603 (Fork Mountain Road SW); POWER: Turbine, briefly, then gasoline engine; STONES: 1 set

History. The mill that Wooster Cockram built around 1913 was powered by water only long enough for him to realize that Mountain Fork Creek was far too small to supply a turbine water wheel. While this is the most dramatic example, there were several other cases in which the turbine originally installed in a mill was unsatisfactory, despite its many advantages over a vertical wheel, because of too little water. All were eventually replaced, usually with an old-fashioned vertical wheel.

Simply put, most turbines require more water than any but the largest streams in the county—Little River and its two main branches, West Fork and Dodd Creek (South Fork Little River); and Greasy Creek, Big Indian Creek, and Burks Fork—could supply under normal conditions, and then only on the lower reaches of the streams. Even these installations employed large mill ponds to build up reserves of water overnight and on weekends. The water consumption of turbines did of course vary with size and design, and at least one, supposed to be 10 inches in diameter, performed tolerably at the German Belcher Mill on Rush Fork (IV, 37). But the site of the Belcher mill was an ideal natural spot for a dam, Rush Fork is a larger stream than Mountain Fork Creek, and Belcher could not grind for long without stopping to let his pond refill.

The size of Cockram's turbine, which he later described as "a thirsty devil," is unknown, but his pond evidently couldn't hold enough water to grind continuously for a worthwhile length of time, so the turbine was removed from service and replaced with a one-cylinder gasoline engine. The site may have lacked the fall required for a vertical wheel, which could be why the turbine, which is well suited to low-head applications, was chosen to begin with. The other side of that coin, that a turbine requires a high volume of water, was either not fully understood, or ignored.

Cockram's mill was powered by the gasoline engine until the building was destroyed by a flood. Wooster Cockram later operated a corn mill in an outbuilding by his home about a mile east of Meadows of Dan on Route 58, using a gasoline engine to power what is presumed to be the same set of mill rocks reserved in the deed dated June 1918. These rocks and the associated framework, shaft, and pulley were bought from Wooster Cockram's son Tommy in July 1992, disassembled, and placed in storage in Indian Valley, VA, by Ricky Cox, Carl Cox, Sammie Phillips, and Frank Webb. Cockram also ran mills for other owners in Floyd and Patrick Counties.

Deeds

Floyd County Deed Book 39, page 68, 2 April 1913. J. T. Agee to J. W. Cockram, 26½ acres on Fork Mountain Creek.

D. B. 39, page 69, 2 April 1913, J. T. Agee to J. W. Cockram, 48 acres on Fork Mountain Creek.

D. B. 43, page 252, 8 June 1918. J. W. Cockram to J. T. Agee and J. W. Agee, 74½ acres on Fork Mountain Creek. One set of mill rocks is reserved.

60. Edwin "Ed" Mabry Mill

Mountain Fork Creek (Tributary of Laurel Fork Creek)

LOCATION: Milepost 176.2, Blue Ridge Parkway. POWER: Wooden overshot wheel: original wheel 16 feet, next wheel 14 feet, 10 inches; 8 HP kerosene engine late '20s to 1936; Currently wooden overshot. STONES: 2 sets, corn, buckwheat; VISIBLE REMAINS: Still standing and in use by the National Park Service as a historical exhibit.

History. One of only two functioning mills remaining in 2010, and the only water-powered mill, Mabry Mill is the most widely known of Floyd County mills, both in and outside the county. As a popular feature of the heavily-traveled Blue Ridge Parkway, it may be the most visited and photographed mill in the United States. Its image has appeared on jigsaw puzzles and calendars, in numerous national magazine advertisements, and on postcards from at least a dozen other states. The visual appeal of a naturally beautiful setting, weathered board siding, and a picturesque overshot wooden wheel have been enhanced by careful landscaping. The result was shaped by, and has in turn shaped, America's vision of its own rural past.

Though it has come to represent the archetypal water-powered mill, Mabry Mill is atypical in at least two ways. Dual wooden flumes or races bringing water from two separate streams were used on only a few other Floyd County mills, most of them, like Mabry Mill, at higher elevations, near the headwaters of mountain streams with substantial "fall," but little volume. And at no other mill in the county was water collected in a pond downstream from the water wheel, where it serves no practical purpose. Officially referred to as the "reflecting pool," the pond that many visitors doubtless perceive as the "millpond" is a pond and it is near a mill, but it is not a millpond in the sense that most millers would recognize, since it neither raises water to the level of the wheel nor stores it for

Mabry Mill, 1947. Following the 1942 replication of the wooden overshot wheel, Mabry Mill, as photographed by Gertrude Harris, probably looked very much as it did during its working life. Constructed as Floyd County's population and farm activity were heading into long declines, Mabry Mill has been a popular working display nearly three times as long as it was operated as a business by Ed and Lizzie Mabry. (Courtesy Gertrude Harris family)

use when the demand for water exceeds the flow of the stream(s) feeding the wheel.

Technical points aside, Mabry Mill offers visitors a first-hand look at water-power and grain-milling technologies that, although already ancient when the mill was initially restored beginning in 1942, were still in use throughout the county no more than twenty years earlier.

Edwin Boston "Ed" Mabry bought the mill property and water rights in a series of purchases between 1903 and 1910 and ran a blacksmith shop here before constructing his mill. In a January 7, 2008, interview, long-time Meadows of Dan resident Coy Lee Yeats stated that the works of the William DeHart (IV, 131) mill, which stood a short distance upstream and was known to him as Scott's Mill, were installed in Mabry's new mill. Deed records for the two mills do not confirm this, but Mabry did buy from the owners of the DeHart/Scott Mill the rights to water he would need for his new mill. Whether or not the sale of these rights means that the older mill was no longer in use when Mabry's mill began operation depends upon the exact location of Scott's Mill and the layout of Mabry's races. But overlaps in dates, locations, and the parties involved in various transactions do leave open the possibility that Mabry purchased stones and other machinery from the owners of the DeHart/Scott Mill in a separate transaction.

Mabry Mill was built about 1910 by Ed Mabry, millwright Alonzo "Newton" Hylton, and others and was operated with water power until 1934, when an eight-horsepower, one-cylinder gasoline or kerosene engine was installed to take the place of a decaying wooden wheel and races. The mill continued under gasoline or kerosene power until 1936 and was sold to the Virginia State Highway Commission by Ed Mabry's widow, Mintory L. "Lizzie" Mabry, in 1938.

The 14-foot, 10-inch wooden wheel replaced by the engine in 1934 was the second at Mabry Mill, and was built in 1922 by Ed Mabry and Newton Hylton to replace the original 16-foot

The most familiar mill in the United States, Labor Day weekend, 2010. (Photograph by Ricky Cox)

wheel, which performed poorly because it "wallowed in," i.e. was too large to run clear of, the tailwaters beneath it. A third wheel was built by Newton Hylton, Bud Goad, and J. C. "Chris" McAlexander in the mid–1940s after the United States Park Service had begun restoring the mill to working order in 1942.

A new wheel installed in 1990 lasted a mere five years before being replaced in 1995. The old wheel was used as a pattern but only the gear teeth and iron gudgeons were reused in the wheel built by Barry Caldwell, Mark Mallette, and George Senne. Other unspecified repairs to the flume and mill were made in 1954, 1974, and 1984.

As operated by Ed and Lizzie Mabry, the grist mill portion of the mill produced corn meal and chop for livestock. The Park Service did away with the chop mill and added a pair of stones for grinding buckwheat. The mill also features a water-powered sawmill and woodworking shop, both built about the same time as the mill. The saw is of the circular type, which was rapidly replacing the slower sash or up-and-down type around the turn of the 20th century. The saw teeth are formed from the saw plate and are integral to it, rather than welded or soldered to it, or inserted into special holders arranged around the perimeter, as are teeth of modern saws. Hence, the "gullets" of the teeth had to be reformed or "gummed" after the saw had been sharpened several times. Encountering a nail in a log or a stone embedded in the bark would have been disastrous in terms of the time required to resharpen a saw of this type. Mabry Mill interpretive staff say that the sawmill required more power than the grist mill, and could be operated efficiently only when the water level was high in the several streams whose combined waters powered the wheel.

Because compliance with FDA guidelines would significantly compromise the authenticity of the mill, nothing intended for human consumption is produced at Mabry Mill, despite the installation of a new oak water wheel and repair of the wall behind it in 1995. Millers have since used one set of stones to crack corn, but no corn meal, grits, or buckwheat flour have been ground there since 1988, according to Frank Webb's notes. Current millers (2010) were told that meal was ground on-site as late as 1996, but this reference could refer to the last time products were ground by Mabry Mill millers, using the rented facilities of Cockram's Mill in nearby Meadows of Dan. Buckwheat Flour, from the King Mill in King, North Carolina, and cornmeal and wheat flour produced by Boonville Flour and Feed Mill in Boonville, NC, are bagged on site and available for purchase at Mabry Mill.

A partial list of Mabry Mill employees begins with Edwin Boston "Ed" Mabry and his wife Mintory L. "Lizzie" Mabry. Foster Dalton, Frank Dalton, and Walter Dalton all worked for Ed Mabry. Foster worked in the late 1920s to 1938 for Lizzie Mabry after Ed Mabry became ill, and later for the National Park Service. Other Mabry Mill employees (with dates of employment), include Vernon Goad (1970–85), Walter Alderman, Walter Harris, Haynie Cockram, Murray Burnett (1966–82), Thomas Belcher (1977–85), Arnold Spangler (1983–92), Leonard Hylton (1988–c. 2006, died in 2007), Jim Sutphin (working in 1996); and Wendell Sutphin (working in 1996). Millers and mill interpreters working in September 2010 were Gray Linville (2004–), Harold Slate (2008–), and Joe Dalton.

Deeds

Floyd County Deed Book "Z," page 490, 22 February 1899. J. D. and V. J. Thompson to M. L. Mabry, wife of E. B. Mabry. Two tracts land, one of 15 acres and one of 26 acres on Mountain Fork Creek.

D. B. 29, page 206, 13 January 1903. C. W. Burwell to Ed Mabry, 46½ acres land.

D. B. 32, page 27, 14 October 1905. C. D. Scott to E. B. Mabry, water rights for mill, north side of Mountain Fork Creek.

D. B. 33, page 108, 19 December 1905. C. W. Burwell to Ed Mabry, 4 acres land.

D. B. 36, page 570, 23 March 1910. C. D. Scott to E. B. Mabry, ¼ acre land and water rights.

D. B. 42, page 453, 2 April 1917. Mary M. Cox and D. W. Cox, her husband, to Emmett Weddle, 100 acres on Laurel Fork Creek. No mill mentioned, but water right is reserved to E. B. Mabry. This deed does not convey the Mabry Mill property, but it does reserve the water right sold to E. B. Mabry by C. D. Scott in 1905.

D. B. 56, page 567, 10 May 1938. Mintory L. "Lizzie" Mabry to Virginia State Highway Commission, 3.01 acres.

D. B. 56, page 574, 31 May 1938. Mintory L. "Lizzie" Mabry, widow of E. B. Mabry, deceased, to Virginia State Highway Commission, 15.74 acres land and mill.

61. John W. "Pob/Pobb/Pop" Boyd Mill/Simon "Wid" Hubbard Mill

Laurel Fork Creek

LOCATION: East side of the creek a few yards upstream (north) from Route 758 (Woodberry Road SW); POWER: Overshot wheel; STONES: 2 sets; VISIBLE REMAINS: A flat expanse built up by silt deposited above the dam drops off sharply at what was apparently the dam site a short distance above Route 758

History. It is believed that John W. "Pobb" Boyd (1854–1940) built this mill and leased it to Joseph Carter, who was running the mill when Boyd sold it to W. J. H. Barringer in 1894. Carter was still (or once again) running the mill when Boyd bought it back in 1904. Barringer was the son and grandson of millers, and during roughly the same ten years owned a modern roller-type flour mill (IV, 40), a sawmill, and a store on Burks Fork. S.R. Conner bought the original one-acre lot and mill, plus an additional 7½ acres from Pobb Boyd in 1916.

The mill and land changed hands frequently over the next 20 years, but were still referred to, collectively, as the Boyd mill property in 1928. Simon H. "Wid" Hubbard bought the "Pob Boyd mill tract" in 1931 and was the last to run the mill. The Commonwealth of Virginia bought it from Hubbard in 1936. It was torn down at an unknown later date by the Blue Ridge Parkway. The mill had a wooden dam on the left side of Route 758 as viewed from the Blue Ridge Parkway. A wooden race about 200 feet long carried water from the dam to the mill.

Carl Belcher's father and uncles amused themselves on hot days by darting between the spokes of the rotating wooden wheel to run in place inside it as it turned, like hamsters on an exercise wheel. The miller, both annoyed and fearful of an accident, would chase them away, but the boys would resume their game as soon as he went back inside.[34]

Deeds

Floyd County Deed Book "S," page 576, 6 October 1884. Daniel R. Conner and Mary J. Conner, his wife, to John W. Boyd, 8½ acres of land on Laurel Fork Creek.

D. B. "W, page 598, 14 March 1894. John W. Boyd and Mary Ann Boyd, his wife, to W. J. H. Barringer, 1 acre on Laurel Fork Creek where Joseph Carter has a grist mill.

For the sake of safety and convenience, mill designers typically avoided exterior stairs of more than two or three steps. The man perched on these nearly ladder-like stairs is thought to be Pobb Boyd, whose mill was the largest of three Floyd County mills that stood along what is now the Blue Ridge Parkway. (Frank Webb collection)

D. B. 32, page 2, 23 July 1904. W. J. H. Barringer and M.E. Barringer, his wife, to John W. Boyd, land "lying on Laurel Fork Creek it being the lot which Joseph Carter has a mill upon the said land it being 1 acre more or less."

D. B. 41, page 585, 24 June 1916. J. W. Boyd and Mary A. Boyd, his wife, to S. R. Conner, 8½ acres of land. Including "grist mill and fixtures which are on the premises."

D. B. 43, page 239, 6 May 1918. S. R. Conner (resident of West Virginia) and Cora, his wife, to J. D. Conner, mill and fixtures and 8½ acres

D. B. 45, page 504, 29 December 1919. J. D. Conner, and Ellen, his wife, and Parriss A. Alderman and Bessie Lee Alderman, his wife, to J. S. Harris and H. E. Thomas, "8½ acres more or less upon which there is a mill erected together with the Machinery belong[ing] to said mill and known as the Pop Boyd Mill ... on waters of Laurel Fork." *The Aldermans had bought the property from J. D. Conner for $600, but had the deed made to Harris and Thomas. Hence the Aldermans are included as parties of the first part.*

D. B. 47, page 524, 14 July 1921. W. P. DeHart to J. A. Conner. Conveys 10½ acres of land and releases "water right on ½ acre of land adjoining this land." *The release of water rights on an adjoining half acre might have been included to resolve some question about what rights had been conveyed earlier, or this deed could be unrelated to this mill since the grantee is J. A. Conner instead of J. D. Conner.*

D. B. 45, page 505, 11 June 1923. H. E. Thomas and wife E. M. Thomas to J. S. Harris, their ½ interest in Pap Boyd Mill property.

D.B. 45, page 560, 5 May 1924. J.S. Harris and M.P. Harris, his wife, to F. J. or F. H. Carter of Draper, North Carolina, 8½ acres and J.D. Conner Mill.

D. B. 47, page 519, 23 November 1925. B. G. Howard, Commissioner for F. J. Carter, deceased, to J. S. Harris and W. G. Burnett, property described in D. B. 45, p. 560. Harris and Burnett bought the property at a public auction forced by a chancery suit among Carter's heirs.

D. B. 46, page 545, 26 January 1928. J. S. Harris and W. G. Burnett to H. E. and E.M. Thomas, "a certain tract of land ... containing 8 acres ... and known as the J. W. Boyd mill tract and conveyed to J. S. Harris and W. G. Burnett by B. G. Howard Jr. and recorded in DB 47, page 519."

D. B. 48, page 49, 20 April 1928. Homer E. Thomas and E. M. Thomas, his wife, to William Edwards, and wife. "Said Thomas grants bargains and sells to said parties of the 2nd part his mill known as William Boyd mill and 8 acres of land more or less."

D. B. 50, page 14, 30 December 1931. W.M. Edwards and Bula, his wife, to S. H. Hubbard, "Land lying on waters of Laurel Fork creek ... containing 8 acres more or less ... also the mill on said land with all appurtenances"

D. B. 53, page 556, 30 April 1936. S. H. Hubbard and Sallie, his wife, to Commonwealth of Virginia (Blue Ridge Parkway), a certain tract or parcel of land lying in Floyd County, Virginia together with all buildings and improvements ... said tract or parcel of land being known as Hubbards Mill."

66. Solomon Slusher Mill/Elijah Turman Mill

Mira Fork Creek

LOCATION: West side of Route 761 (Sugar Tree Road SW) .1 mile from Route 221 (Floyd Highway S); POWER: Overshot wheel; STONES: 2 sets, sawmill; VISIBLE REMAINS:

What appears to be the original "eastern" race may be seen from Route 221 and intersects Route 761 about .1 mile from Route 221. The "southern" race was probably obliterated by improvements to Route 761.

History. The first written record of Solomon Slusher, Sr.'s (1796–1878) mill on the Mirey or Mira Fork of Greasy Creek appeared in 1874 when he applied to the Floyd County Circuit Court for a *writ of ad quod damnum.* The jury assembled by Sheriff F. W. Edwards to conduct an inquest at the proposed site decided in Slusher's favor and compelled the adjoining landowners, John B. Slusher and others, to sell Solomon Slusher one acre for a dam and pond and ¾ acre for a race. The new race carried water from the main branch of Mira Fork northward along the west side of what is now Sugar Tree Road. The wording of the court order (appended below) does not state whether the new dam and race were to serve a brand new mill, or were needed to bring additional water to an existing grist mill and sawmill. The second scenario seems most likely considering that in 1874 Slusher was 78 years old, at which age few people would begin to build a mill from the ground up, and that Slusher's 1875 will makes clear that there were then two races. An interesting twist is that John B. Slusher, whose land Solomon sought to take in his petition for a *writ of ad quod damnum,* had filed his own petition for a *writ of ad quod damnum,* to build a mill on his own land on Mirey Fork, in 14 April 1874. No record of a ruling on his petition was found, but John B. Slusher is listed as a miller on Mira Fork in several editions of the Virginia business directory (see text and endnote 2, Appendix A). Whether he did build his own mill or was running the Solomon Slusher Mill is unknown.

An 1884 deed conveying the land and mills from Solomon Slusher's son Allen to Elijah Turman describes the race acquired in 1874 as approaching from the south. The other "east" race brought water from the branch of Mira Fork that runs westward along the south side of Route 221 between Route 758 (Buffalo Mountain Road SW) and Route 761 (Sugar Tree Road SW). At that time it crossed the lands of A.T. Slusher and Landon Slusher as well as Solomon Slusher's own land. It could be that one race served the grist mill and one the sawmill, or the two could have joined to power a single wheel which could be made to drive either mill.

Developing and maintaining two water supplies would have added considerably to the cost of building or of operating a mill, but it was a workable solution in cases where two streams of low volume but sufficient fall were nearing their confluence, and when the more logical solution of building or rebuilding the mill below that meeting point was impractical. Mabry Mill (IV, 60) is similarly fed by two streams, neither being sufficient on its own to supply the single wooden water wheel that drives either the grist mill or the sawmill at that site.

A second unusual feature of this mill is that one of the leaky wooden races (probably the original eastern race) would have crossed over what is now Route 761 (Sugar Tree Road SW) to the delight or dismay of travelers passing beneath. The same situation existed at the Powhatan Williams Mill (I, 25) that stood across Route 690 (Sandy Flats Road SE) from the Shelor/Nolen/Pine Creek Mill (I, 123), at the Willis Mill that once stood on Route 769 (Reed Lane NW) near its intersection with Route 768 (Firehouse Road NW), and at the Will Stanley Mill (IV, 54) on Tory Creek.

A third point of interest is that Elijah Turman's wife, Adaline, sometimes ran the mill during his lifetime and continued to operate it after his death. This is probably more common than is suggested by written records. Since the operators of many smaller mills worked at farming or sawmilling when no customers were at the mill, they could easily be unavailable when someone showed up with a turn of corn. Josie Moses was also a competent miller, as were Millie Jane Belcher, Ella Thomas, Mattie Weddle, and Luvenia Harris.

According to Arthur Bond, Elijah Turman's son-in-law, Doc Hylton, who was known as a thresher man in the Willis area, lived there at one time and "did away with" the mill.[35] The mill was not mentioned in the deed when the land was sold by Elijah Turman's heirs in 1913.

Deeds and other Public Records

Floyd County Deed Book "A," page 283. 20 October 1834. Christopher Slusher, Sr., to Solomon Slusher, 200 acres on the Miry Fork of Greasy Creek.

D. B. "N," page 330, 1 July 1874. Court decision on *writ of ad quod damnum. Solomon Slusher vs. J. B. Slusher et al.,* 1 acre for dam to be 4 foot high and ¾ acre for race. Court found $21.72 to be fair compensation to J. B. Slusher, et al. *Solomon Slusher's application to the Circuit Court of Floyd County for a* writ of ad quod damnum *indicates that J. B. Slusher, et al., the owners of land along Mira Fork Creek across from the point at which Solomon Slusher wished to build a dam, were unwilling to sell to Solomon Slusher the land needed for his dam, pond, and race, or that the two could not agree on a price. The complete text of the jury's report, and the subsequent court order, both of which are part of a deed recorded in Deed Book N, on pages 330 and 331, are reproduced in full at the end of this mill history.*

Floyd County Will Book "E," page 196–197, 29 March 1875. Solomon Slusher [Sr.] to son, Allen Slusher, "my mills and mill houses [grist and saw] and two and about one fourth or one half acres of land upon which mill is situated and all rights titles and privileges to water power and land acquired under a writ of ad quod damnum in County Court of Floyd for mill purposes and also the land upon which the two races to the mill is situated for ten feet in breadth, the entire length of the two races through my land."

D. B. "R," page 461, 9 July 1884. Allen Slusher, et al., to Christopher Turman, 6½ acres land and mills, land embraces 2 races, one is south race (condemned and acquired per above), and one is east race which runs through A. T. Slusher meadow and then Landon Slusher's land. This race is one rod wide and was deeded by Solomon Slusher and confirmed by will to Allen Slusher.

D. B. "T," page 212, 24 July 1886. Christopher Turman, deceased, by J. P. Turman, Trustee, to Elijah Turman and wife, Adaline, 16½ acres, including 6½ acre and mill tract (10 acres from C. L. Sutphin).

D. B. 40, page 188, 1 October 1913. Adaline Turman and others for Elijah Turman to Jabe H. Turman, 16¾ acres land.

See Table B, *1870 Products of Industry.* Name of corporation or individual: Slusher, Solomon

The following is the verbatim text of Floyd County Deed Book N, Pages 330 and 331. (Boundary descriptions are omitted in the interest of brevity; commas are inserted for clarity.)

> Inquest taken before F. W. Edwards, S. F. C. [assumed to mean "Sheriff of Floyd County"] on 1st day of July 1874 upon the lands of John B. Slusher, Sarah E. Slusher, Z. T. Slusher and Ezra Slusher in Burks Fork Township in County of Floyd in pursuance [or pursuit] of the writ of ad quod damnum awarded on application of Solomon Slusher at June term 1874 of Floyd County Court and hereto amended and made part of this finding, report, and inquest.
>
> The undersigned Jury, after being duly sworn by said sheriff that if the leave asked for in said writ be granted to build the dam and enlarge the canal, that no part of any mansion house, outhouse, yard, garden, or orchard will be taken or overflowed, ordinary navigation will not be obstructed and that the health of the neighbors will not be annoyed by stagnation of waters. There are but few fish in the stream (?. E. Fork of Greasy Creek) the stream being quite small but the building of the dam will obstruct the passage of fish to some extent but this will be avoided by erection of a fish [word, probably "ladder," omitted by clerk]. The land required for the abutment of the dam contains one acre and is bounded as follows, beginning at a stake by the road ... 19 poles to the beginning. The land required for the canal contains ¾ of an acre and is bounded as follows. Beginning at a Spanish oak ... to the beginning, said canal being less than 100 feet in width, to wit 60 feet wide. We are of the opinion that twenty-one dollars and 72 cents would be fair compensation for the land required for abutment of dam and for canal which should be paid by Solomon Slusher, Z. T. Slusher, Sarah E. Slusher, and Ezra Slusher.
>
> Dam to be four feet high will overflow only the one acre of land of said Ezra Slusher, John B. Slusher, Z. T. Slusher and Sarah E. Slusher which is subscribed above. No other lands about or below the dam or mill or canal will be overflowed or deprived of water. Given under our hands and seals July 1st, 1874.
>
> F. W. Edwards, S. F. C.
> Samuel McDaniel
> Greenberry Keith
> S. G. Conduff
> James M. Castle
> P. A. Weeks
> P. H. Fisher
> S. G. Conduff, Jr.
> Early Dickerson
> T. C. Divers
> W. H. Lee
> John L. Caldwell
> Joshua Weddle
> Virginia:
>
> At a County Court held for the County of Floyd on the 14th day of July 1874.
>
> Solomon Slusher vs. John B. Slusher and others, on writ of ad quod damnum for dam and land for a water mill.
>
> This day came the plaintiff and it appearing that the writ of ad quod damnum awarded at last term has been issued duly executed on day mentioned in the order and returned under seal of the Sheriff and Jury and that all the defendants have been properly notified and that the applicant has completed his saw and grist water mill useful to the public, and it appearing by granting the leave asked for in the application that the mansion house, out houses and yards, gardens, orchards of no person will be taken or overflowed, and that the health of the neighbors will not be annoyed, leave to erect the dam and cut the canal is hereby granted for the height, width, and length set out in the inquest returned by the Jury. But this order is not to be effective until the applicant pays to the defendants $21.72, the compensation awarded defendants by the Jury and set out in the inquest, and when applicant does pay such sum to the defendants he shall be seized in fee simple of the land circumscribed by the Jury. The applicant is to erect over the dam a fish ladder similar to ladders required by law over all dams.
>
> It is further ordered that said inquest together with this order be recorded in deed book of the county. A Copy. Teste. J. Godbey[?], Clerk.

89. George W. Slusher and Dr. Will G. Hylton Mill

South Locust Street, Floyd

LOCATION: West side of Route 8 (South Locust Street) Town of Floyd, now 114 S. Locust St.; POWER: 10 H. P. International Harvester Mogul engine; STONES: 1 set (Meadows type mill, with vertical stones), buckwheat mill, and feed mill.

History. This mill might more properly belong on the long list of small mills built (mostly after 1915) with the intention of grinding corn one or two days a week using power from a steam engine, an electric motor, or a gasoline engine rather than water power. It is included here because it was apparently open daily during its short history, and because of the relatively large scale of operations, which included the production and sale of corn meal, cattle feeds, and buckwheat flour.

The reference in the first deed (below) to a "roller mill" raises a question: Did the writer of the deed mistakenly refer to the machinery later advertised for sale by Slusher and Hylton — one set of stones for grinding corn and one for grinding buckwheat — as a roller mill? Or did Slusher and Hylton remove the works of a true roller mill, which is typically a complex arrangement of shafting, pulleys, belts, roller mills, and supporting machinery before going out of business less than six months later? The short life of this milling operation (less than one year) would be a record low among water-powered mills and would be financially disastrous. But since the milling machinery and equipment and the engine that it could be taken down and relocated, the loss to investors could have been relatively small. The owners may even have profited when the South Locust Street building was afterward sold or put to some more lucrative use. In contrast, potential alternative uses and the value of a building constructed expressly for use as a water-powered mill could be limited by its design and would definitely be affected by location.

Will Poff ran the Slusher and Hylton mill for part of its short life. After the mill closed, "In the 1920s [the building] held displays for undertaker C. M. Graham. W. J. Ayers' general store operated here 1938–70. [The building is] now a gift shop."[36] Aside from a single deed, below, three *Floyd Press* advertisements offer the only contemporary written records found for this mill:

> [November 20, 1919] Notice, we have a 16 inch Burr Meadows Corn Mill, a buckwheat mill and bolter, also a 10 H. P. Mogul engine, some belts and shafting, all only used a short time. We will sell to a quick buyer at the right price. Slusher and Hylton, Floyd, Virginia
>
> [December 11, 1919] Notice: I have bought the Slusher and Hylton Corn and Buckwheat mill and will operate same at the old place. Am prepared to do your work in this line. Will keep a good supply of grain on hand at all times also buckwheat flour, meal and feed stuff. Your patronage solicited. John G. West, Floyd, Virginia
>
> [June 10, 1920] For Sale: A genuine French burr buckwheat mill, good as new will sell at a bargain for quick sale. Can be seen at Rakes and Robertson's store. John G. West, Floyd, Virginia

Deeds

Floyd County Deed Book 43, page 474, 27 June 1919. W. A. Sowers to George W. Slusher and Will G. Hylton, a certain lot on South Locust Street upon which there is now situated a roller mill.

126. John W. Epperly Mill

Dodd Creek

LOCATION: Route 720 (Epperly Mill Road SW) .8 miles south of Route 221 (Floyd Highway South); POWER: Turbine; STONES: 2 sets and 4 stand Salem Roller Mill; VISIBLE REMAINS: Mill building still standing in good condition; dug race and stacked stone dam abutments.

History. As one of the most accomplished and sought after Floyd County millwrights of the late 19th and early 20th century, John William Epperly (1856–1931) played some part in the histories of many mills near the Town of Floyd. Of the mills still standing in Floyd County in 2010, this, the second of two mills he built for himself, seems as likely as any to survive the 21st century. This is due in part to luck, and to the care of subsequent owners, but is also a testament to the skill of the mill's designer and builder. Epperly built his first mill, now known as the Homer Roberson mill (I, 27) in the mid-1880s. It, too, still stands, but is in poor condition. Epperly was also a skilled blacksmith and gunsmith. Only 44 years old at the turn of the 20th century, he was one of the last men in Floyd County to fully master any of these trades, let alone all three.

Epperly bought the Ira Howard Mill prop-

Both mills designed and built by master millwright John Epperly still stand. His second mill, constructed around 1901, was much larger and technologically superior to his first mill, farther up Dodd Creek. The space in front of the mill has been filled and raised by highway improvements, but the building itself, aside from the loss of paint and the mill bell and the addition of modern wires, pipes, and fuel tanks, looks much like it did during Epperly's lifetime. (Ricky Cox photograph, 2008)

erty and eleven acres of land in 1899, and in 1901 put up the land, mill, and water rights to secure a debt of $1,670 owed to Salem Machine and Foundry Works of Salem, VA, for equipment they had furnished to him. Epperly evidently ran the mill as long as he was able and still owned it when he died in 1931. However, the mill and 28 acres and several other tracts had been signed over to attorneys Brown G. "Peachy" Howard and B. G. Howard, Jr., as security for a loan in 1928. The debt remained unsettled when Epperly died and his creditors forced a sale of his property in July 1932.

The first in a series of new owners of the mill property was Epperly's son Levi (sometimes spelled "Levy") who was living in Norfolk, VA, when he bought the mill and 28 acres at a public auction in July 1932. Like his father, Levi Epperly was known to be a good carpenter and millwright. Frank Webb's notes say that Levi also ran the C.W and. A.T. Harman Mill (III, 88) on West Fork Creek, but his source may have confused Levi G. Epperly with his younger brother Giles M. Epperly, who owned the Harman mill from December 1938 until it was destroyed by a flash flood in August 1943.

In November 1933 Levi Epperly sold his father's mill to Harry D. Hylton, who in turn sold the mill, 28 acres, and all water rights to Rufus Abraham "Abe" DeHart, who had also succeeded John Epperly as an owner of the Homer Roberson Mill (I, 27) farther up Dodd Creek. Twelve years later DeHart sold the mill to Claude S. Weddle and Douglas Gray. Weddle, who is supposed to have operated the mill, later sold his interest in "Epperly Mill Property with all machinery" to his partner, Gray, who sold most of the machinery for scrap sometime before selling the mill and 2 acres, more or less, to T.L., H.D., J.L., and L.A. Hylton in 1975. The mill has had a number of owners since then, and has served as a home, a food co-op, a coffee shop, and an art gallery.

The building is in remarkably good condition, due in part to its having been remodeled to serve as a home and business. A few pieces of machinery, no longer operable, remained in the building in the mid–1990s.

Millers who worked here but did not own a part of the business include John Epperly's son Cal and Joe Terry, himself a well-known miller and millwright. For earlier history of the site of the John Epperly mill, refer to the Ira Howard Mill (IV, 31).

In its heyday, the Epperly mill was, like many other large mills, both an economic and social centerpiece of the community it served. This bit of society news appeared in the *Floyd Press* on July 2, 1908:

A FISHING PARTY

The young ladies of the town enjoyed a fishing party near Epperly's mill very much last Friday. Those present were: Misses Maidai Howard, Mayday Hall, Eliza Terry, Ola Howard, Lucy Stump, Nancy Howard, Elizabeth Pendleton, Anna Bette Howard, Annie Stump, Kate Howard, Mary Spencer Pendleton, Gretchen Howard, Elizabeth Hall, Elizabeth Howard, and Mrs. Stump chaperoned the young people. As it rained the greater part of the afternoon not much fishing was accomplished, but the rain did not prevent them from having a delightful time in the mill. After a delicious luncheon was served, the merry crowd started for home.

The mill pond also drew turtle hunters, one of whom drowned in 1940. (Typesetting errors in the August 15, 1940, *Floyd Press* article have rendered uncertain the victim's first and last names.)

Clara Burton Carper Drowned

Thirty year old Clary Burton Carper, of near Ferrum, drowned about 3 o'clock August 3rd when he fell into the Epperly Mill pond a mile south of here.

Dr. J. M. Harman, county coroner, said he was informed the man was subject to heart attacks and declared that Carter apparently suffered an attack, fell into the pond and drowned. Carter was seeking turtles with his two brothers, John and Charles, who were some distance away when the mishap occurred. The man's disappearance was noted some 20 minutes after he fell into the pon when one of his brothers found his hat floating on the water's surface.

The body was located by Bobby Shelor, 14, and was brought to the surface by Joseph DeHart, 17, of here.

Deeds and Other Public Records

Floyd County Deed Book 26, page 244, 3 October 1899. Walter S. Tipton and Pattie L. [Howard] Tipton, Maud L. Howard and T. H. Howard, her husband, and Cassie Dobyns to J. W. Epperly, mill property, lands and water rights and privileges known as the Ira Howard Mill property. (Maude L. Howard, T. H. Howard and Cassie A. Dobyns had bought the mill from the Tiptons and paid for it, but no deed had yet been recorded when Epperly bought the property, so the Tiptons are also listed as sellers in the deed.)

D. B. 27, page 496, 8 April 1901. John W. Epperly and wife, Nancy, to J. L. Thompkins, Commissioner, 11 acres land and mill thereon and all water rights to secure debt of $1,670.00 to Salem Machine Works.

D. B. 29, page 405, 30 December 1902. Salem Machine Works to J. W. Epperly, release of Deed of Trust to secure payment of $1,670.00 for mill equipment furnished J. W. Epperly for his mill.

D. B. 48, page 24, 6 April 1928. Deed of Trust. John W. Epperly, widower, to B. G. Howard and B. G. Howard, Jr., trustees, 3 tracts including mill tract of "11 acres and all appurtenances thereto belonging" to secure notes held by Peoples Bank of Floyd County. (Epperly had evidently not repaid the loan secured by this deed of trust when he died in 1931.)

D. B. 50, page 216, 5 August 1932. B. G. Howard, Trustee, to Levy Epperly lot number 4 containing 28 acres and J. W. Epperly mill property and water rights.

D. B. 50, page 218, 5 August 1932. Deed of Trust, Levy G. Epperly to B. G. Howard, Trustee, mill property and 28 acres land with all water rights.

D. B. 51, page 45, 21 November 1933. Levy G. Epperly to Harry Hylton, 28 acres land and roller mill thereon with all water rights.

D. B. 54, page 23, 12 November 1936. Harry Hylton to R. A. DeHart, mill property and 28 acres land on Dodd Creek and all water rights.

D. B. 62, page 510, 26 January 1945. R. A. DeHart to Claude S. Weddle and Douglas A. Gray, parcel of land, 28 acres, and mill locally known as the J. W. Epperly mill.

D. B. 65, page 160, 22 October 1946. Claude S. Weddle to Douglas Gray, his undivided interest in Epperly Mill property with all machinery.

D. B. 114, page 229, 12 June 1975. Douglas Gray to T. L. Hylton, H. D. Hylton, J. L. Hylton, and L. A. Hylton, the Epperly Mill property and 2 acres plus or minus.

D. B. 120, page 201, 4 June 1976. T. L., H. D., J. L., and L. A. Hylton to Harold N. Flinsch, 2 acres and Epperly Mill property.

D. B. 120, page 204, 2 March 1977. Harold N. Flinsch to Floyd County Food Co-op, 2 acres and mill property thereon and all water rights.

D. B. 140, page 552, 31 August 1982. The Floyd Mill Co-op to Harold N. Flinsch, 2 acres mill and water rights.

D. B. 140, page 756, 14 January 1983. Harold N. Flinsch to Anna M. Smith, 2 acres and property known as the Floyd Mill Co-op.

D. B. 170, page 246, 22, September 1989. Anna M. Smith to Kalinda Wycoff, 60 percent interest, and Theresa L. Cook, 40 percent interest, in 1.6727 acres and real estate thereon.

129. Homer Thomas Mill/Eli S. Sowers Mill

Unnamed Tributary of Burks Fork Creek

LOCATION: On tributary of Burks Fork on Route 604 (Halls Store Road SW) just above where road crosses creek, about 1 mile west of Route 799 (Conner Grove Road SW); POWER: Overshot wheel

History. Homer and Ella Thomas, who are said to have built this mill, bought the land on which it stood in two tracts in 1921. The relationship between the seller, H. B. Thomas, and Homer Thomas is unknown, but Louvenia and Rufus Harris, who had sold the land to H. B. Thomas three years earlier, were Ella Thomas's parents. "Eller" is supposed to have run another mill (IV, 41) she and Homer acquired by sale or inheritance from her mother in 1936, and it is likely that she helped with this one as well.

Eli Samuel Sowers (1872–1956), a nephew of the Eli Sowers who built the Rufus Harris mill (IV, 41) on Burks Fork, bought this mill from Homer and Ella Thomas in 1927. The deed recording that sale includes the only deed book reference to this mill. The mill tract was included with land sold to Buren Hall in 1934 or 1944 but is not mentioned in either deed. Hall tore the mill down.

Deeds

Floyd County Deed Book 43, page 349, 27 December 1918. Louvenia Harris and Rufus Harris, her husband, to H. B. Thomas (McDowell County, West Virginia) for $1,500 paid to A. F. Harris, 2 tracts of land, one of 17½ acres and one of 2⁹⁄₁₀ acres on the headwaters of Burks Fork Creek.

D. B. 45, page 275, 7 June 1921. H. B. Thomas to Homer Thomas and Eller Maude, two tracts of land, one of 17½ acres, one of 2⁹⁄₁₀ acres.

D. B. 45, page 275, 11 June 1923. H. E. Thomas to Eller M. Thomas, sells his ½ and entire interest in two tracts of land, 17½ acres and 2⁹⁄₁₀ acres.

D. B. 46, page 344, 16 July 1927. Eller M. Thomas to H. E. Thomas, her ½ interest in two tracts of land, one of 17½ acres and one of 2⁹⁄₁₀ acres.

D. B. 46, page 403, 5 September 1927. H. E. Thomas and E. M. Thomas, his wife, to E. S. and Virginia Sowers, two tracts land one of 17 plus acres and one of 2 plus acres adjoining each other and "mill on the same with all appurtenances."

D. B. 52, page 196, 1 August 1934. E. S. Sowers and wife to B. H. Hall and wife, 17 acres land on the waters of Burks Fork Creek.

D. B. 77, page 182, 27 December 1944. E. S. Sowers to B. H. Hall, 12¾ acres land on headwaters of Burks Fork Creek.

130. A. W. Stanley Mill

Taylor's Creek

LOCATION: Route 603 (McAlexander Road SW), .2 mile west from Route 632 (Shelor Road SW) between the road and Taylor's Creek, on the site of the Chris McAlexander Mill (IV, 53). POWER: Wooden overshot wheel. Water diverted from creek into dug race on south side (opposite the mill) of Taylor's Creek. VISIBLE REMAINS: Dug race

History. None of the deeds by which A. W. Stanley acquired the three tracts of land he owned when he died in 1917 mentions a mill. Nor was there mention of a mill when the 36-acre tract that included the mill site was assigned to his widow, Ella or Ellen Stanley, following his death. The disposition of that land was the object of two chancery suits filed in October 1928 (Floyd County Chancery Court File Nos. 3272 and 3273) but neither the public notices of the upcoming auction nor any other documents on file make reference to a mill. In fact, the original handbills included in the chancery case files point out, using what was probably a standard phrase in rural real estate advertisements of that era, that the location of the property is "convenient to schools, churches, and mills." Neither deed conveying the land to two subsequent owners mentions a mill.

As shown in numerous other cases, the

"silence" of court documents in this regard proves nothing, but it does invite speculation that the A. W. Stanley Mill had fallen into disrepair and was abandoned if not gone by the late 1920s, and so was unworthy of mention in a deed or advertisement. Relatives of the Bowman family who owned the 36-acre mill tract between 1929 and 1932 have stated, somewhat ambiguously, that "the Bowmans owned the mill before the McAlexanders did."[37] The most logical reading of this statement would seem to be "the Bowmans owned the [McAlexander] mill before the McAlexanders did [own the McAlexander mill.]" But since Chris McAlexander built his mill after the Bowmans sold the land in 1932, on the same spot where the Stanley mill had stood, the Bowmans' reference to "the mill" had to mean either the Stanley Mill, in which case the building was still there in 1929, or merely the mill site or mill "seat" on which the McAlexander Mill was afterward built.

The origins of the mill are even more obscure. It may have been there when Stanley bought the land, and for who knows how long before, which is consistent with the idea that it was gone or not worth saving by 1934. Or, since it was a very small, simple mill, it may still have been fairly new and structurally sound in 1934, but configured so differently, in its use of an overshot wheel, as to make it easier to tear the thing down and start from scratch.

Whatever the exact circumstances of its beginning or its end, the mill was unusual in having been served by a dug race that lay on the opposite side of the creek from the mill. A wooden race, a.k.a. box race or sluice, carried the water from the lower end of the dug race across the creek to the mill. Only a few other mills (including IV, 41 and IV, 57) are known to have employed this design. In this instance there was definitely more room to build the mill on the south side of the creek than on the north, and the turbine would have worked as well either way, but putting the mill across the creek from the road would have required the building of a bridge or the maintenance of a ford. It was apparently decided that in the long run it would be easier to get the water from one side of the creek to the other than to do the same with customers.

A. W. Stanley also ran an up and down (sash type) sawmill below his grist mill. Sawing only on the downward stroke, the mill went through a log so slowly that Stanley is supposed to have started a cut, then went to the field to plow corn while the saw worked its way through a log. This rate is glacial by today's standards, but would have seemed marvelous to men who were familiar with a two-man pit saw.

Deeds

Floyd County Deed Book 48, page 271, 25 February 1928. J. E. Proffitt, Commissioner in chancery suit against Ellen Stanley, widow of A. W. Stanley, to Walter Hylton, "36 acres more or less being originally two tracts known as the AW Stanley homeplace."

D. B. 49, page 441, 18 February 1929. Walter Hylton to W. J. Bowman, 36 acres known as the dower lands of Ellen Stanley.

D. B. 50, page 256, 21 September 1932. J. E. Burwell, Commissioner for W. J. Bowman, Deceased, to G.W. McAlexander and assigned to Dollie V. McDaniel, 36 acres more or less.

D. B. 56, page 114, 15 December 1937. Dollie V. McDaniel to Chris McAlexander, 36 acres. "Exempted from and not conveyed by this deed about one acre of said tract of 36 acres ... upon which is situated the corn mill" and "all water rights belonging to and appertaining to the conduct and operation of said mill."

D. B. 63, page 523, 1 February 1946. Dollie V. McDaniel to Chris and Goldie J. McAlexander, two tracts of land, "Tract No. 1 containing 1 acre, more or less, and being the remainder of a 36 acre tract of land ... known as the corn mill lot."

---

131. William DeHart Mill

Laurel Fork Creek

LOCATION: Private road, approximately one-half mile downstream from Blue Ridge Parkway at milepost 175; POWER: Overshot wheel; VISIBLE REMAINS: None

History. Like several other mills in the southwest corner of Floyd County, the DeHart Mill was built on land that became available for development only after the huge holdings of the Lee and later Burwell families were broken up following the death of Nathaniel Burwell, Sr., in the 1870s. His son and executor, Charles W. Burwell, sold 98 acres to William DeHart, his wife Oney, and their sons William B. and Charles T. DeHart in 1880. Natives of Patrick County, the DeHarts appeared in the Floyd County census for the first time in 1890. The deed stipulated that William B. was to have his choice of one acre on Laurel Creek on which to build a mill. In 1883 William B. sold his interest in the Burwell land to his parents and brother, having already sold the one-acre tract selected for a mill site to Isaac Jones. Along with the mill site, Jones acquired the right to flood the upstream land of William DeHart with a mill pond "as long as mill or machine is in operating or in running order."

In 1890 Jones sold the mill to John T. Agee, who, after only one year, sold it to William, Oney, and Charley DeHart, who still owned the original Burwell tract from which the mill lot had been taken. After two years they sold the mill and lot, along with the rest of the original 98 acres, to D. S. and J. W. Scott.

The mill changed hands several more times in the next decade and was still listed as part of the property when Mary M. Cox and her husband, D. W., bought it in 1910. The mill was not mentioned when the Coxes sold the land in 1917, and the rights to water crossing over the property were reserved for E. B. Mabry, who by that time was running Mabry Mill (IV, 60) a short distance downstream.

The reference to a "corn mill" in the last deed, below, indicates that this was a small mill with one run (pair) of millstones and none of the more complex machinery required for producing flour. The size of stream no doubt limited the capacity of the mill, as well. Aside from deed book records, all that was found about this mill is the supposition that Emmett Weddle (see last deed in the following list) tore down the mill and used the lumber to construct or improve other buildings on the property. In a January 7, 2008, interview, long-time Meadows of Dan resident Coy Lee Yeats stated that the William Dehart mill, which he knew as the Scott Mill, closed before 1910, in time for its works to be used by Ed Mabry in constructing his now famous mill.

Deeds

Floyd County Deed Book "P," page 570, 26 July 1880. Charles W. Burwell to William DeHart, William B. DeHart, Charles T. Dehart, and Oney DeHart, 98 acres on Laurel Fork Creek. "Said Wm. B. DeHart is to have the full and complete control fee simple title to one acre of said land on Laurel Creek for purpose of erecting a Mill which is to be selected by him."

D. B. "Q," page 463, 14 April 1883. William B. DeHart to William and Charley T. DeHart, 98 acres. *This deed conveys his interest in the land the three bought together, not the mill.*

D. B. "R," page 417, 11 March 1882. William B. DeHart to Issac Jones, "one acre for a mill site together with rights and privileges to overflow land of William DeHart as long as mill or machine is operating or in running order."

D. B. "U," page 639, 17 January 1890. Issac Jones and Sarah, his wife, to John T. Agee, one acre "on which a mill now stands on the waters of Laurel Fork Creek" and water rights.

D. B. "V," page 12, 22 January 1891. John T. and Mary Agee to William DeHart, Oney DeHart and Charley T. DeHart, one acre "upon which a Mill now stands" and mill, being same lot purchased from Issac Jones for a mill site of William B. DeHart.

D. B. "W," page 366, 15 February 1893. William DeHart and Oney, his wife, to D. S. and J. W. Scott, 98 acres and mill on Laurel Fork Creek.

D. B. 29, page 451, 25 May 1903. V. M. Sowder, Commissioner in Chancery Cause of *W. S. Scott vs. J. W. Scott, party of the first part, and C. D. Scott, of the second part,* to C. D. Scott, 98 acres and mill known as the William DeHart Mill on Laurel Fork Creek.

D. B. 31, page 219, 25 May 1903. C. D. Scott to J. R. Beverly, 1 acre and mill.

D. B. 35, page 603, 14 December 1908. J. R. Beverly to C. D. Scott, 1 acre and mill.

D. B. 37, page 115, Deed Not dated. Notary in Washington County, Alabama, certified signatures on 22 October 1910. C. D. Scott to Mary Cox, 100 acres on the headwaters of Laurel Fork Creek, including one corn mill and all lumber on the place.

D. B. 42, page 453, 2 April 1917. Mary M. Cox and D. W. Cox, her husband, to Emmett Weddle, 100 acres on Laurel Fork Creek. No mill mentioned, but water right is reserved to E. B. Mabry.

134. Shelor Furnace Mill/New Castle Forge Mill

Furnace Creek, Tributary of West Fork of Little River

LOCATION: About .25 mile northwest of the intersection of Route 605 (Happy Valley Road SW) and Route 724 (Old Furnace Road SW). POWER: Vertical wheel, probably overshot. STONES: At least 1 run. VISIBLE

REMAINS: Part of dug race and water wheel pit.

History. The only specific mention of a grist mill on this property appears in a Montgomery County deed written in 1821. The deed refers not only to the grist mill and a sawmill, but also to a furnace for smelting cast iron, which was done at that time by mixing together and burning iron ore, limestone, and charcoal, all typically mined or manufactured nearby. The molten iron could be poured into molds to produce cast iron cookware, wash pots, and other items, or channeled into a series of shallow trenches and allowed to cool into cast iron "pigs" (hence "pig iron").

The on-again, off-again history of the furnace and of the iron and copper mines nearby is recounted in "The Furnace Stack," a three-part series of articles collected in *Mountain Memories, Collection Number 2*, copyright 1983 by Laurel Publications. Even though the high copper content of the iron ore from this property is supposed to have made it especially suitable for the casting of cooking pots, which were less brittle than those made of ordinary cast iron, the furnace seems never to have run for more than a few years at a time.

The use of the word "forge" in the 1821 deed could suggest that there was also a facility here to refine the brittle cast iron pigs into bars of softer, more malleable wrought iron, which can be shaped, cut, or welded as needed by a blacksmith. However, none of the three parts of the article cited above mentions any such facility or activity.

The one sure connection between this water power site and iron smelting is that a water wheel probably drove a bellows which supplied the extra oxygen needed to raise the temperature inside the furnace above the melting point of iron. Water power could also have driven the hammers needed to crush the iron ore before it was poured into the furnace. Whether the same wheel drove the grist mill and sawmill, which were on a different but probably adjoining tract, and whether their years of operation overlapped with those of the furnace, are unknown. Considering that there are no known mills close by, and that cornmeal was always in demand whatever the economic climate, it is likely that the grist mill was a more stable venture than the furnace, albeit no more likely to generate a cash profit for the owner. Yet there is no reference to a mill in any of the four Floyd County parcels acquired by Robert Toncray, nor is there reference to a mill when his lands were sold at auction by commissioner F. A. Winston in September 1871.

The furnace was enjoying one of its brief periods of prosperity when this advertisement appeared on page 1 of the *Floyd Intelligencer* on May 5, 1855:

R. L. Toncray
MANUFACTURER AND DEALER IN
Hollow Ware

Keeps constantly on hand, at his Furnace, a large assortment of Hollow Wares, of every variety suitable for country purposes; which he will dispose of on accommodating terms for cash.

The massive iron-smelting furnace remains largely intact, although the upper portion of the native stone furnace stack was, according to the article cited above, removed without the owner's permission and used by the state highway department to repair a washed out bridge. This and the mill remains described above are on the property known today as the Shelor Park.

Deeds

Montgomery County Deed Book "H," page 44. 1 September, 1821. Castilio Hill and Joseph Townes to Robert T. Woods and Abraham Jones, a parcel of land on West Fork Little River and known by the name of Shelor Furnace, containing 240 acres with all appendages of said furnace; also a tract of land containing 15 acres on which stands a forge, sawmill and grist mill. Known by the name of New Castle Forge.

Epilogue

When weekly or monthly trips to the mill for cornmeal, flour, and livestock feeds were routine events, the local grist and feed mill played a part in the life of every man, woman, and child, and in the lives of most farm animals as well. Even household pets depended on cold cornbread and old biscuits, or on the mice fattened on stored grains and processed feeds. By 1960 most Floyd County farms had stopped raising wheat and "bread corn," but even then our flour and meal producing mills were linked to our dining tables. As long as local mills were competing with Martha White and Gold Medal for shelf space in local stores, and as long as enough buyers believed that the old familiar brands made the best cornbread and biscuits, there was a place for grist and flour mills in Floyd County. But in the fifty years since 1960, a relationship that extends into the past thousands of years has all but ended. In the same innocent way that many of us drive past locally owned retail businesses to embrace the economy and convenience of nationwide chain stores, we have, without malice or conscious intent, abandoned a once indispensable local industry.

In 2010 only one commercial mill is still operating in Floyd County. Vaughn's Mill produces feeds for livestock, using diesel and electric power, and retails various animal feeds. The meal and flour on sale there are made in a Boonville, North Carolina, mill (run by fourth and fifth generation millers from Floyd County, Virginia). Except as a historical demonstration at Mabry Mill, water-powered grain milling in Floyd County is a thing of the past, and the future of milling of any kind is dim, and will stay that way as long as economies of scale and access to transportation make it cheaper to get the things we need from other places than to produce them here. Barring some unforeseen reason to return to the old ways, most of the milling still to be done here will take place in memory or in conversation.

The overview of water-powered grain milling and the individual mill histories in this book draw heavily on impersonal written sources, most of which will probably be around for a good while longer. But building a mill and tearing it down are only the bookends to its history. And a record of its passage, in the meantime, from one owner to the next is only an outline of what happened in between. The work that resulted in this book has been enthusiastic, but not systematic, and many gaps remain. More than twenty years after Frank Webb began this project, there are still many people living nearby who have clear, first-hand memories of "taking a turn" to the mill and waiting to have it ground. But having never formed the habit of writing down their own experiences, they are unlikely to take it up now. And if they did, going to the mill probably isn't what they'd write about. Aside from disaster tales of floods and fires and the occasional suicide or accidental death, stories about mills and millers lack the drama of car wrecks, bear hunts, and moonshine haulers. Whether and how a little boy (or girl) got a half-bushel sack of corn or cornmeal, or him or herself for that matter, astride a patient old horse and on to the mill makes a good tale only in the hands of a skillful teller. But a dozen or more different versions of this childhood adventure can be heard around the county today, and they ought not be forgotten. Many of them will be gone within the next decade, and we will have missed our chance to know and preserve the most human pieces of this story.

The Roman general who conquered ancient Carthage at the end of the Punic Wars is said to have commanded his soldiers to leave no stone stacked atop another. For the water-powered mills that served us so well for so long, that same job is being done more slowly and with less violence, but just as implacably by water, tree roots, frost, fire, and the occasional backhoe or bulldozer. Before it is too late to preserve the stories of these ruins, you, the reader, are urged to stop milling around, put your nose to the grindstone, and set about getting what you can of this history in the sack. There'll be time enough later to separate the wheat from the chaff. Bear in mind that "the mill cannot grind with the water that is past."

Appendix A

Floyd County Mills, Millwrights and Distillers Listed in Virginia Business Directories, 1880–1917

All the information in this appendix comes from the same Virginia business directory. Both the format and the title evolved over its forty-year publication history.

Chataigne's Virginia Business Directory and Gazetteer 1880–81
Richmond, VA: Baughman Brothers
"Floyd County"
Mills: Corn and Flour (* also owned a sawmill, probably at the same site)

Mill Name, Owner or Operator	***Postal/Magisterial District***	***Location in Book (Section, number)***
Boothe's, Geo. B. Boothe	Boothe's Mill	III, 82
*Epperly's, John Epperly	Little River	See Wimmer, II, 105
*Eureka, Matthew Scott	Floyd Courthouse	II, 91
*Graysville, Wertz and Harvey	Graysville	I, 9
*Harman's, Wm. H. Harman	Little River	See Blackburn Akers, II, 102
*Howard's, J. L. Howard	Floyd C. H.	IV, 31
Lancaster's, W. C. Lancaster	Turtle Rock	I, 19
Lee's, Keeth and Shesher [Keith and Slusher]	Burk's Fork	IV, 42
*Martin's, Wm. Martin	Floyd C. H.	IV, 28
*Scott's, Walford Scott	Camp Creek	II, 95
Siner's, Peter Siner	Copper Hill	II, 117
Smith's, Jas. W. Smith	Locust Grove	I, 5
*Sowers,' Caleb Sowers	Alum Ridge	II, 94
*Stuart's, John Stuart	Camp Creek	II, 133
*Williams,' G. M. D. Williams and Bro.	Floyd C. H.	I, 25
Wimmer's, Hosea Wimmer	Brush Creek	II, 119

Chataigne's Virginia Gazetteer and Classified Business Directory 1884–85
Richmond, VA: J.H. Chataigne
"Floyd County"
Population in 1880—White 11,981, Colored, 1,274, Total 13,255.

Distillers

Collins, M. J., Jack's Mill; Cox, F. P., Indian Valley; Cox, W. S., Indian Valley; Dulaney, A., Booth's Mills; Graham, William, Little River; Hall, B. R., Floyd Courthouse; Hall, W. L.,Floyd Courthouse; Harman, William H., Little River; Lain, John, Broad Shoals; Lester, Johnson, Broad Shoals; Lester, W. C., Floyd Courthouse; Mills, J. T., Copper Hill; Martin, C., Floyd Courthouse; Martin, W. P., Floyd Courthouse; Morton, John W., Flint; Phillips, Jefferson D., Mira Fork; Poff, William H., Jack's Mill; Poff, W. J., Copper Hill; Reed, C. D., Broad Shoals; Shortt and Robertson, Floyd Courthouse; Weddle, A.,Floyd Courthouse.

Mills: Corn and Flour (* also owned a sawmill, probably at the same site)

Owner or Operator	*Nearest Community or PO*	*Location in Book (Section, number)*
*Booth, A.	Booth's Mill	II, 93
*Booth, G. B.	Booth's Mill	III, 82
*Duncan, Joshua	Turman	Mill in Carroll County, VA, see end note 1
Epperly, Henry	Little River	II, 103
Fowler, N. T.	Copper Hill	See Huff Mill I, 2
*Harman, William H.	Little River	See Blackburn Akers, II, 102
*Harris, J. W.	Burks Fork	IV, 44
Harvey, R. O.	Turtle Rock	I, 19
Howard, J. L.	Floyd Courthouse	IV, 31
Huff, Byrd	Copper Hill	I, 1
Hurt, J. W.	Alum Ridge	III, Unknown.
Hylton, H. P.	Hylton	III, 69
Lemon, J. H.	Jack's Mill	I, 5
*Link, Albert	Flint	II, Unknown
*Link, Joel H.	Flint	II, Unknown
Martin, Moses	Little River	Later owned II, 107
*Martin, William	Floyd Courthouse	IV, 28
Phillips, H. C. and N.	Turman	III, 63
Phillips, Noah	Showalter	May be the same as III, 63
*Phillips, T. D.	Phillips	III, 65
Potter, C.	Burks Fork	IV, 47
Price, J. T.	Jack's Mill	Later owned I, 5
Ridinger, A.	Basham	II, 118
*Scott, Matthew	Floyd Courthouse	II, 91
*Scott, Walford	Camp Creek	II, 95
Siner, Peter	Copper Hill	II, 117
Slusher, A.	Burks Fork	Possibly Archeleus Slusher. See IV, 44
*Slusher, J. B. and Co.	Mira Fork	See IV, 66; see endnote 2
Smith, Alexander	Phillips	Unknown; see endnote 3
Smith, James W.	Simpson's	Unknown; see endnote 4
Spangler, George	Hylton	III, 69
Spangler, Solomon	Duncan's	III, 72
*Spence, Lewis P.	Indian Valley	III, 73
*Stuart, John	Camp Creek	II, 133
*Turman, C.	Mira Fork	IV, 66
*Turman, George H.	Mira Fork	III or IV, Unknown
Wade, Noah	Alum Ridge	III, 76
Wertz, J. W.	Graysville	I, 9
*Whitlock, Albert	Booth's Mill	II, 92
*Whitlock, D. F.	Camp Creek	II, 92
Williams, G. M. D. and Bro.	Floyd Courthouse	I, 25
*Wimmer, F. D. [F.B.?]	Huffville	Probably II, 105
*Young, Thomas	Indian Valley	III, Unknown

Mills — Woolen

Eller, George,Booth's Mill; Vaughn and Brothers, Burks Fork.

Millwrights

Darnall, R. F., Floyd Courthouse; Duncan, George R., Alum Ridge; Duncan, J., Duncan's; Ekiss, J. D., Little River; Epperly, John, Huffville; Fowler, N. T., Copper Hill; Goodson, T., Camp Creek; Keeth, Zebrana [Zebram], Burks Fork; Mitchell, John H., Indian Valley; Phillips, T. D., Phillips; Purvater, C. T., Flint; Ramon, Y. D., Terry's Fork; Slusher, Allen, Mira Fork; Slusher, J. B., Mira Fork; Sowder, William, Jr., Copper Hill; Stuart, John, Camp Creek; Wall, Daniel, Little River; Whitlock, D. F., Camp Creek; Wimmer, Hosea, Terry's Fork.

Chataigne's Virginia Gazetteer and Classified Business Directory 1893–94
Richmond, VA: J. H. Chataigne,
"Floyd County"
Population in 1890—White 13,183, Colored 1,210, Total 14,393.

Distillers

Basham, M. F., Basham; Booth, A. A., Posey; Cannaday, Z. T., Amos; Custer, W. P., Check; Dulaney, E., Alum Ridge; Earls, John, Weddle; Epperly, William, Huffville; Gerhart, Johnson, Copper Hill; Graham, C. L., Little River; Hale, J. L., Jack's Mill; Hall, B. R., Weddle; Hall, P. P., Turtle Rock; Hall, W. L., Floyd; Hannan, Wm. H., Little River; Harris, J. G., Mira Fork; Lain, John, Broad Shoals; Lawrence, W. B., Flint; Lee, A. J., Floyd; Lester, Johnson, Broad Shoals; Lester, R. E., Posey; Lester, W. C., Floyd; Manning, V. T., Jack's Mill; Martin, C., Floyd; Martin, W. P., Floyd; Montgomery, M. C., Turman; Morton, J. W., Flint; Nester, Andrew, Quoit; Phillips, J. D., Mira Fork; Phillips, Mahlon, Turman; Poff, Joshua, Hylton; Poff, J. J., Check; Pratt, Geo. W., Santos; Quisenberry, Edward, Indian Valley; Richards, Berry, Jack's Mill; Short, J. D., Floyd; Slusher, John, Mira Fork; Thomas, J. T., Camp Creek; Thomas, S. H., Camp Creek; Weddle, John D., Weddle; Williams, Wm., Quoit; Wilson, James W., Copper Hill; Wimmer, Wm. J., Copper Hill.

Mills: Corn and Flour (* also owned a sawmill, probably at the same site)

Owner or Operator	*Nearest Community or PO*	*Location in Book (Section, Number)*
*Akers, Beauford	Beauford	III, Unknown
*Barringer, W. J. H.	Pax	IV, 40
Booth's Mill, A. A. Booth	Posey	II, 93
Bower, Robt.	Howery's	Bro. of O. C. Bower, II, 96
Burnett, A.	Indian Valley	III, 68
Cannaday, Otho	Howery's	Owned I, 16 in early 1890s
Custer, M. H.	Check	I, 8
*Duncan, Beauford	Beauford	See III, 77
Duncan, Joshua	Truman [Turman]	See endnote 1
*Epperly, Floyd	Huffville	II, Unknown; see endnote 5
Epperly, Henry	Little River	II, 103
*Epperly, John	Weddle	I, 27
Hambrick, George	Copper Valley	III, 136
*Harmon, John	Abraham	III, 88
*Harman, Wm. H.	Little River	See Blackburn Akers, II, 102
Harris, B.	Pax	IV, Unknown; see endnote 6
*Harris, J. W.	Burks Fork	IV, 44
Harvey, R. O.	Pizarro	I, 19
Howard and Dobyns	Floyd	IV, 31
*Howell, S. F.	Camp Creek	II, Unknown
Harvey, R. O.	Turtle Creek	I, 19 Same as 3 rows above.
Huff, G. W.	Copper Hill	Possibly I, 1; see also I, 2, and endnote 7
Huff, Robert	Copper Hill	I, 1
Hylton and Co.	Hylton	III, 69
Jennings, John	Huffville	II, Unknown; see endnote 8
Link, Albert	Flint	II, Unknown
Link, Joel	Flint	II, Unknown
Martin, Moses	Little River	II, 107
*O'Neil, J. W.	Mira Fork	Possibly IV, 49; see endnote 9
Phillips, Noah	Turman	III, 63
Phillips, Reed	Indian Valley	III, 65, son of T.D. Phillips
*Phillips, T. D.	Phillips	III, 65
*Poff, J. J.	Check	II, 114
*Poff, W. H.	Jack's Mill	I, 6
*Potter, W. C.	Vaughn	IV, 47
*Price, J. T.	Jack's Mill	I, 5
*Ridinger, A.	Basham	II, 118
*Roberts, J. M.	Flint	II, 107
Scott, M.	Floyd	II, 91
Shelor, R. O.	Alum Ridge	May be Rufus O. Shelor. Owned III, 72 in 1860s

Owner or Operator	Nearest Community or PO	Location in Book (Section, Number)
Short, A. D.	Turtle Rock	I, 22
Simmons, Obadiah	Weddle	IV, 38
Simpson, Wm. A.	Howery's (?)	II, 113
Siner, Peter	Copper Hill	II, 117
Slusher, A.	Burk's Fork	Possibly Archeleus Slusher. See IV, 44
Slusher, Romney	Santos	Unknown; see endnote 10
*Slusher, J. B. and Co.	Mira Fork	See IV, 66 and endnote 2
Smith, Alex	Phillips	See endnote 3
Smith, Eli	Weddle	IV, 33
Smith, J. W.	Simpson	See endnote 4
Sowers, D. L.	Alum Ridge	III, 76
Spangler, Geo.	Hylton	Owned one-half interest in III, 69 in 1883–85
*Spangler, Solomon	Duncan's	Owned III, 72 until c. 1891
Spence, J. F.	Turman	III, Unknown
Turman, Elijah	Mira Fork	IV, 66
*Wade, Eli	Abraham	III, 87
*Wade, Noah	Beauford	III, 76
*Walters, John	Walcott	II, Unknown
Wertz, J. W.	Graysville	I, 9
Whitlock, J. A.	Camp Creek	II, 92
Wimmer, H. A.	Huffville	II, 119
*Woodward, J. H.	Copper Hill	II, 115

Mills — Woolen

Eller, Geo., Posey; Martin, James, Jack's Mill; Morton, W. M., Santos; O'Neal, J. W., Mira Fork; Vaughn, C. P. and Bro., Mira Fork.

Millwrights

Akers, W. B., Huffville; Boone, R. T., Flint; Clover, W. L.,Walcott; Darnell, R. F., Floyd; Dickerson, Riley, Walcott; Duncan, Lynch,Duncan's; Ekiss, J. B., Little River; Epperly, John, Weddle; Goodson, T. M., Camp Creek; Hylton, Henderson, Indian Valley; Hylton, Silas, Santos; Keith, S., Pax; Keith, Z., Hylton; Light, H. W., Flint; Phillips, T. D., Phillips; Ramon, Y. D., Terry's Fork; Sowder, Wm., Jack's Mill; Wall, Daniel, Little River; Wimmer, Hosea, Terry's Fork.

Virginia State Gazetteer and Business Directory 1897–98
Richmond, VA: J. L. Hill Printing
1897
"Floyd County"

Mills: Corn and Flour (* also owned a sawmill, probably at the same site)

Owner or Operator	Nearest Community or PO	Location in Book (Section, Number)
*Akers, Beauford	Beauford	III, Unknown
*Barringer, W. J. H.	Pax	IV, 40
*Booth's Mill, A. A. Booth	Posey	II, 93
Booth, J. H.	Camp Creek	II, 95
Bower, Robt.	Howery's	I, Unknown
Burnett, A. D.	Indian Valley	III, 68
Cannaday, Otho	Howery's	Owned I, 16 in early 1890s
*Carr, William O.	Simpson's	II, 111
Custer, M. H.	Check	I, 8
Dulany, William	Dulany	III, 82
*Duncan, Beauford	Beauford	See III, 77
Duncan, Joshua	Truman [Turman]	See endnote 1
Duncan, N. and M.	Ego	III, Unknown
Duncan, Wesley	Phillips	Mill is in Carroll County, VA See endnote 1

Owner or Operator	*Nearest Community or PO*	*Location in Book (Section, Number)*
*Epperly, Floyd	Huffville	II, Unknown. See endnote 5
Epperly, G. W.	Little River	II, 103, son of Henry Epperly
Epperly, Henry	Little River	II, 103
Epperly, John	Weddle	I, 27
Epperly, W. J.	Narcott	II or III, Unknown
Hambrick, George	Copper Valley	III, 136
Harman, Mary	Abraham	III, 88
Harman, P. Miss	Topeco	Possibly III, 88
*Harman, Wm. H.	Little River	See Blackburn Akers, II, 102
Harris, B.	Pax	IV, Unknown. See endnote 6
*Harris, J. W.	Burk's Fork	IV, 44
Harvey, R. O.	Pizarro	I, 19
Harvey, R. O.	Turtle Rock	I, 19
Helms, G. W.	Santos	Part owner of IV, 42, c. 1912 See IV, 44
Helms, Joseph	Santos	See IV, 44.
*Hylton and Co.	Hylton	III, 69
Jennings, John	Huffville	II, Unknown. See endnote 8
Keeth, G.	Santos	IV, Unknown
Martin, Moses	Little River	II, 107
*O'Neil, J. W.	Mira Fork	Possibly IV, 49. See endnote 9
Phillips, F. R.	Phillips	III, 65
Phillips, H. C. and Co.	Turman	III, 63
Phillips, Reed	Indian Valley	III, 65, Same as 2 rows above
Phillips, T. D.	Phillips	III, 65, Same as 3 rows above
*Poff, J. J.	Check	II, 114
*Potter, W. C.	Vaughn	IV, 47
*Price, J. T.	Jack's Mill	I, 5
Price, Thomas	Simpson's	If J. Thomas Price, could be same as 1 row above
Reed, A. J.	Dulany	III, 82
*Ridinger, A.	Basham	II, 118
Scott, S. and W.	Floyd	II, 91
Shelor, R. O.	Alum Ridge	May be Rufus O. Shelor Owned III, 72 in 1860s
Short, A. D.	Turtle Rock	I, 22
*Simmons, Obadiah	Weddle	IV, 38
Simons, O. Miss	Topeco	IV, 38
Simpson, Wm. A.	Simpson's	II, 113
Siner, P. F.	Copper Hill	II, 117
Slusher, A.	Burks Fork	IV, Unknown
*Slusher, Henry	Santos	IV, 42
*Slusher, J. B. and Co.	Mira Fork	See IV, 66. See endnote 2
Smith, Alex	Phillips	See endnote 3
Smith, Eli	Weddle	IV, 33
Smith, J. W.	Simpson	See endnote 4
*Sowers, Caleb	Narcott	II, 94
Sowers, D. L.	Alum Ridge	Owned III, 76 in 1890s
Spangler, Geo.	Hylton	Sold interest in III, 69 in 1885
Spangler, Solomon	Duncan's	Sold interest in III, 72 in 1893
Spence, J. F.	Turman	III, Unknown
*Turman, Elijah	Mira Fork	IV, 66
*Wade, Eli	Abraham	III, 87
*Wade, E. T.	Topeco	May be same as 1 row above
*Wade, Noah	Beauford	III, 75
*Walters, John	Walcott	II, Unknown
Wertz, J. W.	Graysville	I, 9
Whitlock, J. A.	Camp Creek	II, 92
Wimmer, H. A.	Huffville	II, 119
*Woodward, J. H.	Copper Hill	See Carr Mill, II, 115

Mills — Woolen

Eller, Geo., Posey; Morton, W. M., Santos; O'Neal, J. W., Mira Fork; Vaughn Bros., Mira Fork; Williams, C. E., Flint.

Millwrights

Akers, W. B., Huffville; Boone, R. T., Flint; Clover, W. L.,Walcott; Dickerson, Riley, Walcott; Duncan, Lynch, Duncan's; Ekiss, Jas D., Flint; Ekiss, J. W., Little River; Epperly, John, Weddle; Goodson, T. M., Camp Creek; Hylton, Henderson, Indian Valley; Hylton, Silas, Santos; Keith, S., Pax; Keith, Z., Willis; Light, H. W., Flint; Oneal, I. W., Mira Fork; Phillips, T. D., Phillips; Price, J. T., Jack's Mill; Price, Wm. E., Simpsons; Ramon, Y. D., Terry's Fork; Sowder, Wm. J., Jack's Mill; Sowers, Caleb,Narcott; Turner, E., Mira Fork; Wade, E. T., Topeco; Wall, Daniel, Little River; Wimmer, Hosea, Terry's Fork.

Virginia Business Directory and Gazetteer 1911
Richmond, VA: Hill Directory
"Floyd County"

Mills: Corn and Flour (* also owned a sawmill)

Owner or Operator	***Nearest Community or PO***	***Location in Book (Section, Number)***
Agee, J. T.	Meadows of Dan	IV, 58 or IV, 57
Buffalo Roller Mills, Inc.	Willis	IV, 42
DeHart, A. W.	Copper Valley	III, 136
Duncan, N. J.	Copper Valley	III, 79
Epperly Brothers	Huffville	II, Unknown
*Epperly, J. W.	Floyd, Dodd Creek	IV, 126
Epperly, Levy	Floyd, W. F. L. R.	Bought IV, 126 in 1932. Said to have run III, 88 (No record)
Floyd Milling Co.	Floyd	II, 122
Gillespie, C. W.	Pizarro R. F. D.	I, 18
Goodson, T. M.	Tindall	II, 118
Hylton, Newton	Floyd R. F. D. No. 6	IV, 35
Light, Victor	Carthage	III, 78
Lester, T. S.	Pizarro R. F. D.	I, 121
Penelton, E. M.	Floyd	Unknown
Phillips, I. R. and Co.	Indian Valley	III, 65 Probably F. R. Phillips
Poff, Henry	Huffville	II, Unknown
Poff, J. J.	Check	II, 114
Pratt, A. G.	Floyd R. F. D. No. 6	IV, 36
Price, Jno. T.	Copper Hill	Sold I, 5 in 1910
Ridinger, H. J.	Huffville	II, 119
Simpson, W. A.	Pizarro	II, 113
Sowers, T. D.	Sowers	II, 94
Spangler, Waller	Poff	II, 99
*Vest, C. W.	Terry's Fork	II, 110
Willis, S. P. and G. A.	Willis	III, 69

Mills — Woolen

Vaughn, C. P. and Son, Willis; Wimmer, F. B., Huffville.

Virginia Business Directory and Gazetteer 1917
Richmond, VA: Hill Directory
"Floyd County"

Mills: Corn and Flour

Owner or Operator	***Nearest Community or PO***	***Location in Book (Section, Number)***
Bolt, A. and Son	Willis	Possibly IV, 49 (J. E. Bolt)
Bower, O. C.	Camp Creek	II, 96, sold mill in 1912
Copper Valley Milling Co.	Copper Valley	III, 79

Owner or Operator	*Nearest Community or PO*	*Location in Book (Section, Number)*
DeHarts A. W. Corn Mill	Copper Valley	III, 136
Epperly, J. W.	Floyd	IV, 126
Gillespie, Chas.	Pizarro	I, 18
Martin, W. H.	Simpson's R F D No 1	Unknown
Reed, Harvey	Alum Ridge	Possibly III, 71
Ridinger, H. J.	Huffville	Sold II, 119 in 1916
Thomas Roller Mill	Camp Creek	II, 93
Vest, D. W.	Simpson's R F D No 1	II, 106
Willis Roller Mill Co.	Willis	III, 69
Wimmer, F. B.	Huffville	II, 105

Mills — Saw

Bolt, A and Son, Willis; Epperly and Poff, Huffville; Poff, J. J., Copper Hill; Shelor, Geo. W., Floyd.

Mills — Woolen

Vaughn, C. P., and Dunn, Willis.

Millwrights

Custer, J. M., Copper Hill.

Appendix B

Petitions to Build Water Grist Mills

In What Is Now Floyd County, Virginia, as Listed in the Court Order Books of Montgomery County, Virginia

Year	*Petitioner*	*Stream*	*Court Order Book*	*Corresponds to Section, Number*
1782	William Davies	Beaver Dam Creek by the	Book 2, page 335	
1784	Joshua Howell	South Fork of Little River	?, 13 April 1784	III, 138
1795	John Strunk	South Fork, Little River	Book 9, page 19	
1797	Jonathan Elswick	Indian Creek	Book 10, page 30	III, 136
1802	John Lester	Brush Creek	Book 13, page 71	
1803	John Grayson	Camp Creek	Book 13, page 263	
1805	Martin Weddle	West Fork, Little River	Book 14, page 212	IV, 38
1806	Jacob Nester	?	Book 15, page 105	
1812	Abijah Booth	Beaver Creek	Book 17, page 274	III, 82
1814	Blanch Duncan	Indian Creek	Book 18, page 120	
1814	John Elkins	Camp Creek	Book 18, page 172	
1814	Adam Thompson	Burks Fork	Book 18, pages 221 and 222	
1823	Peter Geurrant	Little River	Book 22, pages 150 and 152	I, 05
1828	Christopher Smith	South Fork, Little River	Book 24, pages 81 and 90	IV, 28

Appendix C

Mills Whose Locations Were Not Found

Name(s)	*Stream and Section of Book*	*Pertinent Deeds (D.B, page, date) C. O. = Court Order*	*Was It Ever Built?*
Ashworth, Henry, and John Thomas	Dodd Creek (S. Fork, Little River)	"F," 263, 10/1850	Unknown. Permission granted, but mill not mentioned in a deed.
Bolt, Charles Sr.	Burks Fork Creek, IV	"A," 111, 9/5/1832	Yes. Mentioned in deed.
Duncan, Annias	Lost Bent Creek, III Said to be about .5 mile from Little River	"M," 237, 1/21/1857 "O", 627, 8/31/1872	Yes. Mentioned in deeds.
Hall, Gordon N., and Edward F.	Goose Creek, II	C. O. Book, 1870–77, 310, 7/5/1873	Yes. Mentioned in deed.
Harrison		"O," 568, 4/26/1877	
Hylton, Valentine, and Jonathon Starr	Burks Fork Creek, IV Said to be near Rufus Harris Mill (IV, 41)	C.O. Book, 1864–70, 258, 7/1867	Yes. C.O. legalizes existing mill. See Table B.
Pepper, Joel	Camp Creek, II	C. O. Book, 7/1849	Yes. Acknowledged by court order.
Shank, Nancy	Meadow Run Creek, I	"R," 243, 12/21/1883	Yes. See Table C.
Thompson, Matthew	Burks Fork Creek, IV	Mont. Co. C. O. Book 18, 221–22, 1814 "A," 423, 6/27/1836 "B," 278, 7/4/1839	Yes. Mentioned in deeds.
Thurman, David, Oil Mill	Boxes Branch of Beaver Dam Creek, II	Floyd Co. Road Book, no date given.	Yes. Referred to in Road Book.
Wall, Daniel, and George Smith	Terry's Creek, II Lasted only 10 years.	"Q," 716, 2/14/1882 "Q," 401, 2/15/1882 "R," 639, 2/16/1882 "V," 218, 11/21/1889 "V," 688, 4/5/1890 "W," 98, 8/22/1892	Yes. Mentioned in deeds.

Appendix D

Cost of Rebuilding Asa D. Shortt Mill, c. 1893

(Section I, Number 22)

Mill Rock	15.00
Lumber	25.00
J.W. Epperly Work	54.78
Machinery Ordered	56.00
Plaster Parris	2.75
Mill Measures	.75
Bolts	1.00
P. Haven Work	5.00
E. Lancaster, Work	1.00
Denist Barton, Work	2.50
Martin Turner and Underwood Work	2.75
30 Days Board	7.50
A.D. Shortt Work	25.00
Total	$199.03
Mortice Wheel Shaft and Step Box	7.00
James Lemon Work	.40
Nails	1.50
Bolting Cloth	10.00
Bolting Chest	10.00
Stove	6.75
Floreing Planks	4.00
Sieve and Measures	1.30
Total	$40.95
J. W. Ferguson Timber	1.45
William Janned Work	.40
Bolts	1.00
Bolts	2.10
Part on Saw Mill	100.00
Postage on Same	.49
Remainder on Saw Mill	49.50
Freight on Same	6.08
Cost of Hauling Same	6.65
Saw and Postage	58.60
Bolts and Paint	2.04
Bolts and Shanking Cogs	3.29
Iron Shaft and Postage	25.12
Belt and Postage	17.13
Freight and Hauling	2.12
Total	$275.97
Sherman Nolin Work	1.00
Z. Guthrey to Lumber	1.31

Work on Race	4.00
Turning Rounds	.80
Iron Bolts and Lock for Gate	2.29
Lase (Lace?) String and Nails	.25
J.W. Epperley Work	85.50
Total	$95.15
Asa D. Shortt Timber	10.00
Fleming Turner Work	3.38
Fleming Janney Timber	2.50
James Martin Timber	1.20
150 Days Board	37.50
Asa D. Shortt Work	50.00
Asa D. Shortt Lumber	5.00
Copering and Weather Boarding	75.44
Total	$185.02
Robert Martin	6.70
James Scott	8.00
James Scott	2.65
William Howell	.50
William Howell	.50
Zebron Guthrey	3.50
Peter Via Col.	.50
Peter Via Col.	1.60
Robert Cropff (Kropff)	1.00
Peter Vest	1.50
Noah Underwood	.40
S. and W. Scott Nails	4.50
John Wertz Mill (?)	10.00
Total	$41.35
Prices for Sawing	20.41
Fleming Janney and Others	1.50
S.T. Turner Nails	2.75
Stuart Cannaday Work	3.75
Stuart Cannaday Work	11.45
Henry Weddle Work	4.25
Henry Weddle Work	8.63
Benjamin Tice Work	4.00
Benjamin Tice Work	25.50
Henry Underwood Timber	11.80
Robert Shelton Work	.50
Total	$94.54
GRAND TOTAL	$932.01

Appendix E

Water-Powered Mills Still Standing in Floyd County, 2010

Mill Name	***Stream***	***Use and Condition***	***Location in Book (Section, number)***
Roberson Mill	Dodd Creek	Complete. Poor.	I, 27
Nolen/Pine Creek Mill	Pine Creek	Race and building restored. Good.	I, 123
*M. Poff Mill	Little River	Converted to house. ?	?
Flint Mill	Beaver Dam Creek	Converted to barn. Good.	II, 106
William Allen Simpson Mill	Pine Branch	Empty. Fair.	II, 113
Ezra Wimmer Mill	Brush Creek	Converted to farm building. Fair-good.	II, 119
O. A. Webb Mill	Dodd Creek	Used for storage. Poor.	II, 122
Vaughn's Mill	Mira Fork Creek	In operation.	III, 64
Greasy Creek Mill	Greasy Creek	Empty, Fair-good.	III, 67
Mabry Mill	Mountain Fork Creek	Operating. Good.	IV, 60
George Shelor Mill	Howell Creek	Converted to farm shop. Good.	IV, 32
John Epperly Mill	Dodd Creek	Converted to residential and retail space. Good.	IV, 126.

**Frank Webb's notes stated that a mill that once belonged to M. Poff on Little River had been moved and made into a house and was still standing in 1996, but this is not consistent with any individual mill record.*

Tables A, B, and C: Products of Industry Reports for Floyd County, VA, 1850, 1870, and 1880

Historical Background

After the first population census of the United States was taken in 1790 it became apparent that there was a need to keep track of the economic growth of the country as well as changes in the size and location of the population. In 1810 Congress directed that an economic census be conducted later that year along with the third census of the population, and afterwards at the same ten-year intervals. This *Economic Census*, and the *Products of Industry* reports that made up one part of it, have an interesting history of their own. Much of that story is told in William M. Micarelli's "*Evolution of the U.S. Economic Census: The Nineteenth and Twentieth Centuries.*" This article first appeared in 1988 on pages 335–377 of *Government Information Quarterly,* Volume 15, Number 3, and is available online. Page numbers in parentheses refer to this article.

It is sufficient here to say that both the *Products of Industry* component and the entire *Economic Census* suffered some growing pains. While both are interesting and useful, neither is regarded as very complete or accurate for most of the 19th century.

Regarding the first three attempts in 1810, 1820, and 1840 (1830 was skipped because of the poor results in 1810 and 1820), Micarelli notes, "Many historians and statisticians consider the economic data tabulated and published for these three censuses to be of little value except as indicators of the gross outlines of manufacturing development" (337). Reasons behind this perception include both the inability of the U.S. Marshals' assistants assigned to this job to ask for detailed information about processes they themselves may not have understood, and the fact that many manufacturers simply didn't keep records of the variety and complexity that would have been required to answer, accurately, all the questions asked by enumerators (347–8).

In the tables that follow, the number of zeros in the last two, three, or even four columns of statistics typically expressed in bushels and pounds suggests significant rounding off and estimating if not outright guessing. Manufacturers and tradesmen, including millers, who were at least as suspicious as 21st century business owners of the motives behind government inquiries, were also no doubt reluctant to reveal the details of their businesses. Even if they did trust the men asking the questions, they must have wondered where the figures they supplied would wind up and what use would be made of them.

Thanks to changes in the administration of the census and in the schedules of questions asked by the enumerators, the 1850 version was much improved, but it was still conceded that undercounting was a problem (339). Both this problem and its immediate cause, at least in

Floyd County, VA, are seen in the *Products of Industry* report for 1850 (Table A). Enumerators were instructed to list only mills whose products had a value of at least $500. As a result, only eight gristmills were described in all of Floyd County. Serving a population of nearly 6,500, this works out to about one mill per 812 people. By contrast, the report compiled in 1870 (using the same rule of $500 minimum value of products) listed 35 mills serving almost 9,500 residents, one mill for only about 270 people. Even a population increase of nearly 50 percent would not likely account for a quadrupling in the number of mills, especially since the Civil War and its aftermath occupied a good portion of the twenty-year period that passed between these two enumerations. The $500 minimum was simply too high for many Floyd County mills to meet, and thereby merit inclusion in the report, even in 1870, let alone twenty years earlier in 1850.

Nationwide, the record for 1860 saw more improvement in accuracy and coverage, but for some reason coverage dropped off sharply in Floyd County. Only one grain mill, Samuel Spangler's on Pine Creek (II, 99), was listed. Because it was almost certainly the largest mill in the county in 1860, it may have been the most productive, and hence the only one that met the $500 minimum. This alone would not seem to account for the exclusion of every other mill in the county since at least seven others had qualified ten years earlier. Other factors, whether systemic, such as a one-time qualification, or localized, such as an overly strict interpretation of a rule or an inept enumerator, must have been at work in 1860.

On the surface, at least, schedules used in 1870 were much like those from 1860, but, as noted, the report for Floyd County was for some reason vastly more inclusive, describing 35 grist and flouring mills. By 1880 the number of questions had been raised from 18 to 44, and for the first time the name of the stream providing water power was included. Another improvement, from the perspective of a local historian, is that administrators hired enumerators to complete separate reports for each magisterial district instead of making one man responsible for the entire county.

Products of Industry reports for 1890 and 1900 were not located, but would be of limited value in confirming the existence of any but the largest and best known Floyd County mills because of a limitation placed on the data collected by the designers of the census. "Under the definitions at the time of the [1890 and 1900] censuses of manufactures, mills that did not produce for sale but only sawed lumber or ground grain for toll did not fall under the factory system" (346). Since virtually all Floyd County mills were self-described as custom rather than merchant mills as late as 1880, a significant portion were probably ignored once again.

Despite these serious shortcomings, the contents of Tables A, B, and C have much to offer even the general reader, such as a look at the types and relative amounts of grains brought to 19th century mills for processing. As the schedules become progressively more specific and tedious (as they were bound to do in the hands of government statisticians), so do they become more interesting to the seeker of more obscure facts about milling in general or descriptive details about the mills that did happen to be enumerated.

Table A, from *Products of Industry*, Floyd County, Virginia, Year Ending June 1, 1850

	Name of Corporation, Company, or Individual, producing articles to the Annual Value of $500.	Name of Business, Manufacture, or Product	Capital Invested in Real and Personal Estate in the Business	Raw Material used, Including Fuel			Kind of motive power, machinery, structure, or resource.	Average no. of hands employed		Wages		Annual Product.		
				Quantities	Kinds	Values		Male	Female	Av. Monthly cost of male labour .	Av. monthly cost of female	Quantities .	Kinds.	Values
Mill No	1	2	3	4	5	6	7	8	9	10	11	12	13	14
I,05	Peter Guerant	Grist Mill	3300	6800 B	Wheat	1100	water power	4		15		1900 barrels	Flower (sic)	
		Distillery	2800	2000 B	Corn & Rye	1050						4000 gallons	Whiskey	4800
IV,28	C. Smith	Grist Mills	3000	4000 B	Wheat	2000						200 barrels	Flower	
				1000 B	Corn	600	water power	1		15		1500 B	Meal	2000* 3000
II,92	D. H. Bird	Woolcarding		3700 lb	Wool	725						3700 lbs	Wool roles	
		Grist Mill	1200	3000 B	Grain	1500	Water power	2		20		3000 barrels	Flower & meal	2620
IV,38	Cory Simmons	Grist Mill	1000	2000 B	Corn	1000						1500 B	cornmeal	
				1000 B	Wheat	750	Water power	1		15		300 ?	Flower	2650

	Name of Corporation, Company, or Individual, producing articles to the Annual Value of $500	Name of Business, Manufacture, or Product	Capital Invested in Real and Personal Estate in the Business	Raw Material Used, Including Fuel			Kind of motive power, machinery, structure, or resource	Average no. of hands employed		Wages		Annual Product		
				Quantities	Kinds	Values		Male	Female	Av. Monthly cost of male labor	Av. monthly cost of female labor	Quantities	Kinds	Values
Mill No.	1	2	3	4	5	6	7	8	9	10	11	12	13	14
II, 94	George Hornsbarger [sic]	Grist Mill	300	1200 B	Corn	600						1000 B	cornmeal	
				800	Wheat	400	Water power	1		15		150 barrels	Flower	1100
III, 81 or 132	James Lester (and Son)?	Grist mill	200	2000 B	Corn	700						1800 B	cornmeal	
				500 B	Wheat	300	Water power	1		15		2000 lb	Flower	350* 1500
III, 138	B. Howell	Wool Carding and Coloring	2000	1,500 lb.	Wool	300						1400 lb	Wool rolls	
			**	**	**	240	Water power	1		25		175 yd	Woolins **	933
I V, 31	Ira Howard	Grist mills	2000	1500 B	Corn	750						2000 B	cornmeal	
			100	1500 B	Wheat	850	Water power	1		12		150 barrel	Flower	1000* 1750

* Written in lightly above the other number in this box (**) Entry Illegible

Table B, from *Products of Industry*, Floyd County, Virginia, Year Ending June 1, 1870

(*) Entry illegible.	Name of Corporation, Company, or Individual Producing to the Annual Value of $500	Name of Business, Manufacture, or Product	Capital (Real & Personal Estate) Invested in Business	Motive Power		Machines		Avg. No. of Hands Employed			Total Amnt. Paid in Wages During Year	No. Mos. in Active Operation Reducing Part Time to Full Time	Materials (Including Mill Supplies and Fuel)			Production (Including all Jobbing and Repairing)		
				Kind	Horsepower	Name or Description	Number of	Males Over 16	Females Over 15	Children and Youth			Kinds	Quantities	Values (omitting fractions of dollar)	Kinds	Quantities	Values (omitting fractions of dollar)
Entry / Mill No.	*1*	*2*	*3*	*4*	*5*	*6*	*7*	*8*	*9*	*10*	*11*	*12*	*13*	*14* *Bushels*	*15* *Dollars*	*16*	*17*	*18* *Dollars*
Page 1 No. 1	Hylton, H. P. & J. Custom Work	Grist and Flouring Mill Capacity 30 Bushels per day	1800	Water One Wheel	20	Breast Wheel Two Run Stones		1			150	10	Corn	1300	1300	Meal	1620 bu	1620
													Rye	1300	1300	"	1730 "	1730
III, 67													Wheat	2750	*	Flour	550 bbl	4600?
													B Wheat	150	150	"	4950 lbs	1710?
Page 1 No. 2	Bair, David Custom	Grist Mill Capacity 25 Bushels per day	1200	Water One Wheel	8	Tub One Run Stone		1				6	Corn	1200	1200	Meal	1500 bu	1500
													Rye	300	300	"	400 "	400
II, 92													Wheat	50	60	Flour	10 bbl	70
													B Wheat	50	50	"	1650 lbs	50
Page 1 No. 3	Spangler, Saml. Private Owners	Grist and Flouring Mill Capacity 40 Bushels per day	3000	Water One Wheel	8	Overshot Two Run Stones		1				10	Corn	1500	750	Meal	1880 bu	940
													Rye	2300	1730	"	30 bu	2290
II, 99													Wheat	1500	9350	Flour	1500 lbs	*
													B Wheat	200	150	"	*	160
Page 1 No. 4	Spangler, Saml. Personal Venture	Grist and Flouring Mill	"	"	"	"							Corn			Meal		
													Rye			"		
II, 99													Wheat	1000	1250	Flour	200 bbl	1400
													B Wheat			"		

Entry / Mill No.	1	2	3	4	5	6	7	8	9	10	11	12	13	14 *Bushels*	15 *Dollars*	16	17	18 *Dollars*
Page 1 No. 5	Earles, John Grist Mill Capacity 36 Bu/day	Grist Mill Capacity 36 Bushels per day	1000	Water One Wheel	10	Breast Wheel Two Run Stones		1				6	Corn	400	400	Meal	500 bu	500
													Rye	150	110	"	200 bu	150
IV, 28?													Wheat	200	250	Flour	40 bbl	300
													B Wheat	100	70	"	3300 lbs	80
Page 1 No. 6	Scott, Matthew Private Owners	Grist and Flouring Mill Capacity 40 * day	2000	Water One Wheel	10	Turbine Two Run Stones		1				10	Corn	1000	1000	Meal	1000 bu	1000
													Rye	800	600	"	900	900
II, 91													Wheat	1500	1880	Flour	300 bbl	2400
													B Wheat	500	380	"	15000 lbs	450
Page 1 No. 7	Scott, Matthew Personal Venture	"	"	"		"		"				"	Corn			Meal		
													Rye			"		
II, 91													Wheat	500	630	Flour	100 bbls	800
													B Wheat	60	50	"	1800 lbs	60
Page 1 No. 8	Barringer, John A. Customers	Grist Mill Capacity 50 Bushels per day	1000	Water One Wheel	10	Overshot Two Run Stones		2				10	Corn	2500	2500	Meal	3120 bu	3120
													Rye	500	500	"	660 bu	660
IV, 45													Wheat	200	250	Flour	40 bbl	280
													B Wheat	100	50	"	3300 lbs	70
Page 1 No. 9	Bower, Christopher Customers	Grist Mill Capacity 36 Bushels per day	1000	Water One Wheel	6	Overshot Two Run Stones		1				10	Corn	1200	960	Meal	1500 bu	1350
													Rye	1500	1200	"	2000 bu	1600
?													Wheat	220	220	Flour	44 bbl	310
													B Wheat	40	20	"	1320 lbs	40
Page 1 No. 10	William H. Harman Customers	Grist and Flouring Mill Capacity 36 Bushels per day	2000	Water One Wheel	30	Overshot Two Run Stones		4			300	11	Corn	2000	2000	Meal	2500 bu	2000
													Rye	2000	2000	"	2670 bu	2670
II, 100													Wheat	2400	3000	Flour	480 bbl	*
													B Wheat	500	500	"	15000 lbs	600
Page 2 No. 1	Lawrence, S.B. Customers	Grist Mill Capacity 30 Bushels per day	1000	Water One Wheel	16	Overshot Two Run Stones		1				11	Corn	2480	2480	Meal	2480 bu	2480
													Rye	1240	1240	"	1240 bu	1240
II, 106													Wheat	500	500	Flour	84 bbls	590
													B Wheat	480	360	"	16600 lbs	390

(*) Entry Illegible

Table B (cont.), from *Products of Industry*, Floyd County, Virginia, Year Ending June 1, 1870

(*) Entry illegible.	Name of Corporation, Company, or Individual Producing to the Annual Value of $500	Name of Business, Manufacture, or Product	Capital (Real & Personal Estate) Invested in Business	Motive Power		Machines		Avg. No. of Hands Employed			Total Amnt. Paid in Wages During Year	No. Mos. in Active Operation Reducing Part Time to Full Time	Materials (Including Mill Supplies and Fuel)			Production (Including all Jobbing and Repairing)		
				Kind	Horsepower	Name or Description	Number of	Males Over 16	Females Over 15	Children and Youth			Kinds	Quantities	Values (omitting fractions of dollar)	Kinds	Quantities	Values (omitting fractions of dollar)
Entry	*1*	*2*	*3*	*4*	*5*	*6*	*7*	*8*	*9*	*10*	*11*	*12*	*13*	*14*	*15*	*16*	*17*	*18*
Mill No.														*Bushels*	*Dollars*			*Dollars*
Page 2 No. 2	Epperly, Henry Custom	Grist Mill Capacity 20 Bushels per day	500	Water One Wheel	12	Overshot Two Run Stones		1				10	Corn	1000	700	Meal	1000 bu	800
													Rye	1100	880	"	1100 bu	900
II, 103													Wheat	250	310	Flour	50 bbl	350
													B Wheat	150	110	"	4500 lb	130
Page 2 No. 3	Lester, James G. Custom	Grist Mill Capacity 20 Bushels per day	800	Water One Wheel	13	Overshot Two Run Stones		1				6	Corn	500	500	Meal	500 bu	500
													Rye	450	360	"	450 bu	360
III, 81													B Wheat	*	50	"	1600 lb	50
Page 2 No. 4	Stuart, John Custom	Grist Mill Capacity 20 Bushels per day	100	Water One Wheel	10	Tub Wheel One Run Stone		1				2	Corn	500	250	Meal	600 bu	300
II, 133																		
Page 2 No. 5	Epperly, J.C. Custom	Grist Mill Capacity 20 Bushels per day	300	Water	10	Tub Wheel		1				6	Corn	500	500	Meal	600 bu	600
													Rye	400	400	"	430 bu	430
II, 98													B Wheat	100	90	"	5000 lb	90

Entry Mill No.	*1*	*2*	*3*	*4*	*5*	*6*	*7*	*8*	*9*	*10*	*11*	*12*	*13*	*14* *Bushels*	*15* *Dollars*	*16*	*17*	*18* *Dollars*
Page 2 No. 6	Lester, Chas. S. or P. Custom	Grist Mill Capacity 20 Bushels per day	500	Water One Wheel	10	Tub Wheel One Run Stone		1				6	Corn	626	470	Meal	780 bu	620
													Rye	240	240	"	320 bu	320
													Wheat	100	130	Flour	20 bbl	160
													B Wheat	40	20	"	1200 lbs	30
Page 2 No. 7	Booth, George H.	Grist and Flouring Mill Capacity 35 Bushels per day	1000	Water One Wheel	8	Overshot Two Run Stones		1				8	Corn	1000	500	Meal	1000 bu	500
													Rye	200	200	"	250	280
III, 82	Custom												Wheat	200	200	Flour	40 bbls	280
													B Wheat	50	30	"	1660 lbs	50
Page 2 No. 8	Duncan, George R.	Grist Mill Capacity 45 Bushels per day	1000	Water One Wheel	8	Overshot One Run Stone		1				8	Corn	500	370	Meal	620 bu	410
													Rye	500	370	"	660 bu	410
III, 78	Custom												Wheat	500	620	Flour	100 bu	800
													B Wheat	150	110	"	4950 lbs ?	140
Page 2 No. 9	Altizer, Elias E.	Grist Mill Capacity 50 Bushels per day	1000	Water One Wheel	1*	Overshot One Run Stone		1				12	Corn	2000	1500	Meal	2000 bu	1500
													Rye	500	370	"	625 bu	400
III, ?	Custom												Wheat	500	630	Flour	100 bbl	700
													B Wheat	80	60	"	2400 lbs ?	70
Page 2 No. 10	Duncan, F. M.	Grist Mill Capacity 30? Bushels per day	500	Water One Wheel	10	Tub Wheel One Run Stone		1				3	Corn	500	500	Meal	620	620
III, 71	Custom																	
Page 3 No. 1	Phillips, Randal	Grist Mill Capacity 30 Bushels per day	500	Water One Wheel	8	Overshot One Run Stone		1				9	Corn	1000	1000	Meal	1250 bu	1200
													Rye	500	500	"	620 bu	620
III, 63?	Custom												Wheat	100	120	Flour	20 bbl	160
													B Wheat	100	80	"	3000 lbs	90
Page 3 No. 2	Slusher, Solomon	Grist Mill Capacity 40 Bushels per day	800	Water One Wheel	8	Overshot One Run Stones		1				9	Corn	900	900	Meal	1000 bu	1000
													Rye	500	500	"	625 bu	625
IV, 66	Custom												Wheat	100	130	Flour	20 bbl	160
													B Wheat	200	80	"	3000 lbs	90

(*) Entry Illegible

Table B (cont.), from *Products of Industry*, Floyd County, Virginia, Year Ending June 1, 1870

(*) Entry illegible.	Name of Corporation, Company, or Individual Producing to the Annual Value of $500	Name of Business, Manufacture, or Product	Capital (Real & Personal Estate) Invested in Business	Motive Power		Machines		Avg. No. of Hands Employed			Total Amnt. Paid in Wages During Year	No. Mos. in active Operation Reducing part time to full time	Materials (Including Mill Supplies and Fuel)			Production (Including all Jobbing and Repairing)		
				Kind	Horsepower	Name or Description	Number of	Males Over 16	Females Over 15	Children and Youth			Kinds	Quantities	Values (omitting fractions of dollar)	Kinds	Quantities	Values (omitting fractions of dollar)
Entry	*1*	*2*	*3*	*4*	*5*	*6*	*7*	*8*	*9*	*10*	*11*	*12*	*13*	*14*	*15*	*16*	*17*	*18*
Mill No.														*Bushels*	*Dollars*			*Dollars*
Page 3 No. 3	Hylton and Starr	Grist Mill Capacity 20 Bushels per day	500	Water One Wheel	6	Tub Wheel One Run Stone		1				3	Corn	700	700	Meal	870	870
	Custom												Rye	200	160	"	250	200
IV, ?													B Wheat	50	30	"	1500 lbs?	50
Page 3 No. 4	O'Neal, A.M. *	Grist Mill Capacity 50 Bushels per day	300	Water One Wheel	6	Tub Wheel One Run Stone		1				6	Corn	360	270	Meal	456	300
	Custom												Rye	360	270	"	486	300
IV, 49													Wheat	40	50	Flour	8 bbl	50
													B Wheat	40	30	"	1320 lbs	40
Page 3 No. 5	Jack, John O.	Grist and Flouring Mill Capacity 50 Bushels per day	2000	Water One Wheel	20	Overshot Two Run Stones		1				10	Corn	750	750	Meal	930 bu	850
	Customer												Rye	750	750	"	900 bu	900
I, 5													Wheat	750	930	Flour	150 bbl	1050
													B Wheat	750	750	"	22500 lbs	880
Page 3 No. 6	Siner, Peter	Grist Mill Capacity 50 Bushels per day	600	Water One Wheel	6	Tub Wheel Two Run Stones		1				6	Corn	500	500	Meal	650	600
	Custom												Rye	300	200	"	250	250
II, 117													B Wheat	*	160	"	6000 lbs	180

Entry / Mill No.	1	2	3	4	5	6	7	8	9	10	11	12	13	14 Bushels	15 Dollars	16	17	18 Dollars
Page 3 No. 7	Huff, Henry	Grist Mill Capacity 30 Bushels per day	1000	Water One Wheel	6	Tub Wheel One Run Stone		1				4	Corn	100	100	Meal	130 bu	130
													Rye	100	100	"	130 bu	130
I, 1	Custom												Wheat	20	25	Flour	5 bbl	30
													B Wheat	300	240	"	9000 lbs	270
Page 3 No. 8	Simpson, James	Grist Mill Capacity 30 Bushels per day	400	Water One Wheel	6	Overshot One Run Stone		1				4	Corn	1000	1000	Meal	1250 bu	1250
													Rye	500	400	"	600 bu	480
II, 113	Custom												Wheat	250	310	Flour	50 bbl	350
													B Wheat	200	150	"	7500 lbs	200
Page 3 No. 9	Shank, P.R.	Grist Mill Capacity 20 Bushels per day	300	Water One Wheel	8	Overshot One Run Stone		1				5	Corn	400	400	Meal	500	500
													Rye	200	160	"	260	200
?	Custom												Wheat	100	130	Flour	20 bbl	160
													B Wheat	100	60	"	2000 lbs	80
Page 3 No. 10	Gray, J. W.	Grist Mill Capacity 60 Bushels per day	800	Water One Wheel	20	Overshot One Run Stone		1				3	Corn	1000	1000	Meal	1200	1200
													Rye	300	240	"	400	320
I, 12?	Custom												Wheat	200	250	Flour	40 bbl	280
													B Wheat	100	60	"	3000 lbs	40
Page 4 No. 1	Carr, H.W.	Grist Mill Capacity 40 Bushels per day	1500	Water One Wheel	8	Overshot Two Run Stones		1			150	10	Corn	1500	1500	Meal	1800 bu	1800
													Rye	800	640	"	900 bu	900
II, 111	Custom												Wheat	1000	1250	Flour	200 bbl	1400
													B Wheat	300	180	"	9000 lbs	220
Page 4 No. 2	Conner, J.P.	Grist Mill Capacity 25 Bushels per day	600	Water One Wheel	8	Overshot One Run Stone		1				6	Corn	100	100	Meal	125 bu	120
													Rye	100	80	"	130 bu	90
II, 120	Custom												B Wheat	100	60	"	3000 lbs	70
Page 4 No. 3	Wimmer, H.	Grist and Flouring Mill Capacity 30 Bushels per day	1500	Water One Wheel	10	Overshot Two Run Stones		1				9	Corn	2000	2000	Meal	2500 bu	2500
													Rye	1000	800	"	1300 bu	1040
II, 119	Custom												Wheat	800	1000	Flour	160 bbl	1200
													B Wheat	200	140	"	6000 lbs	150

(*) Entry Illegible

Table B (cont.), from *Products of Industry*, Floyd County, Virginia, Year Ending June 1, 1870

(*) Entry illegible.	Name of Corporation, Company, or Individual Producing to the Annual Value of $500	Name of Business, Manufacture, or Product	Capital (Real & Personal Estate) Invested in Business	Motive Power		Machines		Avg. No. of Hands Employed			Total Amnt. Paid in Wages During Year	No. Mos. in Active Operation Reducing Part Time to Full Time	Materials (Including Mill Supplies and Fuel)			Production (Including all Jobbing and Repairing)		
				Kind	Horsepower	Name or Description	Number of	Males Over 16	Females Over 15	Children and Youth			Kinds	Quantities	Values (omitting fractions of dollar)	Kinds	Quantities	Values (omitting fractions of dollar)
Entry / Mill No.	*1*	*2*	*3*	*4*	*5*	*6*	*7*	*8*	*9*	*10*	*11*	*12*	*13*	*14 Bushels*	*15 Dollars*	*16*	*17*	*18 Dollars*
Page 4 No. 4	Smith, Eli	Grist Mill Capacity 30 Bushels per day	500	Water One Wheel	10	Overshot One Run Stone		1				10	Corn	1000	1000	Meal	1200 bu	1200
	Custom												Rye	800	640	"	900 bu	720
IV, 33													B Wheat	100	60	"	3000 lbs	70
Page 4 No. 5	Phlegar, Benj.	Grist Mill Capacity 25 Bushels per day	500	Water One Wheel	8	Overshot One Run Stone		1				3	Corn	500	500	Meal	600 bu	600
	Custom												Rye	200	140	"	250 bu	170
II, 90													B Wheat	100	60	"	3000 lbs	80
Page 4 No. 6	Harman, John	Grist Mill	500	Water One Wheel	10	Tub Wheel One Run Stone		1				3	Corn	500	500	Meal	550 bu	500
	Custom												Rye	300	240	"	350 bu	280
III, 88													B Wheat	200	120	"	6000 lbs	180
Page 4 No. 7	Kinser, O.T.	Grist Mill Capacity 30 Bushels per day	500	Water One Wheel	8	Overshot One Run Stone		1				6	Corn	600	600	Meal	600 bu	
	Custom												Rye	300	240	"	300 bu	
II, 140													B Wheat	100	70	"	3000 lbs	

Entry Mill No.	*1*	*2*	*3*	*4*	*5*	*6*	*7*	*8*	*9*	*10*	*11*	*12*	*13*	*14* *Pounds*	*15* *Dollars*	*16*	*17*	*18* *Dollars*
Page 6 No. 5 III, 84	Eller, George Custom	Wool Carding	1000	Water	8	Carding Two Machines	2	2			30	5	Wool	10,000	6000	Rolls	10,000 lbs	6500
Page 6 No. 6 II, 94?	McAlexander, Lester	Wool Carding	750	Water	8	Carding Machine Single	2	1			50	3	Wool	2300 lb	920	Rolls	2116 lbs	1058
Page 6 No. 7 IV, 49	ONeal , A.M & Co.	Wool Carding	500	Water	14	Double Machine	1	1				5	Wool	8500 lb	3400	Wool Rolls		6000
Page 6 No. 8 II, 115	Carr, H. W.	Wool Carding	500	Water	6	Double Machine One Machine	1	1			30	3	Wool	5000 lb	2000	Rolls	5000 lbs	2500
Page 6 No. 9 III, 70	Spangler, E.	Wool Carding	500	Water	6	Single Two Machines	2	1				9	Wool	2000 lb	800	Rolls	2000 lbs	1000
Page 6 No. 10 III, 84	Eller, George	Oil Mill Linseed oil	400	Water	8	Press Six pestles		2			50	2	Flax Seed	200 bu (bushel)	350	Linseed Oil	300 gal	390

Table C, from *Products of Industry*, Flour and Grist Mills in Floyd County, VA, Year Ending May 31, 1880

Mill No.	Corp., Co., or Individual Producing to $500 Annually	Capital (real and personal estate) Invested	Greatest no. of hands employed at any one time during the year.	Avg. No. Hands Employed			Wages and Hours of Labor					Mos. in Operation				Number of runs of stone	Est. max. capacity per day (Bu)	Do you do custom work or all for market?	Is an elevator connected with estab? Capacity in Bu?	Power Used in Manufacture						
							Hrs. in Ordinary Day Of Labor													If Water Power is Used						
																						Wheels				
				Males Above 16	Females Above 15	Children and Youth	May to Nov.	Nov. to May	Average day's wages for a skilled mechanic	Average day's wages for an ordinary laborer	Total amount paid in wages during the year	On full time	On 3/4 time only	On 1/2 time only	Idle					On what river or stream?	Height of fall in ft.	Number	Kind	Breadth in ft.	RPM	Horsepower
	1	2	3	4	5	6	7	8	9	10	11	12	13	14	15	16	17	18	19	20	21	22	23	24	25	26
Burks Fork (District)																										
IV, 49	O'Neal, Alford	1500					12	9				8			4	1	30	All Cstm.	no	Burks Fork	8	1	Tub	3	70	10
IV, 42	Slusher and Keith Co.	2500	2				12	12	1	1	312	12				1	30	All Cstm	no	Burks Fork	10	1	Overshot	6	5	10
III, 69	H.P. Hylton	3500	2				12	10	2	1	394	12				2	50	All Cstm.	no	Greasy Creek	20	1	Overshot	4	6	16
IV, 44	Jackson Harris	1000	1				12	8		1	100	10		2		1	20	All Cstm.	no	Burks Fork	20	1	Overshot	4	5	12
Indian Valley (District)																										
III, 63	Randolph Phillips	400	1	1			10	8		.75	100		8	2	2	1	12	All Cstm.		Mill Cr. of Greasy Cr.	18	1	Overshot	16	5	6
III, 136	Jno. C. Hambrick	1000	1	1			10	8		.50	100		4	8		1	10	All Cstm		Indian Creek	3	1	Tub	3.5	150	5
III, 78	Geo. R. Duncan and Son	600	1	1			10	8		.50	100		12			1	12	All Cstm.		Indian Creek	8	1	Tub	4	100	8

	1	2	3	4	5	6	7	8	9	10	11	12	13	14	15	16	17	18	19	20	21	22	23	24	25	26
III, 72	Solomon Spangler	1200	1	1			10	8		.60	125	9		1	2	1	18	All Cstm.		Indian Creek	18	1	Oversho t	18	7	7
III, 68	Saunders and Hylton	300	1	1			10	8		.75	100		6	4	2	1	18	All Cstm.		Greasy Creek	10	1	Tub	4	140	6
III, 65	Tobias D. Phillips	800	1	1			10	8		.75	125			12		1	10	All Cstm.		Mira Fork	18	1	Tub	3	150	10
Alum Ridge (District)																										
III, 82	Boothe, George B	600	2	1			12	10	1.25	.50	150	12				2	30	Cstm.	100	Beaver Cr/ Little River	18	1	Overshot	4	10	10
Little River Magisterial (District)																										
II, 106	David Hall	1200	1	1			12	12	$1	.75	$250	12				2	45	All Cstm.	No	Beaver Dam Cr.	15		?	4	?	24
II, 103	Henry Epperly	500	1	1			12	10	1.25	.75	115	8			4	1	25	All Cstm.	No	Terry's Creek	13		?	2	10	8
II, 95	Wohlford, Scott	800	1	1			8	8	1.20	.75	115	12				1	50	All Cstm.	No	Camp Creek	21		Overshot	3	15	10
I, 19	Rob. A. Harvey	2000	1	1			12	10	1.25	.75	250	12				1	30	All Cstm.	No	Meadow Creek	20		Overshot	3	15	9
I, 25	P. Williams and Sons	2500	1	1			12	8	1.25	.75	30			12		1	40	All Cstm.	No	Pine Creek	17		Tub Wheel	3	60	6
II, 102	W. H. Harman	2500	2	1			12	8	1.25	.75	375	12				2	96	All Cstm.	10/hr	Middle Creek	17?		Overshot	4	20	15
I, 20	Geo. Akers	1000	2	2			12	8	1.25	.75	100			12		1	40	All Cstm.	No	Beaver Dam Cr.	10		Tub Wheel	3	40	6

Table C (cont.), from *Products of Industry*, Flour and Grist Mills in Floyd County, VA, Year Ending May 31, 1880

Mill No.	Corp., Co., or Individual Producing to $500 Annually	Capital (real and personal estate) Invested	Greatest no. of hands employed at any one time during the year	Avg. No. Hands Employed			Wages and Hours of Labor					Mos. in Operation				Number of runs of stone	Est. max. capacity per day (Bu)	Do you do custom work or all for market?	Is an elevator connected with estab? Capacity in Bu?	Power Used in Manufacture						
							Hrs. in Ordinary Day Of Labor													If Water Power is Used						
																						Wheels				
				Males Above 16	Females Above 15	Children and Youth	May to Nov.	Nov. to May	Average day's wages for a skilled mechanic	Average day's wages for an ordinary laborer	Total amount paid in wages during the year	On full time	On 3/4 time only	On 1/2 time only	Idle					On what river or stream?	Height of fall in ft.	Number	Kind	Breadth in ft.	RPM	Horsepower
	1	2	3	4	5	6	7	8	9	10	11	12	13	14	15	16	17	18	19	20	21	22	23	24	25	26
Jacksonville District																										
II, 91	Scott, Matthew	6000	2	1			12	12	3	.75	100	12				2	50	Mkt. and Cstm.		South Fork Little R.	12	1	Turbine	2.5	84	20
IV, 31	Howard, J.L.	4000	1	1			10	12	3	.75	50	12				2	75	Cstm.		South Fork Little R.	10	1	Brest	7	25	20
III, 88	Harman, John	1500	1	1			8	12	3	.75	50	12				2	40	Cstm.		West Fork Little R.	12	2	Tub	3	50	20
IV, 38	Simmons, O.	1500	1	1			8	7	3	.75	40	12				2	50	Cstm.		West Fork Little R.	12	1	Turbine	3	60	20
II, 99	Spangler, Sam	1500	1	1			8	10	3	.75	30	12				2	50	Cstm.		Pine Creek	10	1	Turbine	3	84	20
IV, 28	Martin, Wm.	1000	1	1			8	10	3	.75	25	12				2	50	Cstm.		South Fork Little R.	10	1	Brest	7	25	20

30th Enumeration District (Locust Grove?)

	1	2	3	4	5	6	7	8	9	10	11	12	13	14	15	16	17	18	19	20	21	22	23	24	25	26
I, 8	Kropf and Custer	800	1	1			10	8	1		375	12				1	40	Cstm.	None	Little River	12	1	Breast	4.5	14	16
I, 9	John W. Wertz	900	1	1			10	8	1		300	12				3	?	Cstm.	100/hr	Little River	9	1	Turbine	8	45	18
I, 11	G. H. Beamer	400	1	1			9	8	1		350	12				1	30	Cstm.	None	Little River*	9	1	Tub	7	50	8
?	Nancy Shank	100	1	1			9	8	1		300	12				1	25	Cstm.	None	Little River	9	1	Overshot	3	15	8
II, 113	Wm. A. Simpson	150	1	1			9	8	1		350	12				1	25	Cstm.	None	Little River*	9	1	Overshot	3.5	14	12
I, 1	Huff, Byrd	1200	1	1			10	8	2		75	12				2	40	Cstm.	100	Little River	12	1	Breast Wheel	5	10	10
II, 117	Siner, Peter	1000	1	1			10	8	1		100	12				2	50	Cstm. Only	50	South Fork Roanoke R	12	1	Breast	6	5	8
?	Fowler and Huff	1400	1	1			8	8	1		313	8			4	4	25	Cstm.	1500	South Fork Roanoke R	14	1	Overshot	4	14	12
II, 119	Wimmer, H. Jr.	2000	2	2			12	12	2	1.00	513	12				2	72	Cstm. Work	2500	Brush Creek	18	1	Overshot	3.5	6	16
II, 105	Epperly, John	150	1	1			5	8	1		25	12				1	40	Cstm. Work	?	Brush Creek	15	1	Overshot	3.5	10	8
I, 5	Smith, Jas. W.	1200	1	1			8	10	1		313	11			1	2	25	Cstm. Work	100	Little?	10	1	Breast Wheel	4	4	16

Table C (cont.), from *Products of Industry*, Flour and Grist Mills in Floyd County, VA, Year Ending May 31, 1880

Mill Number	Corp., Co., or Individual Producing to $500 Annually	Cols. 27, 28, and 29 Pertain to Steam	Materials						Products								
			No. of Bushels of Wheat	Value (Dollars)	No. of Bushels of other Orains	Value (Dollars)	Value of Mill Supplies	Total Value of all materials	Number of barrels of wheat flour	Rye Flour (Barrels)	Number of pounds of buckwheat flour	Barley Meal (Pounds)	Number of pounds of corn meal	Number of pounds of feed	Hominy (Pounds)	Value of all other products (dollars)	Total value of all products (dollars)
	1	27-9	30	31	32	33	34	35	36	37	38	39	40	41	42	43	44
Burks Fork (District) (* = illegible)																	
IV, 49	O'Neal, Alford		50	$50	4000	$2000	50	$2100	10		14000		202000	8700	200		$3130
IV, 42	Slusher and Keith Co.		2600	2600	5600	3360	50	6010	572	14	36400		242***	72500			7117
III, 69	H.P. Hylton		5000	5000	3000	1500	50	6550	1125	150	15000		98200 *	103950	200		7733
IV, 44	Jackson Harris		1500	1500	3500	2100	60	3660	330	75	23000		*	35000			4200
Indian Valley (District) (* = illegible)																	
III, 63	Randolph Phillips		500	$500	2140	$1070	$25	$1595	100	30	10000		94560	15000		$75	$1900
III, 136	Jno. C. Hambrick		500	500	2100	1050	30	1580	94	25	10000		104500	*		75	1850
III, 78	Geo. R. Duncan and Son		600	600	2640	1250	35	1885	122	60	21000		102000	22760		80	2280
III, 72	Solomon Spangler		1200	1200	4650	2250	40	3490	250	68	28000		*	9000		100	4000
III, 68	Saunders and Hylton				1400	700	20	720					76400	2000			1000
III, 65	Tobias D. Phillips		200	200	2500	1250	25	1475	41	31	10800		*	9000		30	*

Alum Ridge (District)																	
	1	27-9	30	31	32	33	34	35	36	37	38	39	40	41	42	43	44
III, 82	Boothe, George B.		500	$500	2000	$1000	$20	$1520	100		2000		106000	11500			$1900
Little River Magisterial (District) (* = illegible)																	
II, 106	David Hall		150	$175	8000	$4000	$25	$4200	30	40	64000		360000	*			$5000
II, 103	Henry Epperly		250	250	5000	2500	20	2770	50	25	6000		256000	*			3900
II, 95	Wohlford, Scott		50	50	2000	1000	10	1060	10	10	*		101960	*			1402
I, 19	Rob. A. Harvey		2400	2400	2000	1000	50	3450	480		2500		105500	40000			*
I, 25	P. Williams and Sons		200	200	2900	1450	30	1680	40	40	10000		133000	13000			*
II, 102	W. H. Harman		3200	3200	2000	1000	100	4300	640	85	12500		*	*			5184
I, 20	Geo. Akers				2000	1000	5	1005			70000		*	*		1050	2050
Jacksonville District (* = illegible)																	
II, 91	Scott, Matthew		9000	$9000	3000	$1500	100	$10600	1700		5000		157000	128000			$10800
IV, 31	Howard, J.L.		4380	4380	4000	2000	50	6430	826		2000		214000	69000			8380
III, 88	Harman, John		3000	3000	3000	1500	50	4550	700		2000		160000	*			7500
IV, 38	Simmons, O.		2000	2000	2000	1000	25	3025	400		1500		*	*			3500
II, 99	Spangler, Sam		3000	3000	2000	1000	25	4025	700		2000		106000	*			4800
IV, 28	Martin, Wm.		2000	2000	2500	1250	25	3275	400		2000		*	*			3580

Table C (cont.), from *Products of Industry*, Flour and Grist Mills in Floyd County, VA, Year Ending May 31, 1880

Mill Number	Corp., Co., or Individual Producing to $500 Annually	Cols. 27, 28, and 29 Pertain to Steam	Materials (* = illegible)						Products (* = illegible)								
			No. of Bushels of Wheat	Value (Dollars)	No. of Bushels of other Orains	Value (Dollars)	Value of Mill Supplies	Total Value of all materials	Number of barrels of wheat flour	Rye Flour (Barrels)	Number of pounds of buckwheat flour	Barley Meal (Pounds)	Number of pounds of corn meal	Number of pounds of feed	Hominy (Pounds)	Value of all other products (dollars)	Total value of all products (dollars)
	1	27-9	30	31	32	33	34	35	36	37	38	39	40	41	42	43	44
30th Enumeration District (Locust Grove?) (* = illegible)																	
I, 8	Kropf and Custer		60	$60	4500	$2250	$30	$2340	12	40	37000		206000	28900			$3000
I, 9	John W. Wertz		1000	1000	2700	1350	75	2425	20	100	22000		91000	19400			3200
I, 11	G. H. Beamer		50	50	1200	600	10	660	10	15	10000		50800	5200			*
?	Nancy Shank				1800	900	10	910	"	25	2200		87500	*			2000
II, 113	Wm. A. Simpson		50	50	1800	900	15	965	10	40	6000		78240	*			2215
I, 1	Huff, Byrd		1200	1000	1500	750	200	1950	240	10	2000			21000			2500
II, 117	Siner, Peter		100	100	1000	500	100	700	20	10	1200		52000	4800			1750
?	Fowler and Huff		1050	1050	1050	525	75	1650	210	20	2000		54000	16800			2400
II, 119	Wimmer, H., Jr.		2000	2000	4000	2000	75	4075	400	50	3330		207000	42000			4538
II, 105	Epperly, John				1600	300	5	305					86400	3200			883
I, 5	Smith, Jas. W.		1460	1460	1600	800	280	2540	252	70	14000		68900	25100			*

Notes

Part One
Historical Overview

1. B.W. Dedrick, *Practical Milling* (Chicago: The Society of Old Mills, 1989), 5.
2. "Genealogy Data: Frequently Occurring Surnames from Census 2000," Census.gov, 29 December 2009, http://www.census.gov/genealogy/www/freqnames2k.html/.
3. Tom Kelleher, "Flour from Mill to Market," *Old Mill News* 37, No. 1 (2009), 19.
4. Dorothy J. Merritts and Robert C. Walter, "Natural Streams and the Legacy of Water-Powered Mills," *Science* 319, issue 5861 (2008), 301–03.
5. William Waller Hening, "An Act for Encouragement of Building Water-mills," in *The Statutes at Large* (New York, Philadelphia, Richmond, 1819–23) 3: 401–04.
6. Betty Fussell, *The Story of Corn: The Myth and Mystery, the Culture and Agriculture, the Art and Science of America's Quintessential Crop*. New York: Alfred A. Knopf, 1992.
7. Hening, "To rectify the great abuse of millers...." 1: 301.
8. Hening, "An Act Prohibiting Millers the Takeing of More Then One Eighth Part of English Graine Toll for Grinding Thereof," 2: 286.
9. Hening, "An Act Concerning Water Mills," 6: 55–60.
10. Hening, "An Act for Encouragement for Erecting Mills," 2:260–61.
11. Hening, "An Act Concerning Mill-dams and Other Obstructions of Water Courses," 12: 187.
12. *Ibid.*, 126–7.
13. Hening, "An Act for Encouragement of Building Water-mills," 3: 401–04.
14. Hening, "An Act Concerning Water Mills," 6: 55–60.
15. Floyd County Deed Book "N" (Floyd County, VA), 331.
16. Melton, *Pittsylvania's Eighteenth Century Grist Mills* (Chatham, VA, 1989), 127.
17. *Ibid.*, 136.
18. Ralph M. Brown, "A Sketch of the Early History of Southwestern Virginia," *William and Mary Quarterly* 17, No. 4 (1937), 512.
19. "County and City Records at the Library of Virginia," *Library of Virginia*, http://www.lva.virginia.gov/whatwehave/local/county_formation/index.htm.
20. Lewis Preston Summers, *Annals of Southwest Virginia, 1769–1800* (Kingsport, TN: Kingsport Press, 1929), 98.
21. Barb Reninger, "Surnames of Families of Floyd County, Virginia," *Ancestry.com*, http://freepages.genealogy.rootsweb.ancestry.com/~barbr/zz_main_surnames.htm. "Goodson."
22. Terry S. Reynolds, *Stronger Than a Hundred Men: A History of the Vertical Water Wheel* (Baltimore: Johns Hopkins, 1983), 347.
23. Russell Dulaney, in discussion with Ricky Cox, 23 December 2009.
24. Oliver Evans, *The Young Millwright and Miller's Guide* (1850; reprint, Salem, NH: Ayer, 1989), 199–200.
25. *Ibid.*, 200.
26. *Ibid.*
27. Dr. James Cox, giving tour of Brookside Mill, Grayson County, VA, August 1996.
28. Jack (M.J.) Spangler, e-mail message to Ricky Cox, Aug. 10, 2008.
29. Rev. Roy Turpin in telephone interview with the author, Ricky Cox, January 2010.
30. *Floyd Press* (Floyd, VA), July 2, 1908.
31. *Floyd County Virginia Heritage*, 35.
32. "Laurel Ridge Items," *Floyd Press* (Floyd, VA), Nov. 3, 1910.
33. *Floyd Press* (Floyd, VA), Oct. 23, 1924.
34. *Floyd Press* (Floyd, VA), Aug. 28, 1902.
35. *Floyd Press* (Floyd, VA), Aug. 08, 1912.
36. *Floyd Press* (Floyd, VA), Aug. 06, 1914.
37. *Floyd Intelligencer* (Floyd, VA), May 5, 1853.
38. Fred E. Coy, Jr., Charles D. Hockensmith, and Jimmie L. Price. *Millstone Manufacture in Virginia: Interviews with the Last Two Brush Mountain Millstone Makers* (Raleigh, NC: Marblehead, 1999).
39. *Floyd Press* (Floyd, VA), March 10, 1909.
40. Jess Carr, *The Second Oldest Profession: An Informal History of Moonshining in America* (Radford, VA: Commonwealth Press, 1972), 72.
41. *Ibid.*, 72–3.
42. "Meal or Whiskey," *Floyd Press* (Floyd, VA), May 07, 1903.
43. Edward Finley, "History of the Meadows Mill Company," *Old Mill News* (January 1983), 4.
44. Jon A. Sass, *The Versatile Millstone: Workhorse of Many Industries* (Knoxville: Society for the Preservation of Old Mills, 1990), 28–9.
45. Thomas Wright, in discussion with the author, Ricky Cox, April 5, 1998.
46. Mail survey of long-time residents of Indian Valley, Floyd County, VA, conducted by Ricky Cox.
47. Jessica Banks, "Mills and Millers in Old and New World Folksong," *Building Community: Medieval Technology and American History*. http://www.engr.psu.edu/mtah/articles/mills_millers.htm. Accessed 3/9/10.
48. E.N. Hartley, *Ironworks on the Saugus* (Norman: University of Oklahoma Press, 1957), 86.
49. Hening, "To rectify the great abuse of millers...." 1: 301.
50. Hening, Act 11, Vol. 1, 347–48.
51. Hening, "An Act for Encouragement of Building Water-mills," 3: 401–04.
52. Jean Haskell Speer, *The Appalachian Photographs of Earl Palmer* (Lexington: University Press of Kentucky, 1990).
53. Floyd County Deed Book "L" (Floyd County, VA), 500.
54. Cavilla McPeak, in discussion with the author, Ricky Cox, Indian Valley, VA, January 2010.
55. Effie King Brown, in discus-

sion with the author, Ricky Cox, Copper Hill, VA, July 7, 2009.

56. R.L. Humbert, *Floyd County, Virginia: Industrial Survey* (Blacksburg: Virginia Polytechnic Institute, 1930), 14.

57. *Ibid.*, 16.

58. Roberta W. Hewett, et al., *Industrial Survey, Floyd County, Virginia* (Floyd, VA: no publisher, 1957), 37.

59. *Ibid.*, 23.

60. Merritts and Walter, 299–304.

Part Two Section I

1. Montgomery County Court Order Book No. 22 (Montgomery County, VA), 150, 152.

2. Hening, "An Act Concerning Water Mills," 6: 89.

3. Amos Wood, *Floyd County: A History of Its People and Places* (Commonwealth Press, 1981), 80.

4. Floyd County Will Book 3 (Floyd, VA), 309.

5. *Floyd Press* (Floyd, VA), July 2, 1908.

6. Glenn Peters, in discussion with Ricky Cox, Copper Hill, VA, July 2008.

7. Sue Proffitt Yeatts, telephone interview by Ricky Cox, November 2009.

8. Thomas Wright, interview by Ricky Cox, Salem, VA, January 1998.

9. Sue Proffitt Yeatts, telephone interview by Ricky Cox, November 2009; Curtis Graham, telephone interview by Ricky Cox, November 2009.

10. *Floyd Intelligencer* (Floyd, VA), May 5, 1854 or 55.

11. *Montgomery County Court Order Book No. 2* (Montgomery County, VA), 335.

12. Floyd County Fiduciary Settlement Book 12 (Floyd County, VA), 33.

13. Marvin Nolen, telephone interview by Ricky Cox, August 2008.

14. *Floyd Press* (Floyd, VA), November 18, 1926.

15. *Floyd Press* (Floyd, VA), June 17, 1909.

16. *Floyd Press* (Floyd, VA), June 25, 1942.

17. Jim Altizer, telephone interview by Ricky Cox, July 4, 2008.

18. Philip E. Gettel, "Roberson's Mill (Epperly)," in "Floyd County Mills and Mills Sites" (unpublished manuscript, Montgomery-Floyd Regional Library, Floyd Branch, Floyd, VA, 1985).

19. Harry Roberson, telephone interview by Ricky Cox, October 17, 2008.

20. *Ibid.*

21. Harry Roberson, telephone interview by Ricky Cox, October 11, 2009.

22. Freeman Cockram, telephone interview by Ricky Cox, August 6, 2008.

23. *Floyd Press* (Floyd, VA), April 14, 1904, 3.

24. *Floyd Press* (Floyd, VA), November 3, 1921, 2.

25. *Floyd Press* (Floyd, VA), November 4, 1926.

26. T.S. Lester Mill Ledger, owned by Mr. and Mrs. E.V. Huff (1990).

27. Effie King Brown, telephone interview by Ricky Cox, December 16, 2007.

28. Reynolds, 18, 20.

29. Judy Nolen Hylton, "Amos 'Cab' Nolen Owns the Mill," in "History of Nolen Mill" (unpublished manuscript), 6.

30. *Ibid.*, 6–7.

31. Reynolds, Chapter 5, "Change: The Emergence of the Iron Industrial Water Wheel, c. 1750–1850," 266–320.

32. Hylton, "The Mill," 3.

33. Ibid, 7.

34. Stefan Klosowski, "Nolen's Mill."

35. Melissa DeVaughn, "A Mill of His Own: Phil Gettel Purchases a Piece of History," *Roanoke Times* (Roanoke, VA). New River Valley edition, Current section, 19.

36. Marvin Nolen, telephone interview by Ricky Cox, August 2008.

37. *Ibid.*

38. Basil Poff Simpson, Sr., "Memoirs of Basil Poff Simpson, Sr.," unpublished manuscript, n.p., n.d.

39. "Electric Feed Mills Becoming Popular in County," *Floyd Press* (Floyd, VA), Sept. 11, 1947.

40. *Ibid.*

41. Kenneth Sewell, telephone interview by Ricky Cox, July 9, 2008; Kyle Lee, telephone interview by Ricky Cox, July 23, 2009.

42. *Floyd Press* (Floyd, VA), April 30, 1950.

43. Kenneth Sewell, telephone interview with Ricky Cox, July 9, 2008.

Section II

1. Mattie Susan Phlegar Brown, "Phlegar Farm 600 Acres" (Floyd, VA: no publisher, no year), 1.

2. *Floyd News* (Floyd, VA), Dec. 21, 1888.

3. *Floyd Press* (Floyd, VA), May 24, 1944.

4. *Floyd Press* (Floyd, VA), March 10, 1910.

5. Marguerite Tise, "Petition to Build a Mill," *Virginia Appalachian Notes* (17, No. 4, 1993), 132–133.

6. *Ibid.*

7. Floyd County Deed Book "P" (Floyd County, VA), 566.

8. Record; Mechanics Liens; Floyd County (Floyd, VA), 32–33.

9. *Floyd Press* (Floyd, VA), Nov. 3, 1910.

10. *Articles of Association of Floyd Electric Power Corporation* (Floyd, VA), Jan. 4, 1912.

11. *Ibid.*; *Floyd Press* (Floyd, VA), June 24, 1937.

12. *Floyd Press* (Floyd, VA), Sept. 8, 1921.

13. Tise, 132.

14. *Floyd Press* (Floyd, VA), Nov. 23, 1922.

15. *Floyd Press* (Floyd, VA), July 22, 1926.

16. *Floyd Press* (Floyd, VA), Aug. 8, 1912.

17. *Ibid.*

18. Hening, "An Act for Encouragement of Building Water-mills," 3: 401–04.

19. Hening, "An Act Concerning Water Mills," Vol. 6: 55–60.

20. M.J. Spangler, *Descendants of Daniel and Mary Spangler, Franklin County, Virginia 1773* (Bountiful, Utah: Family History Publisher, 1997), 3.

21. M.J. Spangler, "Memories of My First 25 Years," unpublished manuscript, 3.

22. Dedrick, 22; Evans, 212–15

23. *Schedule 5. Products of Industry in the County of Floyd, State of Virginia, for the Year Ending June 1, 1860.* Microfilm. Newman Library, Virginia Tech, Blacksburg, VA, 89.

24. *Floyd Press* (Floyd, VA), August 10, 1911.

25. *Floyd Press* (Floyd, VA), Oct. 2, 1924.

26. *Floyd County Court Order Book, 1877–1882* (Floyd County, VA), Nov. 13, 1879.

27. *Floyd Press* (Floyd, VA), July 10, 1913.

28. Gettel, *Tise's Mill*, 2.

29. Anna Hale, "Williams: Charles Emmett Williams," in *Floyd County Virginia Heritage* (Summersville, WV: S.E. Gross, 2001), 173.

30. Mary Williams Willis and Ruby Willis Goodman, "Blue Ridge Mary and Other Manuscripts," n.d., n. pub.

31. *Ibid.*

32. Reynolds, 343.

33. *Floyd Press* (Floyd, VA), Feb. 5, 1903.

34. Humbert, 22, 24.

35. "Montgomery County," *Virginia State Gazetteer and Business Directory, 1897–98* (Richmond, VA: J.L. Hill, 1898).

36. "Montgomery County," *Virginia Business Directory and Gazetteer.* Richmond, VA: Hill Directory, 1917.

37. "Hog Killing Time at Flint Mills," *The Roanoke Times*, New River Valley section, p. 20.

38. Wanda Whitenack Vest, interview by Ricky Cox, July 2009, Check, VA.

39. William "Benny" Conner, telephone interview by Ricky Cox, August 2008.

40. *Virginia Atlas and Gazetteer* (Freeport, Maine: DeLorme Mapping, 1989), 42. "Floyd County, Virginia" (map) (Glen, MS: Southern Engineering, 2005).

41. Daniel Worth Lucas, telephone interview by Ricky Cox, Oct. 27, 2008.

42. J. David McNeil, "A Mill on Goose Creek," in *Floyd County Virginia Heritage* (Summersville, WV: S.E. Gross, 2001), 31–2.

43. Ibid, 31.

44. J. David McNeil, "James Ireson McNeil/McNiel," in *Floyd County Virginia Heritage* (Summersville, WV: S.E. Gross, 2001), 128.

45. McNeil, "A Mill," 31.

46. Anders, 174.

47. McNeil, "A Mill," 32.

48. *Floyd County Virginia Heritage.* McNiel photo, 129; Willis photo, 174; mill and store photos, 32.

49. Effie King Brown, letter to Ricky Cox, June 16, 2010.

50. Effie King Brown, telephone interview by Ricky Cox, December 2007.

51. Effie King Brown, letter to Ricky Cox, December 2007.

52. *Floyd Press* (Floyd, VA), Aug. 31, 1911.

53. Simpson.

54. *Ibid.*

55. *Floyd Press* (Floyd, VA), March 2, 1933.

56. Record; Mechanics Liens; Floyd County (Floyd, VA), 24.

57. Ralph Stump, interview by Frank Webb, n. d.

58. Summers, 843.

59. Floyd County Deed Book "M" (Floyd County, VA, 1871), 223.

60. Sharon Shaver Wood, e-mail message to Ricky Cox, Jan. 2009.

61. Hugh Bowman and Forest Light, interview by Frank Webb, November 1994.

62. *Floyd Press* (Floyd, VA), April 7, 1904.

63. Hugh Bowman and Forest Light, interview by Frank Webb, November 1994.

64. Melvin Light, interview by Ricky Cox, 13 October 2007, Willis, VA.

65. *Floyd Press* (Floyd, VA), November 9, 1933.

66. "Amendment: Scott Roller Mill Company," 10 July 1909. Amendment to charter of corporation formed on 24 January 1902.

67. Joan V. Schroeder, "In Search of Winfield Scott," *Roanoker Magazine* (Roanoke, VA), 12–15, 45.

68. *Floyd Press* (Floyd, VA), August 18, 1910.

69. *Floyd Press* (Floyd, VA), August 20, 1942.

70. Melissa DeVaughn, "It was a Grind, But Floyd County Miller Loved His Work," *Roanoke Times and World News* (Roanoke, VA), October 23, 1994.

71. *Floyd Press* (Floyd, VA), July 11, 1940.

72. *Floyd Press* (Floyd, VA), December 18, 1947.

73. Gettel, 1.

Section III

1. Cox, 9.

2. *Ibid.*

3. Ibid, 25.

4. James "Arthur" Bond, interview by Ricky Cox, Indian Valley, VA, August 2, 1991.

5. *Ibid.*

6. Vinton Gallimore, interview by Ricky Cox, c. 1995.

7. Irene Hollandsworth Gallimore, telephone interview by Ricky Cox, July 4, 2009.

8. Record; Mechanics Liens; Floyd County (Floyd, VA), 34–35.

9. James "Arthur" Bond, interview by Ricky Cox, August 2, 1991.

10. S. Jett Phillips, interview by Ricky Cox, Indian Valley, VA, May 6, 1991.

11. Sammie Phillips, interview by Ricky Cox, Dugspur, VA, July 25, 2008.

12. Sammie Phillips, telephone interview by Ricky Cox, October 17, 2009.

13. Gene Vaughn in a personal interview with author, September 19, 2009.

14. Ottie Padgett, "Vaughn's Mill One of the Few Operating in This Area," *Galax Gazette* (Galax, VA), July 10, 1969, 4B.

15. Floyd County Deed Book 28 (Floyd County, VA), 406.

16. Klosowski, "Greasy Creek Mill."

17. Copies of a 1989 study of the farm compiled by Ricky Cox are held by the Blue Ridge Institute at Ferrum College (includes text, drawings, black and white negatives, and color slides) and the Appalachian Regional Studies Center at Radford University (includes text, drawings, and photos made from slides). ARSC call number MA194.57A, *Blue Ridge Farm Study.*

18. Klosowski, "Greasy Creek Mill."

19. Foster Quesenberry, interview by Ricky Cox, Indian Valley, VA, June 30, 2009.

20. *Floyd Press* (Floyd, VA), February 27, 1908.

21. *Floyd Press* (Floyd, VA), March 12, 1908.

22. Una C. Hylton, "The History of Willis," in *Floyd County Virginia Heritage* (Summersville, WV: S.E. Gross, 2001), 44.

23. James "Arthur" Bond, interview by Ricky Cox, Indian Valley, VA, August 2, 1991.

24. Carlysle Phillips, interview by Ricky Cox, Indian Valley, VA, January 2008.

25. Aaron Altizer, interview by Ricky Cox, Alum Ridge, VA, March 26, 2008.

26. Roy Turpin, interview by Ricky Cox, Indian Valley, VA, January 13, 2010.

27. Vinton Marshall, telephone conversation with Ricky Cox, March 1991.

28. Carlysle Phillips, interview by Ricky Cox, Indian Valley, VA, January 2008.

29. *Ibid.*

30. Barb Reninger, "Surnames of Families of Floyd County, Virginia," *Ancestry.com*, http://freepages.genealogy.rootsweb.ancestry.com/~barbr/zz_main_surnames.htm "Duncan 1."

31. Victor Phillips, response to mailed survey conducted by Ricky Cox, February 28, 1991.

32. Floyd County Deed Book 41 (Floyd, VA), 20.

33. Aaron Altizer, interview by Ricky Cox, Alum Ridge, VA, March 26, 2008.

34. Dale Teel, in discussion with Ricky Cox, at Dulaney Mill site, Indian Valley, VA, December 8, 2008.

35. Floyd County Deed Book 41 (Floyd, VA), 20.

36. Paul Duncan, telephone interview with Ricky Cox, January 3, 2009.

37. Paul Duncan, interview with Ricky Cox, Fairlawn, VA, January 6, 2009.

38. *Ibid.*

39. Aaron Altizer, interview with Ricky Cox, Alum Ridge, VA, February 2009.

40. Montgomery County Court Order Book No. 17 (Montgomery

County, VA), 274. See also Appendix B of this book.
41. Wood, 287.
42. *Ibid.*
43. Leonard Booth, interview with Kathleen Ingoldsby, Floyd, VA, February 17, 1993.
44. *Ibid.*
45. *Chataigne's Virginia Business Directory and Gazetteer* for 1897.
46. *Floyd Press,* (Floyd, VA), February 18, 1908.
47. John Newton Harman, *Genealogy (Southern Branch): With Biographical Sketches, 1700–1924* (Richmond, VA: W.C. Hill, 1925), 299.
48. Wood, 285.
49. Don Conner in an e-mail message to the author, February 2008.
50. "Cloudburst in Topeco Area of Floyd County Caused Heavy Damage: State Highway Department Plans to Rebuild Washed Out Bridges," *The Roanoke Times* (Roanoke, VA), August 15, 1943.
51. Harman, 299.
52. *Floyd Press,* (Floyd, VA), 5 June 1902, 2.
53. Jim Altizer, interview with Ricky Cox, Floyd, VA, July 4, 2008.
54. Table A, 1850 *Products of Industry.*
55. Linwood Altizer, interview with Ricky Cox, Alum Ridge, VA, March 2009.
56. Mary B. Kegley and F.B. Kegley, *Early Adventurers on the Western Waters, Vol. I: The New River of Virginia in Pioneer Days 1745–1800* (Orange, VA: Green, 1980), 213–14.
57. Montgomery County Court Order Book No. 10 (Montgomery County, VA), 30.
58. Jay S. Hoar, *The South's Last Boys in Gray: An Epic Prose Elegy, A Substudy of Sunset and Dusk of the Blue and the Gray* (Bowling Green, OH: Bowling Green State University Popular Press, 1986), 261–62.
59. *Ibid.*, 261.
60. *Ibid.*
61. Olga DeHart Harvill, interview by Ricky Cox, Copper Valley, VA, December 2008.
62. *Ibid.*
63. Aaron Altizer, interview by Ricky Cox, Alum Ridge, VA, December 2008.
64. Olga DeHart Harvill, interview by Ricky Cox, Copper Valley, VA, December 2008.
65. Jay S. Hoar, *The South's Last Boys in Gray: An Epic Prose Elegy, A Substudy of Sunset and Dusk of the Blue and the Gray* (Bowling Green, OH: Bowling Green State University Popular Press, 1986), 261.
66. Linwood Altizer, interview with Ricky Cox, Alum Ridge, VA, December 2008.
67. Hoar, 262.
68. *Slusher Family Organization Newsletter*, January 1974, Vol. 4, No. 2, 5.
69. *Slusher Family Organization Newsletter*, January 1974, Vol. 4, No. 2, 5–6.
70. Foster Quesenberry, in discussion with Ricky Cox, Dugspur, VA, January 17, 2010.

Section IV

1. Wood, 285.
2. Montgomery County (VA) Court Order Book No. 24. 81, 90.
3. Wood, 285.
4. *Ibid.*
5. Our Correspondents, "Howell's Creek," *Weekly News* (Floyd, VA), Dec. 6, 1895.
6. *Floyd Press,* (Floyd, VA), Feb. 18, 1932.
7. Speer, 3–15.
8. *Chataigne's Virginia Business Directory and Gazetteer* for 1911.
9. *Floyd Press* (Floyd, VA), February 8, 1940.
10. Montgomery County (VA) Court Order Book No. 14, 212.
11. *Weekly News* (Floyd, VA), December 6, 1895.
12. *Floyd Press* (Floyd, VA), October 9, 1902, 3.
13. *Floyd Press* (Floyd, VA), August 28, 1902, 3.
14. "Local News," *Floyd Press* (Floyd, VA), December 21, 1908.
15. Floyd County (VA) Will Book "C," page 596, 15 February 1862.
16. Gerald Via, e-mail message to Ricky Cox, January 10, 2008.
17. *Ibid.*
18. Floyd County (VA) Deed Book "R," page 254.
19. Charter, Buffalo Roller Mills, Inc., 4 October 1906.
20. Schroeder.
21. Supplements to *Floyd Press* (Floyd, VA), Jan. 14, 1909; Feb. 04, 1909.
22. *Floyd Press* (Floyd, VA), January 23, 1909.
23. Ronald Weddle, in discussion with Frank Webb and Ricky Cox, Willis, VA, June 3, 1995.
24. Source not named by Frank Webb.
25. Pauley, 80.
26. *Floyd County*, 30.
27. *Floyd Press* (Floyd, VA), May 29, 1902, 2.
28. Mary Bland Armistead, "Water Wheel Provides Energy Independence," *Roanoke Times and World News* (Roanoke, VA), February 1, 1981.
29. Calvin McAlexander, in discussion with Ricky Cox, July 19, 2010.
30. Reynolds, 285–86.
31. "Old Agee Mill Sold," *Floyd Press* (Floyd, VA), November 20, 1947.
32. Dan Hubbard, telephone interview with Ricky Cox, December 3, 2009.
33. Floyd County (VA) Deed Book "O," Page 258.
34. Carl Belcher, in discussion with Ricky Cox, Radford, VA, October 27, 2009.
35. James "Arthur" Bond, interview by Ricky Cox, August 2, 1991.
36. Floyd County Historical Society.
37. Debra McAlexander, e-mail message to Ricky Cox, August 2, 2010.

Appendix A

1. This mill was probably in Carroll County, VA, less than two miles west of the Floyd/Carroll border. The Carroll County mill Joshua Duncan inherited from his father, Greenberry Duncan, in 1857 may have been at the Greasy Creek site later known by the name of J. Wesley "Wes" Duncan, Joshua Duncan's son. The appearance of both a Wesley Duncan mill and a Joshua Duncan mill, with references to different post office addresses, in the 1897 list in Appendix A may mean that they were two separate mills, or could simply be a redundancy, of which similar examples exist in this source. Inclusion of the Duncan mill(s) does show that the lists in the appendix may include mills that were patronized by Floyd Countians, but actually stood in adjoining counties.
2. John Slusher applied to the Circuit Court of Floyd County for a *writ of ad quod damnum* on April 14, 1874. No record of the court's decision or of an inquest finding was found in Circuit Court records. Slusher's inclusion on this list suggests that Slusher did build a dam and mill with or without permission, but no other evidence of his having done so was found.
3. Alex Smith appears in Carroll County, VA, census records. He may have worked in a Floyd County mill, or, like Joshua Duncan, owned a Carroll County mill that served customers from the northwestern edge of Floyd County.
4. James W. Smith owned the John O. Jack Mill (I, 5) in 1881, but none of the subsequent listings (1885, 1894, 1897) of his name in Appendix A can be matched with a

specific mill history. The number of deeds made to and from Smith suggest that he was a land speculator and investor, hence he may have been a silent partner in other mills in the eastern end of Floyd County.

5. Daniel Floyd Epperly (1833–1910) appears in the 1860 census as a miller living in Huffville, VA.

6. Possibly Bennet L. Harris, aka Benjamin L. Harris, who described himself as a 23-year-old mechanic when he married in 1896.

7. If this is Giles W. Huff, he may have been assisting his aging father-in-law, Robert Huff, who died in 1895. Giles Huff died in 1893 and his widow inherited the Byrd Huff mill property, on which the Irvin Huff Mill was later built.

8. John Jennings appears in the 1870 census as a 49-year-old miller in Locust Grove Township.

9. O'Neal is an unusual name in 19th century Floyd County. No J. W. O'Neal appears in census records in 1890 or 1900. If this is instead I. W. (Isaac) O'Neal, this reference is to the O'Neal/Bolt Mill (IV, 49) on Burks Fork Creek.

10. Romney Slusher (1888–1961) of Santos, VA, was only 6 years old in 1894. In an e-mail to Ricky Cox (7/5/10), Slusher family historian Alice Slusher surmises that the submission of his name as a mill owner was a joke played by someone who resented the perceived meddling of Richmond bureaucrats. Known as a joker, himself, throughout his adult life (Hanley and Keith, 50–54), Rom Slusher probably would have appreciated it.

Selected Bibliography

Anders, Betty Willis. ""Willis: Roscoe Willis's Store." *Floyd County Virginia Heritage.* Summersville, WV: S.E. Gross, 2001. 174.

Brown, Ralph M. "A Sketch of the Early History of Southwestern Virginia," *William and Mary Quarterly* 17, No. 4 (1937): 512.

Carr, Jess. *The Second Oldest Profession: An Informal History of Moonshining in America.* Radford, VA: Commonwealth Press, 1972.

Cox, Elza B. *The Genealogy of William Tobias Phillips, nee Alice (Elsie) B. Henson, c. 1770–1989, Plus Fourteen Generations 1653–1989.* Jefferson City, MO: Brown, 1989.

Coy, Fred E., Jr., Charles D. Hockensmith, and Jimmie L. Price. *Millstone Manufacture in Virginia: Interviews with the Last Two Brush Mountain Millstone Makers.* Raleigh, NC: Marblehead, 1999.

Dedrick, B.W. *Practical Milling.* 1924; reprint, Chicago: The Society of Old Mills, 1989.

Evans, Oliver. *The Young Mill-Wright and Miller's Guide.* 1850; reprint, Salem, NH: Ayer, 1989.

Finley, Edward. "History of the Meadows Mill Company," *Old Mill News,* January 1983. 4–5.

Floyd County Historical Society. "Floyd Historic Walking Tour: A Virginia Crossroads Town" (brochure), n.p., n.d.

Floyd County Virginia Heritage. Summersville, WV: S.E. Gross, 2001.

Fries, Robert. "Millstone Memories," *Roanoke* (VA) *Times and World News.* February 5, 1995, New River Current, 17–19.

Fussell, Betty. *The Story of Corn: The Myth and Mystery, the Culture and Agriculture, the Art and Science of America's Quintessential Crop.* New York: Alfred A. Knopf, 1992.

Gettel, Philip E. "Floyd County Mills and Mills Sites." Unpublished manuscript. Montgomery-Floyd Regional Library, Floyd Branch, Floyd, VA. 1985.

Hale, Anna. "Williams: Charles Emmett Williams," in *Floyd County Virginia Heritage.* Summersville, WV: S.E. Gross, 2001. 173.

Hartley, E.N. *Ironworks on the Saugus.* Norman: University of Oklahoma Press, 1957.

Hening, William Waller. *The Statutes at Large: Being a Collection of All the Laws of Virginia, From the First Session of the Legislature in the Year 1619,* 12 vols. New York, Philadelphia, Richmond. 1819–23.

Hewett, Roberta W., et al., *Industrial Survey, Floyd County, Virginia.* Floyd, VA: No publisher, 1957.

Humbert, R.L. *Floyd County, Virginia: Industrial Survey.* Blacksburg: Virginia Polytechnic Institute, 1930.

Hylton, Judy Nolen. "History of Nolen Mill." Unpublished manuscript.

Kelleher, Tom. "Flour from Mill to Market," *Old Mill News* 37, No. 1 (2009): 19–22.

Klosowski, Stefan. "Greasy Creek Mill." Unpublished manuscript. c. 1976.

Klosowski. "Nolen's Mill." Unpublished manuscript. c. 1976.

Klosowski. "Pine Creek Mill." Unpublished manuscript. c. 1976.

Melton, Herman. *Pittsylvania's Eighteenth Century Grist Mills.* Chatham, VA, 1989.

_____. *Pittsylvania's Nineteenth-Century Grist Mills.* Chatham, VA, 1991.

McNeil, J. David. "James Ireson McNeil/McNiel," in *Floyd County Virginia Heritage.* Summersville, WV: S.E. Gross, 2001. 128.

_____. "A Mill on Goose Creek," in *Floyd County Virginia Heritage.* Summersville, WV: S.E. Gross, 2001. 31–2.

Merritts, Dorothy J., and Robert C. Walter, "Natural Streams and the Legacy of Water-Powered Mills," *Science* 319, issue 5861 (2008): 299–304.

Micarelli, William M. "Evolution of the U.S. Economic Census: The Nineteenth and Twentieth Centuries" *Government Information Quarterly* 15, 3 (1998): 335–377.

"Montgomery County." *Virginia Business Directory and Gazetteer.* Richmond, VA: Hill Directory, 1917.

"Montgomery County." *Virginia State Gazetteer and Business Directory, 1897–98.* Richmond, VA: J.L. Hill, 1898.

"Pauley, E. Catherine Vaughn. "Barringer/Alderman/Vaughn," in *Floyd County Virginia Heritage.* Summersville, WV: S.E. Gross, 2001. 80–82.

Products of Industry reports for 1850, 1860, 1870, and 1880. These enumerations of manufacturing businesses in each locality were made alongside the decennial population census of the United States. Microfilm copies of the reports from Floyd County, VA, for these years are available in the Newman Library at Virginia Tech, Blacksburg, VA. Records for grist and flour mills are transcribed in Tables A, B, and C, which represent, respectively, the years 1850, 1870, and 1880. Records for other business, such as wool carding mills, may or may not have been transcribed. The record for 1860 included only one grain mill (Spangler, Section II, 99) and was not transcribed.

Reynolds, Terry S. *Stronger Than a Hundred Men: A History of the Vertical Water Wheel.* Baltimore: Johns Hopkins, 1983.

Sass, Jon A. *The Versatile Millstone: Workhorse of Many Industries*. Knoxville: Society for the Preservation of Old Mills, 1990.

Schroeder, Joan V. "In Search of Winfield Scott," *Roanoker Magazine*, February/March 1993, 12–15, 45.

Simpson, Basil Poff, Sr. "Memoirs of Basil Poff Simpson, Sr." Unpublished manuscript, n.p, n.d.

Slusher, R.O. Interview by Alice Slusher. "Rom: A Little Bit of Humor," in *250 Years in the Blue Ridge: The Legacy of the Slushers and Other Settlers in Floyd County* by Nancy Slusher Hanley and Janet Slusher Keith (n.p.: 1993). 50–4.

Spangler, M.J. *Descendants of Daniel and Mary Spangler, Franklin County, Virginia 1773*. Bountiful, Utah: Family History Publisher, 1997.

_____. "Memories of My First 25 Years." Unpublished manuscript.

Speer, Jean Haskell. *The Appalachian Photographs of Earl Palmer*. Lexington: University Press of Kentucky, 1990.

Summers, Lewis Preston. *Annals of Southwest Virginia, 1769–1800*. Kingsport, TN: Kingsport Press, 1929.

Tise, Marguerite. "Petition to Build a Mill," *Virginia Appalachian Notes* 17, No. 4 (1993): 132–133.

Virginia Business Directory and Gazetteer. This guide to industries, post offices, and services in Virginia localities appeared under various titles between 1877 and 1917. See Appendix A.

Willis, Mary Williams, and Ruby Willis Goodman. "Blue Ridge Mary and Other Manuscripts," n.d., n. pub.

Wood, Amos. *Floyd County: A History of Its People and Places*. Commonwealth Press, 1981.

Index

Page numbers in **bold** indicate sections on a particular mill.
Numbers in ***bold italic*** indicate illustrations.

www.ingramcontent.com/pod-product-compliance
Ingram Content Group UK Ltd.
Pitfield, Milton Keynes, MK11 3LW, UK
UKHW051929100926
8556IPUK00022B/1452

9 780786 437238